Behavioral Mechanisms in Evolutionary Ecology

BEHAVIORAL MECHANISMS *in* EVOLUTIONARY ECOLOGY

Edited by Leslie A. Real

The University of Chicago Press
CHICAGO AND LONDON

Leslie A. Real is Professor of Biology, Indiana University.

The University of Chicago Press, Chicago 60637
The University of Chicago Press, Ltd., London
© 1994 by The University of Chicago
All rights reserved. Published 1994
Printed in the United States of America
03 02 01 00 99 98 97 96 95 94   1 2 3 4 5

ISBN: 0-226-70595-1 (cloth)
ISBN: 0-226-70597-8 (paper)

Library of Congress Cataloging-in-Publication Data

Behavioral mechanisms in evolutionary biology / edited by Leslie A.
Real.
      p.   cm.
   Includes bibliographical references and index.
   1. Animal behavior.   2. Behavior evolution.   3. Animal
intelligence.   I. Real, Leslie.
   QL751.B346   1994
   591.51—dc20   94-14131
                                                    CIP

⊚ The paper used in this publication meets the minimum requirements of the American
National Standard for Information Sciences—Permanence of Paper for Printed Library
Materials, ANSI Z39.48-1984.

To my wife, Damienne Palazzola Real
and my son, Daniel Jordan Real

CONTENTS

PREFACE

Mechanistic analyses of population, community, and ecosystem processes have become the norm throughout much of the evolutionary and ecological literature. The great advances that have been made over the last two decades in ecology have largely been the result of attention to the mechanisms that generate ecological patterns and determine ecological organization.

A similar interest in mechanism is growing in the field of behavioral ecology. The functional interpretation of behavior within specific ecological contexts is only part of a sufficient explanation for a set of actions undertaken by an organism. How are such actions and behaviors implemented and affected by physiology, morphology, and ontogeny? To what degree are physiological processes shaped by natural selection to produce appropriate behaviors within specific environments? To what degree do existing organismal adaptations and genetic systems limit the range of behavioral responses of organisms to changing environments? The answers to these questions require articulating the fundamental mechanistic processes that generate behavioral responses both within current ecological time and over a time frame sufficient for the evolution of behavioral traits.

The contributors to this book have each addressed one or more of these general questions in an attempt to outline the mechanistic foundations for a behavioral evolutionary ecology. The volume originated in a symposium organized for the American Society of Naturalists during summer 1992. The symposium was published as a supplement to the *American Naturalist*, and the five original contributions to the symposium are reprinted here (chaps. 3, 5, 11, 13, and 14). The present volume represents a greatly expanded version of the *American Naturalist* supplement, including additional areas that were not represented in the original symposium. The essays cover a wide variety of topics and should provide readers with a general introduction to the major research topics associated with mechanistic approaches to behavioral ecology.

I wish to thank the contributors for their patience and forbearance during editing and production and for the high quality of the articles they have contributed. Much of the research reported here has been supported by the National Science Foundation and the National Institutes of Health, and I wish to acknowledge and encourage their continued support for research on the biological bases of behavior. Lastly, I would like to thank my wife, Damienne Palazzola Real, and my son, Daniel Jordan Real, for their patience, kindness, encouragement, and enthusiasm throughout this project.

LESLIE A. REAL

Bloomington, IN

ONE

How to Think About Behavior: An Introduction

Leslie A. Real

THE STUDY OF BEHAVIOR is unique in providing and promoting connections across different levels of biological organization. Anyone who has studied contemporary issues in behavior knows that no successful study can be undertaken without working across multiple levels of organization. The organism's behavior stands at the interface between the set of internal physiological and psychological processes that give rise to particular actions and the set of ecological and evolutionary consequences of individual actions. The principal force that links the internal domain of causes with the external domain of consequences is the force of natural selection.

Many ecological phenomena and many aspects of biological community organization (e.g., predation, mating, dispersal, and a variety of types of interference competition) are the immediate consequence of individual organisms' actions and behaviors. By emphasizing and assessing individual actions and behaviors, we can characterize the magnitude of variation among individuals with respect to important ecological traits. While determining individual variation within populations has been characteristic of evolutionary analyses, such variation is often obscured in traditional ecological studies. The chapters in this book suggest that all ecological phenomena and patterns of community organization can be viewed as the immediate consequence of individual actions and behaviors, and that consequently, an analysis of specific biological mechanisms at the level of the individual should increase our capacity to predict and account for patterns and structures at higher levels of ecological organization. The only successful way to think about behavior is to unite its internal mechanistic foundations with its external effects in specific ecological settings, and to recognize that the internal and external worlds of the organism are united through the action of natural selection and evolution.

The set of internal processes that form the mechanistic basis of behavior can be divided roughly into five interacting subcomponents (fig. 1.1). The most immediate precursors to a given behavioral response depend on particular decision-making rules that may have evolved for the efficient exploitation of resources or the efficient processing of information (e.g., optimal foraging policies or sampling strategies), and on a variety of psychophysical constraints that may limit the capacity of the organism to perceive or process information (e.g., constraints on memory, learning, or perception). Either of these two broadly defined determinants of behavior could be further partitioned into specific hierarchies. For example, cognitive functions are often divided into lower-level and higher-level tasks and abilities (fig. 1.2). The details of each of these components are discussed in the chapters that follow.

The physiological processes represented by the remaining three components—

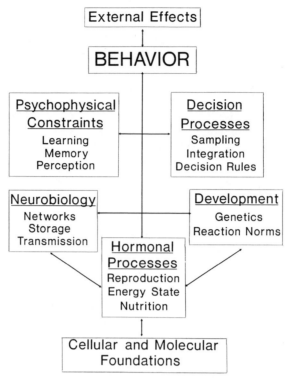

Fig. 1.1   Schematic representation of the principal components accounting for the generation of particular behavioral actions. This scheme represents the internal environment of the animal that leads to external ecological consequences and effects.

neuronal, hormonal, and developmental/genetic—rest upon a variety of cellular and molecular processes. Often the greatest gap in our understanding occurs between the level of cellular processes and whole-organism behavior. However, recent advances in neuroethology, such as the current work on the molecular basis for learning and memory (Kandel and Schwartz 1982; Alkon 1987; Cotman and Lynch 1990), may prove a means of closing this gap.

If we wish to understand the evolution of behavior and its foundation in the internal processes and mechanisms that give rise to particular actions, then we must link particular behaviors to their ecological consequences. The ecological consequences of any behavioral act can be roughly partitioned into four categories (fig. 1.3). Foraging ecology involves the acquisition of both energy and essential nutrients and has been a major focus of traditional behavioral ecology. Species interactions (including predation, competition, herbivory, pathogen/parasite systems, and mutualism) constitute the traditional domain of population and community ecology. The importance of predatory functional response in determining the stability of predator-prey interactions indicates the necessity for incorporating behavior into models of ecological interactions. Many of the most important types of interactions involve one or more aspect of organismal behavior, and individual-based approaches to population theory that incorporate a variety of behaviors are increasing (DeAngelis and Gross 1992). Social organization involves aspects of individual movement, aggregation, territoriality, and dispersal, all of which influence patterns of resource

Cognitive Function

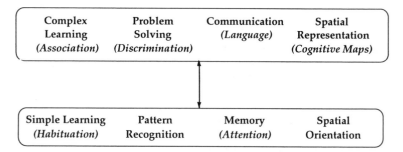

Higher Level Cognition

Complex Learning (Association)	Problem Solving (Discrimination)	Communication (Language)	Spatial Representation (Cognitive Maps)

Simple Learning (Habituation)	Pattern Recognition	Memory (Attention)	Spatial Orientation

Lower Level Cognition

Fig. 1.2   Each component of behavior represented in figure 1.1 can be further partitioned into hierarchical components contributing to operations at that specific level. For example, cognitive functions can be partitioned into lower-level cognitive operations (such as spatial representations) that form the basis for more complicated cognitive tasks (such as the formation of cognitive maps).

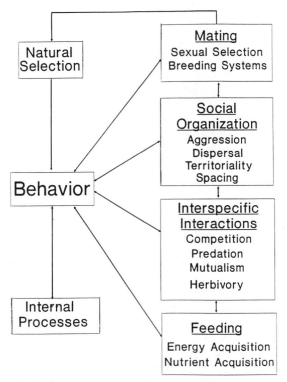

Fig. 1.3   Schematic representation of the principal categories of ecological effects that result from specific behavioral actions. These ecological effects constitute the arena within which natural selection can operate to shape the organism's internal processes and behavioral responses.

use and availability. Finally, mating ecology has obvious links to behavioral studies and has been of great interest to evolutionary biologists since Darwin's speculations on sexual selection. Mating ecology, however, has received little attention from ecologists (compared with, say, the physiological requirements for survival), which may seem ironic considering that births constitute one of the two main determinants of local population demography.

The first two parts of this book explore the psychological and cognitive foundations of behavior that pertain to perceiving information in the environment (e.g., spatial relations, sounds, probabilities, the actions of other members of the population) and integrating that information into ecologically meaningful aspects of learning and decision making.

Alan Kamil (chap. 2) explores traditional approaches in experimental animal psychology and calls for a reformulation of these approaches to embrace a more ecological and evolutionary perspective on animal behavior. Kamil's primary emphasis is on aspects of learning. The call for an ecological and evolutionary analysis of learning is echoed in the chapter by John Krebs and Alastair Inman (chap. 3). Krebs and Inman extend the discussion of learning to include issues such as how organisms should sample their environments for information, and how learning to exploit patchily distributed resources affects the spatial distribution of individuals across patches.

Fred Dyer examines how foragers use spatial perception to orient over landscapes (chap. 4). His experiments question many of the current conjectures on the formation of cognitive maps in honeybees, and suggest that simpler mental functions can explain the observed patterns of resource exploitation. My chapter also uses bees (albeit bumblebees). I examine the nature of animal decision making and choice behavior (chap. 5). Throughout my chapter I adopt a cognitive view of behavior and suggest that much of the structure of the internal mental operations that inform decisions can be viewed as the product of evolution and natural selection.

Issues in perception and decision making are applied to the specific problem of communication in the chapters by Haven Wiley (chap. 7) and Michael Ryan (chap. 8). Wiley applies concepts from signal detection theory to problems in mating behavior and the organization of social interactions in birds. Ryan examines the neural basis of sound perception in anurans and suggests that preexistent neuronal mechanisms may have constrained the direction of mating behavior and sexual selection. Both chapters underscore the importance of exploring the mechanistic foundations of behavior alongside their evolutionary origins and implications.

The two broad categories of processes that form the most immediate precursors of individual behavioral actions—the behavioral objectives embodied in decision rules and their psychophysical constraints—are themselves linked to three interconnected physiological processes: neurobiological, hormonal, and genetic/developmental (see fig. 1.1).

At the neurobiological level, one may ask, for example, how integrated sensory information is stored as a specific memory, and how the representation of specific memories and events is constrained by neural machinery. The current analysis of the relation between hippocampal structure, spatial memory, and the ecology of food caching is one such study that has obvious ecological importance (reviewed in Sherry and Schacter 1987). Ryan's analysis of the neuronal basis of mate choice and of the constraints imposed by neural machinery provides another illustration. In this volume, Arthur Arnold (chap. 9) examines the interaction between neural mechanisms and the development of bird song. He demonstrates that certain critical aspects of the bird's phenotype can only be understood after appreciating the underlying developmental processes. The developmental context of phenotypic expression of bird song is reiterated in the chapter by Meredith West, Andrew King, and Todd Freeberg (chap. 10). In this study, the developmental context is not so much the underlying neuronal mechanisms but the set of interactions with conspecific members of the population during the ontogeny of song.

The past ten years have seen a tremendous increase in interest in modeling the nervous system and in applications of neural modeling to problems of perception, learning, and memory. Much of this new interest is attributable to the rejuvenation of neural network modeling through the emergence of the connectionist/parallel distributed processing (PDP) paradigm (Rumelhart and McClelland 1986). Daniel Papaj (chap. 6) applies some of these new concepts to ecological learning theory and demonstrates the implications of optimizing learning in a model neuronal system for the evolution of insect behavior.

When discussing animal behavior it seems natural to focus on the nervous system, since actions are always tied to brain and nervous system function in some way. However, individual actions, including many aspects of brain function, are influenced by hormonal condition, and often hormones act as intermediaries of brain and nervous system responses. Hormones affect and reflect reproductive condition and behavior, nutritional state, and motivation, and can influence the outcome of ecologically important forms of interaction, such as mating attempts, aggressive encounters, and territorial defense. Recent methodological advances allow investigators to manipulate hormonal processes directly and thereby influence the range and relative frequency of potential phenotypes in a population. By extending the range of phenotypic expression beyond that found under natural conditions, investigators can assess the selective forces acting on physiological regulation in defined ecological settings. For example, artificially adjusting testosterone levels in individual birds by using testosterone implants extends the range of phenotypic behavior expressed by males in natural populations. This approach has been elegantly exploited by Ellen Ketterson and Val Nolan (chap. 14) through their inquiry into the evolution of avian reproductive behavior and life history tactics. Such manipulations may lead to a greater understanding of the role of physiological processes in determining the outcome of selection on mating behaviors and, consequently, the long-term evolution of mating systems and life history phenomena.

Behavioral ecology has few overriding general principles that have survived empirical investigation for very long. One of the more persistent claims is that females will generally be more choosy than males in their selection of mates. Male fitness will thus be limited by access to females (leading to increased competition among males), while female fitness will be limited by resources available for offspring production and development. This general claim has been elevated to the status of a law and often appears in the literature as "Bateman's principle," named after A. J. Bateman (1948). Recent developments in the behavioral ecology of mate choice have suggested that greater female selectivity may not only be driven by resource allocation but may also be associated with the assessment of disease status and/or the prevention of disease transmission (Hamilton and Zuk 1982). The features of males, for example, that are often used as choice criteria (bright coloration, vigor, etc.) are often affected by disease, and therefore may be used to distinguish males of different quality and/or disease status. The direct effects of disease on male signals will often be mediated by the immune system. The loss of bright coloration in diseased male birds may be a general phenotypic response resulting from the immune response. Marlene Zuk (chap. 15) explores the general problem of mate choice, disease, and the implications of the immune response for the evolution of behavior. The role of disease agents in controlling biotic interactions has been a relatively underexplored aspect of ecology, and research on disease ecology has been targeted as a major research priority over the next decade (Lubchenco et al. 1991). Zuk's chapter stands as a model of how an appreciation of the physiological mechanisms of disease response can affect our interpretation of population-level phenomena.

Behavior can be viewed as an exceedingly plastic aspect of the organism's phenotype. As such, the evolution of the range of phenotypic response can be treated as a special problem in the evolution of reaction norms and phenotypic plasticity. Steve Arnold (chap. 11) summarizes our current understanding of how genetic constraints (revealed through variance-covariance matrices) determine both the rate and direction of phenotypic evolution. He then applies these general principles to the evolution of behavior. One of his more interesting conjectures is that behavior is not only constrained by genetic correlations but can also be an important force in the disassociation of already existent genetic correlations. In an analysis that principally relies on years of fieldwork on the checkerspot butterfly, Michael Singer (chap. 12) shows how the application of genetic correlations and constraints can influence our understanding of oviposition behavior and our interpretation of the evolution of host plant selection.

The timing of developmental stages and ontogenetic transitions reflects the constraints imposed upon development by the genetic program, the historical and social context of development, and specific adaptations for changing behavior to increase the efficiency of resource acquisition. Earl Werner (chap. 13) explores the importance of changes in habitat use associated with changes in body size and timing of metamorphosis, and documents how these ontogenetic changes can mediate competitive interactions among species. Werner attributes much of the organization of anuran communities to adaptations for dealing with temporal changes in competitive interaction associated with the changing scale of interactions dependent on body size. The influence of the developmental context on the expression of specific behaviors is further explored by Meredith West, Andrew King, and Todd Freeberg (chap. 10). The ontogeny of bird song is apparently influenced by the particular constellation of the social group during early song learning. This dependence of song development on social interaction during early development may lead to important effects on social organization and the mating behavior of adults. Recent interest in the relations between phylogeny, evolution, and behavior may strengthen interest in development as a precursor to behavior (Brooks and McLennan 1991). Nonetheless, the role of development and ontogeny in the evolution of behavior is a relatively underexplored topic, and considerably more research is needed in this area.

The behavior of an individual organism changes with the social context not only during early development but at all stages of life. The last four chapters of this book examine the social context of behavior during the organism's mature stages and in fully developed animal societies.

In a provocative and ingenious application of principles from cognitive science, Robert Seyfarth and Dorothy Cheney, (chap. 16) examine the psychological mechanisms that underlie interindividual and intergroup interactions in primate societies.

They find much evidence suggesting that monkeys use cognitive mental operations that "represent" particular classes of social interactions. Their findings and their approach challenge the traditional limitations on mental ability that have been imposed on the interpretation of primate behavior. This chapter stands as an example of how evolutionary and ecological approaches to behavior can contribute substantially to other fields, such as the philosophy and psychology of mind.

Throughout the history of science, layperson and scientist alike have marveled at the complex organization of insect societies. Many of the most complex features of such organization must result from the aggregate interactions of the individuals that make up the colony. How do aggregate properties of social organization emerge from underlying individual dynamics, and can our understanding of individual dynamics contribute to an understanding of the evolution of sociality? Nigel Franks and Lucas Partridge (chap. 17), Deborah Gordon (chap. 18), and Blaine Cole (chap. 19), tackle this problem using slightly different perspectives and methods. Cole uses the techniques developed for chaos theory and nonlinear dynamics to show that individual ants appear to follow chaotic activity patterns while ant colonies as a whole show regular periodicities in activity. Cole goes on to speculate on how aggregate regularities may emerge from the underlying nonlinear dynamics of individual behavior.

Franks and Partridge (chap. 17) address the problem of how individual members of complex social species organize for collective combat when successful predation depends upon the cooperative effort of many individuals. They examine the self-organization of army ants and cellular societies and show that the organizational principles first advanced by Lancaster as a normative model for the deployment of human combat forces appear to apply equally to the organization of simpler species (such as ants and myxobacteria). Franks and Partridge argue that it is the very restricted cognitive field of nearest neighbors influencing one another, rather than global organization and centralized control, that seems to account for the appearance of well-organized societies in ants and slime molds. Their emphasis on the emergence of organized collective activity through an analysis of individual behaviors echoes the themes advanced by Cole (chap. 19) and Gordon (chap. 18). I suspect that much of what has been attributed to centralized organization in "superorganisms" and societies will be revealed as the product of self-organizing decentralized interactions among the individual components of these complex systems.

Gordon (chap. 18) addresses two important issues in the social organization of behavior: first, how social organisms monitor changes in their environment and communicate information about changing conditions, and second, how the group responds to changing conditions based on this shared information. Do groups respond adaptively to alterations in the environment? Gordon focuses on the integration of information within ant colonies using neural network models of information storage; however, her approach may prove generally applicable to information exchange and storage within other animal societies.

In this book I have tried to represent various paths that lead from specific behavioral mechanisms to important ecological interactions—from the psychology of learning to the spatial distribution of individuals within populations, from choice behavior to pollinator-plant interactions, from hormonal condition to reproductive success and population demographics, and so on. This book represents only a small sample of the potential pathways that must be explored if we are to produce a unified and mechanisitc understanding of biological and ecological organization.

REFERENCES

Alkon, D. L. 1987. *Memory traces in the brain*. Cambridge: Cambridge University Press.

Bateman, A. J. 1948. Intrasexual selection in *Drosophila*. *Heredity* 2:349–68.

Brooks, D. R., and D. A. McLennan. 1991. *Phylogeny, Ecology and Behavior*. Chicago: University of Chicago Press.

Cotman, C. W., and G. S. Lynch. 1990. The neurobiology of learning and memory. In P. D. Eimas and A. M. Galaburda, eds., *Neurobiology of cognition*, 201–41. Cambridge, Mass.: MIT Press.

DeAngelis, D. L., and L. J. Gross, eds. Individual-based models and approaches in ecology: Populations, communities, and ecosystems. London: Chapman and Hall.

Hamilton, W. D., and M. Zuk. 1982. Heritable true fitness and bright birds: A role for parasites? *Science* 218:384–87.

Kandel, E. R., and J. H. Schwartz. 1982. Molecular biology of learning: Modulation of transmitter release. *Science* 218:433–43.

Lubchenco, J., A. M. Olson, L. B. Brubaker, S. R. Carpenter, M. M. Holland, S. P. Hubbell, S. A. Levin, J. A. MacMahon, P. A. Matson, J. M. Mellilo, H. A. Mooney, C. H. Peterson, H. R. Pulliam, L. A. Real, P. J. Regal, and P. G. Risser. 1991. The sustainable biosphere initiative: An ecological research agenda. *Ecology* 72:371–412.

Rumelhart, D. E., and J. L. McClelland. 1986. *Parallel distributed processing*. Cambridge, Mass.: MIT Press.

Sherry, D. F., and D. L. Schacter. 1987. The evolution of multiple memory systems. *Psychological Review* 94:439–54.

Part I

PSYCHOLOGICAL *and* COGNITIVE FOUNDATIONS

TWO

A Synthetic Approach to the Study of Animal Intelligence

ALAN C. KAMIL

TWO ANECDOTES

IT IS 7:00 A.M. The sun has just risen over the botanical gardens, and my research team and I are about to give up our attempt to catch a male Anna's hummingbird with a discrete white spot behind his left eye. "Spot" has been defending a small, flower-rich territory, and we want to put a colored plastic band on his leg as part of our study of nectar-foraging patterns. To catch Spot we had arrived before sunrise and strung a mist net, 5 feet high and 18 feet long, across the middle of his territory. Mist nets, made of very thin black nylon thread, are designed to entangle any bird that flies into them. Unfortunately, a heavy dew at sunrise had collected on the strands of the net, and Spot saw it immediately. He had flown along it and even perched on it. Experience has taught us that once a hummingbird has done this, it will never fly into the net. So we were about to take down the net, but first we were having a cup of coffee. Spot was sitting on his favorite perch, overlooking the territory from its southwest edge. Suddenly an intruding hummingbird flew into the territory from the northeast and began to feed.

Male Anna's hummingbirds are extraordinarily aggressive animals. Usually they will utter their squeaky territorial song and fly directly at an intruder, chasing it out of the territory. But that is not what Spot does. He silently drops from his perch and flies around the perimeter of the territory, staying close to the ground, until he is behind the other bird. Then he gives his song and chases the intruder—directly into the mist net. Spot pulls up short, hovers over the bird, utters another burst of song, and returns to his perch.

This anecdote raises many questions with interesting implications. For example, did Spot have a "cognitive map" of his territory that allowed him to understand that if he moved to a point behind the other bird he could force the intruder into the net? Since this is only an anecdote, it provides no definitive answer. But many more mundane empirical studies of nectar-feeding birds offer systematic data showing that they do possess considerable knowledge about spatial and temporal patterns of food production on their territories (Gass and Montgomerie 1981; Gill and Wolf 1977; Gill 1988; Kamil 1978).

Consider this observation of chimpanzees reported by Goodall:

The juvenile female Pooch approaches high-ranking Circe and reaches for one of her bananas. Circe at once hits out at the youngster, whereupon

This chapter was previously published in a slightly different form in *Nebraska Symposium on Motivation, 1987: Comparative Perspectives in Modern Psychology*, edited by D. W. Leger, 257–308, © 1988 by the University of Nebraska Press. Reproduced by permission.

Pooch, screaming very loudly indeed, runs from camp in an easterly direction. Her response to the rather mild threat seems unnecessarily violent. After two minutes, the screams give way to waa-barks, which get progressively louder as Pooch retraces her steps. After a few moments she reappears; stopping about 5 meters from Circe, she gives an arm-raise threat along with another waa-bark. Following behind Pooch, his hair slightly bristling, is the old male Huxley (who had left camp shortly before in an easterly direction). Circe, with a mild threat gesture towards Pooch and a glance at Huxley, gets up and moves away. Pooch has used Huxley as a "social tool." This little sequence can be understood only because we know of the odd relationship between the juvenile and the old male who served on many occasions as her protector and was seldom far away (Goodall 1986, 567).

There are many objections to the use of anecdotes such as these. As Thorndike (1898) pointed out, hundreds of dogs get lost every day and nobody pays much attention except the unfortunate dogs' owners. But let one dog find its way from Cambridge to London, or Boston to New Haven, and it becomes a famous anecdote. Anecdotes cannot provide definitive evidence about animal intelligence (or anything else). But it may be a serious mistake to completely ignore their implications, which can provide interesting hypotheses for rigorous test.

Furthermore, the two anecdotes related above are not isolated examples. Most fieldworkers have similar stories from their own experience. Books such as Goodall (1986) and Smuts (1985) are replete with them (see also Kummer 1982; Kummer and Goodall 1985). Much more important than the number of these anecdotes, however, is the fact that empirical data are being amassed to support their specific implications. The main point is that these anecdotes and supporting data suggest that the traditional psychological approach to the study of animal learning is too limited.

Psychologists have been studying animal learning for about a century. This century of experimental and theoretical work has produced some remarkable successes, particularly in understanding basic conditioning processes. However, these successes are limited in two major ways. First, they have been confined to a narrow domain. Recent research from a variety of settings has demonstrated that animals have mental abilities far beyond what they were given credit for just a few years ago, We must dramatically expand the range of phenomena addressed by the study of animal learning. Second, there has been an almost complete failure to place animal learning in any kind of comparative evolutionary framework, primarily because of a failure to develop any detailed understanding of how animals use their ability to learn outside the laboratory. Recent developments in psychology and biology are beginning to suggest how this gap may be filled.

The expansion of the range of phenomena under study is already well under way, with the emergence of the cognitive approach to animal learning (Hulse et al. 1978; Roitblat et al. 1984) and diverse new techniques for exploring the capacities of animals (Griffin 1976, 1978). The development of a meaningful comparative approach is also beginning to emerge, thanks to developments in both psychology and biology.

My purpose in this chapter is to outline the beginnings of a new way to study animal intelligence. I have labeled this the synthetic approach because it represents an attempt to synthesize the approaches of psychology, ethology, and behavioral ecology. I have used the term intelligence, rather than more specific terms such as learning or cognition, to emphasize the breadth of the phenomena to be included.

The synthetic approach builds upon previous successes but is much broader and more biological than the predominantly psychological approaches of the past. Its goal is to develop a full understanding of the intellectual abilities of animals, with particular emphasis on psychological mechanisms and functional significance.

What I am proposing is not a new theory. Rather, it is an attempt to outline a new scientific research program (Lakatos 1974). According to Lakatos, research programs consist of two parts: a central core of laws, principles, and assumptions that are not subject to direct empirical test, and a protective belt of "auxiliary hypotheses" that relate the central core to observations and can be tested and perhaps rejected. The central core and its auxiliary hypotheses function to direct research toward certain problems and away from others. In these terms, I am urging two changes in the central core of the psychological approach to animal learning: a broadening of the discipline's domain and the adaptation of a biological and ecological approach to the study of learning. These changes could redirect attention to important and interesting facets of animal learning that have been ignored by the traditional psychological approach.

THE TRADITIONAL APPROACH

The purpose of this section is to identify the central core of the psychological study of animal learning. There are two difficulties. First, the programs Lakatos discusses are from the history of physics, with explicit, usually mathematical, specifications of their central core. In the case of animal learning, the central core is less formalized and more difficult to specify. Another difficulty is that although it is easy to talk and write about "the traditional approach" to animal learning in psychology, in fact there have been a number of different approaches. Nonetheless, a few assumptions have been widespread, if not universal. Some of these assumptions formed the central core of the scientific research programs that have dominated animal learning psychology and have directed attention away from important phenomena and issues.

General Processes

One basic assumption has been that one or a very few general principles can account for all of animal learning. A variety of principles have been proposed, but the two dominant ones have been associationism and reinforcement theory.

Rescorla (1985) provides an extremely coherent overview of the associationist approach that is remarkable in the extent to which it agrees, in form, with Lakatos's description of a research program. The central core is the assumption that virtually all learning can be understood as the formation of an association between two events. The associationist approach then attempts to explain the diversity and richness of an animal's knowledge of its world not by hypothesizing a richness and diversity of learning mechanisms, but by weaving a web of auxiliary hypotheses around the central learning mechanism. Rescorla identifies three types of auxiliary hypotheses that serve this function: the complexity of the conditions that govern the formation of associations, a wide range of elements that can be associated, and multiple mechanisms by which associations can affect behavior. These auxiliary hypotheses have made associationism a powerful force for understanding some aspects of learning in animals, a force that is often underappreciated by those working in other areas.

The central core of reinforcement theory is that behavior can best be understood in terms of the strengthening or weakening effects of reinforcers and punishers on the responses that have preceded them. This was first clearly formulated by Thorndike (1911) and has been elaborated in many ways by others (Herrnstein 1970; Skinner 1938). Like associationism, reinforcement theory attempts to account for the richness and diversity of behavior by using a single principle with a web of auxiliary hypotheses. Among these hypotheses are the complexity of the effects of schedules of reinforcement and alterations in the definition of what constitutes a reinforcer. The study of reinforcement has made many important contributions to our understanding of learning.

Although associationism and reinforcement theory have proved to be powerful concepts, they have often been overemphasized. There are too many phenomena they cannot easily account for, including those studied by many cognitive animal psychologists and those beginning to be revealed by naturalistic studies of intelligence. The learning of associations between events and the effects of reinforcement must be investigated as part of any study of animal learning and intelligence. But these two principles in themselves cannot completely account for how animals adapt their behavior on the basis of experience.

Radical Behaviorism

Two kinds of behaviorism need to be distinguished. Methodological behaviorism simply recognizes that behavior is what we must measure in experiments. Its central tenet is that all the mechanisms we may theorize about are known to us only through behavior.

Radical behaviorism goes beyond stating that it is behavior we seek to understand. According to the radical behaviorist, any theoretical constructs, especially about cognitive structures animals may possess, are not just unnecessary, but dangerous (Skinner 1977); behavior can best be understood in terms of the functions that relate stimulus events to responses.

Radical behaviorism has been unremitting in its concentration on the similarities between species. For example, an often-quoted comment of Skinner's (1959) accompanies the cumulative records from several species: "Pigeon, rat, monkey, which is which? It doesn't matter . . . once you have allowed for differences in the ways in which they make contact with the environment, and in the ways in which they act upon the environment, what remains of their behavior shows astonishingly similar properties" (374–75).

The interesting aspect of this quotation is that it acknowledges the existence of differences between species but relegates them to the realm of the uninteresting. It provides a clear case of Lakatos's concept of a negative heuristic, directing research away from certain topics. For the synthetic approach, these differences are of interest. If they had been of more interest to the radical behaviorist, phenomena such as autoshaping and instinctive drift (see below) would have come as less of a surprise.

Another problem with the radical behaviorist position has been that it tends to be radically environmentalistic, regarding the organism as a tabula rasa upon which experience writes. This emphasis ignores the potential importance of the effects of genetics and evolutionary history. The emerging current view, particularly apparent in the cognitive approach to animal learning, is that organisms bring certain processes, such as attention and memory, to bear on problems. This in turn has serious implications for evolutionary analyses of animal intelligence.

Comparative Generality

Another traditional assumption has been that the basic properties of animal learning are the same in a wide variety of organisms, which has justified the use of relatively few species in animal learning research. The logic underlying this assumption may have been that many of the psychologists studying animal learning were not primarily interested in the species they studied, but were using these species as convenient substitutes for humans. Therefore the only learning processes of real interest were those that could be generalized to our own species. This is a coherent, sensible approach, but it suffers from a basic flaw. The animals under investigation are biological entities, with their own evolutionary history. The way that evolutionary history might influence the outcome of learning experiments was not considered by most psychologists.

As reviewed below, there are special and substantial logical and methodological problems confronting the comparative analysis of learning and intelligence in animals. But to assume the absence of such differences, or at least their relative unimportance, has some major drawbacks because it places the study of learning outside the realm of modern evolutionary theory. Suppose there are, in fact, no important differences in the processes of learning among a wide variety of species—say, all vertebrates. This would imply that learning plays no adaptive role at all for vertebrates. Indeed, a number of ethologists (e.g., Lorenz 1965) and psychologists (Boice 1977; Lockard 1971) have suggested that learning is relatively unimportant to animals in their natural environments. But more recent data have clearly demonstrated that learning and memory do function in crucial ways for foraging animals (Kamil et al.1987; Kamil and Sargent 1981; Shettleworth 1984) and animals in social situations (Cheney et al. 1986; Kummer 1982). As is explained in more detail below, evidence for the functional significance of learning is evidence that there must be significant variation in intelligence between species.

Empty Methodological Sophistication

Bolles (1985a) suggested that an angry god put a terrible curse on psychology: "You will never discover anything about underlying causal processes, and you will never ever understand the overlying functional significance of anything. You will be forever doomed to be methodologists. You will content yourselves with teaching each other how to do experiments, and you will never know what they mean" (137). According to Bolles, because of this curse psychologists have become more caught up with their procedures than with the animals they study.

Another way to express this problem is to say that psychologists have concentrated disproportionately on internal validity and ignored the issue of external validity. Internal validity refers to the internal logic of the experiment, including factors such as the absence of confounding conditions and the adequacy of controls. External validity refers to the extent to which the results of laboratory studies can be generalized beyond the laboratory situation. When one designs a single experiment, there tends to be a trade-off between internal and external validity. Well-designed and well-controlled experiments are generally carried out under highly artificial or constrained conditions, which limits external validity. But at some point, any area of scientific endeavor must be concerned with the issue of external validity.

For example, consider the study of language acquisition by children. At one time this field was dominated by laboratory research in highly constrained situations and theoretical work on transformational grammars. But at some point researchers began to ask whether the ideas developing from this laboratory work could deal accurately with the actual process of language acquisition as it occurs in normal circumstances. This in turn led to many naturalistic studies of language acquisition, whose results have had a large impact on theoretical ideas and laboratory research (Gardner 1978).

The only external referent for animal learning research has been applied research with humans and animals. The applied work with animals immediately suggested problems (Breland and Breland 1961), but they were largely ignored. The applied work with humans has had some success, but this too has been limited (Schwartz 1984). What we need are additional external referents against which to judge the generality and importance of the information we have gained about animal learning and intelligence. As we shall see below, the absence of external criteria has caused particularly serious problems for comparative analyses of animal learning.

In summary, then, there are several problems with the traditional approach: a concentration on just a few general processes, with the possible elimination from consideration of many others; a concentration on behavior, ignoring the processes with which animals are endowed; the lack of an evolutionary, comparative framework; and the lack of substantial measures of external validity. These problems with the traditional approach have had particularly serious implications for the comparative analysis of learning and intelligence.

THE COMPARATIVE ANALYSIS OF INTELLIGENCE

The traditional psychological approach to animal learning has largely ignored comparative questions, concentrating research on just a few species. This tendency has been documented and criticized many times over the past 35 to 40 years (Beach 1950; Bitterman 1960). Despite this, most learning research in psychology is still conducted with just a few species. Why has this criticism had so little effect?

One reason is the commitment to general processes. The assumption has been that just a few general processes can explain most learning in many species. If that were true, there would be no reason not to concentrate on a few available species. And of course the general principles of association and reinforcement have been demonstrated (but not studied in depth) in a wide range of species. One must wonder, however, to what extent the emphasis on general processes has restricted the view of the animal learning psychologist.

Another important reason for the lack of comparative work among traditional animal learning psychologists is the substantial methodological and theoretical problems presented by any comparative analysis of learning. The major methodological problem involves the difficulty of measuring species differences in learning because of the learning/performance distinction. The major theoretical problem is due to the logical status of the so-called mechanisms of learning.

The Learning-Performance Distinction

As Bitterman (1960, 1965) has so clearly articulated, the performance of a species in a particular situation is a joint function of its abilities and the particulars of the task presented. Thus the failure of a species (or an individual) to perform well on a particular test does not necessarily mean the species lacks the ability for which it is supposedly being tested. Rather, it may be that the situation is in some way inappropriate. A species may fail to solve a problem, for example, not because it is incapable of solution in a general way, but because the experiment was improperly conducted. In Bitterman's terms, some contextual variable, such as motivational level or response requirement, may have been inappropriate.

Bitterman's (1965) solution to this problem is "control by systematic variation," in which one systematically varies the contextual variables in an attempt to find a situation in which the species will perform well on the task. So, for example, one might vary motivational level, the intensity and nature of the stimuli, the response required, and so on. The problem, of course, is that control by systematic variation can never prove that a species difference exists. It is impossible to prove that there are no circumstances in which a species will learn a particular type of problem. Some untested combination of variables may produce positive results in the future.

This leaves a curious asymmetry in the interpretation of comparative learning research. The meaning of similar results with different species is supposedly clear: the species do not differ in the learning ability being tested. The meaning of different results with different species is never clear. No matter how many failed attempts there have been, the skeptic can always claim, with impeccable logic, that the apparent difference may be due to something other than a species difference in learning abilities.

"Mechanisms" of Learning

The second problem that presents substantial challenges to the comparative analysis of intelligence is the logical status of what are commonly called the "mechanisms" of learning. In normal language a mechanism is machinery, like gears in a clock. The machinery is physical and can be observed directly. In comparative anatomy and physiology the mechanisms are also physical; respiration has a physically observable and measurable basis in trachea, lungs, and hemoglobin. In principle, learning mechanisms also have a physical basis in the brain. But that physical basis is as yet unknown in any detail, especially for more complex forms of learning. In any case, the way psychologists define learning (or cognitive) mechanisms is independent of the physical basis of these mechanisms.

The "mechanisms" of learning are known in terms of input-output relationships. That is, models are constructed that accurately predict output, behavior, from the input, previous experience. A successful model is then called a learning mechanism. The things we call learning mechanisms are not really mechanisms at all but hypothetical constructs, models that accurately predict behavior. What does it mean to say that the same hypothetical construct correctly predicts learning in two different species?

It certainly does not mean that the mechanisms of learning, in the physical sense, are identical in the two species. It is instructive, in this context, to look at an example from comparative physiology. There is considerable variety in the physical mechanisms of respiration, even among air-breathing vertebrates. The mechanisms (e.g., the lungs) are not inferred, they are directly observable. It is hard to imagine comparative physiologists arguing much about whether the differences between bird and mammal lungs are quantitative or qualitative, or whether we should use one mathematical model with changeable parameters or two different mathematical models. The differences are there to be directly observed and measured. In other words, some of the arguments about comparative interpretation of possible species differences in learning have their origin in the hypothetical nature of learning "mechanisms," not in the logic of comparative analysis per se.

Given the hypothetical nature of mechanisms of animal learning or intelligence, one of the central arguments of the traditional approach, that of qualitative versus quantitative differences, will often be impossible to resolve, and it misses the point in any case. For example, consider the argument over long-delay taste aversion learning. Baron et al. (1969) have shown that as the delay between a bar press and a shock increases from 0 to 60 sec, the extent of suppression of bar pressing decreases. Andrews and Braveman (1975) have shown that as the delay between saccharin consumption and poisoning increases from a few minutes to 25 hours, the suppression of saccharin intake decreases. In describing these results, Mazur (1986) concludes that they "do not require the postulation of a different law to replace the principle of contiguity; they merely require the use of different numbers in describing the relationship between contiguity and learning" (228). Although this statement is literally true—a single model can describe both sets of results with a change in parameter value—what does the word "merely" imply?

Clearly, it implies that the difference is "only" quantitative and therefore not of much interest (to Mazur). But how large does a quantitative difference have to be before it can escape the description "merely"? A difference between seconds and hours is a difference of more than a thousandfold. As Bolles (1985a) points out, a thousandfold difference in a biological system is never just quantitative. One can find

on the skeletons of some snakes little bumps on certain vertebrae where the legs might be if the snake had legs. They are pelvic bumps, and it is my understanding that these bumps may be 1 or 2 mm in size . . . although a 1- or 2-mm leg is not much of a leg, it is actually about 1/1000th of the length of the legs of a race horse. So the difference in legs between a snake and a race horse is really only a matter of degree. (393)

From a biological point of view, it does not matter whether one chooses to call the differences between taste aversion learning and bar press suppression qualitative or quantitative. The difference can be accounted for by postulating a single "mechanism" with a parameter or two whose values can be changed to accommodate the temporal differences. It can also be accounted for by postulating two "mechanisms." What matters is that there are differences, and these raise a large number of issues that need empirical attention. Since these issues are primarily evolutionary and functional in nature, the traditional approach is not likely to pursue them.

An analogy that may be useful in thinking about this problem is to compare learning mechanisms to computer programs. Suppose one were given two programs that solved arithmetic problems in compiled form, so that the programs could not be listed. How would one go about determining whether these programs were based on the same underlying algorithms? One would have to study the input-output relationships—give each program a set of standard problems and compare the speed and accuracy with which they solved the different problems. If the results for both programs were identical, it would seem highly likely that the programs were the same, although one could not be positive. Perhaps some other arithmetic test would produce results that were different for the two programs.

What would happen if the programs differed in some systematic way? For example, suppose that one program always took longer than the other, but only when division was involved. One would naturally be led to conclude that the programs used different algorithms for division. But wait! A theorist could claim that the difference was only quantitative—perhaps the slower program used the same algorithm but had a pause statement added to its division subroutine.

No analogy should be pushed too far. But my general point is that it would be very hard to know with certainty whether the two programs used the same algorithm. Furthermore, it would probably be impossible to tell the "evolutionary" relationship between the programs—whether they had been independently written or one had led to the other. This, of course, is the problem of homology versus analogy in the evolutionary study of traits.

There is one final point to milk from this analogy. One approach to the problem of comparing the two programs would be to attempt measurement at the molecular level and measure the activities of the microprocessor itself. Thinking about this brings out some interesting implications for the relationship between behavioral mechanisms and the physical processes instantiating them. At one level the mechanism for the two programs would be identical—the same processor, and so on, would be involved, even if the programs were written in different languages. But I am sure suitable measurements could be made that would reveal any difference. This suggests that knowledge of the events in the central nervous system that underlie the intellectual capacities of animals will be useful in understanding these processes. But it will have to be information of a certain type. I suspect it will be a long time before the neuroanatomy and neurophysiology underlying the complex processes involved in animal intelligence are understood at all. Behavioral work needs to proceed. The issues are too important to wait on the assumption that the physiological level of analysis will eventually solve these problems. In addition, without good understanding of the way mental processes function at the behavioral level, it is unlikely that physiological work can succeed (Kamil 1987).

THE NULL HYPOTHESIS

Many of the problems that the traditional approach encounters in the comparative realm can be seen quite clearly by examining the methods and conclusions of Macphail (1982, 1985), who conducted an extensive critical survey of the literature on the comparative study of learning in vertebrates. His conclusion was that there was no compelling reason to reject the null hypothesis "that there are no differences, either quantitative or qualitative, among the mechanisms of intelligence of non-human vertebrates" (Macphail 1982, 330), and he has reaffirmed this more recently (1985). How does Macphail reach this conclusion?

One approach to this question would be to take each of the phenomena Macphail examined and decide how plausible his conclusions are. However, that would probably take a book as long as his. In any event, I want to raise a more crucial point. Does Macphail's basic approach to the comparative study of vertebrate intelligence have some basic flaw (or flaws) that calls his conclusion into question. One can argue that his logic forced the final conclusion.

The first problem with Macphail's analysis is his definition of intelligence. In his opening chapter, he avoids any explicit definition. In particular, he states that it would be best to leave open the question "whether intelligence is some unitary capacity, or better seen as a complex of capacities, each of which might be independent of the others" (1982, 4). Macphail says that a decision about this issue might bias his review. However, his review is in fact biased toward the unitary view. For example, in discussing the results of a comparative research program on reversal learning in birds conducted by Gossette and his associates (Gossette 1967; Gossette et al. 1966), Macphail dismisses their findings. The reason for the dismissal is that different patterns of reversal learning between species were found with spatial and nonspatial cues. Macphail states, "If the ordering of species in serial reversal performance can be changed by altering the relevant dimension, it seems clear that serial reversal in itself cannot give a reliable measure of general intelligence" (Macphail 1982, 223). In the concluding discussion of his last chapter, Macphail talks extensively in terms of general intelligence.

A second contributor to Macphail's conclusion is an extreme willingness to believe in the untested intellectual capacities of animals. If some apparently complex learning ability has been demonstrated in two distantly related species, Macphail is willing to assume it can be found in all species. For example, win-stay, lose-shift learning in object-discrimination learning set is best tested by looking for transfer from object-reversal learning to learning set. This phenomenon has been demonstrated in relatively few species (blue jays: Kamil et al. 1977; rhesus monkeys: Warren 1966; chimpanzees: Schusterman 1962), and tests for such transfer have failed in at least two cases (cats: Warren 1966; squirrel monkeys: Ricciardi and Treichler 1970). The failure with cats is dismissed as apparently due to contextual variables, the failure with squirrel monkeys is not cited. The major implication of the discussion is that though most species have not been tested, they would show the phenomenon.

Another, perhaps more egregious, example is drawn from Macphail's (1985) discussion of languagelike behavior. Such behavior has been demonstrated in some primates using sign language or artificial language (e.g., Gardner and Gardner 1969; Rumbaugh 1977). Pepperberg (1981, 1983) has recently demonstrated similar behavior in an African gray parrot using "speech." Although the parrot has not achieved the level of performance shown by the primates (at least not yet), he has demonstrated capacities beyond what anyone (except Pepperberg) might have expected. Macphail (1985) concludes by saying, "As the single avian subject yet exposed to an appropriate training schedule, he [the parrot] gives good support to the view that the parrot's talent for language acquisition may not be significantly different from the ape's" (Macphail 1985, 48). Macphail seems to be implying that the same would be true of every vertebrate species if only suitable testing procedures could be devised. This exceptional willingness to assume that species possess abilities for which they have not even been tested stands in marked contrast to Macphail's extreme unwillingness to accept apparent species differences that have been revealed.

The most important reason for Macphail's conclusion of no species differences among vertebrates in learning or intelligence is his extensive use of the contextual stimulus argument (Bitterman 1960, 1965). As discussed above, whenever an explicit comparison of two species in the same learning task turns up differences, one can always argue that they reflect some performance factor (the effects of a contextual variable) rather than a difference in intelligence. Proving that there is no set of circumstances in which an animal can learn a particular task (e.g., that frogs cannot acquire language like behavior) is impossible.

Thus Macphail's argument leaves us with two competing null hypotheses. One is the null hypothesis of no differences in intelligence among vertebrates. Macphail holds that this null hypothesis should be maintained unless clear, convincing evidence against it is obtained. But clear, convincing evidence must prove the second null hypothesis, that no contextual variable is responsible for the proposed species differences. This logic essentially makes it impossible ever to demonstrate that there are species differences in intelligence.

Macphail would probably say I have overstated his argument. He does not require absolute proof of the second null hypothesis through systematic variation, only some reasonable attempt at evaluating contextual variables. But who is to determine what constitutes reasonable? In fact, the problem of contextual variables can never be completely dealt with through control by systematic variation.

Macphail has performed a valuable service. His arguments have clearly demonstrated that the traditional approach to the comparative study of learning can never succeed. One can never be certain that a species lacks a particular learning ability. This lesson applies not just to the study of learning, narrowly defined, but to the study of animal intelligence in general. An alternative approach that avoids the problem of contextual variables must be found. As described later in this chapter, there are compelling biological reasons to believe that species differences in intelligence do exist. Given that Macphail's approach can never successfully demonstrate such differences, it is crucial to find an alternative approach that avoids the problem of contextual variables.

THE SYNTHETIC APPROACH TO ANIMAL INTELLIGENCE

In this section I will outline an alternative approach to the study of the mental capacities of animals. I have labeled this the synthetic approach because it represents an attempt to synthesize the approaches of psychologists and organismal biologists. The synthetic approach has three major aspects: (1) a broad definition of the phenomena of interest, and (2) a comparative, evolutionary orientation, which leads to (3) an emphasis upon the importance of studying learning and its effects both in the laboratory and in the natural environment of the species being studied.

Broad Definition of the Phenomena of Interest

Using the term *animal intelligence* is a calculated gamble. It has the substantial advantages of communicating the general topic of interest to a wide audience in many different fields and of emphasizing the broad range of phenomena to be included. But it also carries a substantial disadvantage. It is a term that has been used and abused in many ways in the past. When technical discussion begins, then, there is a risk of misunderstanding based on people's assuming different definitions of animal intelligence.

I want to be explicit about the definition of animal intelligence I am using. The synthetic approach defines animal intelligence as those processes by which animals obtain and retain information about their environments and use that information to make behavioral decisions. Several characteristics of this definition need to be emphasized.

First of all, this is a broad definition. It includes all processes that are involved in any situation where animals change their behavior on the basis of experience. It encompasses the processes studied with traditional methods, such as operant and classical conditioning. It also includes processes such as memory and selective attention, which animal cognitive psychologists study (Roitblat 1986). It includes processes involved in complex learning of all sorts, including that demonstrated in social situations. It also includes the study of more "specialized" learning, such as song learning and imprinting.

Second, the definition emphasizes the information-processing and decision-making view of animals. This makes it very consistent with the approach of animal cognitive psychologists. It also makes the synthetic approach consistent with behav-

ioral ecology (Krebs and Davies 1978, 1984), which emphasizes the adaptive significance of the behavioral decisions of animals.

Third, this definition assumes that animal intelligence is multidimensional, not
unidimensional, in accordance with recent thinking about human intelligence (Gardner 1982). It also prohibits any simple ordering of species in terms of general intelligence. Species that are very good at some problems may be bad at others.

Fourth, this definition offers the possibility of conceptually integrating environmental and genetic influences on behavior, thus avoiding the nature/nurture controversy. It is generally recognized that no behavior is determined completely by either
genetic or environmental variables alone. However, this realization does not seem
to have had much effect on animal learning research in psychology, which still tends
to ignore the idea that the learning abilities of animals are part of their biological
heritage. The synthetic approach regards learned behavior as the result of experience. But these effects of experience are determined by the intellectual capacities
of the organism, which in turn depend upon the expression of genetically and
ontogenetically determined abilities.

This focus on processes instantiating behavior obviously entails rejecting most
types of behaviorism, but not methodological behaviorism. The primary way to
learn about these processes is by studying behavior. There is no desire to throw
away the considerable methodological sophistication that has been developed over
the past century, only to redirect that sophistication.

Comparative, Evolutionary Orientation

There has been considerable disagreement and confusion about the importance,
role, and purpose of comparative research on animal learning. Some have viewed
animal learning research as primarily a way of understanding basic mechanisms that
would, at least in the long run, lead to fuller (or even complete) understanding
of our own species. For these scientists, comparative research has been relatively
unimportant. Others have viewed comparative research as important but have
adopted approaches in conflict with evolutionary theory (Hodos and Campbell
1969). For example, Yarczower and Hazlett (1977) have argued in favor of anagenesis, the linear ranking of species on a trait. But given the complexity of relationships
among existing species, it is hard to see how such linear ranking would be useful,
though it is possible.

The synthetic approach adopts a view of comparative research on animal intelligence that is based upon modern evolutionary theory. The essence of the approach
is to assume that the various processes composing animal intelligence have adaptive
effects and to use this assumption as a starting point for research, particularly comparative work. In this framework the goal of research is to develop a full understanding of animal intelligence at all relevant levels of explanation, including developmental, mechanistic, physiological, phylogenetic, and ecological levels. For
comparative work, this sets the goal of understanding patterns of similarities and
differences among species. The evolutionary framework offers several new research
strategies for the study of animal intelligence, discussed in the last section of this
chapter.

One important implication of the synthetic approach is that both qualitative and
quantitative differences between species are of interest. This is important for two
reasons. First, the distinction between qualitative and quantitative differences is
often a matter of individual judgment. Second, examining the comparative study
of morphological traits clearly shows that the distinction between qualitative and

quantitative differences is blurred. Understanding qualitative differences, particularly the relationship between qualitative differences and the ecology of the species in question, is a crucial part of developing a full understanding of the phenomena of interest.

For example, consider once more the comparative physiology of respiration. Those writing about the comparative study of learning often use respiration, or some other physiological system, as an analogy that may offer some guidance (e.g., Bolles 1985a; Revusky 1985). At some levels the respiratory system is the same in a wide variety of animals. For example, fish, amphibians, reptiles, birds, and mammals all use various hemoglobins to bind oxygen and transport it through the circulatory system. But at other levels respiratory systems differ dramatically. Many amphibians utilize a positive-pressure ventilation system to move air through the lungs. Mammals utilize negative-pressure ventilation in which pressure in the thoracic cavity is slightly lower than atmospheric pressure. Birds, in contrast, have a flow-through lung ventilation system that requires two respiratory cycles for the complete passage of a breath of air. These differences are related to various ecological correlates of the different niches of these organisms (Hainsworth 1981). Revusky (1985) uses the analogy between learning and respiration to argue for the existence of a general learning process. But the substantial variation in the respiratory systems of different animals can be used to reach another conclusion: that full understanding requires the analysis of differences among species as well as similarities.

The Emphasis on Both Laboratory and Field

Because the synthetic approach is evolutionary in orientation, it necessarily views events in the field, under natural conditions, as crucial. That is, it is assumed that the intellectual capacities of animals serve important biological, adaptive functions. Therefore studies of learning, memory, and so on under natural conditions can throw considerable light upon animal intelligence. In most cases coordinated laboratory and naturalistic research will be the most informative.

This coordinated approach to laboratory and field research on animal intelligence is important for two reasons. First, it addresses the problem of external validity raised earlier. If the principles of animal intelligence derived from laboratory research prove useful in the field, this will increase our confidence that important mechanisms of animal behavior have been successfully identified. Second, it is important for theoretical reasons. Since the synthetic approach depends heavily on identifying the specific ways animal intelligence affects biological success, field research will be necessary. These issues will permeate the rest of this chapter.

The Place of General Processes in the Synthetic Approach

The emphasis on general learning processes has been so pervasive that explicit discussion of their place in the synthetic approach could be valuable. Two extreme views about general processes can be identified (Bitterman 1975). The extreme general process view is that a single general process is responsible for all learning. The extreme anti–general process view, perhaps best exemplified by Lockard (1971), holds that there is no generality, that learning in each species is unique.

The synthetic approach views both these positions as too extreme. On the one hand, the available evidence, especially the research of Bitterman and his colleagues with honeybees (e.g., Abramson and Bitterman 1986; Bitterman et al. 1983; Couvillon and Bitterman 1984) clearly demonstrates impressive similarity in basic associative learning among diverse species. On the other hand, the demonstration of a general learning process present in many species does not rule out the possibility of important, significant species differences, both qualitative and quantitative.

Assume that animals use a host of processes to obtain environmental information and that some of these are quite general across species, others widespread but less general, and others very limited in distribution. A research program based upon the assumption of general processes would appear successful—general processes would be found. However, the less general processes would remain undiscovered. Furthermore, and more important for any comparative, evolutionary study of animal intelligence, differences among species and the adaptive role of cognitive processes outside the laboratory would remain unknown.

ARGUMENTS FOR INCREASED BREADTH

The synthetic approach calls for two broad changes in the traditional psychological approach to animal learning: increasing the breadth of phenomena being studied, and placing these phenomena in an evolutionary, ecological framework. In this section I will present the arguments for increased breadth.

Cognitive Processes in Animals

Perhaps the greatest challenge to the traditional approach from within psychology has been the emergence of the cognitive approach to animal learning. This development has been thoroughly documented in a number of publications (Hulse et al. 1978; Riley et al. 1986; Roitblat 1986; Roitblat et al. 1984). The cognitive approach emphasizes the internal states and processes of animals.

Organisms are assumed to have internal cognitive structures that depend on their individual development as well as their evolution. External objects cannot enter directly into an organism's cognitive system, and so they must be internally encoded—that is, "represented." Accordingly, much cognitive research involves techniques for studying the representations used by an organism, the processes that produce, maintain, and operate on them, and the environmental and situational factors that affect them (Roitblat et al. 1984, 2).

One important area of cognitive research focuses upon the "memory codes" animals use. For example, in a symbolic matching-to-sample task, the animal is first presented briefly with a single stimulus, the sample. Then it is presented with an array of test stimuli. Choice of one of the test stimuli will be reinforced. Which stimulus is correct depends upon which sample stimulus was previously presented. There are at least two ways the animal could code the sample information: retrospectively, by remembering the sample itself, or prospectively, by remembering which test stimulus would be correct. Roitblat (1980) found that errors tended to be directed toward test stimuli resembling the to-be-correct test stimulus, implying a prospective code. Cook et al. (1985) have obtained data in the radial maze implying that rats use both retrospective and prospective memory in this spatial task.

Another cognitive issue that has received a great deal of attention is animals' ability to time the duration of events. One procedure that has been used to study timing is the "peak procedure" of Roberts (1981). On most trials, rats receive food for bar pressing after a signal has been present for a fixed duration. On occasional probe trials, the signal remains on for a much longer period. When the rate of bar pressing on these probe trials is analyzed as a function of time into the trial, the response rate is highest at that point in time when food is usually presented on nonprobe trials. The process underlying this ability to gauge time appears to have many of the properties of a stopwatch. For example, the clock can be stopped or reset (Roberts 1983).

Another cognitive ability that has been extensively studied is counting. The major methodological problem facing research on counting, or sensitivity to numerosity, is how to demonstrate that behavior can be brought under the discriminative control of number and not any of the many other attributes that may correlate with number. Although not every study has addressed this problem, it has long been recognized (Koehler 1950; Thorpe 1956). Fernandes and Church (1982) presented rats with sequences of either two or four short sounds. If there were two sounds, the rat was reinforced for pressing a lever on the right. If there were four sounds, the rat was reinforced for pressing the level on the left. Not only did the rats perform accurately, but they maintained this accuracy when nonnumerical aspects of the sequences, such as stimulus duration and interstimulus intervals, were varied.

Davis and Memmott (1983) demonstrated sensitivity to sequentially presented stimuli with a much different procedure. Rats were trained to respond on a variable-interval food reinforcement schedule until they were responding steadily. They were then exposed to three unsignaled shocks during each session. Responding was initially suppressed, but after some time responding accelerated after the third shock, even though there was considerable variation in when during the session the shocks could occur. For example, in control sessions in which there were only two shocks, one early and one late, there was no acceleration of responding after the second shock, which came near the end of the session.

The existence of cognitive abilities such as counting, timing, and memory coding clearly challenge the traditional approach, especially radical behaviorism. The nature and implications of this challenge have been discussed in many places in the literature (e.g., Roitblat 1982, and replies: Riley et al. 1986). The cognitive approach is an alternative research program to radical behaviorism and also can be claimed to include associationism, since modern theories of association are very cognitive in nature. Furthermore, as I will discuss below, the various aspects of the cognitive approach fit very well with the synthetic approach, particularly when it comes to comparative, evolutionary issues.

Complex Learning in Animals

The cognitive approach has begun to emphasize more complex forms of animal learning, but many examples of research on complex learning remain to be integrated within the cognitive approach. In some cases these areas of research predate the emergence of the cognitive approach by many years.

One clear example of this is provided by the literature on object-discrimination learning set (Bessemer and Stollnitz 1971). In an object-discrimination learning set (ODLS) experiment, animals are given a series of discrimination problems to solve. Each problem is defined by the introduction of a new pair of stimuli, one arbitrarily designated as correct. Of main interest is an improvement in the speed of learning new problems, especially above chance choice on the second trial of new problems. Many primate species (Bessemer and Stollnitz 1971), as well as several avian species (Hunter and Kamil 1971; Kamil and Hunter 1969), have been shown to reach high levels of performance on the second trial of new problems.

The model that best accounts for ODLS performance in primates is a cognitive model. The basic idea is that the animals learn a pattern of choices descriptively labeled "win-stay, lose-shift." That is, on trial 2 of a new problem, they remember two aspects of what happened on trial 1: which stimulus was chosen and whether they received reinforcement. Then if they remember reinforcement (win) on trial 1, they choose the same stimulus on trial 2. If they remember nonreinforcement (lose) on trial 1, they shift their choice on trial 2. The results of many experiments on long-term and short-term memory, on the effects of switching stimuli between trials 1 and 2, on positive transfer from reversal learning to ODLS, and on stimulus preferences on trial 1 are all consistent with this model.

Despite this impressive literature, the ODLS phenomenon has been largely ignored by those working on animal learning. It apparently lies outside the realm of phenomena traditional workers are willing to consider. Given the apparent involvement of long- and short-term memory and strategy learning, it is particularly surprising that animal cognitive psychologists have ignored ODLS.

There are many other examples of complex learning in animals that are generally ignored, in the sense that no consistent attempt has been made to integrate these phenomena into a systematic cognitive-based scheme. These include evidence for categorical learning by pigeons (Herrnstein 1985), detailed spatial representational systems in a variety of organisms (bees: Gould 1987; primates: Menzel and Juno 1982, 1985), and various forms of reasoning in chimpanzees (Gillan et al. 1981).

These phenomena suggest that the cognitive approach needs to be expanded. At least to an outsider like me, it appears that many of the issues of central concern for animal cognitive psychologists originate in procedures used in the past. A good example of this point is provided by research on selective attention in animals. Some psychological work on selective attention has attempted to determine whether attention could account for certain phenomena such as reversal learning (Bitterman 1969; Mackintosh 1969). Other research has attempted to demonstrate attention to abstract dimensions, such as color or line orientation, in matching-to-sample tasks (e.g., Zentall et al. 1984). These types of research are very different and perhaps in the long run less informative than direct attempts to study selective attention and its characteristics. One area in which selective attention and its effects have been examined is research focused upon the detection of cryptic, hard-to-see prey. Selective attention appears to play a substantial role in prey detection (Bond 1983; Dawkins 1971a, 1971b; Pietrewicz and Kamil 1981). Animal cognitive psychology needs to broaden its scope and focus more directly on the information-handling processes of animals, with less focus on the particular issues generated by methodological developments of the past. The broad definition of intelligence offered by the synthetic approach would hasten this process.

Evidence from the Field: Social Knowledge

The emergence of behavioral ecology in the past 20 years has led to a dramatic increase in our knowledge of the behavior of individual animals in the field (see Krebs and Davies 1978, 1984). This literature contains many examples of data demonstrating that animals know a great deal about their environments, especially in two contexts: foraging and social behavior. In this section I discuss some of the data on social relationships. Data on foraging behavior will be reviewed later.

As I indicated at the very beginning of this chapter. many anecdotes based on observations in the field suggest that animals possess considerable knowledge about their world, particularly social interactions. Because anecdotes have generally been regarded as scientifically unacceptable, they are most often unreported. As Kummer

(1982) has observed, this is unfortunate. It has left each fieldworker aware only of his or her own observations.

My own experience confirms this. After observing the behavior of "Spot" described at the beginning of this chapter, I filed the incident away and for a long time never discussed it with anyone. One night, with some hesitation, I told the story to a group of fieldworkers. It turned out that another hummingbird researcher had seen a similar incident in another territorial species. Every fieldworker present that evening had stories that suggested animals possess more knowledge of their environment than typically considered by the laboratory researcher.

Although these are only anecdotes and their scientific validity is limited, it is time to take their implications seriously and begin to design experiments to test those implications. For example, Goodall (1986) reports many observations of the chimpanzees at Gombe that indicate these animals are acutely aware of the social relationships of their group. In Goodall's terminology, animals manipulate others and assess others' interactions. Are there any more systematic data to support these implications?

Kummer and his associates have tested some of these ideas in their research program with hamadryas baboons. Hamadryas baboons have a single-male, multiple-female social system in which males "appropriate" females. Kummer et al. (1974) found that if a male was allowed to watch another male with a female, this inhibited the tendency of the observing male to attempt to take over the female, even if the observing male was dominant to the other male. Something analogous to a concept of "ownership" appears to be present.

Even more intriguing, Bachmann and Kummer (1980) found that male hamadryas baboons assess the relationship between another male and a female. They tested twelve baboons, six of each sex. In the first stage they tested all possible different-sex pairs for grooming preference. This allowed the experimenters to construct a hierarchy of preference of each animal for each of the opposite-sexed animals. They then allowed males to watch another pair for 15 minutes. At the end of the 15-minute observation period, they gave the observer a graded set of opportunities to attempt to appropriate the female. They found that the observer assessed the relationship between the male and female he had been observing. The probability of the observer's attempting to appropriate the female depended on the female's preference for the original male. If that preference was weak, appropriation was more likely.

The research program of Cheney and Seyfarth is generating similar kinds of data for vervet monkeys. Cheney and Seyfarth (1980) conducted playback experiments in the field during which the scream of a juvenile was played through a hidden loudspeaker to groups of females that included the juvenile's mother. Mothers responded more strongly to these calls than the other females did. More surprisingly, the other females often responded by looking at the mother before the mother herself had reacted. This indicates that the females recognized the relationships of other females and young.

More recent data indicate that vervets have knowledge about other social relationships. Cheney and Seyfarth (1986) recorded the probabililty of agonistic encounters between members of a vervet group as a function of recent social interactions. There were two main findings. First, they found that individuals were more likely to behave aggressively toward other group members who had recently fought with their own kin, indicating that they know their own kin. Kin recognition is well known in many species. Second, Cheney and Seyfarth found that individuals were more likely to interact aggressively with others whose close kin had recently fought with their own kin. This indicates that vervet monkeys know about the relationships of other monkeys in their group. This appears to be learned, since monkeys under 3 years of age did not show the effect. How the relationships are learned is unknown.

Cheney and Seyfarth (1985) have argued that primate intelligence may have evolved primarily to deal with social relationships. Monkeys and apes clearly recognize social relationships and remember recent affiliative and aggressive interactions. But when tested for similar nonsocial knowledge, the monkeys appear surprisingly unresponsive. In various field experiments, vervets failed to respond to signs of predators. Cheney and Seyfarth's (1985) argument seems premature because these experiments on nonsocial knowledge may have failed to produce positive results for many reasons other than the monkeys' lack of knowledge. Nonetheless, their more general point about the importance of cognition in social settings deserves careful attention, not only in primates but in many group-living animals.

Conclusions

It is clear that a trend toward studying more complex forms of animal learning is well under way. It is important that this trend continue. Many unanticipated intellectual abilities have been revealed, and this implies that there are more waiting to be discovered.

Griffin (1976, 1978) has argued that interspecies communication offers an important tool for investigating the knowledge animals possess about their world. This is certainly true, and it is encouraging to see the technique being used with more species, including not only apes (Savage-Rumbaugh 1988) but birds (Pepperberg 1981, 1983), dolphins (Herman et al. 1984), and sea lions (Schusterman and Krieger 1986).

There are two general suggestions about how this search for complex processes in animals should proceed that I would like to make at this point. First, some research should concentrate primarily on what animals know, without worrying too much, for the time being, about how they acquire the knowledge. For example, the research of Premack and his associates with Sarah, a chimpanzee trained to use plastic symbols as a medium for communication, indicates that Sarah understands many relationships among stimuli. Although this research tells us little about how Sarah acquired this knowledge, it begins to tell us some of the things any complete theory of animal intelligence will have to be able to explain.

Second, it is important to continue to test animals in relatively unconstrained situations. It is quite possible that by restricting attention to experimental situations in which animals had few response alternatives and had to deal only with a few simple stimuli, psychologists have underestimated the abilities of their subjects. For example, the research of Menzel and Juno (1982, 1985) has demonstrated one-trial discrimination learning and extensive long-term memory for the spatial location of many objects in group-living marmosets, in marked contrast to the relatively poor performance of marmosets in more traditional experimental settings (e.g., Miles and Meyer 1956). The distinguishing features of the procedures of Menzel and Juno (1982, 1985) were probably the lack of constraints on the behavior of the marmosets and the use of knowledge about the natural foraging environment of these marmosets in designing the problems. These two characteristics were probably crucial to making it possible for the animals to demonstrate what they knew about their environment.

ARGUMENTS FOR A MORE BIOLOGICAL APPROACH

In this section I will review three areas of research—biological constraints on learning, "specialized" learning, and learning under natural conditions. The results of research in these three areas, considered together, provide convincing evidence that learning must be considered in a biological, evolutionary framework.

Biological Constraints on Learning

The phenomena that are usually called biological constraints on learning indicated the intrusion of biological factors into standard, traditional conditioning situations. Breland and Breland (1961) were the first to recognize the importance of constraints in operant conditioning situations. They observed what they called instinctive drift, a tendency for "natural behaviors" of animals undergoing operant conditioning to intrude upon and interfere with the emission of the response being reinforced. The Brelands clearly recognized the fundamental importance of their observations, which they viewed as a "demonstration that there are definite weaknesses in the philosophy underlying these [conditioning] techniques" (Breland and Breland 1961, 684). However, their findings had little effect at the time. The later discoveries of taste aversion learning, autoshaping, and species-specific defense reactions had more impact.

Taste aversion learning was first reported by Garcia and Koelling (1966). In essence, taste aversion learning suggests that some stimuli are more associable than others, challenging the often implicit assumption of associationists that stimuli are generally equipotential (Seligman 1970). These studies show that many animals are more likely to associate intestinal illness with gustatory (or olfactory) stimuli than with external stimuli. Garcia and Koelling (1966) proposed that these results demonstrate that rats may have a genetically coded hypothesis: "The hypothesis of the sick rat, as for many of us under similar circumstances, would be 'it must have been something I ate'" (Garcia and Koelling 1966, 124).

The phenomenon of autoshaping was first reported by Brown and Jenkins (1968). Brown and Jenkins found that if they simply illuminated a light behind a pecking key for a few seconds, then presented food, the pigeons began to peck the key even though these pecks had no effect on the presentation of the reinforcer. Although they felt that an appeal to some species-specific disposition was necessary, and though Breland and Breland reported many similar findings in less constrained situations, Brown and Jenkins do not cite the Brelands. The implication that species-specific predispositions affect the key peck has been confirmed. Jenkins and Moore (1973) showed that the topography of the pigeon's key peck depends on the reinforcer used. Mauldin (1981; Kamil and Mauldin 1987) found that three different passerine species each used species-specific response topologies in an autoshaping situation.

The concept of species-specific defense reactions originated in a seminal paper by Bolles (1970). Bolles argued that many of the results of avoidance conditioning experiments could best be understood in terms of the innate species-specific responses of the species being tested, such as fighting and fleeing. The opening sentence of his abstract was, "The prevailing theories of avoidance learning and the procedures that are usually used to study it seem to be totally out of touch with what is known about how animals defend themselves in nature" (Bolles 1970, 32).

I have been brief in describing these developments because there are already so many extensive reviews of biological constraints available in the literature (e.g., Seligman and Hager 1972; Hinde and Stevenson-Hinde 1973). And there is still considerable controversy about the extent to which these phenomena require abandoning any of the central assumptions of the traditional approach. For example, Revusky (1985) argues against radical behaviorism but also contends that taste aversion learning can be encompassed in a general associationist approach (see below).

There can be no doubt that these "biological constraints" on learning demonstrate that the evolutionary history of the species being studied can affect the outcome of a conditioning experiment. Whether the differences between taste aversion learning and other aversive conditioning are considered qualitative or quantitative, differences that seem most explicable in functional grounds do exist. The form of the response in a Skinner box depends on the natural repertoire of the animal, as do the results of avoidance learning experiments. However, the impact of these findings on the psychological study of animal learning has been limited.

The very label given to these phenomena, biological constraints on learning, reveals this limited impact. The label implies that there is some general process, learning, that is occasionally constrained by the biology of the organism (Kamil and Yoerg 1982). Surely a broader view is justified. The animal comes to the learning situation with a set of abilities that determine what behavioral changes will occur. These abilities are part of the animal's biological endowment. (I do not imply that they are completely genetically determined—clearly ontogenetic factors play an important role.) In that case a functional, evolutionary approach is necessary.

"Specialized" Learning

The value of a functional approach to the study of learning can be seen clearly in the literature on specialized learning. Specialized learning appears in specific biological contexts and plays very specific roles. Examples include song learning, imprinting, and homing/migration. In each of these cases, available data demonstrate that the phenomena in question meet any reasonable definition of learning—changes in behavior based on experience. The data also show important species differences in learning, which can often be related to differences in the natural history of species.

Naturalistic studies of nest and egg recognition by gulls and terns suggest the existence of important differences in learning among closely related species that correlate meaningfully with natural history (Shettleworth 1984). Royal terns nest in dense colonies where it is difficult to discriminate among nest sites. Their eggs are highly variable in appearance, and they learn to recognize their own eggs. Herring gulls build elaborate nests that are spaced farther apart, and they learn to recognize their nests but not their eggs. By the time the chicks are old enough to wander from the nest, the parents have learned to recognize them (Tinbergen 1953). Yet another pattern is shown by kittiwakes. These birds nest on cliff ledges, and their chicks do not (cannot) wander from the nest site. Parent kittiwakes recognize only their nest sites and do not discriminate their own eggs or young from those of others (Cullen 1957).

As Shettleworth (1984) has pointed out, these kinds of differences do not necessarily result from differences in learning *ability*. It may be that all the species have the same ability to learn to recognize their eggs, young, and nest sites, but natural circumstances of the species vary so as to favor one type of learning. For example, kittiwakes might learn to recognize their eggs if their eggs varied as much in appearance as do those of royal terns. The necessary experiments, such as placing eggs that vary in appearance in kittiwake nests, have not been carried out. However, this consideration does not apply to all examples of specialized learning.

In the case of song learning, at least some of the necessary experiments exploring differences in learning abilities have been done. Many male passerine birds sing songs that function both to attract a mate and to defend a territory against other males (Kroodsma 1982). In many species these songs are acquired through experience. Chaffinches, marsh wrens, white-crowned sparrows, and many other species must hear adult song when young to sing appropriately when mature. In many cases there are "dialects" of bird songs—different versions are observed in the same species in different geographical areas. The dialect an adult male sings often depends upon which dialect he heard during development.

The findings of Kroodsma and his associates on differences in song learning between eastern and western marsh wrens (currently classified as two subspecies) provide particularly clear evidence on differences in song learning between these two populations of marsh wrens. Kroodsma and Verner (1987) found that the normal repertoire size—the number of different songs sung by a single individual—varied considerably between the two populations. Eastern birds had repertoire sizes of about 30–60 songs while western birds had repertoire sizes of 120–220. While this could represent a difference in learning ability, it could also be the result of differences in early experience. It seem likely that the eastern wrens hear fewer songs when young than do western birds.

Kroodsma and Canady (1985) have performed the experiment necessary to distinguish between these possibilities. They raised eastern and western marsh wrens in identical laboratory environments. All subjects heard 200 tutor songs during development. Eastern birds learned 34–64 different songs, while the western wrens learned 90–113 songs under identical conditions. Furthermore, Kroodsma and Canady found significant differences in the size of the song control nuclei in the brains of the two groups. Eastern birds had smaller song control areas. The differences in song learning ability and neuroanatomy appear to be associated with several ecological differences between the populations, including year-round residency and high population densities in the western population.

Thus the evidence on song learning among passerine birds clearly demonstrates that species differences in ability exist. Many such differences are known, and they appear to correlate with natural history and ecology (Kroodsma 1983; West and King 1985). The finding that two subspecies of wrens learn different things from the same experience is particularly noteworthy. There can be important differences in specialized learning among extremely closely related animals. The question is whether such differences can be expected in more general types of learning.

The discussion of general and specific adaptations by Bolles (1985a) provides a good framework for this discussion. He points out that some adaptations are

> common, but unrelated, evolutionary adjustments to common circumstances. The phenomenon is called convergence, and color vision is an illustration of it. Full spectrum color vision pops up here and there in the evolutionary tree . . . it appears in some mammals, in most birds, in some fish, and in some of the arthropods. Animals in between are more or less color-blind . . . One way to think of color vision is that it has been discovered or invented several times independently (394).

Bolles (1985a) contrasts these reversible adaptations with others that apparently are not reversible, such as feathers:

Only birds have feathers. But the feather idea was apparently stupendously successful, because there are no birds without feathers. Once feathers came upon the scene, that was it, all descendants were stuck with feathers. Some birds (e.g., penguins) have funny feathers. . . . [Feathers] may change shape and size and color and waxiness and so on, but evidently if you have feathers you can depend upon all your descendants having feathers. . . . Is associative learning like feathers? Is the ability to learn such a stupendous advantage that once in possession of it, there is no way back? (394–95)

There can be little doubt that some specialized forms of learning are like color vision. Song learning appears scattered, albeit fairly widely, among passerines, varying significantly in its characteristics. The same may be said of imprinting. But are there forms of learning that are like feathers?

Bolles suggests that associative learning may be like feathers. The similarity in basic conditioning processes among widely different species suggest that this is so. The same argument can be made about the law of effect. The effects of reinforcement have also been demonstrated in many species. However, several points must be made about the analogy between feathers and learning.

First of all, even if some kinds of learning are like feathers, this does not mean there are not important differences between species in the learning. Although all feathers have certain features in common, they also vary. They are different at different stages of a bird's life and on different parts of a bird's body. And there are substantial variations between avian species. A large part of understanding feathers is understanding this variation. We need to examine even the most general kinds of learning for significant variation. To do so will require knowledge of the function of learning. I will return to this point later.

Second, even if there are general kinds of learning, this does not necessarily settle the question of homology and analogy. These concepts are labels for two very different possible evolutionary reasons for similarity between species. Homology is similarity through common evolutionary origin or descent. In contrast, analogy is similarity despite separate evolutionary origin because of similar adaptive pressure (see Atz 1970 for a discussion of the difficulties of applying these concepts to behavior). General forms of learning, unlike feathers, may have arisen two or more times during evolutionary history. For example, the similarities Bitterman and his co-workers have found between associative learning in honeybees and mammals may be the result of analogy, or convergence (Abramson and Bitterman 1986). It can be argued that the world is structured in such a way that any learning mechanisms that accurately and efficiently predicted events would have to have certain characteristics, namely, those that associative learning shows. (Dennett [1975] has argued that the law of effect must be part of any adequate and complete psychological theory. This philosophical argument implies that evolution may have invented the law of effect any number of times.)

Third, it would be premature at this time to attempt to decide whether any particular kind of learning is general. Biological variation, whether in general adaptations like feathers or in more specialized adaptations like color vision, requires some understanding of the function of the trait in question. Variation in feathers and in color vision relates to adaptive functioning. For example, in the case of color vision one can hypothesize that honeybees have color vision because they feed from colorful flowers (and this is exactly what made von Frisch [1954] so sure that honeybees did have color vision).

The problem is that in the case of possibly general processes of learning, we have little idea of their specific functions. One can reasonably speculate that associative learning is useful for an animal because it allows accurate prediction of future events. One can reasonably argue that the law of effect is useful because it allows the animal to obtain resources like food or water. But these are very general arguments and do not easily lead to the selection of particular species for study on ecological grounds. What is needed is some more definitive and specific idea of how learning and cognitive processes actually function under natural conditions. Fortunately, for the first time recent developments in behavioral ecology are making data relevant to this problem available in a substantial way.

Learning in the Field

Certain kinds of learning have long been known to occur in the natural world of animals: song learning and imprinting are the outstanding examples. But these are specialized forms of learning. Is there any evidence that the types of learning psychologists have typically been interested in occur outside the laboratory?

Many have maintained that learning in a more general sense is not important to animals under natural conditions (Boice 1977; Lockard 1971). This presented a problem to anyone attempting an evolutionary, adaptive approach to learning. If learning is unimportant in the field, why is it so evident in the laboratory? Do animals carry around what Boice called surplusage—unneeded and unnecessary abilities?

The problem appears to have been methodological, at least in part. Learning is much more difficult to observe than is learned behavior. Imagine a bird eating a monarch butterfly and subsequently throwing up. After that experience, it will simply avoid eating monarchs (Brower 1969). The scientist watching birds would have to see the brief first encounter to understand that later avoidance of monarchs was learned. This raises the second problem. The identification of learning requires documenting changes in the behavior of individuals over time. Until relatively recently, there were very few extended field studies of known or marked individuals. In the past 20 to 30 years such studies have become much more common, thanks in part to the emergence of behavioral ecology. These studies have revealed that animals in their natural environments face many problems that they appear to solve through learning and cognition (see Krebs and Davies 1978, 1984 for reviews of behavioral ecology; Shettleworth 1984 for an explicit discussion of the behavioral ecology of learning). For example, bumblebees learn how to handle different flower species and which flowers are most profitable (Heinrich 1979); nectar-feeding birds remember which flowers they have emptied (Gass and Montgomerie 1981; Kamil 1978); food-caching birds remember the locations of their stored food (Kamil and Balda 1985; Shettleworth and Krebs 1982) as well as the contents of the caches (Sherry 1984); and young vervet monkeys learn the social relationships among members of their groups (Cheney and Seyfarth 1986). In light of the accumulating evidence, it is difficult to conceive of anyone's believing that learning is not important in the natural world of animals outside the laboratory.

In addition to these empirical developments, important theoretical developments in behavioral ecology have emphasized the potential biological importance of learning. A variety of models have shown that if animals are sensitive to many of the features of their environment, they can increase the efficiency of their behavior. For example, the original "diet" selection model of MacArthur and Pianka (1966) assumes that predators know the nutritional value and density of their prey and the time required to handle it. Given that they possess this information and that they can rank prey types in terms of the ratio of nutritional value to handling time, a relatively simple rule can determine which prey types should be eaten whenever encountered and which ones should never be eaten in any given set of circumstances. Although this model has not been completely successful in predicting selection among prey types, it has had considerable success (for recent reviews see Krebs et al. 1983; Schoener 1987). Studies stimulated by this model have shown that animals respond adaptively to changes in the density of their prey (e.g., Goss-Custard 1981; Krebs et al. 1977) and learn to rank different prey types as the model predicts (Pulliam 1980). Other models have similarly predicted learning effects that have been confirmed by subsequent experiments (see Kamil and Roitblat 1985 for review; see Stephens and Krebs 1986 for detailed presentation of foraging theory, especially chap. 4).

There can be no doubt animals use learning to modify their behavior under natural conditions and that such learning can have very important adaptive implications. This is good news for the student of animal learning: the phenomena we have been interested in are biologically significant. However, we must also recognize the implications of this conclusion, the most central being that the study of learning must be placed in a biological context, and we must deal with the thorny problems this outlook raises.

In summary, then, three types of research indicate the need for a biological approach to learning: (1) studies of biological constraints, which clearly show that the evolutionary history of the species can affect the outcome of conditioning experiments in a variety of ways; (2) studies of specialized learning, which indicate that there can be significant variation in learning mechanisms that correlate with the ecologies of the species being studied; and (3) evidence from behavioral ecology, which shows that general forms of learning are of adaptive significance and may also, therefore, vary in ways that correlate with ecology.

THE IMPLICATIONS OF AN ADAPTIVE APPROACH
TO INTELLIGENCE

In earlier sections of this chapter, I argued that learning is adaptive and proposed that the synthetic approach should operate under that assumption. This assumption has important comparative implications, primarily that there *must be* significant variation in intelligence among species. Why is this a necessary implication?

Let us return to the feather analogy used by Bolles (1985a). He pointed out that learning might be like feathers—such a stupendously successful adaptation that, once developed, it could not be lost. Some might be tempted to use this analogy to argue that some adaptations are so successful that they simply do not vary significantly among species that possess them. This conclusion is not supported by available evidence on successful adaptations.

Feathers represent an extremely successful adaptation. But not all feathers are the same. Different types of feathers serve different functions and have different structures. Some feathers, such as down, serve as insulation. Other feathers function primarily in flight. Still others, the filoplumes, apparently serve as sensory organs, sensitive to the position of other feathers. Furthermore, within a feather type there can be considerable between-species variation in structure that is related to special adaptations. For example, the underside of an owl's wings has a velvety pile produced by special processes of the barbules, which reduces the sound of the wings when the owl swoops down on prey. Birds of the open sky have long primary flight feathers best suited to fast, straight flight, whereas woodland birds have shorter primaries that increase maneuverability. Diving birds have overlapping feathers that reduce drag (Lucas and Stettenheim 1972; Spearman and Hardy 1985). The list of functional variations of feathers is extremely long, even without mentioning perhaps the biggest source of variation, the evolution of brightly colored feathers for intraspecific display.

The point of this discussion of more than you (or I) ever wanted to know about feathers is that traits with adaptive functions vary between species, in ways that make sense in terms of the ecology and adaptations of the organisms they serve. If animal intelligence is adaptive—and as I have already stated, ample evidence of this is emerging—then intelligence must vary between species. The variation may be qualitative or quantitative; intelligence may consist of a complex of processes. But differences there must be. I cannot think of a single adaptive trait that does not vary in some way between species, often closely related species—the structure of the eye, the forelimb or hind limb, the stomach, the lungs. Why should animal intelligence be any different?

One reason animal intelligence could be different has been proposed by Shettleworth (1982, 1984): the distinction between function and mechanism. Shettleworth argues that because natural selection selects only among outcomes, not among the processes that produce them, any of a number of different mechanisms may be selected in any given situation. While this is true in global terms, it may well be false when examined in detail. Different mechanisms are unlikely to produce exactly the same outputs. In fact, as long as we are limited to input-output studies of the mechanisms of intelligence, we will classify two mechanisms producing the same results as the same mechanism (as would evolution).

However, as in the computer program example explored earlier, different mechanisms are likely to have different input-output relationships. If the input-output relationships differ, detailed analysis may prove that one mechanism is more functional than the other for problems the species faces. In that case natural selection will favor the more functional mechanism.

Returning to the main argument, my analysis of Macphail's approach to the evolution of intelligence among vertebrates suggests that his analysis is based upon prevailing, but unproductive, assumptions and definitions. Macphail recognized this possibility when he pointed out that "even the tentative advocacy of [the null] hypothesis is in effect a *reductio ad absurdum* which merely indicates that comparative psychology has followed a systematically incorrect route" (Macphail 1982, 334). That is exactly my contention. The challenge is to devise an alternative approach that can be used to investigate the evolution of animal intelligence while avoiding the snares that entangled Macphail and others.

Another potential problem with the literature upon which Macphail's analysis is based must also be noted. It is quite conceivable, perhaps even likely, that some mechanisms of intelligence are widespread throughout broad segments of the animal kingdom while others are not. Indeed, one could argue that the literature on classical conditioning demonstrates that the associationistic mechanisms involved are widespread whereas the literature on song learning, for example, demonstrates narrow distribution of song-learning mechanisms. It may be that the psychological study of animal learning has concentrated upon general mechanisms, ignoring those with more limited distribution. But some of these mechanisms of limited distribution may be more general, across tasks, than very specific forms of learning like song learning. In particular, some more complex forms of learning—so far little studied outside a few primate or avian species—deserve comparative attention (Humphrey 1976).

RESEARCH STRATEGIES

The purpose of this section is to propose a set of research strategies to further our knowledge of animal intelligence. In outlining these strategies I have been guided by the two criticisms of the traditional approach developed earlier: that we know relatively little about the intellectual capacities of animals and that we understand very little about how these capacities function or evolved. I have also sought to develop a set of strategies that will avoid the problems revealed by analysis of Macphail's review of the comparative literature on animal learning.

There are two components to any strategy for studying animal intelligence: selecting the procedures to be used and selecting the species to be studied. These are not unrelated problems. Research will proceed most readily if there is a good match between the task employed and the species under study.

These suggested research strategies originate from several considerations: (1) the characteristics of research that has produced good evidence for complex intelligent processes in animals; (2) the decision-making processes that are being revealed by laboratory and field research in behavioral ecology; and (3) an examination of the biological approach to comparative research.

Developing a Natural History of Animal Intelligence

One important step to developing a new approach to the comparative study of animal intelligence will be to develop a natural history of animal intelligence. This would consist of a detailed study of intelligence under natural conditions. The focus would be upon the problems animals are faced with in the field and how they use their mental capacities to solve them. In many cases field experiments or laboratory work closely coupled to natural history would be necessary.

I have already referred to many examples of field data that demonstrate or suggest how intelligence is used to solve the problems nature presents to animals. These include timing in hummingbirds, spatial memory in food-storing animals, and knowledge of social relationships in primates. The two major arenas for the operation of animal intelligence are foraging and social behavior. These areas need to be examined much more closely, and in a wider variety of species, from the point of view of the functional significance of animal intelligence.

Using Natural History to Choose Species and Design Procedures

Once the study of natural history has revealed a particular problem that is (or might be) solved by learning in the field, this knowledge can be used to select species for study and to design experimental procedures for testing. This is a strategy ethologists have used with considerable success in studying "specialized" learning such as song learning, imprinting, and migration. There are also a number of examples of this approach dealing with processes that may be more general: these include the detection of cryptic prey (Bond 1983; Pietrewicz and Kamil 1981); spatial memory in food-caching parids (Sherry 1984; Sherry et al. 1981; Shettleworth and Krebs 1982) and nutcrackers (Balda 1980; Kamil and Balda 1985); and pitch perception in starlings (Hulse et al. 1984).

Another approach has been to design experimental situations to test models of natural behavior, particularly optimal foraging models. For example, there have been tests of patch selection (Smith and Sweatman 1974), within-patch persistence (Cowie 1977; Kamil and Yoerg 1985; Kamil et al. 1988), and collecting food to be brought to a central place (Kacelnik 1984; Kacelnik and Cuthill 1987). One problem with some of these studies is that researchers sometimes fail to consider whether the species they choose to study are appropriate for the model they wish to test.

This raises the general point of evaluating ecological validity. It is relatively easy to argue that laboratory tasks should reflect the problems animals normally face in nature. But it is not so easy to judge how well any particular task meets that requirement. The best way to address this issue is to collect laboratory data that can be compared with effects known to occur in the field. For example, when Pietrewicz and I were first developing our procedure for studying cryptic prey detection by training jays to detect cryptic moths in slides, we collected data that could be checked against phenomena known to occur in the field. We found that the moths in the slides were least detectable by the jays when shown in their species-typical body orientation (Pietrewicz and Kamil 1977). The jays slow their search immediately after finding a moth (unpublished data), a result identical to the "area-restricted search" often observed in the field (Croze 1970). They also search more slowly when the prey are more cryptic (Getty et al. 1987), an effect also analogous to data collected in the field (Fitzpatrick 1981). These isomorphisms between laboratory and field mean that when we investigate parameters that cannot be studied under natural conditions, there is some reason to believe the results are applicable to the field.

We hope that adopting this research strategy based upon natural history will have two effects: first, that it will lead to a clearer and fuller understanding of animal intelligence; second, that it will change the focus of research on animal learning and cognition, making it more animal oriented and less process oriented. This will allow greater integration with organismal biology. It will also focus more attention on a crucial evolutionary issue, the adaptive significance of animal intelligence. But it will not solve the problem of contextual stimuli and the difficulty of establishing that species differences in learning or cognition even exist. However, the synthetic approach does suggest some ways around this problem.

Using External Criteria to Make Comparative Predictions

One way to minimize the problem of using contextual stimuli as an alternative explanation for species differences is to have some external criterion that predicts differences among a number of species. For example, Rumbaugh and Pate (1984) have used an index of encephalization to predict species differences among seven nonhuman primate species on a complex learning task. The encephalization index accurately predicts the performance of the species. Since there are many predictions, supported in detail by the comparative data, contextual stimuli do not provide a likely alternative explanation. The probability that contextual stimuli will produce a ranking of nine species by chance is exceedingly small. Thus the use of an external criterion to make a priori specific predictions provides an explicit alternative to the null hypothesis of no species differences. If this alternative makes many predictions and these are supported, then contextual stimuli cannot be taken seriously as an explanation.

Indexes of brain size or encephalization provide one source of external predictions. These indexes may be particularly useful for comparing closely related species, as in Rumbaugh's research program. Natural history and ecological considerations can provide another source of a priori predictions of species differences in animal intelligence. If some animals face specific foraging or social problems that others do not face, and if learning is used to solve these problems, then a comparative prediction is at least implicit. For example, do food-storing birds have a greater ability to remember spatial locations than other birds? Are animals that utilize food resources that are renewed on a strong temporal schedule, like traplining hummingbirds, better at timing? Are animals that live in stable, long-lasting social groups better able to learn about social relationships either between themselves and others or among others?

The key to overcoming the problem posed by contextual variables is generating multiple predictions about species differences. The ecological approach leads to such multiple predictions because of the processes of convergence and divergence. Divergence refers to differences between closely related species owing to differences in their ecologies. The differences in the beaks of the Galápagos finches are the classic case. Convergence refers to similarities between distantly related species because of similar ecological pressures and adaptations. For example, nectar feeding has evolved independently among many groups of birds, including the hummingbirds of North and South America, the honeycreepers of Hawaii, and the sunbirds of Africa and Asia. Many of these birds have decurved beaks that are well suited to extracting nectar from flowers.

The ways convergence and divergence can be used to generate multiple predictions can be seen by considering a specific example. Suppose one hypothesized that nectar feeders should have particularly good spatial memory (Kamil 1978) or timing ability (Gill 1988). This hypothesis could be tested by comparing closely related animals, only some of which feed on nectar, such as the Hawaiian honeycreepers, which vary enormously in foraging specializations. Any supporting evidence could then be tested with other groups of nectar-feeding birds. It could also be tested by doing comparative research with other groups that include nectar feeders, such as bats.

The strategy of selecting species for study based upon convergence and divergence can be applied to many aspects of animal intelligence. For example, if the social context has been crucial for the evolution of learning, as Cheney and Seyfarth (1985) suggest, then at least some of the phenomena observed in group-living primates should be found in some avian species. Many birds have long life spans spent in stable groups with established genealogies (e.g., Florida scrub jays: Woolfenden and Fitzpatrick 1984; bee-eaters: Emlen 1981). Some of these groups have been studied for as long as 20 years. The findings reported suggest that these birds may be making judgments of the sort described for primates, but the appropriate data have not been collected. It would be important to collect them.

Using Specific Processes to Generate Multiple Predictions

Another way to minimize the interpretive problems posed by contextual variables is to design several experimental procedures, each measuring the same intellectual ability, and test two or more species with all the procedures. The species tested should be chosen with some external criterion so that specific predictions are made in advance. Then if the results of each of the procedures indicate the same ordering of the species, contextual variables are unlikely to be responsible.

One example of this strategy can be found in ongoing research on spatial memory in Clark's nutcrackers. These birds are known to use spatial memory in recovering their caches (Balda 1980; Balda et al. 1986; Kamil and Balda 1985). This memory is remarkable in at least two ways: it is long lasting and of large capacity. We have found that nutcrackers perform better than pigeons in an open field analogue of the radial maze (Balda and Kamil 1988). Data collected by Olson (1991) indicate that the nutcrackers also perform better than pigeons in a spatial operant task. As data from different settings accumulate and are consistent in showing that nutcrackers remember spatial locations better than pigeons, our confidence that there is a species difference in cognitive ability increases.

CONCLUSIONS

In this chapter I have argued for a new, broader approach to studying the evolution of the cognitive capacities of animals. This synthetic approach is based upon several arguments. (1) Data from the natural world of animals as well as from the laboratory clearly show that the intellectual capacities of animals are greater than previously thought. This means that we need to use a broad definition of animal intelligence. (2) The traditional psychological approach to the study of animal learning has been defined too narrowly, and its logic has prevented meaningful comparative, evolutionary analysis. (3) The literature on several phenomena, including constraints on learning and "specialized" learning, indicates that an approach based on research strategies drawn from biology and behavioral ecology can be useful in analyzing the evolution of animal intelligence. (4) As a prerequisite to engaging in a meaning comparative analysis of animal cognition, we must develop hypotheses that make multiple and detailed predictions about species differences in intelligence. Natural history and behavioral ecology are important sources of such hypotheses.

We have a great deal yet to learn about the cognitive abilities of animals. If we adopt a broad approach, using the best of what psychology and biology have to offer, we are most likely to succeed in our efforts to understand these abilities and their evolution. The next 20 years of research on these problems should be very exciting.

ACKNOWLEDGMENTS

The ideas presented in this chapter have undergone a long, and still incomplete, development. During this time support has been received from the National Science Foundation (BNS 84-18721 and BNS 85-19010 currently), the National Institute of Mental Health, and the University of Massachusetts. I have also been stimulated by discussions, conversations, and arguments with many individuals. I would particularly like to thank Robert L. Gossette for first igniting my interest in the comparative study of learning, Daniel S. Lehrman and Robert Lockard for first directing my attention toward biology and ecology, and Charles Van Riper III for this guidance during my first research experience outside the cloisters of the laboratory. I would also like to thank Sonja I. Yoerg, Kevin Clements, and Deborah Olson for their comments and suggestions on a previous version of this chapter.

REFERENCES

Abramson, C. I., and M. E. Bitterman. 1986. Latent inhibition in honey bees. *Animal Learning and Behavior* 14:184–89.

Andrews, E. A., and N. S. Braveman. 1975. The combined effects of dosage level and interstimulus interval on the formation of one-trial poison-based aversions in rats. *Animal Learning and Behavior* 3:287–89.

Atz, J. W. 1970. The application of the idea of homology to behavior. In L. R. Aronson, E. Tobach, D. S. Lehrman, and J. S. Rosenblatt, eds., *Development and evolution of behavior.* San Francisco: W. H. Freeman.

Bachmann, C., and H. Kummer. 1980. Male assessment of female choice in hamadryas baboons. *Behavioral Ecology and Sociobiology* 6:315–21.

Balda, R. P. 1980. Recovery of cached seeds by a captive *Nucifraga caryotactes. Zeitschrift für Tierpsychologie* 52:331–46.

Balda, R. P., and A. C. Kamil. 1988. The spatial memory of Clark's nutcrackers (*Nucifraga columbiana*) in an analog of the radial-arm maze. *Animal Learning and Behavior* 16:116–22.

Balda, R. P., A. C. Kamil, and K. Grim. 1986. Revisits to emptied cache sites by nutcrackers. *Animal Behaviour* 34:1289–98.

Baron, M., A. Kaufman, and D. Fazzini. 1969. Density and delay of punishment of free-operant avoidance. *Journal of the Experimental Analysis of Behavior* 12:1029–37.

Beach, F. A. 1950. The snark was a boojum. *American Psychologist* 5:115–24.

Bessemer, D. W., and F. Stollnitz. 1971. Retention of discriminations and an analysis of learning set. In A. M. Schrier and F. Stollnitz, eds., *Behavior of nonhuman primates,* vol. 4. New York: Academic Press.

Bitterman, M. E. 1960. Toward a comparative psychology of learning. *American Psychologist* 15:704–12.

Bitterman, M. E. 1965. Phyletic differences in learning. *American Psychologist* 20:396–410.

Bitterman, M. E. 1969. Habit reversal and probability learning: Rats, birds and fish. In R. Gilbert and N. S. Sutherland, eds., *Animal discrimination learning.* New York: Academic Press.

Bitterman, M. E. 1975. The comparative analysis of learning. *Science* 188:699–709.

Bitterman, M. E., R. Menzel, A. Fietz, and S. Schafer. 1983. Classical conditioning of proboscis extension in honeybees (*Apis mellifera*). *Journal of Comparative Psychology* 97:107–19.

Boice, R. 1977. Surplusage. *Bulletin of the Psychonomic Society* 9:452–54.

Bolles, R. C. 1970. Species-specific defense reactions and avoidance learning. *Psychological Review* 77:32–48.

Bolles, R. C. 1985a. The slaying of Goliath: What happened to reinforcement theory. In T. D. Johnston and A. T. Pietrewicz, eds., *Issues in the ecological study of learning.* Hillsdale, N.J.: Lawrence Erlbaum Associates.

Bolles, R. C. 1985b. Short term memory and attention. In L. Nilsson and T. Archer, eds., *Perspectives on learning and memory.* Hillsdale, N.J.: Lawrence Erlbaum Associates.

Bond, A. B. 1983. Visual search and selection of natural stimuli in the pigeon. *Journal of Experimental Psychology: Animal Behavior Processes* 9:292–306.

Breland, K., and M. Breland. 1961. The misbehavior of organisms. *American Psychologist* 16:681–84.

Brower, L. P. 1969. Ecological chemistry. *Scientific American* 220 (2):22–29.

Brown, P. L., and H. M. Jenkins. 1968. Auto-shaping of the pigeon's key peck. *Journal of Experimental Analysis of Behavior* 11:1–8.

Cheney, D. L., and R. M. Seyfarth. 1980. Vocal recognition in free-ranging vervet monkeys. *Animal Behaviour* 28:362–67.

Cheney, D. L., and R. M. Seyfarth. 1985. Social and non-social knowledge in vervet monkeys. In L. Weiskrantz, ed., *Animal intelligence.* Oxford: Clarendon Press.

Cheney, D. L., and R. M. Seyfarth. 1986. The recognition of social alliances by vervet monkeys. *Animal Behaviour* 34:1722–31.

Cheney, D. L., R. Seyfarth, and B. Smuts. 1986. Social relationships and social cognition in nonhuman primates. *Science* 234:1361–66.

Cook, R. G., M. F. Brown, and D. A. Riley. 1985. Flexible memory processing by rats: Use of prospective and retrospective information in the radial maze. *Journal of Experimental Psychology: Animal Behavior Processes* 11:453–69.

Couvillon, P. A., and M. E. Bitterman. 1984. The over-learning extinction effect and successive negative contrast in honeybees (*Apis mellifera*). *Journal of Comparative Psychology* 98:100–109.

Cowie, R. J. 1977. Optimal foraging in great tits (*Parus major*). *Nature* 268:137–39.

Croze, H. J. 1970. Searching image in carrion crows. *Zeitschrift für Tierpsychologie*, Suppl. 5:1–85.

Cullen, E. 1957. Adaptations in the kittiwake to cliff-nesting. *Ibis* 99:275–302.

Davis, H., and J. Memmott. 1983. Autocontingencies: Rats count to three to predict safety from shock. *Animal Learning and Behavior* 11:95–100.

Dawkins, M. 1971a. Perceptual changes in chicks: Another look at the "search image" concept. *Animal Behaviour* 19:566–74.

Dawkins, M. 1971b. Shifts in "attention" in chicks during feeding. *Animal Behaviour* 19:575–82.

Dennett, D. C. 1975. Why the law of effect will not go away. *Journal of the Theory of Social Behaviour* 5:169–87.

Emlen, S. T. 1981. Altruism, kinship and reciprocity in the white-fronted bee-eater. In R. D. Alexander and D. W. Tinkle, eds., *Natural selection and social behavior: Recent research and new theory*. New York: Chiron Press.

Fernandes, D. M., and R. M. Church. 1982. Discrimination of the number of sequential events by rats. *Animal Learning and Behavior* 10:171–76.

Fitzpatrick, J. W. 1981. Search strategies of tyrant flycatchers. *Animal Behaviour* 29:810–21.

Frisch, K. von. 1954. *Dancing bees: An account of the life and senses of the honey bee* (trans. Dora Ilse). London: Methuen.

Garcia, J., and R. A. Koelling. 1966. Relation of cue to consequences in avoidance learning. *Psychonomic Science* 4:123–24.

Gardner, H. 1978. *Developmental psychology*. Boston: Little, Brown.

Gardner, H. 1982. *Frames of mind: The theory of multiple intelligence*. New York: Basic Books.

Gardner, R. A., and B. T. Gardner. 1969. Teaching sign language to a chimpanzee. *Science* 165:664–72.

Gass, C. L., and R. D. Montgomerie. 1981. Hummingbird foraging behavior: Decision-making and energy regulation. In A. C. Kamil and T. D. Sargent, eds., *Foraging behavior: Ecological, ethological, and psychological approaches*. New York: Garland Press.

Getty, T., A. C. Kamil, and P. G. Real. 1987. Signal detection theory and foraging for cryptic or mimetic prey. In A. C. Kamil, J. R. Krebs, and H. R. Pulliam, eds., *Foraging behavior*. New York: Plenum.

Gill, F. B. 1988. Trapline foraging by hermit hummingbirds: Competition for an undefended, renewable resource. *Ecology* 69:1933–42.

Gill, F. B., and L. L. Wolf. 1977. Nonrandom foraging by sunbirds in a patchy environment. *Ecology* 58:1284–96.

Gillan, D. J., D. Premack, and G. Woodruff. 1981. Reasoning in the chimpanzee. 1. Analogical reasoning. *Journal of Experimental Psychology: Animal Behavior Processes* 7:1–17.

Goodall, J. 1986. *The chimpanzees of Gombe: Patterns of behavior*. Cambridge, Mass.: Harvard University Press.

Goss-Custard, J. D. 1981. Feeding behavior of red-shank, *Tringa totanus*, and optimal foraging theory. In A. C. Kamil and T. D. Sargent, eds., *Foraging behavior: Ecological, ethological, and psychological approaches*. New York: Garland Press.

Gossette, R. L. 1967. Successive discrimination reversal (SDR) performance of four avian species on a brightness discrimination task. *Psychonomic Science* 8:17–18.

Gossette, R. L., M. F. Gossette, and N. Inman. 1966. Successive discrimination reversal performance by the greater hill myna. *Animal Behaviour* 14:50–53.

Gould, J. L. 1987. Landmark learning by honey bees. *Animal Behaviour* 35:26–34.

Griffin, D. R. 1976. *The question of animal awareness: Evolutionary continuity of mental experience*. New York: Rockefeller University Press.

Griffin, D. R. 1978. Prospects for a cognitive ethology. *Behavioral and Brain Sciences* 1:527–38.

Hainsworth, F. R. 1981. *Animal physiology*. Reading, Mass.: Addison-Wesley.

Heinrich, B. 1979. *Bumblebee economics*. Cambridge, Mass.: Harvard University Press.

Herman, L. M., J. P. Wolz, and D. G. Richards. 1984. Comprehension of sentences by bottle-nosed dolphins. *Cognition* 16:1–90.

Herrnstein, R. J. 1970. On the law of effect. *Journal of the Experimental Analysis of Behavior* 13:243–66.

Herrnstein, R. J. 1985. Riddles of natural categorization. In L. Weiskrantz, ed., *Animal intelligence*. Oxford: Clarendon Press.

Hinde, R. A., and J. Stevenson-Hinde. 1973. *Constraints on learning.* New York: Academic Press.

Hodos, W., and C. B. G. Campbell. 1969. *Scala naturae:* Why there is no theory in comparative psychology. *Psychological Review* 76:337–50.

Hulse, S. H., H. Fowler, and W. K. Honig. 1978. *Cognitive processes in animal behavior.* Hillsdale, N.J.: Lawrence Erlbaum Associates.

Hulse, S. H., J. Cynx, and J. Humpal. 1984. Cognitive processing of pitch and rhythm structures by birds. In H. L. Roitblat, T. G. Bever, and H. S. Terrace, eds., *Animal cognition.* Hillsdale, N.J.: Lawrence Erlbaum Associates.

Humphrey, N. K. 1976. The social function of intellect. In P. P. G. Bateson and R. A. Hinde, eds., *Growing points in ethology.* Cambridge: Cambridge University Press.

Hunter, M. W., and A. C. Kamil. 1971. Object-discrimination learning set and hypothesis behavior in the northern bluejay. *Psychonomic Science* 22:271–73.

Jenkins, H. M., and B. R. Moore. 1973. The form of the autoshaped response with food and water reinforcers. *Journal of the Experimental Analysis of Behavior* 20:163–81.

Kacelnik, A. 1984. Central place foraging in starlings (*Sturnus vulgaris*). 1. Patch residence time. *Journal of Animal Ecology* 53:283–300.

Kacelnik, A., and I. C. Cuthill. 1987. Starlings and optimal foraging theory: Modelling in a fractal world. In A. C. Kamil, J. R. Krebs, and H. R. Pulliam, eds., *Foraging behavior.* New York: Plenum.

Kamil, A. C. 1978. Systematic foraging by a nectar-feeding bird, the amakihi (*Loxops virens*). *Journal of Comparative and Physiological Psychology* 92:388–96.

Kamil, A. C. 1987. Sensory biology and behavioral ecology. In A. M. Popper and J. Atema, eds., *Sensory biology of aquatic animals.* New York: Springer-Verlag.

Kamil, A. C., and R. Balda. 1985. Cache recovery and spatial memory in Clark's nutcrackers (*Nucifraga columbiana*). *Journal of Experimental Psychology: Animal Behavior Processes* 11:95–111.

Kamil, A. C., and M. Hunter. 1969. Performance on object-discrimination learning-set by the greater hill myna (*Gracula religiosa*). *Journal of Comparative and Physiological Psychology* 73:68–73.

Kamil, A. C., and J. E. Mauldin. 1987. A comparative-ecological approach to the study of learning. In R. C. Bolles and M. D. Beecher, eds., *Evolution and learning.* Hillsdale, N.J.: Lawrence Erlbaum Associates.

Kamil, A. C., and H. L. Roitblat. 1985. The ecology of foraging behavior: Implications for animal learning and memory. *Annual Review of Psychology* 36:141–69.

Kamil, A. C., and T. Sargent. 1981. *Foraging behavior: Ecological, ethological and psychological approaches.* New York: Garland Press.

Kamil, A. C., and S. I. Yoerg. 1982. Learning and foraging behavior. In P. P. G. Bateson and P. H. Klopfer, eds., *Perspectives in ethology,* vol. 5. New York: Plenum.

Kamil, A. C., and S. I. Yoerg. 1985. Effects of prey depletion on patch choice by foraging blue jays (*Cyanocitta cristata*). *Animal Behaviour* 33:1089–95.

Kamil, A. C., T. B. Jones, A. T. Pietrewicz, and J. Mauldin. 1977. Positive transfer from successive reversal training to learning set in blue jays. *Journal of Comparative and Physiological Psychology* 91:79–86.

Kamil, A. C., J. R. Krebs, and H. R. Pulliam. 1987. *Foraging behavior.* New York: Plenum.

Kamil, A. C., S. I. Yoerg, and K. C. Clements. 1988. Rules to leave by: Patch departure in foraging blue jays. *Animal Behaviour* 36:843–53.

Koehler, O. 1950. The ability of birds to "count." *Bulletin of Animal Behaviour* 9:41–45.

Krebs, J. R., and N. B. Davies. 1978. *Behavioural ecology: An evolutionary approach.* Oxford: Blackwell Scientific Publications.

Krebs, J. R., and N. B. Davies. 1984. *Behavioural ecology: An evolutionary approach.* 2d ed. Oxford: Blackwell Scientific Publications.

Krebs, J. R., T. J. Erichsen, M. I. Webber, and E. L. Charnov. 1977. Optimal prey selection in the great tit (*Parus major*). *Animal Behaviour* 25:30–38.

Krebs, J. R., D. Stephens, and W. Sutherland. 1983. Perspectives in optimal foraging. In A. H. Brush and G. A. Clark, Jr., eds., *Perspectives in ornithology.* Cambridge: Cambridge University Press.

Kroodsma, D. E. 1982. Song repertoires: Problems in their definition and use. In D. E. Kroodsma and E. H. Miller, eds., *Acoustic communication in birds,* vol. 2. New York: Academic Press.

Kroodsma, D. E. 1983. The ecology of avian vocal learning. *BioScience* 33:165–71.

Kroodsma, D. E., and R. A. Canady. 1985. Differences in repertoire size, singing behavior, and associated neuroanatomy among marsh wren populations have a genetic basis. *Auk* 102:439–46.

Kroodsma, D. E., and J. Verner. 1987. Use of song repertoires among marsh wren populations. *Auk* 104:63–72.

Kummer, H. 1982. Social knowledge in free-ranging primates. In D. R. Griffin, ed., *Animal mind–human mind.* Berlin: Springer-Verlag.

Kummer, H., and J. Goodall. 1985. Conditions of innovative behaviour in primates. In L. Weiskrantz, ed., *Animal intelligence.* Oxford: Clarendon Press.

Kummer, H., W. Gotz, and W. Angst. 1974. Triadic differentiation: An inhibitory process protecting pair bonds in baboons. *Behaviour* 49:62–87.

Lakatos, I. 1974. *The methodology of scientific research programs.* Cambridge: Cambridge University Press.

Lockard, R. B. 1971. Reflections on the fall of comparative psychology: Is there a message for us all? *American Psychologist* 26:168–79.

Lorenz, K. 1965. *The evolution and modification of behavior.* Chicago: University of Chicago Press.

Lucas, A. M., and P. R. Stettenheim. 1972. *Avian anatomy: Integument.* Agricultural Handbook 362. Washington, D.C.: United States Department of Agriculture.

MacArthur, R. H., and E. R. Pianka. 1966. On optimal use of a patchy environment. *American Naturalist* 100:603–9.

Mackintosh, N. J. 1969. Habit-reversal and probability learning: Rats, birds and fish. In R. Gilbert and N. S. Sutherland, eds., *Animal discrimination learning.* New York: Academic Press.

Macphail, E. M. 1982. *Brains and intelligence in vertebrates.* Oxford: Clarendon Press.

Macphail, E. M. 1985. Vertebrate intelligence: The null hypothesis. In L. Weiskrantz, ed., *Animal intelligence.* Oxford: Clarendon Press.

Mauldin, J. E. 1981. Autoshaping and negative automaintenance in the blue jay (*Cyanocitta cristata*), robin (*Turdus migratorius*) and starling (*Sturnus vulgarius*). Ph.D. dissertation, University of Massachusetts, Amherst.

Mazur, J. E. 1986. *Learning and behavior.* Englewood Cliffs, N.J.: Prentice-Hall.

Menzel, E. W., Jr., and C. Juno. 1982. Marmosets (*Saguinus fuscicollis*): Are learning sets learned? *Science* 217:750–52.

Menzel, E. W., Jr., and C. Juno. 1985. Social foraging in marmoset monkeys and the question of intelligence. In L. Weiskrantz, ed., *Animal intelligence.* Oxford: Clarendon Press.

Miles, R. C. and D. R. Meyer. 1956. Learning sets in marmosets. *Journal of Comparative and Physiological Psychology* 49:219–22.

Olson, D. 1991. Species differences in spatial memory among Clark's nutcrackers, scrub jays and pigeons. *Journal of Experimental Psychology: Animal Behavior Processes* 17:363–76.

Pepperberg, I. M. 1981. Functional vocalizations of an African gray parrot (*Psittacus erithacus*). *Zeitschrift für Tierpsychologie* 55:139–51.

Pepperberg, I. M. 1983. Cognition in the African gray parrot: Preliminary evidence for auditory/vocal comprehension of the class concept. *Animal Learning and Behavior* 11:179–85.

Pietrewicz, A. T., and A. C. Kamil. 1977. Visual detection of cryptic prey by blue jays (*Cyanocitta cristata*). *Science* 195:580–82.

Pietrewicz, A. T., and A. C. Kamil. 1981. Search images and the detection of crypic prey: An operant approach. In A. C. Kamil and T. D. Sargent, eds., *Foraging behavior: Ecological, ethological, and psychological approaches*. New York: Garland Press.

Pulliam, H. 1980. Do chipping sparrows forage optimally? *Ardea* 68:75–82.

Rescorla, R. A. 1985. Associationism in animal learning. In L. Nilsson and T. Archer, eds., *Perspectives on learning and memory*. Hillsdale, N.J.: Lawrence Erlbaum Associates.

Revusky, S. 1985. The general process approach to animal learning. In T. D. Johnson and A. T. Pietrewicz, eds., *Issues in the ecological study of learning*. Hillsdale, N.J.: Lawrence Erlbaum Associates.

Ricciardi, A. M., and F. R. Treichler. 1970. Prior training influences on transfer to learning set by squirrel monkeys. *Journal of Comparative and Physiological Psychology* 73:314–19.

Riley, D. A., M. F. Brown, and S. I. Yoerg. 1986. Understanding animal cognition. In T. J. Knapp and L. C. Robertson, eds., *Approaches to cognition: Contrasts and controversies*. Hillsdale, N.J.: Lawrence Erlbaum Associates.

Roberts, S. 1981. Isolation of an internal clock. *Journal of Experimental Psychology: Animal Behavior Processes* 7:242–68.

Roberts, S. 1983. Properties and function of an internal clock. In R. L. Mellgren, ed., *Animal cognition and behavior*. New York: North-Holland.

Roitblat, H. L. 1980. Codes and coding processes in pigeon short-term memory. *Animal Learning and Behavior* 8:341–51.

Roitblat, H. L. 1982. The meaning or representation in animal memory. *Behavioral and Brain Sciences* 5:353–72.

Roitblat, H. L. 1986. *Introduction to comparative cognition*. San Francisco: W. H. Freeman.

Roitblat, H. L., T. Bever, and H. Terrace. 1984. *Animal cognition*. Hillsdale, N.J.: Lawrence Erlbaum Associates.

Rumbaugh, D. M. 1977. *Language learning by a chimpanzee: The LANA project*. New York: Academic Press.

Rumbaugh, D. M., and J. L. Pate. 1984. The evolution of cognition in primates: A comparative perspective. In H. L. Roitblat, T. G. Bever, and H. S. Terrace, eds., *Animal cognition*. Hillsdale, N.J.: Lawrence Erlbaum Associates.

Savage-Rumbaugh, S. 1988. A new look at ape language: Comprehension of vocal speech and syntax. In D. Leger, ed., *Comparative Perspectives in Modern Psychology* (Nebraska *Symposium on Motivation, 1987*, vol. 35), 201–56. Lincoln: University of Nebraska.

Schoener, T. W. 1987. A brief history of optimal foraging ecology. In A. C. Kamil, J. R. Krebs, and H. R. Pulliam, eds., *Foraging behavior*. New York: Plenum.

Schusterman, R. J. 1962. Transfer effects of successive discrimination-reversal training in chimpanzees. *Science* 137:422–23.

Schusterman, R. J., and K. Krieger. 1986. Artificial language comprehension and size transposition by a California sea lion (*Zalophus californianus*). *Journal of Comparative Psychology* 100: 348–55.

Schwartz, B. 1984. *Psychology of learning and behavior*. 2d ed. New York: Norton.

Seligman, M. E. P. 1970. On the generality of the laws of learning. *Psychological Review* 77: 406–18.

Seligman, M. E. P., and J. L. Hager. 1972. *Biological boundaries of learning*. New York: Appleton-Century-Crofts.

Sherry, D. F. 1984. Food storage by black-capped chickadees: Memory for the location and contents of caches. *Animal Behaviour* 32:451–64.

Sherry, D. F., J. Krebs, and R. Cowie. 1981. Memory for the location for stored food in marsh tits. *Animal Behaviour* 29:1260–66.

Shettleworth, S. J. 1982. Function and mechanism in learning. In M. Zeiler and P. Harzen, eds. *Advances in analysis of behavior*. vol. 3, *Biological factors in learning*. New York: Wiley.

Shettleworth, S. J. 1984. Learning and behavioral ecology. In J. R. Krebs, and N. B. Davies, eds., *Behavioral ecology*, 2d ed. Oxford: Blackwell Scientific Publications.

Shettleworth, S. J., and J. Krebs. 1982. How marsh tits find their hoards: The roles of site preference and spatial memory. *Journal of Experimental Psychology: Animal Behavior Processes* 8:354–75.

Skinner, B. F. 1938. *The behavior of organisms.* New York: Appleton-Century-Crofts.

Skinner, B. F. 1959. A case history in scientific method. In S. Koch, ed., *Psychology: The study of a science.* New York: McGraw-Hill.

Skinner, B. F. 1977. Why I am not a cognitive psychologist. *Behaviorism* 5:1–10.

Smith, J. N. M., and H. P. A. Sweatman. 1974. Food searching behavior of titmice in patchy environments. *Ecology* 55:1216–32.

Smuts, B. B. 1985. *Sex and friendship in baboons.* Hawthorne, N.Y.: Aldine.

Spearman, R. I. C., and J. A. Hardy. 1985. Integument. In A. S. King and J. McLelland, eds., *Form and function in birds,* vol. 3. New York: Academic Press.

Stephens, D. W., and J. R. Krebs. 1986. *Foraging theory.* Princeton: Princeton University Press.

Thorndike, E. L. 1898. Animal intelligence: An experimental study of the associative processes in animals. *Psychological Monographs* 2 (whole No. 8).

Thorndike, E. L. 1911. *Animal intelligence: Experimental studies.* New York: Macmillan.

Thorpe, W. H. 1956. *Learning and instinct in animals.* Cambridge: Harvard University Press.

Tinbergen, N. 1953. *The herring gull's world.* London: Collins.

Warren, J. M. 1966. Reversal learning and the formation of learning sets by cats and rhesus monkeys. *Journal of Comparative and Physiological Psychology* 61:421–28.

West, M. J., and A. P. King. 1985. Learning by performing: An ecological theme for the study of vocal learning. In T. D. Johnston and A. T. Pietrewicz, eds., *Issues in the ecological study of learning.* Hillsdale, N.J.: Lawrence Erlbaum Associates.

Woolfenden, G. E., and J. W. Fitzpatrick. 1984. *The Florida scrub jay: Demography of a cooperative-breeding bird.* Princeton: Princeton University Press.

Yarczower, M., and L. Hazlett. 1977. Evolutionary scales and anagenesis. *Psychological Bulletin* 84:1088–97.

Zentall, T. R., D. E. Hogan, and C. A. Edwards. 1984. Cognitive factors in conditional learning by pigeons. In H. L. Roitblat, T. G. Bever, and H. S. Terrace, eds., *Animal cognition.* Hillsdale, N.J.: Lawrence Erlbaum Associates.

THREE

Learning and Foraging: Individuals, Groups, and Populations

John R. Krebs and Alastair J. Inman

WHAT CAN BEHAVIORAL ECOLOGISTS contribute to the study of animal learning? Learning is traditionally the domain of experimental psychologists working with animals in the laboratory, and they have developed sophisticated tools for both experimental and theoretical analysis of questions such as the role of and nature of representations in learning (Mackintosh 1983; Roitblat 1987; Gallistel 1990); the relationship among learning, performance, and reinforcement (Dickinson 1980; Commons et al. 1987); and the nature of associations formed during learning (Rescorla 1988).

Meanwhile, ethologists and behavioral ecologists have worked on "specialized" forms of learning such as song learning, imprinting, and food storing (Krebs and Horn 1991). These studies raise important questions about whether there are specialized memory systems (Sherry and Schacter 1987; Weiskrantz 1991), but they are often difficult to relate to mainstream issues in animal learning because they involve different experimental procedures, time scales, and measures of performance from those used by psychologists. For example, the question of whether the memory of food-storing birds for storage sites (Shettleworth 1990) is akin to working memory or to reference memory in typical animal memory tasks (Pearce 1987; Roitblat 1987) is not clear. However, the links between learning as studied by behavioral ecologists working on foraging and mainstream animal learning, both operant and Pavlovian, are more obvious (Shettleworth 1989). An animal learning to exploit which of a number of patches in the environment has the highest density of food is solving a problem in essence identical to that of a pigeon in a Skinner box learning to choose which of a number of keys offers the highest rate of reinforcement. Foraging theorists became interested in learning for two reasons: the adoption of techniques, especially from operant psychology, for application to foraging problems, and an interest in whether existing theories of learning could help to predict how foraging animals track changes in the spatial and temporal distribution of food in the natural environment.

For behavioral ecologists, a major function of learning is to enable animals to track such spatial and temporal changes. Certainly the application of techniques from animal learning has led to more refined experiments by behavioral ecologists, and often (but not always: see, e.g., Houston 1987) it has proved possible to account for results of foraging experiments in the theoretical framework of either operant or

This chapter was previously published in a slightly different form in *American Naturalist* 140, Supplement (November 1992), S63–S84. © 1992 by The University of Chicago.

Pavlovian conditioning. A possible contribution by behavioral ecologists to animal learning theory is to consider from a Darwinian perspective of design how animals ought to learn about spatial and temporal changes in their food supply to maximize fitness and predict properties of learning from this functional perspective (Krebs et al. 1978).

In this chapter we will describe models and experiments based on this line of thinking, starting off with a discussion of how individual foragers might track temporal changes in food supply, before going on to look at how individual learning strategies might be affected by foraging in a group. In the final section we will switch from considering how ecology might affect learning to discussing how learning might affect ecology, in particular the population interactions between predators and their prey.

TRACKING TEMPORAL CHANGES IN THE ENVIRONMENT

Descriptive Analysis

Two approaches have been taken to how foraging animals track temporal changes in their environment: descriptive (Pulliam and Dunford 1980; Kacelnik and Krebs 1985; Dow and Lea 1987; Kacelnik et al. 1987) and optimality (McNamara and Houston 1985; Krebs et al. 1978; Stephens 1987; Tamm 1987; Shettleworth et al. 1988). An example of the descriptive approach is the work of Cuthill et al. (1990), who studied how starlings in the laboratory respond to changes in travel time between patches. In their experiments, food intake rate within a patch declined with time, so that the relationship between the rate-maximizing patch residence time and average travel time between patches could be predicted from the marginal value theorem (Charnov 1976). For fixed values of travel time, the predicted relationship is known to hold for starlings in both the field (Kacelnik 1984) and the laboratory (Kacelnik and Cuthill 1987). The experiments of Cuthill et al. differed from these earlier studies in that the responses of the birds to fluctuations of travel time within a single foraging bout were studied. Starlings were presented with random changes in the value of travel time between two values, long and short, with a 0.5 probability of change on each successive patch visit. The patch residence time of birds following different recent experience of long and short travels was used to analyze the extent to which starlings remember the past. If, for example, the bird was strongly influenced by the most recent travel experience, its patch residence time would be largely determined by, and therefore predictable from, the last travel. If, at the other extreme, the animal remembered a very long run of travel experiences, the last travel would have virtually no effect on patch residence time.

Cuthill et al. (1990) found that starlings behaved as if they responded only to the most recent travel time: they appeared to have a very short "memory window" for past events. However, other experiments show that starlings do in fact have a longer memory for the past (Cuthill et al., in press). When, for example, starlings were exposed to many hours of long travel times and then a transition to short travel times (or vice versa), the transient patch residence time following the step change in travel time was not just one "patch visit cycle," as predicted if the memory window is one travel. Instead, the transient time lasted for about seven patch visit cycles. It appears, therefore, that starlings are able to remember farther back into the past, but use only information from the last travel to control the next patch visit in the environment in which there were random oscillations of travel time. This "short hindsightedness" may be adaptive in certain situations. Cuthill et al. show that if, for example, changes in travel time occur independently of the animal's activity and are not serially autocorrelated, the short memory window is a rate-maximizing strategy. Frequent interruptions to foraging may also favor a short memory window.

Optimality Analysis: Theory

The study of Cuthill et al. (1990) did not seek to predict a priori from functional considerations the properties of a learning algorithm. Instead, it sought a functional interpretation of an observed pattern of performance. In contrast, Stephens (1987), Tamm (1987), Shettleworth et al. (1988), and Inman (1990) predict from an optimality analysis how foraging animals should perform in a particular task involving tracking a changing environment. The predictions do not relate to a particular mechanism of learning but rather to the behavioral output, so the model is not an alternative to a mechanistic model of learning, although it may reveal unexpected aspects of learning.

The problem analyzed by these authors is illustrated in figure 3.1. A predator is faced with a choice of two feeding sites. One offers a constant, low reward rate, whereas the other fluctuates between a "good" state, which offers a higher reward rate than the constant alternative, and a "bad" state, which offers a lower reward rate (fig. 3.1A). The transitions between the good and bad states follow a first-order Markov chain, so that on each visit to either patch there is a fixed probability that the fluctuating patch will change state. The model considers probabilities of change less than 0.5, so the fluctuating patch alternates between bouts in the good state and bouts in the bad state, with each bout being unpredictable in length. An omniscient predator (one with complete information about changes in the state of the fluctuating patch) would achieve its maximum intake over a long time period by exploiting the poor, stable patch when the fluctuating patch is in its bad state, and then switching to the fluctuating patch whenever it is good (fig. 3.1B). For an animal that is unable to tell the state of the fluctuating patch except by visiting it, however, the best strategy is to feed in the stable patch and occasionally "sample" the fluctuating patch (fig. 3.1C). When, as a result of sampling, the fluctuating patch is found to be good, the predator should immediately switch to this patch until its reward rate drops as a result of the patch's becoming bad, when it should switch back to the stable patch.

Given this general scenario, the optimality question can be phrased as follows: How frequently should the animal sample in order to maximize overall intake rate? Stephens (1987) analyzed this problem, making specific assumptions about the structure of the environment (see legend of fig. 3.1). He has shown that the rate-maximizing frequency of sampling depends on how often the fluctuating patch changes state and the ratio $R_g:R_s$, where R_g is the reward rate of the fluctuating patch in its good state and R_s is the reward rate of the stable alternative patch. (For simplicity in this illustrative example, the probability of reward in the fluctuating patch in its bad state, R_b, is assumed to be zero.) These effects can be understood intuitively as follows. If the fluctuating patch never changed state, sampling would not pay. If, on the other hand, changes in state occurred every time the animal visited one or another patch, sampling would also not pay, because there would be no potential benefit derivable from using the information obtained while sampling. Sampling, therefore, pays off with an intermediate frequency of change in the fluctuating patch. The values of R_g and R_s reflect the benefits and costs associated with

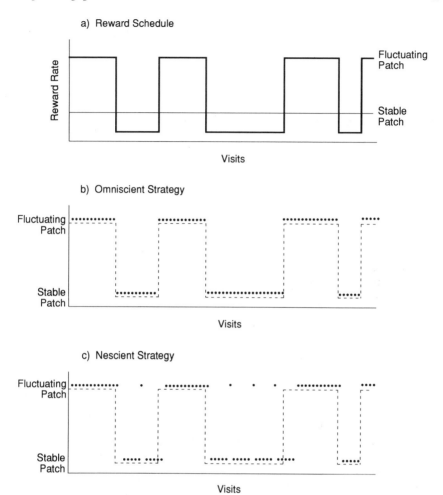

a) Reward Schedule

b) Omniscient Strategy

c) Nescient Strategy

Fig. 3.1 A summary of the sampling problem modeled by Stephens (1986). (*A*) The pattern of change in reward rate with time. The stable patch offers a low, constant probability of reward, while the fluctuating patch changes at random between good and bad states. In the good state it is much better than the stable patch; in the bad state it offers no rewards. Note that "time" is represented by "visits" in which the animal goes to one or the other patch. (*B*) An omniscient forager would follow the exploitation pattern indicated by the dotted line. The dashed line traces the pattern of change in the fluctuating patch as indicated in (*a*). (*C*) A nescient forager has to visit the fluctuating patch in its bad state to detect the change to a good state. These visits are referred to as *sampling*.

sampling. The benefit ($R_g - \bar{R}_s$) is the extra reward obtained by switching to the fluctuating site in its good state, while the cost of sampling is the lost opportunity to forage in the stable site. If the fluctuating site has no food when in its bad state, each time the animal visits this site to sample it pays a cost of R_s. When the animal samples at too low a frequency, it is slow to detect changes in the fluctuating site and misses opportunities to gain $R_g - R_s$ extra food on each foraging trial. But if sampling is too frequent, the benefits of rapid detection of changes in state of the fluctuating site are more than offset by the cost of missing food (R_s).

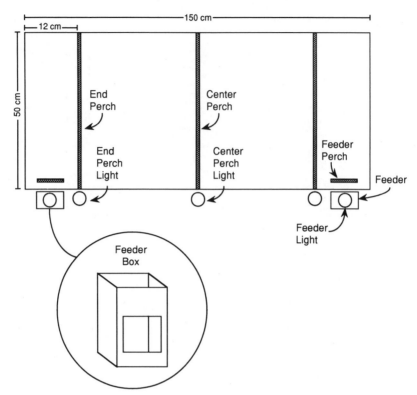

Fig. 3.2 The experimental apparatus. A trial starts every 30 seconds with a center perch cue light coming on. If the bird lands on the center perch within 5 seconds, it illuminates both end perches and turns the center light off. At the stable patch, the end perch light goes off and the feeder light comes on. At the fluctuating patch, the feeder light comes on only if the patch is in its good state.

Optimality Analysis: Experiment

Inman (1990) has tested predictions of this model relating to the effects of R_g and R_s on the optimal sampling rate in a laboratory experiment with starlings (see also Tamm 1987; Shettleworth et al. 1988). The experimental setup, designed to reflect the structure of the model, is illustrated in figure 3.2. An experimental session consisted of a series of choices or trials. The bird initiated each trial by sitting on the center perch in response to a cue light, after which lights on the two perches at either end of the cage were illuminated, which indicated to the bird that it could make a choice. One of the end perches was associated with a fluctuating patch and the other with a stable patch. If the bird chose to go to the stable patch, it obtained rewards from the feeder at that end of the cage with a probability of 0.2 or 0.4, depending on the experimental treatment. If, on the other hand, the bird chose to go to the fluctuating patch, it obtained either rewards with a probability of 1.0, or no rewards at all, depending on the state of the fluctuating patch. Changes in state of the patch were signaled to the bird by a cue light on the fluctuating feeder after the bird had made a decision to visit the fluctuating site and landed on the end feeder perch. Thus, in the experimental setup, the bird had to sample by visiting the fluctuating site in order to detect its state, but once it had sampled, the state of the fluctuating site was recognizable instantaneously from the feeder light. After obtaining the appropriate reward, the bird could initiate the next trial by returning to the center perch in response to the center perch light.

This experimental arrangement matched the assumptions of the model in that, upon visiting the fluctuating patch, recognition of changes in state was in principle instantaneous. There was no cost of sampling other than the lost opportunity to forage in the stable patch, because the birds started each trial equidistant from the two sites, and changes in state of the fluctuating patch occurred as a function of the number of trials that the bird had been through. The experiment involved manipulating the values of R_g (either one or two 20 mg measures of reward, delivered with probability of 1.0) and R_s (as mentioned earlier, either 0.2 or 0.4 probability of obtaining 20 mg of reward). The length of runs in the good and bad states of the fluctuating patch was not varied as an experimental treatment; in all conditions the mean length of the bad states was 75 trials with a geometric distribution about this mean, while the good states were fixed at 40 visits.

Figure 3.3A shows the average record of choice in one experimental treatment for one bird. When the fluctuating patch was in its bad state, the bird visited this patch at a low frequency. When the fluctuating patch changed to its good state, the bird switched to almost exclusive foraging in this patch. There was a slight but significant decrease in the proportion of trials spent foraging in the fluctuating patch in its good state until the patch returned to its bad state, at which point the bird switched back to the stable patch. In other words, the bird succeeded in tracking changes in the environment by sampling. But did the pattern of sampling fit the predictions of the optimality model? The answer is "Qualitatively, yes, but quantitatively, no."

Figure 3.3B shows the mean data for four birds compared with the predictions. The model failed to predict the observed sampling rates of each bird in at least two of the four conditions. Qualitatively, however, the predicted effects were seen. Birds sampled more when R_s was 0.2 than when it was 0.4, as predicted from the reduced cost of sampling when the stable patch provided a lower rate of reward. They also sampled more when R_g was 2.0 than when it was 1.0, predicted from the increased benefit of early detection of changes to the good state of the fluctuating patch. Although the model predicts equivalent effects of changing R_s and R_g, in the observed data the effect of R_g was weaker than that of R_s. While the effect of R_s was significant for all birds at both levels of R_g, the effect of R_g was significant for all birds at $R_s = 0.2$, but only for two of the four birds at $R_s = 0.4$.

A further prediction of the optimal sampling model is that sampling should be regularly spaced. Intuitively one can see that two samples taken one after the other are less efficient in terms of detecting change than samples that are regularly spaced. Figure 3.3C shows that starlings tended to space their sampling regularly. This is demonstrated by calculating the delay to detect the change in state of the fluctuating patch expected from random and from regular sampling. The observed delays, given the observed average levels of sampling, show that the birds did significantly better than if they sampled at random, and in some treatments they approximated the optimum of regular sampling.

A qualitative feature of the results that does not fit the predicted pattern is the slight but significant decline in percentage of visits to the fluctuating patch in its good state as a function of time since the start of the good state. We will return to this pattern later.

The most detailed previous study of the sampling model is that of Shettleworth

A

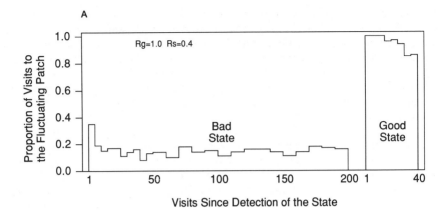

Visits Since Detection of the State

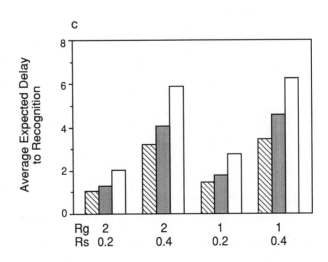

et al. (1988). They found that R_s had an effect on the sampling frequency, R_g had no effect, and that samples were distributed at random. The difference between the two sets of results may be the result of a species difference (Shettleworth et al.'s study was done on pigeons) or subtle variations in procedure. Shettleworth et al. argued that their data, although deviating from the qualitative predictions of the sampling model in several respects, are consistent with a mechanistic model of memory derived by Gibbon et al. (1988). This model assumes that the birds choose the option with the shorter remembered delay to food. The bird stores in memory experienced delays to food with a variance that is proportional to the remembered interval. It then draws a sample at random from the two distributions associated with the two patches and chooses the site associated with the shorter delay. This model, however, fails to account for certain qualitative features of the starling results. It does not, for example, explain the tendency to regularly spaced sampling, the effect of R_g on sampling frequency, and the decline in proportion of responses to the good patch.

Inman (1990) has suggested an alternative model that might account for the present results and those of Shettleworth et al. (1988). It is based on the argument that, rather than maximizing intake rate, the birds may be trading off intake against other benefits. If the benefit of food intake is a decelerating function of amount, as seems very likely, changes in the value of R_g may be of less significance to the animal than changes in R_s, because, when foraging in the fluctuating patch in its good state, the birds are close to the upper limit of benefit derivable from food intake. Hence increasing the value of R_g from 1.0 to 2.0 in the present experiments has less effect than changing the value of R_s. If Shettleworth et al.'s pigeons were already at the ceiling of benefit, then changes in the value of R_g would not be expected to have any effect at all. This functional argument could also be expressed in mechanistic terms by saying that the animals are near or at a constraint on rate of food intake when foraging in the good patch.

Three observations are consistent with this view. First, the latency to respond to the trial-initiating cue light in Inman's (1990) experiments was longer when R_g was 2.0 than when it was 1.0, suggesting that the birds were less motivated to feed. Second, the birds missed an increasing proportion of trials as a function of time in the good state. A trial was missed if the bird did not respond within 5 seconds to

---

Fig. 3.3 (*A*) An average record of the proportion of visits to the fluctuating patch for one bird in one experimental treatment. The *x*-axis shows time, represented by discrete trials since the bird has visited the fluctuating patch after a change in state. The first block shows the results for the bouts in the bad state, and the second block shows the results for bouts in the good state. The *y*-axis shows the proportion of visits to the fluctuating patch, which has been calculated across five-visit intervals. When the fluctuating patch is in the bad state, it is visited at a frequency of about 20% (sampling). After a change to the good state has been detected, the forager visits the fluctuating patch almost exclusively. Logistic regressions of each bird's proportion of visits to the fluctuating patch against the number of visits since detecting a change to the good state revealed a slight but significant decline with time. (*B*) Predicted and observed levels of sampling in the four different treatments. *G*-tests comparing each bird's sampling rate with the model's predictions revealed significant differences in at least two of the conditions for each bird. Logistic regressions of each bird's sampling frequency against the level of R_g and R_s, however, found that the birds sampled more when R_s was lower (0.2 versus 0.4) and when R_g was higher (2.0 versus 1.0), as predicted. (*C*) One bird's delay to detect changes to the good state of the fluctuating patch. Delays predicted from a regular (*cross-hatched bars*) and random (*open bars*) spacing of the observed level of sampling are compared with the observed mean delay (*shaded bars*). A randomization technique revealed that each bird's observed delays were significantly lower than would be predicted for a random spacing of samples in all four treatments.

the center perch cue light at the start of a trial. Finally, as mentioned earlier, the proportion of visits to the good patch declined with time since the start of the good state. All of these observations could be explained by postulating that in mechanistic terms, the birds become satiated in the good state and switch to activities other than foraging, such as resting or sampling. Equally, in functional terms, they could be explained by arguing that the birds are balancing a trade-off between intake and other activities.

GROUP FORAGING IN A PATCHY ENVIRONMENT

The idea that animals foraging in groups might do better when exploiting patchy, ephemeral food was developed by authors such as Crook (1964), Lack (1968), Krebs et al. (1972) and Ward and Zahavi (1973). A central idea in these articles is that, by foraging in a group, individuals might learn about temporary local abundances of food, either by going to sites where others were already foraging or by learning about new foraging techniques. These potential benefits of group foraging must be weighed against its costs, such as competition for resources once they have been discovered. Crook and Lack, noting that seed- and fruit-eating birds tend to forage in flocks while insect eaters do not, argued that the balance of costs and benefits favors flocking in the former group because food tends to occur in large, ephemeral clumps. This increases the benefit of information acquisition about the location of clumps while simultaneously decreasing the costs of foraging in a group in terms of competition.

The extent to which flocking birds are sensitive to the costs and benefits of group foraging is illustrated by a field experiment in which starlings were observed feeding in a group of four small patches (0.35×0.35 m) containing variable densities of mealworms hidden in sand (Inman 1990). A multiple regression analysis of the factors influencing whether starlings landed in a particular patch showed that the birds tended to avoid landing in patches that already contained a large number of individuals (presumably avoiding competition) but that they were attracted to patches in which other birds had already discovered food and patches containing a high proportion of the foraging birds in the area (independent of absolute number). Thus the stimuli to which starlings respond in choosing foraging patches may be quite subtle and include not just the presence of other birds but also their behavior and spatial distribution (Inman 1990).

Sampling in Groups: Theory

Inman (1990) has developed a more quantitative analysis of the costs and benefits of group foraging in relation to tracking ephemeral food, based on the sampling model described earlier. Imagine, for the sake of simplicity, a group of two individuals sampling in the environment described above for a single bird (see fig. 3.1). Each individual samples independently and can still detect changes in the fluctuating patch as a result of its own sampling, but it is also able to detect changes revealed by the other bird's sampling. This was incorporated into the model by assuming that when one individual detected a change to the good state, the other individual automatically became aware of that change. In the field, the reactor might detect subtle cues related to foraging success, as in the field experiment described in the previous paragraph.

Figure 3.4 shows the payoff to individual A in relation to the amount of sam-

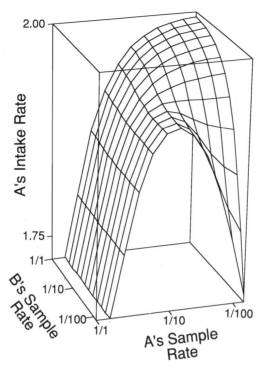

Fig. 3.4 The intake rate of A in relation to different amounts of sampling by itself and B. As B's sampling rate increases, A's optimal (intake rate–maximizing) sampling rate decreases. $R_g = 3$, $R_s = 1$; $L_g = L_b = 200$ (length of run of good and bad states respectively).

pling done by itself and by B. If B samples very little, A has the highest payoff by sampling as if it were alone. In the case of the parameters used in figure 3.4, this is a sampling rate of 7% of trials. If, on the other hand, B samples on every trial, A does not need to sample at all, because it is able to reap the benefits of B's sampling by detecting changes in the fluctuating patch without paying the costs of sampling. In between these two extremes, A's optimal sampling frequency decreases as B's frequency of sampling increases. Of course exactly the same argument can be applied in relation to B. In other words, both individuals benefit the most when the other bird does all the sampling.

Figure 3.5 shows a graphical analysis of the equilibrium sampling rate of A and B together. The surface represented in figure 3.4 has been reduced to a single line by plotting the optimum (intake rate–maximizing) sampling rate for bird A for all possible sampling rates of bird B (fig. 3.5A) and vice versa (fig. 3.5B). Figure 3.5C superimposes the optimum sampling rate lines for both individuals. Points of intersection represent points of equilibrium, where both are simultaneously at their optima. There are three points of intersection, but only two are stable equilibria. Either B samples at the optimum for a single forager and A does not sample at all (equilibrium 1), or B does no sampling and A samples at the single-forager optimum (equilibrium 2). At these two equilibria, the sampler has a lower payoff than the "copier" because the latter does not have to pay the cost of sampling. Equilibrium 3 in figure 3.5C is not stable because if either individual slightly changes its sampling rate, the

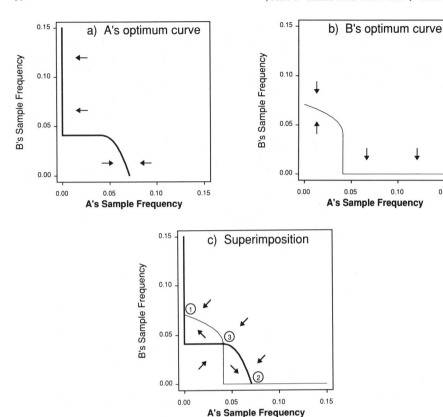

Fig. 3.5 (A) The intake-maximizing sampling rate for A plotted against the sampling rate adopted by B. This is derived from the surface plotted in fig. 3.4 by choosing A's optimum (intake rate–maximizing) sampling rate for all possible sampling rates of bird B. (B) The intake rate–maximizing sampling rate for B plotted against the sampling rate adopted by A. (C) Superimposition of the optimal sampling strategies of A (*bold line*) and B (*thin line*). The three points of intersection are equilibria, but only 1 and 2 are stable. The arrows indicate directions of movement toward/away from the equilibria. $R_g = 3$, $R_s = 1$; $L_g = L_b = 200$.

pair will quickly move to either equilibrium 1 or equilibrium 2. If, for example, A slightly increased its sampling rate at equilibrium 3, B would immediately decrease its own rate, leading to A's increase up to equilibrium 2.

So, if animals adopt a simple rate-maximizing strategy—adapting their sampling rate to achieve the highest possible intake rate given the present sampling rate of their partner—the only stable outcome is that one animal samples as if alone and the other animal does not sample at all. The animal that does not sample at all experiences the higher intake rate of the pair because it takes advantage of its opponent's sampling without paying the cost of sampling itself. The animal that does not sample exploits the animal that does.

Could animals adopt strategies other than simple rate maximization that might ensure that they avoid being exploited and do not end up being the one to do all the sampling? Inman (1990) considers a population consisting of a mixture of three strategies. The first of these is "maximizer," as we have already described. The second is "coercer," a strategy that avoids outcomes in which it samples alone by sampling at a rate low enough to ensure that its opponent maximizes its intake rate by sampling at an equal or higher rate. In figure 3.5 this would correspond to sampling at a rate equal to or below that of intersection 3. If such a strategy is adopted against a maximizer, the maximizer ends up sampling as if alone, enabling the coercer to stop sampling. The only situation in which a coercer will fare more poorly than a maximizer is when playing against an individual that does not sample at all. In this case, an individual that samples as if alone, like the maximizer, will achieve a higher intake rate than one that samples at the lower rate adopted by coercers. Thus, a third strategy is included in the population: "obstinate," an individual that never samples.

Through a series of computer simulations, Inman (1990) demonstrated that the evolutionarily stable strategy (ESS) in this game is coercer. As described earlier, when playing against maximizers, coercers force their opponent into sampling as if alone, allowing the coercer to play the nonsampling role in one of the two stable equilibria described in figure 3.5. How does the inclusion of obstinates affect the outcome? Both coercers and maximizers can invade a population of obstinates. Although obstinates achieve a higher intake rate than their opponent when playing against either a coercer or a maximizer, obstinates fare very poorly when playing against another obstinate, much more poorly than either a coercer or a maximizer. This is because when two obstinates play against each other, neither samples at all, so they miss all opportunities to exploit the fluctuating patch when it is in the good state. As both maximizers and coercers sample when playing against an obstinate, they continue to exploit the good states of the fluctuating patch. As described earlier, maximizers actually fare better against obstinates than do coercers because the maximizers respond to the obstinates' lack of sampling by adopting the rate-maximizing sampling rate. Coercers, which are limited to a lower sampling rate, do not do quite as well. As a result, when invading a population of obstinates, maximizers initially have a slight advantage over coercers. As the proportion of maximizers and coercers in the population increases, however, the advantage quickly switches to coercers, as they fare better than maximizers in playing against either other coercers or maximizers. Thus, any population will eventually evolve to all coercers.

So far we have assumed that the reactor detects changes in the fluctuating patch at the same time as the sampler. The addition of a "recognition delay" has the effect of decreasing the amount of information an animal can derive from its partner's sampling. The reactor experiences a delay to detect the change, whereas its partner does not. As recognition delay increases, each bird receives progressively less information from its partner's sampling and so ends up being better off if it samples more itself. The intake-maximizing lines for each bird in figure 3.5 move upward and outward until, with a long enough recognition delay, there is only one intersection, with both individuals sampling at the same frequency. At this point, coercer and simple maximizer become indistinguishable, as each individual is better off to continue sampling even if its opponent also samples as if alone.

Further modifications of the model considered by Inman (1990) include the introduction of asymmetries between the two individuals. If, for example, one individual is hungrier than the other, it should sample more because the potential benefit to it of discovering the fluctuating patch in its good state is greater. This is analogous to risk-prone foraging (Caraco et al. 1980; McNamara and Houston 1985). If there is competition for food (this has not been included in the model so far) and an asymmetry in dominance, the extent to which the dominant and subordinate individuals should sample depends on the recognition delay. If there is a long delay, the subordinate should sample more because it benefits more from a period of competition-free foraging in the good state of the fluctuating patch. With short recognition delays, however, this benefit no longer holds, so it does not pay the subordinate to sample. In fact, as a result of the dominant's sampling, the subordinate has the benefit of foraging in the stable patch without competition. In other words, the subordinate has a greater disincentive to sample and therefore samples less than the dominant.

Sampling in Groups: Experiment

Inman (1990) describes the results of an experiment with starlings in which the effects of being in a group on sampling behavior were investigated. Four pairs of birds were tested in the following sequence of treatments: sampling alone, sampling in a group of two, sampling alone again, and sampling with a cue visible from the central perch to indicate the state of the fluctuating patch. (In this last condition, the birds did not need to sample in order to detect changes in the fluctuating patch, so it was predicted that they should stop sampling.) All treatments were carried out in the same apparatus, consisting of two side-by-side cages like the one shown in figure 3.2. In the sampling alone and cued conditions, each bird had a separate schedule of rewards, and an opaque screen separated the two. In the group sampling treatment, the screen was removed so that the birds could see each other, and the reward schedules of the two individuals were yoked: both birds experienced changes in the state of the fluctuating patch simultaneously.

Figure 3.6A shows the results of the four treatments for one of the pairs of birds. In all four pairs, one of the birds sampled at a lower rate in the group condition than when alone, whereas the other did not change its sampling rate in the group condition. The bird in each pair that reduced its sampling rate was nevertheless able to detect changes in the fluctuating patch. In fact, its latency to detect the changes was only slightly longer than that of the other bird, although it was significantly shorter than it would have been had the latency been determined entirely by the bird's own sampling. In other words, the reduced sampler in each pair was tracking changes by following its partner. This pattern is analyzed in more detail in figure 3.6B, in which the latencies of each bird in a pair to detect individual switches from bad to good state are plotted against each other. The latencies were closely correlated, with the reduced sampler taking on average 1.5 trials longer than the other bird to detect the change.

Why did only one bird in each pair reduce its sampling rate? The ESS model suggested that, within the strategy set considered, coercer was an ESS and that both birds should therefore reduce their sampling rate. Asymmetries in dominance and/ or hunger could modify this conclusion, although the experiment was set up in such a way that both birds experienced the same daily food intake and there was no competition, so dominance interactions should have been unimportant. An alternative explanation is that the birds are incapable of playing the coercer strategy but instead behave as simple maximizers, in which case the equilibrium is for one bird to sample and the other not. The coercer strategy requires the birds to forgo immediate gain in order to force the other individual to increase its sampling rate. There is an extensive literature on the response of animals to reinforcement after long and short delays that shows that, on the whole, animals maximize in the very short term and do not choose long delays to reinforcement even if the overall reward rate as a result of doing so would be greater (Logue 1988). Thus, coercer may be an unrealistic strategy in psychological terms.

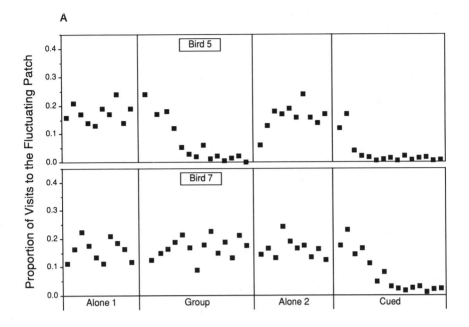

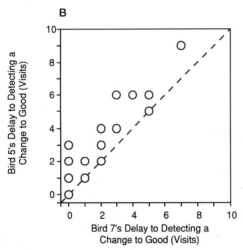

Fig. 3.6 (A) Results of the group sampling experiment. The average data for one pair of birds are displayed, representative of the four pairs that were tested. Each point is the average of three or more cycles of change from good to bad, based on a minimum of 30 visits to the fluctuating patch. The y-axis shows the amount of sampling (proportion of visits that are to the fluctuating patch in its bad state). Both birds sample at a rate of about 20% in the alone condition, but bird 5 almost stops sampling in the group condition while bird 7 does not change. Both birds stop sampling when a cue is provided. (B) The delay to detect a change in the fluctuating patch from bad to good state for each member of the pair of birds, plotted against each other. The rate-maintaining bird is plotted on the x-axis and the rate-reducing bird on the y-axis.

There is yet another explanation, and perhaps this one is the most likely. The asymmetry in sampling behavior may have arisen simply because one of the two birds happened to learn before its partner that information could be acquired by copying. As soon as this individual started to learn that it could copy its partner, it decreased its sampling rate so that the other individual had no opportunity to learn. Hence the asymmetry in sampling in the group condition arises as a result of an initial difference in learning about the possibility of copying.

The results of the cued treatment indicate that when sampling does not pay, the birds stop visiting the fluctuating patch in its bad state (fig. 3.6A). This supports the view that the behavior labeled by us as sampling is indeed, in functional terms, related to acquisition of information about the state of the fluctuating patch.

LEARNING AND POPULATIONS

In this final section we will describe models of the possible effects of learning by individual predators on the spatial distribution of predator and prey populations. The models refer primarily to specialist predators feeding on a particular species of prey.

Two contrasting approaches have been taken in the past to modeling the distribution of predators among prey patches. Hassell and May (1973) used an aggregate function to describe the distribution of predators. This approach is essentially phenomenological and so does not allow one to predict the distribution of predators a priori. Fretwell and Lucas (1970), on the other hand, derive predictions about the distribution of predators among patches of resource from a consideration of how individuals might maximize their intake rates, the so-called ideal free distribution (table 3.1).

This approach has the advantage of making a priori predictions, but it has the disadvantage that many field studies appear not to support its predictions. Numerous experimental studies of continuous input systems support the numerical predictions (e.g., Milinski 1979, 1984; Harper 1982; Abrahams and Dill 1989), but the results may be open to other interpretations. The numerical predictions are not supported by most field studies (Parker and Sutherland 1986). In addition, some studies find equal average intake rates across patches (e.g., Whitham 1980) as predicted, but

Table 3.1   The Classic Ideal Free Distribution

ASSUMPTIONS	PREDICTIONS
1. Resources are in patches	1. Numerical: Ratio of predators to prey is equal between patches
2. All competitors are equal	
3. Intake rate is proportional to number of prey per number of predators	2. Intake: Average intake rate in all patches is equal, and individuals have equal intake rates[a]
4. There is no travel time or restriction on patch choice	3. Risk: Percentage prey mortality is the same in all patches
5. Predators are omniscient	
6. No depletion occurs	
7. Predators are rate maximizers	

[a] The second prediction is treated by some authors as an assumption.

others do not (Parker and Sutherland 1986). There is a consistent lack of support for the prediction that all competitors should have equal intake rates (Parker and Sutherland 1986). As for the risk prediction, field studies of insect parasitoids and their hosts do not find an equal percentage of prey mortality across patches (Lessells 1985; Stiling 1987; Walde and Murdoch 1988).

One reason why the predictions of the ideal free distribution may not hold up in the field is that the "classic version" of the model makes too many simplifying assumptions. Various attempts have been made to modify the original model to make its assumptions more realistic. One such attempt is that of Bernstein et al. (1988, 1991a, 1991b), who described a model in which the consequences of individual behavior for the distribution of populations of predators among patches are explored. Our particular interest in the present context is their evaluation of the effects of learning by the predators. In their model, Bernstein et al. assume that within a patch, the number of prey eaten per predator per unit time follows a "disc" equation (Holling 1959), which takes into account the effects of handling time and of interference, the (usually) negative effect of one predator on the searching efficiency of others (Rogers 1972). Individual predators move between patches according to the rule, "migrate if current intake rate is less than the expected value for the rest of the environment." Each individual predator obtains an estimate of the value of the environment as a result of its experience, according to a learning algorithm, the so-called linear operator (Bush and Mosteller 1951). In different versions of their model, Bernstein et al. made various assumptions about the cost of migration (measured in time cycles of the model), the pattern of migration (e.g., random or viscous—movement only to neighboring places), and the rate of prey depletion. Two versions of the model track the distribution of predators in two different kinds of environments. The first is a matrix of forty-nine patches in which prey are initially distributed at random, and the second is a "continuous" habitat in which prey density changes smoothly with distance in a sinusoidal function.

The aim of Bernstein et al.'s (1988, 1991a, 1991b) model is twofold. First, it asks which behavioral properties of individuals would lead to deviations from the ideal free distribution. The reason for asking this question is to try to explain the variety of patterns of distribution observed in field studies. The second aim is to explore the implications of different predator behaviors for the spatial pattern of prey mortality. These two objectives are closely related because, as Bernstein et al. (1991a, 1991b) explain, spatial prey mortality is linked to whether the predators have an ideal free distribution.

Some important results of the Bernstein et al. (1988, 1991a, 1991b) models are as follows. First, if there is no or very slow prey depletion, predators following the individual learning and migration rules described above eventually stabilize in an ideal free distribution. For the range of parameters studied by Bernstein et al., this implies spatial density dependence of prey mortality. With rapid prey depletion, however, the predator's estimate of the environment as a whole tends to lag behind changes in prey density, and there is thus no stable equilibrium distribution of predators.

Second, the cost of migration has an effect on the fit to the ideal free distribution and hence the spatial mortality of prey. In general, as migration costs increase, the predators become more reluctant to leave their present patch (because the cost of migration detracts from the benefit of moving), and, hence, the predators equilibrate farther away from the ideal free distribution. With very high migration costs predators often fail to leave their starting patch, and since the model starts with a random distribution of prey and predators, mortality of prey is proportionately higher in patches where, by chance, at the beginning there were relatively many predators and few prey (inverse density dependence). With intermediate migration costs, only predators in very poor patches migrate, and the resulting density dependence is domed, with a peak at intermediate values.

Finally, in the continuous environment, the equilibrium distribution of predators depends on the pattern of migration and the grain of the environment (i.e., how quickly the density of prey changes in space, reflected in the period of the sinusoidal change). If predators are constrained to migrate to neighboring patches and if the environment is coarse-grained, predators equilibrate far from the ideal free distribution. If the environment is fine-grained or if predators migrate at random, the equilibrium distribution is close to the ideal free.

The explanation of these apparently puzzling results relates to how predators learn about their environment. Recall that the migration decision of an individual depends on its estimate of the expected gain from the environment as a whole, which is derived from past experience. Predators with a viscous movement rule in a coarse-grained environment tend to build up an inaccurate picture of the world as a whole. Take as an example individuals starting off in a poor area. Their first few moves are to other poor patches, and their estimate of the world thus rapidly moves toward a low value. As a result, these individuals give up migrating and stay in poor patches. Hence, poor patches are overexploited relative to good ones. In the fine-grained environment, or if predators move at random from one patch to another, the predators rapidly build up an accurate estimate of the average value of the world. Migration decisions thus tend to be "correct" and the population ends up in an ideal free distribution.

To summarize, the models of Bernstein et al. (1988, 1991a, 1991b) show that a small number of behavioral properties of individual predators, especially learning and migration rules, can have a dramatic effect on the spatial distribution of predators and hence the mortality they impose on their prey. Further, the observed range of patterns of spatial density dependence (Lessells 1985; Stiling 1987; Walde and Murdoch 1988) concurs with the range accounted for by the models.

CONCLUSION

To forage efficiently, an animal must keep track of changes that occur in its environment. In some cases, the information the animal requires to track its environment can be derived directly while foraging, as was the case when Cuthill et al.'s (1990) starlings monitored changes in the travel times between patches. Foraging required the birds to travel between patches, so monitoring changes in travel time could be accomplished during the birds' normal foraging activity. Other kinds of information require that an animal actively sample, investing time and energy specifically in tracking the environment. Such was the case when Inman's (1990) starlings sampled the fluctuating patch to keep track of its state when foraging in Stephens's (1987) two-patch environment. In both cases, what the animals learned about the state of their environment affected their subsequent foraging decisions.

Keeping up-to-date in a changing environment also requires that animals integrate different sources of information. In the case of Cuthill et al.'s starlings, information about the state of the environment in the past had to be integrated with information that the animal gained from its present foraging activity to derive an estimate of the present state of the world. The learning rules modeled by Bernstein et al. (1988, 1991a, 1991b) used linear operators to incorporate a similar kind of integration. Inman's (1990) experiments with pairs of starlings showed that animals can also derive information from watching conspecifics, however, and that this information has to be integrated with information that the animals have gained from their own foraging experience, past and present. In Inman's experiments this integration was relatively simple. Evidence of a change to the good state could come either from the bird's sampling the fluctuating patch itself and receiving food there or from its watching its partner receive food from that patch. As either of these events provided unambiguous evidence of a change to the good state, the animal should have switched to foraging in the fluctuating patch as soon as *either* of these events occurred. In other situations the integration of information derived from an animal's own foraging activity and information derived by watching conspecifics may be more complicated. For example, if an animal needs to estimate the absolute profitability of a patch, rather than just whether the patch is providing food or not, the animal may need to integrate information derived from its own intake rate with information it derives from watching the pecking rate of conspecifics. This may involve the same kind of relative weighting that linear operator learning rules use in integrating past and present foraging experience. Little is known about how animals actually integrate these different sources of information.

The models of Bernstein et al. (1988, 1991a, 1991b) demonstrate how the decision rules of individuals can affect population distributions. Population distributions can also affect an individual's decision rules, however, as was demonstrated by the effect of the distribution and behavior of conspecifics on the patch choice of starlings foraging in the field (Inman 1990). How this would affect the outcome of Bernstein et al.'s model has not been examined. It would seem likely that if individuals can gain additional information about their environment by observing conspecifics, they should have a more accurate estimate of the distribution of food in their world, and as a result the population should more closely approach an ideal free distribution.

ACKNOWLEDGMENTS

We thank A. Houtman and L. Real for their valuable comments on earlier versions of this chapter.

REFERENCES

Abrahams, M. V., and L. M. Dill. 1989. A determination of the energetic equivalence of the risk of predation. *Ecology* 70:999–1007.

Bernstein, C., A. Kacelnik, and J. R. Krebs. 1988. Individual decisions and the distribution of predators in a patchy environment. *Journal of Animal Ecology* 57:1007–26.

Bernstein, C., A. Kacelnik, and J. R. Krebs. 1991a. Distribution of birds amongst habitats: Theory and practice. In C. M. Perrins, J.-D. Lebreton, and G. Hirons, eds., *Bird population studies: Relevance to conservation and management*, 317–45. Oxford: Oxford University Press.

Bernstein, C., A. Kacelnik, and J. R. Krebs. 1991b. Individual decisions and the distribution of predators in a patchy environment. II. The influence of travel costs and the structure of the environment. *Journal of Animal Ecology* 60:205–25.

Bush, R. R., and F. Mosteller. 1951. A mathematical model of simple learning. *Psychological Review* 58:313–23.

Caraco, T., S. Martindale, and T. S. Whittam. 1980. An empirical demonstration of risk-sensitive foraging preferences. *Animal Behaviour* 28:338–45.

Charnov, E. L. 1976. Optimal foraging: The marginal value theorem. *Theoretical Population Biology* 30:45–75.

Commons, M. L., S. J. Shettleworth, and A. Kacelnik, eds. 1987. *Quantitative analyses of behavior.* Vol. 6, *Foraging.* Hillsdale, N.J.: Lawrence Erlbaum Associates.

Crook, J. H. 1964. The adaptive significance of avian social organisations. *Symposia of the Zoological Society of London* 14:181–218.

Cuthill, I. C., A. Kacelnik, J. R. Krebs, P. Haccou, and Y. Iwasa. 1990. Patch use by starlings: The effect of recent experience on foraging decisions. *Animal Behaviour* 40:625–40.

Cuthill, I. C., P. Haccou, and A. Kacelnik. In press. Starlings exploiting patches: Response to long-term changes in travel time. *Animal Behaviour.*

Dickinson, A. 1980. *Contemporary animal learning theory.* Cambridge: Cambridge University Press.

Dow, S. M., and E. G. L. Lea. 1987. Foraging in a changing environment: Simulations in the operant laboratory. In M. L. Commons, S. J. Shettleworth, and A. Kacelnik, eds., *Quantitative analyses of behavior,* vol. 6, *Foraging,* 89–113. Hillsdale, N.J.: Lawrence Erlbaum Associates.

Fretwell, S. D., and H. J. Lucas, Jr. 1970. On territorial behaviour and other factors influencing habitat distribution in birds. *Acta Biotheoretica* 19:429–512.

Gallistel, C. R. 1990. *The organisation of learning.* Cambridge, Mass.: MIT Press.

Gibbon, J., R. M. Church, S. Fairhurst, and A. Kacelnik. 1988. Scalar expectancy theory and choice between delayed rewards. *Psychological Review* 95:102–14.

Harper, D. G. C. 1982. Competitive foraging in mallards: "Ideal free" ducks. *Animal Behaviour* 30:575–84.

Hassell, M. P., and R. M. May. 1973. Stability in insect host-parasite models. *Journal of Animal Ecology* 42:693–726.

Holling, C. S. 1959. Some characteristics of simple types of predation and parasitism. *Canadian Entomologist* 91:385–98.

Houston, A. I. 1987. Control of foraging decisions. In M. L. Commons, S. J. Shettleworth, and A. Kacelnik, eds., *Quantitative analyses of behavior,* vol. 6, *Foraging,* 41–61. Hillsdale, N.J.: Lawrence Erlbaum Associates.

Inman, A. J. 1990. Foraging decisions: The effects of conspecifics and environmental stochasticity. D. Phil. thesis, University of Oxford.

Kacelnik, A. 1984. Central place foraging in starlings (*Sturnus vulgaris*). I. Patch residence time. *Journal of Animal Ecology* 53:283–300.

Kacelnik, A., and I. C. Cuthill. 1987. Starlings and optimal foraging theory: Modelling in a fractal world. In A. C. Kamil, J. R. Krebs, and H. R. Pulliam, eds., *Foraging Behaviour,* 303–33. New York: Plenum.

Kacelnik, A., and J. R. Krebs. 1985. Learning to exploit patchily distributed food. In R. M. Sibly and R. H. Smith, eds., *Behavioural ecology: 25th symposium of the British Ecological Society,* 189–205. Oxford: Blackwell Scientific Publications.

Kacelnik, A., J. R. Krebs, and B. Ens. 1987. Foraging in a changing environment: An experiment with starlings (*Sturnus vulgaris*). In M. L. Commons, S. J. Shettleworth, and A. Kacelnik, eds., *Quantitative analyses of behavior,* vol. 6, *Foraging,* 63–87. Hillsdale, N.J.: Lawrence Erlbaum Associates.

Krebs, J. R., and G. Horn, eds. 1991. *Behavioural and neural aspects of learning and memory.* Oxford: Clarendon Press.

Krebs, J. R., M. MacRoberts, and J. Cullen. 1972. Flocking and feeding in the great tit *Parus major*: An experimental study. *Ibis* 114:507–30.

Krebs, J. R., A. Kacelnik, and P. Taylor. 1978. Test of optimal sampling by foraging great tits. *Nature* 275:27–31.

Lack, D. 1968. *Ecological adaptations for breeding in birds.* London: Methuen.

Lessells, C. M. 1985. Parasitoid foraging: Should parasitism be density-dependent? *Journal of Animal Ecology* 54:27–41.

Logue, A. W. 1988. Research on self-control: An integrating framework. *Behavioral and Brain Sciences* 1:665–709.

Mackintosh, N. J. 1983. *Conditioning and associative learning*. New York: Academic Press.

McNamara, J. M., and A. I. Houston. 1985. Optimal foraging and learning. *Journal of Theoretical Biology* 89:200–219.

Milinski, M. 1979. An evolutionarily stable feeding strategy in sticklebacks. *Zeitschrift für Tierpsychologie* 51:36–40.

Milinski, M. 1984. Competitive resource sharing: An experimental test of a learning rule for ESSs. *Animal Behaviour* 28:521–27.

Parker, G. A., and W. J. Sutherland. 1986. Ideal free distributions when individuals differ in competitive ability: Phenotype-limited ideal free models. *Animal Behaviour* 34:1222–42.

Pearce, J. M. 1987. *An introduction to animal cognition*. Hove and London: Erlbaum.

Pulliam, H. R., and C. Dunford. 1980. *Programmed to learn: An essay on the evolution of culture.* New York: Columbia University Press.

Rescorla, R. A. 1988. Behavioural studies of Pavlovian conditioning. *Annual Review of Neuroscience* 1:329–52.

Rogers, D. J. 1972. Random search and insect population models. *Journal of Animal Ecology* 52:821–28.

Roitblat, H. L. 1987. *Introduction to comparative cognition*. New York: W. H. Freeman.

Sherry, D. F., and D. L. Schacter. 1987. The evolution of multiple memory systems. *Psychological Review* 94:439–54.

Shettleworth, S. J. 1989. Animals foraging in the lab: Problems and promises. *Journal of Experimental Psychology: Animal Behavior Processes* 15:81–87.

Shettleworth, S. J. 1990. Spatial memory in food-storing birds. *Philosophical Transactions of the Royal Society of London*, Series B 329:143–51.

Shettleworth, S. J., J. R. Krebs, D. W. Stephens, and J. Gibbon. 1988. Tracking a fluctuating environment: A study of sampling. *Animal Behaviour* 36:87–105.

Stephens, D. W. 1987. On economically tracking a fluctuating environment. *Theoretical Population Biology* 32:15–25.

Stiling, P. D. 1987. The frequency of density dependence in insect host-parasitoid systems. *Ecology* 68:844–56.

Tamm, S. 1987. Tracking varying environments: Sampling by hummingbirds. *Animal Behaviour* 35:1725–34.

Walde, S. J., and W. W. Murdoch. 1988. Spatial density dependence in insect parasitoids. *Annual Review of Entomology* 33:441–66.

Ward, P., and A. Zahavi. 1973. The importance of certain assemblages of birds as 'information centres' for food finding. *Ibis* 115:517–34.

Weistkrantz, L. 1991. Problems of learning and memory: One or multiple memory systems? In J. R. Krebs and G. Horn, eds., *Behavioural and neural aspects of learning and memory,* 1–10. Oxford: Clarendon Press.

Whitham, T. G. 1980. The theory of habitat selection examined and extended using *Pemphigus* aphids. *American Naturalist* 115:449–66.

FOUR

Spatial Cognition and Navigation in Insects

Fred C. Dyer

INVESTIGATIONS OF HOW ANIMALS meet complex challenges to survival have, over the past 30 or so years, undergone a cognitive revolution. A cognitive approach to behavior grants an important causal role to internal mechanisms whereby sensory data are organized to form more or less detailed representations of events and relationships in the outside world. The ascendancy of this cognitive perspective comes at the expense of both the classical ethological view that much complex behavior reduces to mindless circuits of innate releasing mechanisms and fixed action patterns (Tinbergen 1951), and the radical behaviorist view that most behavior is forged from environmentally conditioned stimulus-response associations (Skinner 1971). Much research in animal psychology is now explicitly concerned with understanding the nature of cognitive representations. Interest focuses on the contents of such representations (i.e., what features of the environment are encoded internally), how they are acquired and encoded, how they are used to control behavior, and how they are implemented in the nervous system (Roitblat 1982; Gallistel 1989, 1990; Yoerg and Kamil 1991; Kamil, chap. 2). Behavioral ecologists, meanwhile, though generally not explicit in referring to the cognitive abilities of their subjects, often imply that animals possess sophisticated abilities to assess the fitness costs and benefits of alternative actions and then to calculate the best action to undertake (Kacelnik et al. 1990; Krebs and Kacelnik 1992). Thus, as in cognitive psychology (see Roitblat and von Fersen 1992), the animal is viewed as an active processor of information about its environment, rather than as an essentially passive "effect" of its innate or conditioned control mechanisms.

One of the great contributions of the early ethologists and behavioristic psychologists was to show that apparently intelligent behavior may often be the result of quite simple processes. Similarly, true to their roots in ethology, behavioral ecologists often suggest that decision making in animals is based upon more or less noncognitive "rules of thumb." Rules of thumb are (usually hypothetical) mechanisms that are relatively unsophisticated in the way they process and use information about the outside world, but well designed to produce appropriate behavior under natural conditions (Charnov 1976; Waage 1979; Cheverton et al. 1985; Schmid-Hempel 1984; Stephens and Krebs 1986; Wehner 1987; Kipp et al. 1989; Real 1990, 1991; Janetos and Cole 1981). For example, some animals behave as if they monitor their net rate of energetic gain while foraging. Under a cognitive interpretation (e.g., Gallistel 1990) one might assume that such an animal is actually equipped with the neural circuitry to measure its rate of energy gain in a literal sense, performing the appropriate computations on internally represented measures of net energy gain and time. Alternatively, perhaps the animal uses something as simple as a stretch receptor monitoring the distension of the gut. Because it depends on the rates of filling and emptying of the gut, gut distension might be well correlated with foraging rate under normal conditions (Charnov 1976; see also Roitblat 1982), thus eliminating the need for computation.

The increasing tendency of behavioral ecologists to look for cognitive explanations has been shaped by two related factors. First, there is the realization that the natural behavior of animals is often difficult to reduce to simple noncognitive rules. For example, even some hypothesized rules of thumb, simple though they may be, postulate information processing abilities (such as counting, measuring time intervals, or tracking fluctuations in environmental conditions) that are considered by psychologists (e.g., Gallistel 1990) to involve at least a low level of cognitive sophistication. The second factor is the notion that an understanding of the computational and representational mechanisms whereby animals process information about their environment may illuminate the adaptive design of the behavior in which these processes participate (Yoerg 1992). Research that has incorporated these perspectives has focused on a diverse array of topics, including spatial learning by food-storing animals (Sherry 1987; Krebs 1990; Shettleworth 1990; Kamil, chap. 2), learning by foragers of time intervals (Gill 1988; Kacelnik et al. 1990; Kacelnik and Todd 1992), learning by foragers of the patterns of reward offered by different patches (Shettleworth et al. 1988) or different food types (Real 1991, see also chap. 5), learning of social relationships (Cheney and Seyfarth 1990); and processing of signaled information in the context of communication (Guilford and Dawkins 1991). Note well that these possible cases of animal cognition largely fall outside the narrow research agenda of so-called "cognitive ethology," which concerns the possibility of consciousness and other humanlike mental events in animals (for reviews see Ristau 1991; Griffin 1992).

The cognitive bandwagon rolls on, but the notion that cognitive representations play an important role in animal behavior is still not universally accepted, either because it is rejected on principle (Amsel 1989; Kennedy 1992), or because it is simply ignored (see Yoerg 1992 for review). A common criticism is that an enthusiasm for cognitive explanations may lead researchers to overlook simpler explanations, even when simpler explanations are appropriate (Wehner 1987; Amsel 1989; Kennedy 1992; Poucet 1993). Cognitive explanations also arouse skepticism because they are about internal events that are observable only indirectly, through their behavioral effects (reviewed by Wasserman 1993).

This chapter evaluates the possibilities and pitfalls of a cognitive approach to behavior by reviewing recent work on the role of spatial memory in insect navigation. Insects provide an interesting opportunity to explore questions about the importance and complexity of internal representations in animal behavior. Although often viewed, no doubt with considerable justification, as rigidly programmed automata, many insects exhibit behavioral flexibility that is quite astounding. Inspired by such behavior, or by the hope that animals with simple nervous systems could serve as models for understanding general principles of cognition, several authors have espoused the heresy that a cognitive approach to insect behavior is both desirable and possible (Churchland 1986, 1988; Lloyd 1989; Gallistel 1989, 1990; Gould 1990; Real 1991). This proposal hinges on the assumption that cognitive processes (defined in some meaningful way) have reality in the behavior of animals with simple nervous systems, and on the further assumption that these processes can be carefully defined in operational terms and studied rigorously. The spatial orientation of insects offers an unusual opportunity to evaluate these assumptions. Some of the most tantalizing hints as to the cognitive sophistication of the insect brain come from studies of how the brain stores and uses information about spatial relationships in the outside world. At the same time, questions about the processing of spatial information are particularly tractable to experimental analysis.

In the following section of this chapter I review in more detail the concept that internal representations play an important role in animal behavior. Then I provide an overview of the navigational problems faced by central-place foraging insects, then review in detail a body of experimental work that has explored how these problems are solved. This review casts a fine-meshed net over a relatively restricted body of research in order to illustrate principles that have emerged only through ever more probing experimental investigations of the mechanisms underlying the behavior.

COGNITION AND REPRESENTATION IN ANIMAL BEHAVIOR

What is it that makes a particular account of behavior cognitive rather than not? Cognitive approaches to behavior share with noncognitive approaches a concern with the causal processes that intervene between sensory input and motor output. Cognitive and noncognitive approaches are distinguished, to a first approximation, by the degree of complexity that is attributed to these processes. Noncognitive interpretations of behavior usually postulate a relatively pared-down set of intervening processes, such that the animal's brain registers and responds to the raw sensory impressions of external events (or relationships among events) as experienced. Learning can be noncognitive, as in simple associative learning (see below) or imprinting. Cognitive hypotheses, on the other hand, postulate processes that organize sensory information into coherent internal models of external events, allowing the animal to respond appropriately to important stimuli even when experiencing them in novel combinations or contexts (Markl 1985; Cheney and Seyfarth 1990; Gallistel 1990).

Beyond this, further distinctions depend upon which of the many cognitive approaches one is considering. In some cases, the actual distinctions between cognitive and noncognitive interpretations of the same behavioral phenomena seem slight. For example, the snapping of a frog at an object that looks sufficiently like a bug could be viewed in the noncognitive terms of classical ethology (Tinbergen 1951) as a fixed action pattern released by a sign stimulus. This same behavior is sometimes cited, however (e.g., Lloyd 1989), as evidence of a simple form of cognition: the eye-brain forming a representation of an external event that is relevant to the animal. The sophistication is presumed to lie in the mechanisms whereby the brain recognizes the particular combination of size, shape, and motion stimuli that characterizes an object as possible prey. It could be argued, however, that identifying such recognition mechanisms as "cognitive" only invites confusion.

To take another example, simple Pavlovian conditioning, traditionally interpreted as a reflexlike variety of associative learning, is now interpreted in cognitive terms. In the classical view, a dog that salivates upon hearing a bell that has previously been rung prior to the delivery of food is said to have formed an association between an initially neutral stimulus (the sound) and the scent of food, to which salivation is an unconditioned (innate) response. Underlying this learning could be a mechanism as simple as a synaptic connection, strengthened through experience, between cells that receive inputs from the sound and the food. In the more modern cognitive interpretation, the strength of the associative bond is taken to be a representation of experience that the animal can actively evaluate, and then either use or ignore, depending upon its needs or upon other associations that it has learned (Rescorla 1988; Baker and Mercier 1989). Experimental evidence supports this shift in perspective, but for someone new to this literature, it appears to involve a rather modest realignment of thinking about the behavior.

Another important source of ideas about the mental abilities of animals is the emerging field of "comparative cognition" (see reviews by Roitblat 1987; Roitblat and von Fersen 1992; Wasserman 1993). Having grown from roots in experimental psychology, most research in this field is "comparative" only in a limited sense, focusing on rats and pigeons studied in artificial laboratory settings. The research questions and explanatory models are often not inspired by a consideration of the biology of the species in question, but by human cognitive abilities. Thus, research focuses on such problems as the structure and performance of "working memory" and "reference memory," the learning of abstract concepts (e.g., same/different, familiar/novel, oddity), categorization of objects, counting, and "language" (symbol manipulation). Roitblat (1982, 1987) places the study of internal representations at the core of comparative cognition. However, this view is not universal: many researchers in this field regard speculations about the nature of internal processes with suspicion (reviewed by Wasserman 1993). Moreover, in many studies the term "representation" could be replaced by "memory" (in the sense of a direct record of experience resulting in a change in behavior) with no loss of meaning, as if the cognitive revolution had affected mainly terminology.

Recently, Gallistel (1989, 1990) has provided a bolder and more comprehensive vision of animal cognition, advocating what he calls a "computational-representational" approach to behavior. At the heart of his argument is his definition of a cognitive representation as a "functioning isomorphism" between a set of objects, locations, or events in the environment and a set of states of the nervous system (see also Churchland and Sejnowski 1992). According to this definition, not only are there mechanisms in the brain for registering different states of the environment, but there are also computational operations that preserve and recover information about the formal (mathematical) relationships among different environmental states. The result is an internal model of a part of the animal's world, including some relationships that the animal derives through computation, rather than by detecting them directly. Furthermore, whereas traditional associative learning theories assume that a small set of general processes underlies most instances of learning (see Amsel 1989), Gallistel argues that distinct neural mechanisms have evolved to handle information about different classes of environmental stimuli. Thus, a cognitive representation of foraging rate, as in the example cited earlier, literally implies the existence of a neural module that divides a stored measure of intake by a measure of time, and then takes on distinct states and produces distinct outputs corresponding to different foraging rates. A spatial representation implies an ability to measure and manipulate geometric relationships among locations in the environment (see Gallistel 1990 for details). A representation of social relationships might entail capacious memory for the identities of individuals and histories of interactions, as well as rules for estimating the probability of different outcomes of future interactions (see Cheney and Seyfarth 1990). A corollary of this view is that we should expect the cognitive abilities of animals to have been shaped by species-specific selection pressures (see also Kamil, chap. 2).

Important theoretical and practical difficulties arise in characterizing the behavior of animals in terms of internal representations, leading many to doubt their reality or their explanatory value (Amsel 1989; Kennedy 1992). A general problem is that a given hypothesized representation, however descriptive of what an animal seems to be doing, may actually play no role in the hierarchy of processes that cause its behavior, either because the actual representation has different properties from those hypothesized, or because information about the outside world is not preserved internally in any meaningful sense. The source of this problem is that hypotheses about representations and other internal processes generally refer not to observed neural mechanisms but to systematic relationships between sensory input and behavioral output, which could be based on a variety of different mechanisms (reviewed by Churchland 1986; Amsel 1989; Kamil, chap. 2).

Two examples illustrate this point. First, as in the case cited earlier, a foraging animal that seems to monitor its foraging rate might merely measure the distension of its gut as a noncognitive correlate of foraging rate (Charnov 1976). At least in this case there is a simple isomorphism whereby the output of a stretch receptor encodes a "nominal representation" (Gallistel 1990) of foraging rate. However, this representation presumably would be less flexible than a rich functioning isomorphism—that is, a neural module that computed foraging rate by dividing a measure of net energy gained by a measure of time, thus preserving information that could be used for other calculations.

The second example illustrates that an adaptive response to changing environmental states might involve no isomorphism whatsoever between sets of internal and external states. Waage (1979), studying oviposition decisions by a hymenopteran parasitoid (*Nemeritis canescens*), showed that wasps stay longer in patches that have a higher density of hosts (insect eggs); this might suggest that they can measure, and hence represent, host density. However, Waage accounted for the wasps' behavior with a model that assumed no ability to represent host density or any correlate of it. He hypothesized an internal "level of responsiveness" that has a certain value when a wasp enters a patch, and then increases by a fixed increment as each host is encountered, and decays at a fixed rate as hosts are not encountered. The wasp stays in a patch as long as its current level of responsiveness is not below a certain threshold; thus in patches with high host density the wasp stays longer. In the model, this hypothesized internal parameter does not even nominally represent host density, since it may have the same instantaneous value for a wasp entering a patch, a wasp that has recently encountered a few hosts in an otherwise bad patch, and a wasp that has recently run into a bad spell in an otherwise good patch. This may not matter to the wasp, but it would matter to an investigator trying to explain the animal's behavior in terms of the presumed processes whereby representations are formed and used.

Some critics of representational approaches conclude from examples such as these that it is inherently impossible to decide, on the basis of behavioral observations alone, among competing hypotheses about the properties of representations and other intervening processes; this is known as the "indeterminacy problem" (Uttal 1982). Since behavioral observations often provide the only evidence we have, this criticism cannot be dismissed lightly, although, as we shall see, it does not seem likely to prove fatal to a representational approach (see also Roitblat 1982).

In this chapter I discuss a series of case studies that provide qualified support for a computational-representational approach to spatial navigation in insects. These studies deal with processes that clearly seem to involve the acquisition, manipulation, and use of relatively complex internal models of spatial relationships experienced by the animal. Furthermore, in contradiction to the indeterminacy objection, they exemplify strategies for studying these representations experimentally. These studies also reveal, however, that the processes that underlie insect spatial memory often turn out to be less sophisticated than the behavior first implies. Indeed, as is often noted, cognitive explanations, especially those patterned after human information processing abilities, have the potential to be grossly misleading as to the nature of the mechanisms really operating.

VISUAL NAVIGATION IN INSECTS: BACKGROUND AND OVERVIEW

For animals that cover large distances in their day-to-day movements, the major navigational tasks are visual ones. This is especially true of the species on which I will focus my attention: hymenopteran insects that provision a central nest, and which go on foraging trips that take them well out of direct visual or olfactory contact with the nest. The main species that have been studied are honeybees (*Apis mellifera*), sphecid digger wasps (*Philanthus triangulum, Ammophila campestris, Bembix rostrata*), and desert ants (*Cataglyphis* spp.). A major task these animals face as foragers is to find their way repeatedly back to a single point in the landscape—the site of the nest. Depending upon their foraging ecology, they may also need to find their way repeatedly to rich and temporarily stable patches of food. As I will discuss in considerable detail, learning of visual features of the environment plays an important role in the solution of these problems. This raises the general question of how visual information is represented internally and then used to guide behavior.

It is common to divide the tasks faced by long-distance navigators into those involving orientation to visual cues directly associated with a goal (e.g., a food source or the nesting site), and those involving orientation across distances over which cues associated with the goal are not visible or recognizable from the starting point (Schöne 1984; Collett 1992). In orienting visually at close range, insects may use spatial features of the goal itself, for example, the shape of a flower, or surrounding landmarks that fix the location of the goal. Starting with Tinbergen's classic studies of how a digger wasp (*Philanthus triangulum*) learns the location of its nest entrance (Tinbergen and Kruyt 1938), considerable work has focused upon the learning of visual features on this spatial scale. This is also the scale on which most studies of spatial memory are carried out in animal psychology labs (reviews by Gallistel 1990; Leonard and McNaughton 1990; Poucet 1993).

In orienting over long distances to a familiar but unseen goal, an animal must rely upon references visible at its starting point and along its route; for many species these references are provided by both celestial cues and landmarks. This is the challenge facing the many insects that forage over distances of hundreds or even thousands of meters from the nest (see Wehner 1981 for review). It is useful, if sometimes misleading, to recognize that a navigator trying to set a course for a distant goal needs to extract two kinds of information from the environment (Keeton 1974; Gallistel 1990). First, it needs the equivalent of a compass: the ability to discriminate directions. But since the direction that leads to a particular site depends upon the location of the starting point, the animal also needs the information that a map provides: its current position relative to its destination. A moment's reflection will make it clear that both directional and positional information are necessary. If you were wandering in unfamiliar terrain, you would be utterly lost if you had only the knowledge that your position was exactly 50 km south of your goal, but no compass to determine the direction to your goal (north). You would be equally lost if you had an accurate compass but no map or other source of information about whether you were north, south, east, or west of your goal. While it is important to recognize that an animal would be lost if it could only discriminate directions or could only determine its position, this distinction need not imply that directional and positional information are derived from different features of the environment, or that the "compasses" and "maps" of animals exhibit consistent properties across species. Indeed, the metaphor of the "map" in animal orientation is confused by the variety of meanings given to the term (Tolman 1948; Keeton 1974; O'Keefe and Nadel 1978; Wehner et al. 1983; Gould 1986).

The review that follows is selective in its emphasis on recent studies of spatial orientation in just a few well-investigated insect species. More exhaustive treatments, including reviews of the earlier literature, can be found in von Frisch (1967), Wehner (1981, 1982), Schöne (1984), and Gallistel (1990). In the following section I deal with how insects (especially desert ants) use celestial cues as a reference for determining their position relative to home during a foraging trip. Then I discuss the use of landmarks for orientation on various spatial scales, addressing the question of whether insects (especially honeybees) acquire some sort of "cognitive map" of the landscape.

PATH INTEGRATION: PYTHAGOREAN ANTS?

How can a foraging insect determine her position relative to her nest when she is ready to set a course for home? Observations from a variety of species (reviews by Mittelstaedt 1985; Wehner and Wehner 1986, 1990) suggest that during their outward path from a central place, animals can obtain information that will allow them to return directly home, even if they have arrived at a site that offers no other information about their location relative to home. A particularly well-studied example is the Saharan desert ant (*Cataglyphis* spp). These ants live in a desert habitat, scavenging for other insects that have succumbed to the midday heat (Wehner et al. 1983). Foragers never lay odor trails to recruit one another to food, but are solitary in their activities and rely entirely upon vision to find food and to navigate.

While searching for food an ant may follow an extremely circuitous outward path, but once she finds food she can set a course straight to her nest, which may be 100 m away. To demonstrate that the ant's ability to determine her position relative to home is not dependent upon familiar landmarks or other information obtained at the starting point, one can displace homing ants to a new location (fig. 4.1). On release they will run in the compass direction that would have led them home from the point of capture; they also run the appropriate distance before beginning a systematic search for the nonexistent nest entrance (Wehner 1982; Wehner and Srinivasan 1981; Wehner and Wehner 1990). The directional reference by which the homing ant sets her course is the celestial compass (the sun and sun-linked patterns of polarized light), as indicated by manipulations of artificial celestial cues presented to freely moving ants (Wehner 1982; Müller and Wehner 1988). The more difficult question is how, based on her outward journey, the ant determines her current position; that is, the appropriate direction and distance she should run to reach home.

It is clear that the ants on the outward path do not simply acquire a tracing in memory of the route actually followed, or a simple rewarded response. Instead, the ants behave as if they acquire a cognitive representation of the homeward vector, a

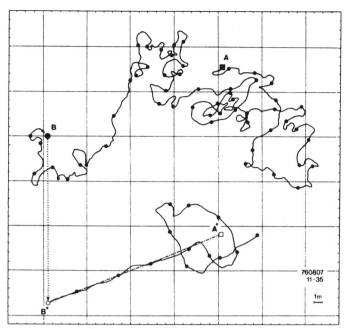

Fig. 4.1   Path integration by desert ants (*Cataglyphis albicans*), as shown by the searching path (above) and the homing path (below) of an individual forager. A, nest opening; B, point at which ant finds a morsel of food; B*, point of release after displacement; A*, location of hypothetical nest, i.e., position at direction and distance relative to B* that is equivalent to position of nest (A) relative to B. Dots along the animal's path show positions at 10-second time intervals. (From Wehner 1982.)

vector that differs from any segment of the circuitous outward path. Since, as Gallistel (1990) notes, a searching ant can set a homeward course from wherever she encounters food, even if she is in unknown terrain, the computation that generates this representation must be performed continuously in real time, prior to any reinforcement provided by the food or a successful trip home. Moreover, the representation compiled to a given point in the trip is stable, in that an ant captured on discovering food can set a homeward course even after several hours of captivity (Wehner 1982). Similar behavior is seen in honeybees (von Frisch 1967) and in a variety of other taxa, and is closely analogous to the dead reckoning calculations performed by human navigators (Wehner and Wehner 1986; Gallistel 1990).

Experiments in which ants are constrained to follow a specific path reveal additional insights into the accuracy of their integrative mechanism and the information on which it is based. Wehner (1982) trained foragers of *Cataglyphis bicolor* to find food by running through a narrow channel, which offered the ants a view of the sky but none of surrounding landmarks. In one experiment the ants had to travel out from the nest and back again in channels that formed the perpendicular sides of a right-angled isosceles triangle; each leg of this path was 192 cm in length. Ants trained in this way were captured as they started home from the food and were transferred to one of two testing channels. One channel lay parallel to the first leg of the homeward path, which the ants were about to travel when caught. The other channel lay parallel to the hypotenuse of the triangle. Quite remarkably, in each channel the ants walked the distance that would be appropriate for the direction

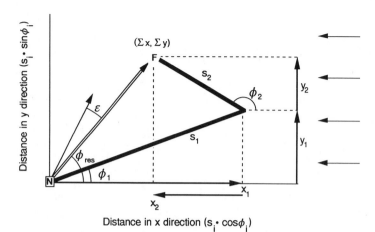

Distance in x direction ($s_i \cdot \cos\phi_i$)

Fig. 4.2   Graphical explanation of vector summation model of path integration. An animal travels a two-leg outward path (heavy line) from nest (N) to food (F). Each leg is decomposed into the sine and cosine projections in a Cartesian reference system defined by a compass reference (arrows at right). The resultant vector of displacement is computed from the summed sine and cosine components of the outward path. In this example, with $\phi_1 = 20°$, $s_1 = 20$ m; $\phi_2 = 150°$, and $s_2 = 10$ m, an animal performing vector summation should compute $\phi_{res} = 49.4°$ and $s_{res} = 15.6$ m. By comparison, an animal that computes the arithmetic average of the two angles traveled, weighted by the total length of the outward path, would compute $\phi_{res} = \Sigma(\phi_i \cdot s_i)/\Sigma s_i = 63.3°$, producing an error of $\epsilon = 13.9°$. (Adapted from Mittelstaedt 1985.)

they were traveling. In the first, the ants walked approximately 192 cm and then began turning back and forth as if searching for the perpendicular leg that would lead them home. In the channel parallel to the hypotenuse, the ants walked about 272 cm before turning and searching; this was exactly the length of the straight-line path from the food to the nest, which the ants had never before traveled. Whatever computational mechanism the ants have, it allows them to integrate the distances and directions traveled relative to the celestial reference over successive segments of a path. At least in this experiment, this mechanism allows the ants to behave as if they can precisely calculate the resultant of the two vectors experienced on the outward path.

But what actually is the nature of this calculation? Do the neural processes that underlie the ant's representation of the homeward vector really transform sensory measures of distance and direction in a way that is isomorphic to the trigonometric calculations underlying vector summation? Historically, this question has been answered in the affirmative, since trigonometry provides the simplest exact solution to the problem of finding a resultant vector. Mittelstaedt (1985) and, more recently, Gallistel (1990) have clearly laid out the sorts of computational abilities that this would require. A minimal model requires the animal to decompose the direction (ϕ) and distance (s) traveled on each segment of a path into its sine and cosine projections ($y_i = s_i \cdot \sin \phi_i$ and $x_i = s_i \cdot \cos \phi_i$) onto a common reference system (fig. 4.2). The reference system could be defined by the current solar meridian (and its perpendicular) or by the local celestial meridian as established by the insect's memory of the pattern of solar movement over the day (Wehner and Lanfranconi 1981; Dyer 1987). As they move, the ants would have to sum the contributions of each successive path segment to update their current position (Σx_i, Σy_i) in this reference system. From these coordinates they could compute their current angle of displacement (ϕ_{res} = $\tan^{-1}$ [$\Sigma y_i / \Sigma x_i$]) and, by the Pythagorean theorem, the net distance of displacement (s_{res} = [(Σy_i)^2 + (Σx_i)^2]$^{1/2}$). This hypothesis assumes no remarkable sensory abilities—many insects can measure angular and linear distances. The question is whether this hypothesis accurately describes the internal operations by which these sensory data are integrated into a representation of the homeward vector.

An alternative hypothesis, which has been set forth in various versions (Jander 1957; Mittelstaedt 1985; Wehner and Wehner 1986; Müller and Wehner 1988), is that animals perform path integration by computing a distance-weighted arithmetic average of the directions traveled on the outward path. In its simplest form, as explained by Mittelstaedt (1985), this hypothesis assumes that an animal obtains the net angle of displacement (ϕ_{res}) by continuously summing the angles traveled on successive path segments, with each angle weighted by the distance traveled on that segment. Then the animal divides this sum by total path length. Hence: ϕ_{res} = $\Sigma (s_i \cdot \phi_i)/\Sigma s_i$. One might argue that since this is computationally simpler than the vector summation model, it is more plausible on the grounds of parsimony. However, this method of path integration has generally been dismissed a priori because it would produce substantial errors for paths containing sharp turns (Mittelstaedt 1985) (fig. 4.2).

Müller and Wehner (1988) found that a variant of the distance-weighted mean direction hypothesis, complete with such errors, actually better accounts for the path integration behavior of desert ants (*Cataglyphis fortis*) than does the vector summation hypothesis. These authors trained ants to run a two-leg outward path in channels, and then transferred them to a featureless landscape where they could choose their own homing direction (fig. 4.3A). The first leg of the outward path was twice as long as the second leg, and the angle the ants had to turn to begin the second leg was varied from 60° to 180°. If the ants could perform vector summation to measure their displacement, they should always set the bearing home without error. However, the bearings chosen by the ants actually deviated from the true homeward bearing in a way that suggested that the ants had computed a distance-weighted mean of the directions traveled on the outward path (fig. 4.3B).

Because a weighted-mean model of the sort described above did not exactly account for the errors, Müller and Wehner explored two other computational models. Both models share some common features that distinguish them from the earlier model. In the first place, an ant is assumed to measure the direction (δ) of each path segment not relative to a fixed directional reference, but relative to her own computation of her direction of displacement (ϕ_n) from the nest at her current position. Furthermore, she weights this angle not by total path length, but by the net distance (l_n) of displacement computed to that position (see fig. 4.3C). Also, Müller and Wehner hypothesized a method for computing net distance of displacement, which the original weighted-mean model does not do. Specifically, each unit path segment traveled by the ant is assumed to add an increment to her computed net distance of displacement that is proportional to δ; this increment varies uniformly from 1 (for δ = 0°) through 0 (for δ = ±90°) to −1 (for δ = 180°). The difference between Müller and Wehner's two models is in how ants are assumed to update their computed net direction of displacement.

In their first model, ants were assumed to add to the current computed direction (ϕ_n) the angle (δ) by which each successive path segment deviated from this direction, weighted by l_n. Thus: $\phi_{n+1} = \phi_n + \delta \cdot l_n^{-1}$. The weighting by l_n guarantees that the contribution of each path segment to the computed net angle is smaller at greater

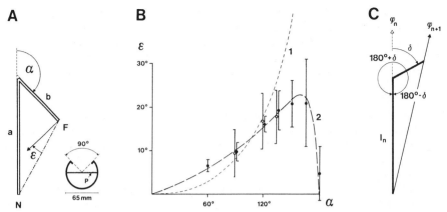

Fig. 4.3   Path integration in desert ants (*Cataglyphis fortis*): Experimental results and interpretation. (A) Ants were trained to run a two-leg outward path in narrow channels which turned through a variable angle α. The lengths of the two legs were constant (a = 10 m, b = 5 m). Inset: Cross-section of training channels. Ants were caught at F and released in open terrain, and the error angle (ε) between their homing bearing and the angle expected for vector summation (see Fig. 4.2) was measured. (B) Error angles (with 99% confidence intervals) plotted for different turning angles (α). Some ants had a view of the sun (open circles), while others could see only blue sky, hence sun-linked patterns of polarized skylight (solid circles). Dashed lines give error predicted by distance-weighted mean angle models 1 and 2. (C) Basis for predicted error angles. ϕ_n and ϕ_{n+1} give the ant's mean direction from the nest after its nth and (n + 1)th path segment, respectively; l_n denotes the distance computed after the nth segment; δ is the angular difference between ϕ_n and the direction traveled during the ant's (n + 1)th segment. According to model 1, the ant determines ϕ_{n+1} by adding δ, weighted by l_n, to ϕ_n; according to model 2, the ant determines ϕ_{n+1} by adding [k(δ + 180)(δ − 180)δ] · l_n^{-1} to ϕ_n (see text). (From Müller and Wehner 1988; see also Wehner 1991.)

distances from the nest. This model predicts that the amount of directional error in the ants' calculation of their direction of displacement increases steadily as the turning angle between two successive path segments increases. On the contrary, in the two-leg experiments the ants actually made smaller directional errors for turns of 180° than for turns of 90°–135° (fig. 4.3B).

Müller and Wehner's second model, a relatively simple extension of the first, provided a much closer fit to the data. In this model, as in the first, the ant's representation of her displacement from home is updated in an iterated fashion, with each successive path segment producing a distance-weighted change in ϕ_n. The equation summarizing this effect is a bit more complicated, however: $\phi_{n+1} = \phi_n + [k(180 + \delta)(180 - \delta)\delta] \cdot l_n^{-1}$. Here, the contribution of each successive path segment to ϕ_n is assumed to scale with the product (180 + δ)(180 − δ)δ. This has the effect that the change in δ_n is greatest when the ant is traveling roughly perpendicular to the estimated homeward direction. A normalization factor, $k = 4.009 \times 10^{-5}$ deg^{-2}, scales this product so that the maximum contribution of each path segment to ϕ_n (before weighting by l_n) is 90°. This model was quite successful at accounting for the errors in the two-leg experiments (fig. 4.3B). It also accurately predicted errors in the homeward bearings chosen by ants that had been constrained to run a three-leg outward path, and by ants that had followed long, unconstrained searching paths in a natural landscape. Finally, Wehner (1991) has used the second model, without modification, to reanalyze path integration data from various other anthropod species, including honeybees and spiders, which have long been assumed to perform accurate vector summation (Mittelstaedt 1985). In all cases the animals actually exhibit errors in their computed homeward bearings that are indistinguishable from the predictions of the model.

How do ants cope with the errors that result from their path integration mechanism? Müller and Wehner provide an answer in three parts. First, all weighted-mean models predict that turns opposite in direction will produce errors opposite in sign, and therefore will cancel out. Since, while searching for food, ants turn left about as often as they turn right, and seldom make very sharp turns (see Wehner and Wehner 1990), their path integration system will often compute a mean direction of displacement with little or no error. Furthermore, when turning biases in the outward path do produce errors, ants can rely upon conspicuous landmarks near the nest to pinpoint its location if they are close (Wehner and Räber 1979). Finally, if they are totally lost, ants can fall back on a systematic search strategy designed to maximize the chance of finding the nest entrance (Wehner and Srinivasan 1981).

In summary, the evidence clearly indicates that ants do not do vector summation, and suggests instead that they perform a type of distance-weighted arithmetic calculation of their displacement during an outward path. Müller and Wehner (1988) argue further that the apparent computational simplicity of this alternative hypothesis is to be expected given the modest size of the ant's nervous system. Of course, neuroscience has not yet provided us with a theoretical framework to evaluate the assumption that trigonometric operations are inherently more difficult to implement in small nervous systems (or in large ones for that matter) than are the operations involved in computing a distance-weighted sum. An important conclusion, however, is that natural selection has engineered neural mechanisms that carry out not the exact mathematical solution but an approximation, and yet the animal gets by quite well. The point has been made before (e.g., Wehner 1987), but it is worth emphasizing that hypotheses about how animals solve a particular problem should not be excluded merely because the proposed solution would produce a certain degree of systematic or random error.

The observational and experimental evidence clearly supports a computational-representational interpretation of the ant's behavior. The animal's homing is guided by a representation of spatial position formed as a result of internal computations on sensory data. Although the mechanism results in an approximate solution to the problem, this does not make it noncognitive; nor would anything be gained by dismissing it as a mere "rule of thumb." It is also hard to explain the ant's path integration system in terms of standard associative learning theories, which usually assume the formation of noncomputational linkages between specific stimuli and responses. A further important lesson, which flies in the face of the indeterminacy objection, is that the computations underlying path integration have proved open to experimental investigation via manipulations of behavior. Müller and Wehner's (1988) experiments, which were designed in a way that explored the limits of the ant's path integration mechanism, virtually eliminate one long-standing computational model (vector summation) as an explanation for path integration. Even if their alternative model fails some future test, it represents progress.

SCALES OF REPRESENTATION IN SPATIAL MEMORY: FROM FLOWER SHAPES TO "MENTAL MAPS"

Honeybees and other insects face the problem of learning spatial relationships in an environment at several different spatial scales, ranging from the scale of individual flowers in a patch to the scale of the landscape. Since the 1930s, studies of how insects use noncelestial visual features of their environment have focused on two related problems: first, what spatial information is stored in memory? and second, how is the stored information subsequently used to guide orientation? The results of this work have consolidated around a general model, which applies, with some variations, to the learning of spatial relationships both for orientation at close range and for long-distance navigation along routes. As the animal learns a spatial pattern, it is hypothesized to store in memory an "eidetic image," the neural equivalent of a snapshot, fixed relative to retinal coordinates (Collett and Cartwright 1983). Subsequently it guides itself by matching this memory image pixel by pixel against what it currently sees, and maneuvering to the location or along the path where the match is best. This model allows for considerable behavioral flexibility without requiring the insect to learn complex geometric relationships among landmarks or locations in the environment. My discussion centers on honeybee spatial learning on a range of spatial scales, starting with flower learning.

Flower Shape

Honeybee foragers can learn the shape (as well as the color and odor: Menzel 1990) of flowers from which they have obtained food. In using what she has learned, a bee approaching a flower behaves as if she has recorded a two-dimensional image of the rewarding shape relative to retinal coordinates, and then fits this memory image to what she currently sees to discriminate rewarding from novel patterns. This learning has been explored in experiments with simplified flowerlike patterns (radial gratings, which look like pie charts with alternating black and white sectors) presented side by side on a vertical screen (Wehner 1981, 469ff.). Bees can discriminate a rewarded pattern from the same pattern rotated by exactly one-half of the spatial period, so that, for example, a white sector instead of a black sector is in the upward position. This suggests that their memory of the shape does not encode a set of parameters such as spatial frequency or the distribution of edge orientations, since such parameters would be identical for the rotated and unrotated patterns. Also, bees evidently do not form an image concept of the sort people use to recognize a familiar shape regardless of its orientation relative to retinal coordinates (Biederman 1987; Kosslyn 1988). To an insect, a shape in a different orientation, or seen by a different eye from the one that saw it originally (Wehner and Müller 1985), is not recognized as the same shape.

The behavior of trained bees performing this discrimination provides further insights. Bees fixate visually on a given shape from a characteristic distance in a way that would ensure that light and dark portions of the currently detected pattern could be brought into register with the parts of the retina that had recorded them previously. Experiments in which parts of the eyes of trained bees were covered with paint bear out the conclusion that the same part of the eye is used to learn and to match a given part of a familiar shape (Wehner and Flatt 1977).

Recent studies (van Hateren et al. 1990; Srinivasan et al. 1993) have suggested that bees can be trained to learn the dominant orientation of edges in a pattern as a parameter associated with food reward. Thus, two patterns that differ in their overall shape may be recognized as similar if they are both dominated by edges of a particular orientation relative to the vertical. Bees therefore must not require a global match to recognize a pattern as familiar. Because of the way the bees are tested in these studies, this result is not necessarily incompatible with Wehner's suggestion that patterns are recognized via retinally localized comparisons of current and remembered images. Bees perform the discrimination not while hovering in front of the target patterns, as in Wehner's earlier work, but while flying into the choice point of a Y-maze where they can see test patterns at the ends of the two arms. The image moving across the eyes of a flying bee presumably allows a given retinal sector to be stimulated by a common shape element (line orientation) in many different patterns. This could produce a partial match even if other elements of the patterns differ.

Thus, insects, unlike humans (Biederman 1987), seem to represent visual images in a way that is strongly dependent on the way they project to retinal coordinates. This may considerably simplify the encoding of a visual pattern for storage in memory. Meanwhile, the ability of bees to direct their responses to the flower shapes that provide food rewards can be interpreted as a straightforward case of associative learning. Thus, so far we need not assume the involvement of very sophisticated cognitive operations in the learning of visual images.

Landmark Arrays

Pictorial images are also thought to underlie the extreme accuracy with which many insects may search for a familiar goal in the environment, such as a nest, a patch of food, or a mating site (reviewed by Wehner 1981). In this case it is a snapshot of the array of landmarks surrounding the goal that the animal stores and then uses to guide its return. The goal itself may play a relatively minor role in guiding the return, a point dramatically illustrated by Tinbergen and Kruyt's (1938) classic experiments with digger wasps (*Philanthus triangulum*). A wasp was presented with a circle of pine cones or some other such array around her nest, and when the array was displaced by a half meter or so, the wasp searched persistently in the array rather than near the actual location of the nest.

Most recent work on how insects learn small-scale landmark arrays has focused on honeybees (Cartwright and Collett 1983, 1987; Cheng et al. 1986, 1987; Gould 1987b; Collett 1992), although important insights have also come from studies of desert ants (Wehner and Räber 1979) and hoverflies (Collett and Land 1975). Experiments that explore the contents of the bee's spatial memory have followed an approach developed by Cartwright and Collett (1983). Individually labeled honeybees are trained singly to find an inconspicuous source of sugar syrup placed in a particular location relative to an array of vertical cylinders or other shapes. To reach this feeding site, a bee must enter a white-painted room, which simplifies the experimental manipulation of the visual surroundings. After the bee has made a number of visits and can find the food quickly, she is tested to see what she has learned about the landmark array. When the bee has finished feeding and has left for the hive, the feeder is removed and some critical feature of the array is changed (e.g., size, spatial arrangement). Then the bee's searching behavior on her next visit is observed. The supposition is that if the bee's behavior is affected by the change in the landmarks, then the memory that she used to guide her search must have encoded information about the feature that was changed.

In such experiments, bees searched during a test at the location where the landmarks appeared the same as they did during previous training visits. This suggested that bees might learn the spatial features of the array as a two-dimensional image representing the angular (apparent) size of each landmark as seen from the food (Cartwright and Collett 1983, 1987; Collett and Cartwright 1983; Gould 1987b), akin to the snapshot of a flower's shape. However, the complexity of the array

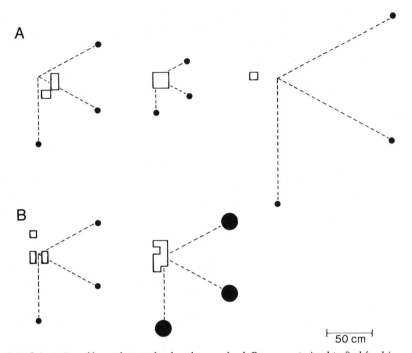

Fig. 4.4   Orientation of honeybees to landmarks near food. Bees were trained to find food (sucrose) next to an array of three black cylinders (each measuring 4 × 40 cm). From the food (at the intersection of the dashed lines), the angular separation between two adjacent cylinders was 60°. The leftmost array in both (A) and (B) shows the configuration to which bees were trained; in the test arrays the sizes of the landmarks or their distances from the position of the food were changed. The shaded areas cover the points where the amount of time spent by the searching bees exceeded 80% of the time spent in the most visited area. (Adapted from Cartwright and Collett 1983.)

seems to influence just how individual landmarks are used, and hence, perhaps, how they are stored (Cartwright and Collett 1983). When bees were trained to search for food at a given distance from a single black cylinder, they behaved as if they used only the size of the image that the landmark projected to their eyes; when tested with a larger or smaller cylinder, the bees searched at the distance where the test landmark subtended the same angle as the training landmark. If, however, three black cylinders were placed in an array to one side of the food (fig. 4.4), changes in the size of each landmark, and hence the angle it subtended at a given distance, did not cause the bees to search in a different location, so long as the angles separating the landmarks were the same. To reconcile these two results we must assume that the angular information provided by a panorama of landmarks is more important than the angular information provided by any one landmark (see Gallistel 1990).

Both the one-landmark and three-landmark experiments suggested that bees were using references other than the cylinders to help them narrow their search. They searched only in one restricted location relative to the single cylinder, even though such a landmark should have looked the same from every point at the same distance around it. As Cartwright and Collett (1983) suggest, bees probably used other features visible in the room, such as the window through which they had entered, to restrict their search; that is, their snapshot contained these features as well as the cylinder. In support of this interpretation, if the three-cylinder array was rotated relative to these features, the bees substantially changed their distribution of searching relative to the array.

These early experiments revealed much about the angular information that a bee encodes in memory. A further wrinkle in the story is the discovery that bees may encode the distances of landmarks from the food. When visually distinct landmarks were placed at various distances from the food during training and then rearranged during testing, bees were somewhat more likely to search at the position specified by the landmarks that had been near the food, even if distant and nearby landmarks subtended the same visual angle (Cheng et al. 1987). Thus, the snapshot acquired by bees has depth, in that the angular information provided by nearby elements of an array has slightly more weight than does the information provided by distant landmarks. Bees are thought to use motion cues to discriminate the relative distances of objects, independently of image size (Lehrer et al. 1988).

Turning to the question of how a bee uses her memory of a landmark array to guide her return to the site of the food, we encounter some intriguing puzzles. Presumably some sort of matching process is involved, such that the bee maneuvers to the place where her current view of the spatial pattern most closely matches what she has stored previously in memory. This matching process must, however, be somewhat more complicated than that used in flower learning, in which the bee inspects the flower's shape while hovering at a fixed point. An insect flying through an array of landmarks has many degrees of freedom in her orientation and position relative to the pattern she is trying to match, and thus many possible ways in which the images of the landmarks may fall onto her retina. Whatever the mechanism whereby current visual input is matched against memory, it must explain the observation that bees and other insects can set a course for their goal from any arbitrary starting point (and initial orientation) relative to an array of landmarks (Collett and Land 1975; Cartwright and Collett 1983).

To explore this problem, Cartwright and Collett (1983) considered several computational models. All of the models assume that a bee searching for food relative to a landmark array guides herself using a snapshot that was previously recorded at the position of the food. When the bee is not yet at her goal, she should see a discrepancy between the snapshot and her current view of the landmark array. The models examine various mechanisms whereby a bee might recognize this discrepancy and adjust her angle of flight to reduce it. Using computer simulations, Cartwright and Collett excluded several models, all of which shared the assumption that bees acquire and use a single snapshot that is fixed relative to retinal coordinates, like a snapshot of a flower's shape. The models with this feature could consistently compute a course toward the food only when the bee happened to be aiming in the same direction in which she recorded the snapshot. Since, as mentioned, insects did not seem constrained in this way, Cartwright and Collett constructed a more sophisticated model. This model assumed that at every location in the environment a bee can match current visual input against a snapshot that is in the same compass orientation as the actual array, no matter what her position and orientation relative to the array (fig. 4.5A). This implies an ability to use references external to the array itself (for example, distant landmarks) to align the snapshot with the array, independent of body orientation, before the match is made.

It remains quite perplexing how this might be accomplished. Collett and Cartwright (1983) offered two possibilities. First, a single snapshot of the array might

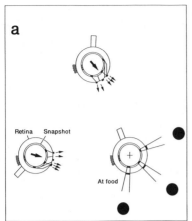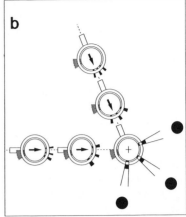

Fig. 4.5   Models that explain how bees might use memory images ("snapshots") of a familiar array of landmarks to find a feeding site. (*A*) Cartwright and Collett's (1983) hypothesis that bees match the current image on the retina to a single snapshot of a landmark array previously taken at food. Dark and light sectors of the snapshot are compared with the nearest comparable sectors on the retina (long axis of body is denoted by rectangle). Discrepancies in the bearings and angular extents of the compared sectors generate a set of unit vectors which, when summed, lead the bee in the direction that will reduce the discrepancies. At the food, there are no discrepancies. For this model to behave like a real bee, the snapshot used at a given location has to be in the same compass orientation as the represented array, as determined by references external to the array. Cartwright and Collett suggest that a distant landmark, shown by the large hatched sector, could be used to activate a snapshot taken at the food in an orientation parallel to the bee's current direction of travel. (*B*) Alternative hypothesis that bees memorize a sequence of snapshots along each of several routes leading to the food (two routes are shown). Each snapshot is specified relative to retinal coordinates, is encoded with information about the direction to fly when that pattern is seen, and is activated when that pattern is seen.

counterrotate within the bee's head to compensate for changes in her orientation relative to compass references, analogous to the way people can mentally rotate an internal image to assess whether two objects in different orientations have the same shape (Kosslyn 1988). This hypothesis attributes considerable computational powers to the insect, though it demands that she store in memory only a single visual image for a given landmark array. The second possibility, which Collett and Cartwright regarded as more plausible, is that the bee records several egocentric (i.e., retinally fixed) snapshots of the landmark array at the position of the food, each with her body in a different orientation (for example, during previous arrivals and departures: Opfinger 1931; Couvillon et al. 1991). When subsequently searching, she relies upon compass cues or other landmarks to activate the snapshot taken in the orientation parallel to her current direction of flight. This hypothesis requires that several images of the array be stored in memory, but the computations involved in their use might be simpler.

Illustrating the challenges inherent in studying these sorts of processes, the possibility remained that completely different models, making quite different assumptions from those made by Cartwright and Collett, might work just as well as theirs did. For example, Cartwright and Collett might have been wrong in their assumption that bees searching for the food matched their current visual input against one or more snapshots previously taken at the goal. Perhaps, instead, bees take snapshots at several points along each of several routes previously flown relative to the array, and encode with each snapshot information about the direction of travel that leads to the food when they see that particular view of the landmarks (fig. 4.5B). This hypothesis could work with snapshots that are fixed relative to retinal coordinates, like the snapshots of flower shape; hence it dispenses with the need to assume that snapshots are encoded and used in an orientation-independent, earth-based frame of reference. It also dispenses with the assumption that bees have a generalized ability to "compute" a course based on a discrepancy between the current visual image and one stored previously at the location of the goal. On the other hand, this hypothesis requires that bees store multiple snapshots in memory during their previous arrivals and departures from the food. Obviously, since bees would be able to steer toward the food only from locations on or near routes they had previously traveled, this sort of mechanism would be less flexible than that proposed by Cartwright and Collett. However, with adequate experience (which the bees in their experiments might well have had), it might produce identical behavior.

A recent study offers yet another alternative to Cartwright and Collett's conclusions, by suggesting that one of the models that they rejected might actually be applicable. Through video analysis, Collett and Baron (1994) found that bees searching for food relative to a landmark usually faced southward (apparently using a magnetic compass) during a certain portion of their approach. Thus, as in Cartwright and Collett's simpler models, bees might use a single snapshot fixed relative to retinal coordinates. It is sobering that their more complex model, and much of the foregoing discussion, may be based on a false premise: that insects are indifferent to how images of landmarks fall on the eye. Of course, we do not know whether the mechanism hypothesized by Collett and Baron is by itself sufficient to explain the learning of landmark arrays by insects.

In discussing the contents and use of the honeybee's representation of a landmark array, I have postponed a couple of fundamental questions, which I will now address. The first question concerns the geometric "power" of the bee's memory: this refers to the degree to which the bee's memory preserves the actual metric relationships among objects in the environment. Metric relationships, or the angles and distances separating points in space, define the shape of space in Euclidean geometry. It is possible, however, that animals encode spatial relationships in weaker geometric systems that preserve only qualitatively the relative positions of objects (e.g., that landmark B is somewhere between landmarks A and C) and distort the metric relationships among them (Gallistel 1990; Poucet 1993). The second question concerns the coordinate system within which the positions of landmarks are represented (Gallistel 1990; Collett 1992). One possibility is that the coordinate system is egocentric, defined by the animal's body axes. Alternatively, the coordinate system may be allocentric or geocentric, defined by references external to the observer. Such a system would be computationally more demanding than an egocentric reference system because it would need to be constructed from images projected to (egocentric) retinal coordinates.

Gallistel (1990), in reviewing the work of Cartwright and Collett (1983), argued that bees form the richest possible cognitive representation of the landmarks in an array: a metric map in geocentric coordinates, analogous to a "plan view" of the array. The behavior of Cartwright and Collett's bees (see fig. 4.5A) closely resembles the behavior of rodents in analogous tests, which many animal psychologists take as evidence of such maps. In one widely used experimental paradigm, rats are placed in a circular pool of murky water and then trained to find a submerged platform where they can rest (Morris 1981). Rats quickly learn the location of the escape platform relative to surrounding landmarks that are visible from every point in the pool. Like bees, rats can travel directly to their goal from any arbitrary starting point, and are confused when surrounding landmarks are placed into new arrangements, suggesting that they use a global maplike memory of the landmarks.

The bees' behavior encourages the interpretation that their egocentric snapshots of landmark arrays represent metric information: the angles separating landmarks as seen from particular local vantage points and, to the extent that nearby landmarks are weighted differently (Cheng et al. 1987), the distances of landmarks from particular vantage points. It is debatable, however, whether this information is also encoded in a common allocentric coordinate system. Cartwright and Collett's (1983) results suggest that a bee searching amidst a landmark array might use one or more snapshots taken at the food, each linked to a common external directional reference (fig. 4.5A). If this qualifies as a allocentric representation, it is a far cry from Gallistel's notion of an internal plan view. Its output is only an angle to be flown given a certain view of the landmark array; there is no recoverable information about distance to the goal or the relative positions of different landmarks. Weaker still is the model illustrated in figure 4.5B, in which the bee is assumed always to use the egocentric snapshot that best matches her current local view of the array. Leonard and Mc-Naughton (1990) make a similar argument regarding the ability of rats to find the escape platform in a water pool. Finally, Collett and Baron (1994), while providing evidence that bees use an external compass to orient to landmarks, suggest that the compass serves merely to steer the bee so that she can more easily match an egocentric snapshot to her current view of the landmarks.

The point of all this is that the highly flexible behavior of searching bees (and that of rodents in a water pool) may be based upon relatively simple, although still poorly understood, geometric operations. Given how well bees do, it is hard to imagine what they (or any animal) might gain from a more detailed representation of the spatial relationships among landmarks. Nor is it obvious what hardware improvements (e.g., in brain size) might be required to form something like a plan view of a landmark array in memory. By providing insights into what the bee's spatial memory does (and especially into what it does not do), the computational-representational approach developed by Collett and others promises to illuminate these and other basic questions about how nervous systems store and use spatial information.

Learning of Multiple Landmark Arrays

Moving up one level of spatial scale, Collett and Kelber (1988) have considered how bees might be equipped to learn the landmark arrays around different feeding sites encountered on the same familiar route. This problem presumably has a natural analogue, since bees might generally not obtain a full load of nectar or pollen in a single restricted area (Janzen 1971; Heinrich 1976). If two rich food patches are sufficiently far apart, the insects will be confronted with the task of learning the location of food relative to two arrays of landmarks. Compounding the problem, some successively encountered arrays of landmarks (e.g., clumps of vegetation) might differ little in their visual appearance, and yet the location of the food relative to each array could be distinct. This might require the animal to rely upon other contextual cues to adopt the correct learned response to an array with a given appearance.

Collett and Kelber brought this problem under experimental control by training

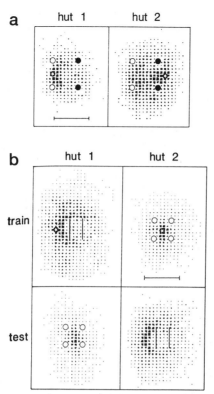

Fig. 4.6   Search distributions of honeybees trained to find food (sucrose) relative to landmark arrays inside two huts; each bee was videotaped from above during a visit on which no food was provided for the first 2 minutes. Sizes of black squares show the relative time spent in each 4.5 cm × 5 cm cell of an imaginary grid superimposed on the floor of the hut. Scale: 50 cm. (*A*) Single bee that was trained to forage in different locations relative to two identical arrays of landmarks (open circles, yellow cylinders; solid circles, blue cylinders; stars, positions of food during training). (*B*) Single bee that was trained to find food relative to two different arrays of landmarks that were then swapped between the two huts (open circles, yellow cylinders; thick lines, blue triangles standing on edge; stars, positions of food during training). (From Collett and Kelber 1988.)

bees to enter a small hut to obtain half a stomach full of sugar syrup, then to fly to a different hut to complete their load. Inside each hut was an inconspicuous feeder that was always placed in the same position on the floor relative to an array of landmarks (colored cylinders or other shapes). As soon as the bee entered, the door of the hut was closed behind her, so that the landmark array provided the only visual reference for finding the food. As in the studies described above, the information stored in memory by trained bees was assessed by observing where they searched during a visit in which the food was removed and the landmarks were altered in some way.

The study suggested that several distinct spatial and contextual cues can trigger retrieval of spatial information from memory. One experiment (fig. 4.6A) addressed whether the sequence in which bees visited the two huts provided them with information about where to search relative to the landmarks inside each hut. For this experiment, each hut had an identical array of landmarks, two yellow cylinders and two blue cylinders arranged in a square pattern. In the first hut the bees visited (hut

1), the food was placed between the yellow cylinders, on one side of the square. In hut 2 the food was placed between the blue cylinders. During the test trial, with the food removed and appropriate measures taken to exclude the use of odor cues, the bees searched predominantly between the yellow cylinders in hut 1, and between the blue cylinders in hut 2. This result strongly suggests that the bees learned each response to these identical landmark arrays according to the order in which they were used.

Next, Collett and Kelber asked whether bees could learn to respond appropriately to different landmark arrays even if they encountered them out of order (fig. 4.6B). For this experiment, the bees were trained in hut 1 to find the food to one side of a pair of blue triangles standing on edge parallel to each other, while in hut 2 the food was placed in the middle of a square array of four yellow cylinders. During test trials, the arrays were swapped between the two huts, so that the bees encountered the cylinders in hut 1 and the triangles in hut 2, and no food in either hut. The bees searched for the food mainly in the place that was appropriate for each landmark array, even though the arrays were presented out of sequence. Thus the appearance of a distinctive array is sufficient to elicit the learned response to that array.

These two experiments, together with other work, suggest that the spatial content of each snapshot is linked in memory not only with information about where to search relative to the corresponding landmark array, but also with additional information about the context in which the learned response is ordinarily used. The order in which a response is used provides one sort of contextual information. Other experiments (Collett and Kelber 1988) showed that distant landmarks may also be used to trigger an appropriate response to an array; this lends support to Cartwright and Collett's (1983) suggestion that distant landmarks could cue the retrieval of an appropriately oriented snapshot of a landmark array (see above). Finally, Bogdany (1978) and Gould (1987a) have shown that bees can use time of day as a contextual cue in using stored information about attributes of feeding places.

Collett and Kelber (1988) suggest that honeybee spatial memory might be an example of "content-addressable memory" (Rumelhart and McClelland 1986), in which exposure to any of several linked features associated with an event is sufficient to provoke the retrieval of other features in the set. Content addressability need not imply the existence of a rich set of cognitive processes for storing and combining information in memory. Instead it could involve the formation of a set of rather simple associative links among stimuli associated with an object or event. Some of these stimuli may be contextual cues that modulate an animal's use of other associations (Rescorla 1988).

However the underlying processing is best interpreted, it is clear that bees store spatial information in a way that guarantees its accessibility. In particular, I want to emphasize one key result that is relevant to the other sections of this chapter: although contextual cues influence the way bees draw upon stored spatial information, bees readily respond in an appropriate way to familiar spatial patterns encountered out of context. Thus, as Collett and Kelber (1988) point out, the sight of a familiar pattern must in some way trigger the retrieval of the spatial memory that the bees use to respond to that pattern.

Long-Distance Navigation: Route-Based Memory Sequences and Mental Maps

Honeybees and other insects, like many vertebrates, can use landmarks to navigate among familiar sites separated by long distances. This implies that they can obtain both directional and positional information from familiar landmarks. Given that such insects may routinely travel over distances of hundreds or even thousands of meters (see Wehner 1981 for review), this would seem quite a formidable task, since it requires the animals to learn something about spatial relationships that they could have experienced only over time, as a result of their movement through the environment.

The first evidence that honeybees can learn directions relative to landmarks was provided by experiments in which bees were trained to find food by flying along a conspicuous line of trees. When their hive was displaced to a landscape with a similar landmark running in a different direction, many of the trained bees ignored their sun compass and followed the trees (von Frisch and Lindauer 1954). Similar fidelity to a familiar route in the landscape has been reported in various insects (Baerends 1941; Janzen 1974; Heinrich 1976; Wehner et al. 1983; Collett et al. 1992).

Evidence that insects can use landmarks to determine their position in the landscape is provided by experiments in which individuals are displaced to distant sites in their foraging range, and hence deprived of any opportunity to use a path integration mechanism to determine their location relative to home. Honeybees (see Wehner 1981 for review), carpenter bees (Rau 1929), and orchid bees (Janzen 1971) can return to their nests after displacements over distances on the order of kilometers. In all cases direct homing is possible only from familiar sites, where the insect can recognize landmarks encountered on previous foraging trips (Wehner 1981).

How can bees, or any animal that performs analogous feats of navigation, store in memory information about landmarks experienced on these large spatial scales? In no animal has the question been more intensively studied than in human beings (reviewed by Byrne 1982). People rely heavily upon landmarks for navigation in a familiar environment, and experimental studies have documented our ability to acquire and use a mental representation of the lay of the land. Specifically, we can acquire at least a rough "metric map" (Gallistel 1990) or "vector map" (Byrne 1982) in memory. Such a representation is analogous to a survey map, on which the directions and distances among charted sites are proportional to directions and distances measured over the ground. The construction of this sort of map requires integration of observations taken in different parts of the landscape into a common geometric frame of reference. In operation, our mental map allows us to estimate the position and direction of each familiar site relative to any other site, even along shortcut routes never directly traveled, or to imagine what it is like to look down upon a familiar landscape from above.

Until recently, it has been almost universally assumed that insects do something rather less complicated, in that they learn the visual features of familiar routes in the landscape, but do not store the equivalent of a survey map in memory. As first articulated by Baerends (1941) and further elaborated by Wehner (1981, 1983), this hypothesis assumes that landmarks encountered in successive stages of a route are stored in memory as a series of "snapshots" (in the terminology of Cartwright and Collett 1983). On subsequent flights the insect retraces this route by matching familiar landmarks sequentially to the stored images. The animal must be able to head in the correct direction along the route, but it need not obtain any additional information about its position and direction of movement in the landscape. Analogously, one can use a route map to take a city bus from one station to another without knowing how the route fits into a network of larger-scale spatial relationships. The use of such a route map would account for the ability of insects to follow a conspicuous line of landmarks out from the nest, or to use familiar landmarks to set a homeward course in the absence of information from path integration. Baerends (1941), in his study of homing by displaced digger wasps (*Ammophila campestris*), showed that wasps could learn multiple stereotyped routes through a landscape, but seemed not to derive from their experience on separately traveled routes a global knowledge of the lay of the land. Similar conclusions have been drawn for desert ants (*Cataglyphis bicolor* and *C. fortis*) (Collett et al. 1992).

Gould (1986) challenged this view of insect spatial memory by providing experimental evidence that honeybees, at least, learn more than just the familiar routes between the nest and feeding sites. He proposed instead that a bee can store in memory an accurate metric landscape map, which she can use to set a shortcut route from one familiar site to any other familiar site, even if the shortcut route is unfamiliar and the bee can see no landmarks previously used on other routes to her goal. To form such a map, a bee would have to record in a common frame of reference the directions and distances traveled to sites on different familiar routes, and then use this same frame of reference to compute the directions and distances among separately visited sites.

This map hypothesis is not utterly implausible. Bees have a directional reference (their celestial compass) and a way to measure flight distance (see von Frisch 1967 for review), which could be used to chart the relative positions of different sites in a common frame of reference, and they can learn certain relationships between their celestial reference and large-scale features of the terrain (Dyer and Gould 1981; Dyer 1987; Cartwright and Collett 1987). The question, however, is whether Gould's evidence for mental landscape maps, which turns out to be hard to replicate (Wehner and Menzel 1990; Wehner et al. 1990; Menzel 1990; Menzel et al. 1990), could be interpreted more parsimoniously by other hypotheses that do not attribute to bees the computational abilities that would be needed to assemble this information into a mental map (Dyer and Seeley 1989).

In his experiment (fig. 4.7), Gould captured bees as they left the hive for a feeding station, and then displaced them to a location off the route they were about to travel. On release the bees flew straight toward the food, as if plotting the shortcut route with a map that relates the position of the release site to the current foraging route. Two problems with this experiment undermine it as convincing evidence for maps (Dyer and Seeley 1989), and in turn suggest alternative interpretations of the results. First, since the prior flight experience of the bees was neither known nor controlled, it is possible that the bees were already familiar not only with the release site but also with the path between the release site and the goal. Following a familiar path through a landscape would not require a bee to have a map—that is, a knowledge of how her current position and direction of movement relates to a larger framework of spatial relationships. Second, it is possible that the bees could see from the release site some of the same landmarks that they were about to use on their flight from the hive to the food. These landmarks need not be ones marking the location of the food itself. Instead, the bees might recognize their displacement relative to a large-scale panorama that is visible both at the release site and at points along the foraging route. If so, then they could correct for their displacement using the same sort of image-matching process that bees use to approach a familiar feeding site relative to a small-scale landmark array (Cartwright and Collett 1983). As we have seen above, this ability would not necessarily require a maplike knowledge of the spatial arrangement of landmarks and locations in the environment. By analogy, a ship's pilot would not need a map to steer toward a familiar headland from a novel direction.

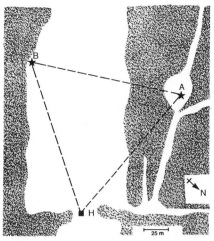 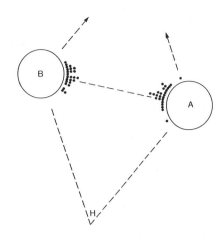

Fig. 4.7 Experiment put forward as evidence for large-scale mental maps in honeybees. Trees are shown by shaded pattern. Bees that were trained to fly from the hive (H) to feeding site A or B were captured upon departure for the food and then displaced to the other site. Dots on polar graphs give vanishing bearings of released bees. Bees consistently departed toward the food, rather than flying in the compass direction they were about to travel when captured or flying toward the hive. (Adapted from Gould 1986.)

I have replicated Gould's experiments with improved control over the bees' visual experience during training and testing (Dyer 1991; Dyer et al. 1993). My experiments provided evidence for both of these alternative hypotheses, but no evidence for metric cognitive maps (fig. 4.8). I found that bees can indeed compensate quite effectively for displacement by setting a direct route to their current goal, but only if they are released at a site from which they can see landmarks specifically associated with previously traveled routes to the food. They can use either landmarks associated with the route they were about to travel (fig. 4.8A), or landmarks associated with the shortcut route itself, if the shortcut is familiar (fig. 4.8B: right polar graph). When released at a site that offered no view of landmarks associated with their current foraging route, bees that had never flown along the shortcut to the food flew in other directions, apparently unable to compute this course (fig. 4.8B: left polar graph). Even when thoroughly familiar with such a release site (because they had previously fed there), bees could not draw upon their experience to set the shortcut path, but instead flew home (fig. 4.8B: center graph).

These results are strong evidence that bees do not form metric cognitive maps that permit them to compute a novel route from one familiar site to another. The data are instead consistent with Baerends's (1941) original hypothesis that insects orienting to landmarks are constrained to follow routes along which they can see a sequence of familiar visual features (see also Wehner et al. 1990). Metaphorically, we can think of the honeybee's large-scale spatial memory as a set of route maps, each consisting of a series of landmarks linked together like beads on a string. Bees can progress from one bead to the next along the same string, but, unlike human beings (Byrne 1982), they cannot find their way from one string to another, as if they cannot perform the computations that would allow the strings corresponding to separately traveled routes to be tacked down to a common frame of reference. In this view, the spatial memory that bees use for orientation on the scale of the

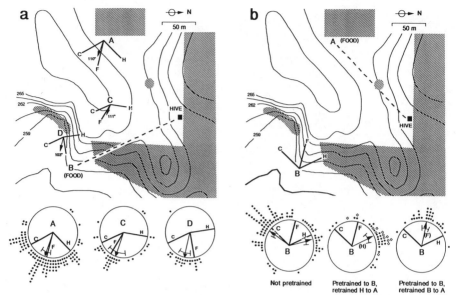

Fig. 4.8   Experiments by Dyer (1991) and Dyer et al. (1993), who used the design of Gould's (1986) map experiments (see fig. 4.7), but trained and tested bees in a landscape in which the bees' visual experience could be better controlled. Contour lines give elevations in meters above sea level; trees are shown by hatching. Polar graphs show vanishing bearings of individually released bees, mean bearings, and 99% confidence intervals of the means. (A) Bees trained to find food at site B were released at three sites from which they should have been able to see landmarks previously experienced along the foraging route. Sites A and C provided a view of landmarks only experienced early in this route; site D, landmarks only experienced near the end of the route. (B) Bees trained to site A were released at site B, from which, because of its lower elevation, they should have had no view of any landmarks previously seen on the route from the hive to site A. Bees that had been trained and tested according to Gould's conditions, and according to the conditions used in (A), failed to set a shortcut path to the food (left graph), but instead flew either toward home or in the compass direction they were about to travel when caught. Bees that had been pretrained for several days to a feeder at B, then retrained to site A along a straight path from the hive (center graph), also did not fly the shortcut to the food, but flew toward home (filled circles, sky clear; open circles, sky overcast). Bees could, however, fly the shortcut from side B to side A if they had previously been trained along this route (right graph).

landscape involves processes of associative learning, with visual images being linked with particular destinations, and perhaps with other images experienced on the same route. It does not involve a cognitive ability to derive geometric relationships (among routes) that have not been directly experienced.

This is not to deny, however, the remarkable flexibility with which a bee can use familiar landmarks to set a course to a goal not currently in view. Indeed, the suggestion that bees learn a route as a series of visual "beads on a string" underrates how flexibly they can respond to familiar route-based landmarks seen from a different perspective or in an unexpected sequence. As Gould's results showed, and my experiments confirmed, displaced bees can quickly set an appropriate course relative to route-based landmarks seen from a starting point off the route they were about to travel. They can even correct for a displacement that delivers them directly to a site near the end of a familiar route (fig. 4.8A); thus, they need not experience landmarks normally encountered early in a sequence to use landmarks that are familiar from late in a sequence (Dyer et al. 1993). Furthermore, bees can respond appropriately to landmarks associated with familiar routes other than the one they were about to travel when displaced (Dyer 1991). This point is illustrated in figure 4.8B. Bees caught as they left the hive for one familiar site (site A) were displaced to a different familiar site (site B) from which they could not see any landmarks associated with their current foraging route. Bees known to have been at this site before responded in a familiar and appropriate way to the landmarks there. They followed the shortcut to site A if they had previous experience along this route; without such experience, bees flew home. These results are strongly reminiscent of Collett and Kelber's (1988) finding that honeybees can retrieve a remembered response to a familiar landmark array even when they encounter the array in an unexpected context (see previous section; fig. 4.6B). Finally, bees that move to a new nesting site within a landscape can rapidly reorganize their responses to familiar landmarks, returning exclusively to the new nest while remembering how to find the old nest (Robinson and Dyer 1993), and flying directly to feeding sites to which they have previously traveled only on routes from the old nest (Dyer 1993). Thus, the underlying spatial memory must not encode merely the specific routes traveled between the nest and a set of known feeding sites.

Representation of Noncelestial Visual Information in Memory: Summary

Taken together, the studies reviewed in this section reveal some interesting common features of the spatial representations that honeybees use for navigation on different spatial scales. Most important, the suggestion that the spatial features of the environment are encoded as a sort of snapshot applies well at each of these scales. At each level, the experimental evidence is consistent with the notion that an encoded image is referenced to retinal coordinates—that is, as originally projected to the eyes—although the ability of flying bees to match remembered images to current images that move across the eye (Cartwright and Collett 1983) certainly warrants keeping an open mind about just how an image is encoded. A further feature exhibited on different scales is that, although additional information may be encoded with a snapshot about the context in which it is used, the retrieval of appropriate spatial information from memory is not driven entirely by context. Instead, the represented image itself can trigger the retrieval of the snapshot, or at least of the correct response to a given pattern (Collett and Kelber 1988; Dyer 1991). Finally, at multiple spatial scales, experiments have suggested that bees at best have a poor ability to compute geometric relationships among images projected to the eyes. For example, identical spatial patterns are not recognized as the same flower shape if they are in different orientations; also, the use of landmarks on both small and (especially) large spatial scales can be explained without assuming that bees compile a rich representation of the geometric relationships among landmarks and locations in the environment.

Thus, the sophistication of honeybee spatial memory appears to lie primarily in its capacity for encoding multiple visual images during a foraging trip and in the flexibility with which a response associated with a given visual image can be elicited, and not in an ability to compute spatial relationships among familiar objects and locations (see also Collett 1992). What is impressive is that, in spite of this constraint, bees exhibit behavioral flexibility that is similar to what an animal with a knowledge of these relationships should exhibit. This raises the fascinating question of whether, and if so then why, the computations required for such abilities as recognizing an object in an arbitrary orientation (Biederman 1987) or constructing a metric map in memory (Byrne 1982; Gallistel 1990) are inherently more difficult to implement in a small brain than is the capacity to store an album of visual images.

CONCLUSIONS

An insect such as a honeybee or a desert ant, like any animal that needs to find its way on long foraging trips from a central place, faces navigational problems whose difficulty human observers can readily appreciate. The ability of the insects to solve these problems has long been a source of wonder (Darwin 1873; Fabre 1879; Rau 1929). Given that the behavior of the insect may closely resemble the behavior of a human navigator confronted with an analogous problem, it is perhaps understandable that insects have been credited with computational abilities analogous to those employed by human navigators (Mittelstaedt 1985; Gould 1986, 1990; Gallistel 1990). The case studies I have discussed here, however, illustrate a well-worn lesson: that the solutions animals have evolved may actually be somewhat less sophisticated and less exact than their behavior first implies. Thus, foraging ants under some circumstances behave as if they can perform trigonometry to sum vectors corresponding to the successive stages of an outward path (Wehner 1982), but they actually carry out an apparently simpler computation that delivers an approximate representation of their net displacement (Müller and Wehner 1988). Similarly, honeybees under some circumstances behave as if they can acquire and use a cognitive map of a landscape, in which the relationships among previously traveled routes are organized by computation into the geometric equivalent of a topographic map (Gould 1986, 1990). Further studies showed, however, that bees encode a set of "route maps," each corresponding to a separately traveled route, and do not learn the relationships among routes (Dyer 1991; Dyer et al. 1993).

Although these case studies make a case for exercising parsimony in the analysis of insect spatial navigation, they also underscore the richness and flexibility of the underlying spatial representations. The work on path integration has confirmed that foraging ants continuously update an internal representation of their displacement relative to home (Wehner 1991; Wehner and Wehner 1986; Müller and Wehner 1988). Simple associative learning is utterly insufficient to explain the animal's behavior; instead we must assume the existence of a computational mechanism that generates a homeward vector based on measurements taken during the twists and turns of the outward path. The studies of spatial memory in honeybees reveal a well-organized set of processes for storing and retrieving spatial patterns experienced on a range of spatial scales. The data are largely consistent with the view that the bees' spatial memory consists of a set of "snapshots" associated with particular goals. But this view may not do justice to the richness and flexibility of the underlying processes. Encoded in (or with) the snapshot is not only the pattern of light and dark registered by the retina but also directional and contextual information that allows a bee to respond appropriately to familiar landmark panoramas whenever they are encountered (Cartwright and Collett 1987; Collett and Kelber 1988; Dyer 1991). Clearly the memory incorporates a rich network of associations among the various elements of a bee's spatial experience. Thus, the lesson is not that insects do not form relatively rich internal representations of spatial relationships in the environment; it is that these representations work so extremely well that we are readily led to overestimate their capacities.

These case studies also demonstrate that the representations formed by insects, though observable only through their behavioral effects, are nevertheless accessible to experimental analysis. In each instance it has been possible to state a hypothesis about the representation underlying the animal's behavior in operational terms, and to devise an experiment which, albeit with varying degrees of success, puts the hypothesis to a test. It is easier to be certain about what a representation is not then to establish exactly what it is. Thus, Müller and Wehner (1988) decisively eliminated a role for true vector summation in path integration by desert ants, but their alternative computational model could be revealed by future work to have missed some important element of the actual computations that underlie the behavior. Similarly, Cartwright and Collett (1983) showed that several fairly simple computational models could not account for the ability of bees in their experiments to match a current to a remembered image, but they left many questions open about the nature of the mechanism the bees do use. Dyer (1991) excluded a role for large-scale metric maps in honeybee navigation, but this result only partially demystifies the ability of bees to use stored information to travel long distances in a landscape. These advances, though still incomplete, pave the way for studies that seek to specify further the behavioral capacities of the animal and the mechanisms underlying them.

I would like to end with a word about how these insights might be relevant to evolutionary ecologists (see also Yoerg 1992). The first point to make is that the growth of interest in cognitive processes among evolutionary ecologists studying behavior has been spurred in part by the desire to understand how species-specific information processing mechanisms might have constrained the strategic design of the behavior in which these mechanisms participate (Cheverton et al. 1985; Guilford and Dawkins 1991; Real 1991; Kacelnik and Todd 1992). The studies I have discussed all have investigated competing cognitive explanations that differ markedly in the constraints that are assumed to apply to the animal's ability to find its way and learn important locations in the environment. The insights from these studies may be of direct relevance to those persons interested in how the capacity and organization of spatial memory constrains the foraging decisions of species that face similar navigational problems. On a more general level, the lesson is that careful experimental investigations may be needed to characterize the cognitive processes underlying any animal's behavior, and hence the ways in which the behavior is constrained by these processes.

A second point is that detailed studies like those reviewed in this chapter provide a firmer basis for exploring the adaptive design of the mechanisms themselves. The most detailed work on adaptive specializations of learning mechanisms and memory systems has been done with vertebrates (reviews by Sherry 1987; Sherry and Schachter 1987; Harvey and Krebs 1990; Krebs 1990; Shettleworth 1990), although a number of studies of insects have appeared in recent years (Menzel 1985, 1987; Rosenheim 1987; Laverty and Plowright 1988; Dukas and Real 1991). The spatial information processing capabilities of insects raise compelling questions not only about how the underlying mechanisms work, but also about why they have the properties that they do. It may be possible to explore such functional questions either through optimality analyses or through comparative studies of closely related species that differ in traits such as the abiltiy to learn spatial locations (e.g., Rosenheim 1987). An important challenge, which is illustrated by the studies reviewed in this chapter, will be to explore the limits of an animal's abilities fully, so as not to be misled by superficial similarities to human information processing abilities.

ACKNOWLEDGMENTS

The author's research on honeybee spatial memory has been supported by NSF Grants BNS 8820010 and IBN 9222015. Mark E. Rilling provided helpful suggestions on an earlier draft of the manuscript.

REFERENCES

Amsel, A. 1989. Behaviorism, neobehaviorism, and cognitivism in learning theory: historical and contemporary perspectives. Hillsdale, N.J.: Lawrence Erlbaum Associates.

Baerends, G. P. 1941. Fortpflanzungsverhalten und Orientierung der Grabwaspe *Ammophila campestris* Jur. *Tijdschrift voor Entomologie* Deel 84:68–275.

Baker, A. G., and P. Mercier. 1989. Attention, retrospective processing and cognitive representations. In S. B. Klein and R. R. Mowrer, eds., *Contemporary learning theories: Pavlovian conditioning and the status of traditional learning theories*, 85–116. Hillsdale, N.J.: Lawrence Erlbaum Associates.

Biederman, I. 1987. Recognition-by-components: A theory of human image understanding. *Psychological Review* 94:115–47.

Bogdany, F. J. 1978. Linking of learning signals in honey bee orientation. *Behavioral Ecology and Sociobiology* 3:323–36.

Byrne, R. W. 1982. Geographical knowledge and orientation. In A. W. Ellis, ed., *Normality and pathology in cognitive functions*, 239–64. London: Academic Press.

Cartwright, B. A., and T. S. Collett. 1983. Landmark learning in bees. *Journal of Comparative Physiology* 151:521–43.

Cartwright, B. A., and T. S. Collett. 1987. Landmark maps for honeybees. *Biological Cybernetics* 57:85–93.

Charnov, E. L. 1976. Optimal foraging: Attack strategy of a mantid. *American Naturalist* 110:141–51.

Cheney, D. L., and R. M. Seyfarth. 1990. *How monkeys see the world.* Chicago: University of Chicago Press.

Cheng, K., T. S. Collett, and R. Wehner. 1986. Honeybees learn the colours of landmarks. *Journal of Comparative Physiology* A 159:69–73.

Cheng, K., T. S. Collett, A. Pickhard, and R. Wehner. 1987. The use of visual landmarks by honeybees: Bees weight landmarks according to their distance from the goal. *Journal of Comparative Physiology* A 161:469–75.

Cheverton, J., A. Kacelnik, and J. R. Krebs. 1985. Optimal foraging: Constraints and currencies. In B. Hölldobler and M. Lindauer, eds., *Experimental behavioral ecology and sociobiology*, 109–26. Fortschritte der Zoologie 31. Stuttgart: Gustav Fischer Verlag.

Churchland, P. S. 1986. *Neurophilosophy.* Cambridge, Mass.: MIT Press.

Churchland, P. S. 1988. The significance of neuroscience for philosophy. *Trends in Neurosciences* 11:304–307.

Churchland, P. S., and T. J. Sejnowski. 1992. *The computational brain.* Cambridge, Mass.: MIT Press.

Collett, T. S. 1992. Landmark learning and guidance in insects. *Philosophical Transactions of the Royal Society of London,* series B 337:295–303.

Collett, T. S., and J. Baron. 1994. Biological compasses and the coordinate frame of landmark memories in honeybees. *Nature* 368:137–40.

Collett, T. S., and B. A. Cartwright. 1983. Eidetic images in insects: Their role in navigation. *Trends in Neurosciences* 5:101–5.

Collett, T. S., and A. Kelber. 1988. The retrieval of visuo-spatial memories by honeybees. *Journal of Comparative Physiology* A 163:145–50.

Collett, T. S., and M. F. Land. 1975. Visual control of flight behaviour in a hoverfly, *Syritta pipiens* L. *Journal of Comparative Physiology* 99:1–66.

Collett, T. S., E. Dilmann, A. Giger, and R. Wehner. 1992. Visual landmarks and route following in desert ants. *Journal of Comparative Physiology* A 170:435–42.

Couvillon, P. A., T. G. Leiato, and M. E. Bitterman. 1991. Learning by honeybees (*Apis mellifera*) on arrival at and departure from a feeding place. *Journal of Comparative Psychology* 105:177–84.

Darwin, C. 1873. Origin of certain instincts. *Nature* 7:417–18.

Dukas, R., and L. A. Real. 1991. Learning foraging tasks by bees: A comparison between solitary and social species. *Animal Behaviour* 42:269–76.

Dyer, F. C. 1987. Memory and sun compensation in honey bees. *Journal of Comparative Physiology* A 160:621–33.

Dyer, F. C. 1991. Bees acquire route-based memories but not cognitive maps in a familiar landscape. *Animal Behaviour* 41:239–46.

Dyer, F. C. 1993. How honey bees find familiar feeding sites after changing nesting sites with a swarm. *Animal Behaviour* 46:813–16.

Dyer, F. C., and J. L. Gould. 1981. Honey bee orientation: A backup system for cloudy days. *Science* 214:1041–42.

Dyer, F. C., and T. D. Seeley. 1989. On the evolution of the dance language. *American Naturalist* 133:580–90.

Dyer, F. C., N. A. Berry, and A. S. Richard. 1933. Honey bee spatial memory: Use of route-based memories after displacement. *Animal Behaviour* 45:1028–30.

Fabre, J. H. 1879. *Souvenirs entomologiques*. 1. Série. Paris: C. Delagrave.

Frisch, K. von 1967. *The dance language and orientation of bees*. Cambridge, Mass.: Belknap Press of Harvard University Press.

Frisch, K. von, and M. Lindauer. 1954. Himmel und Erde in Konkurrenz bei der Orientierung der Bienen. *Naturwissenschaften* 41:245–53.

Gallistel, C. R. 1989. Animal cognition: The representation of space, time, and number. *Annual Review of Psychology* 40:155–89.

Gallistel, C. R. 1990. *The organization of learning*. Cambridge, Mass.: MIT Press.

Gill, F. B. 1988. Trapline foraging by hermit hummingbirds: Competition for an undefended, renewable resource. *Ecology* 69:1933–42.

Gould, J. L. 1986. The locale map of honey bees: Do insects have cognitive maps? *Science* 232:861–63.

Gould, J. L. 1987a. Honey bees store learned flower-landing behaviour according to time of day. *Animal Behaviour* 35:1579–81.

Gould, J. L. 1987b. Landmark learning by honey bees. *Animal Behaviour* 35:26–34.

Gould, J. L. 1990. Honey bee cognition. *Cognition* 37:83–103

Griffin, D. R. 1992. *Animal minds*. Chicago: University of Chicago Press.

Guilford, T., and M. S. Dawkins. 1991. Receiver psychology and the evolution of animal signals. *Animal Behaviour* 42:1–14.

Harvey, P. H., and J. R. Krebs. 1990. Comparing brains. *Science* 249:140–46

Hateren, J. H. van, M. V. Srinivasan, and P. B. Wait. 1990. Pattern recognition in bees: Orientation discrimination. *Journal of Comparative Physiology* A 167:649–54.

Heinrich, B. 1976. The foraging specializations of individual bumblebees. *Ecological Monographs* 46:105–28.

Jander, R. 1957. Die optische Richtungsorientierung der roten Waldameise (*Formica rufa* L.). *Zeitschrift für vergleichende Physiologie* 40:162–238.

Janetos, A. C., and B. J. Cole. 1981. Imperfectly optimal animals. *Behavioral Ecology and Sociobiology* 9:203–10.

Janzen, D. H. 1971. Euglossine bees as long-distance pollinators of tropical plants. *Science* 171:203–5.

Janzen, D. H. 1974. The deflowering of Central America. *Natural History* 83:48–53.

Kacelnik, A., and I. A. Todd. 1992. Psychological mechanisms and the marginal value theorem: Effect of variability in travel time on patch exploitation. *Animal Behaviour* 43:313–22.

Kacelnik, A., D. Brunner, and J. Gibbon. 1990. Timing mechanisms in optimal foraging: Some applications of scalar expectancy theory. In R. N. Hughes, ed., *Behavioural mechanisms of food selection*, 61–82. Berlin, Heidelberg: Springer-Verlag.

Keeton, W. T. 1974. The orientational and navigational basis of homing in birds. *Advances in the Study of Behavior* 5:47–132.

Kennedy, J. S. 1992. *The new anthropomorphism*. Cambridge: Cambridge University Press.

Kipp, L. R., W. Knight, and E. R. Kipp. 1989. Influence of resource topography on pollinator flight directionality of two species of bees. *Journal of Insect Behavior* 2:453–72.

Kosslyn, S. M. 1988. Aspects of a cognitive neuroscience of mental imagery. *Science* 240: 1621–26.

Krebs, J. R. 1990. Food-storing birds: Adaptive specialization in brain and behaviour? *Philosophical Transactions of the Royal Society of London* B 329:153–60.

Krebs, J. R., and A. Kacelnik. 1992. Decision making. In J. R. Krebs and N. B. Davies, eds., *Behavioral ecology: An evolutionary approach*, 3d ed., 105–36. Oxford: Blackwell Scientific Publications.

Laverty, T. M., and C. Plowright. 1988. Flower handling by bumblebees: A comparison of specialists and generalists. *Animal Behaviour* 36:733–40.

Lehrer, M., M. V. Srinivasan, S. W. Zhang, and G. A. Horridge. 1988. Motion cues provide the bee's visual world with a third dimension. *Nature* 332:356–57.

Leonard, B., and B. L. McNaughton. 1990. Spatial representation in the rat: Conceptual, behavioral, and neurophysiological perspectives. In R. P. Kesner and D. S. Olton, eds., *Neurobiology of comparative cognition*, 363–422. Hillsdale, N.J.: Lawrence Erlbaum Associates.

Lloyd, D. 1989. *Simple Minds*. Cambridge, Mass.: MIT/Bradford Books.

Markl, H. 1985. Manipulation, modulation, information, cognition: Some of the riddles of communication. In B. Hölldobler and M. Lindauer, eds., *Experimental behavioral ecology and sociobiology*, 163–94. Fortschritte der Zoologie 31. Stuttgart: Gustav Fischer Verlag.

Menzel, R. 1985. Learning in honey bees in an ecological and behavioral context. In B. Hölldobler and M. Lindauer, eds., *Experimental behavioral ecology and sociobiology*, 55–74. Fortschritte der Zoologie 31. Stuttgart: Gustav Fischer Verlag.

Menzel, R. 1987. Memory traces in honeybees. In R. Menzel and A. Mercer, eds., *Neurobiology and behavior of honeybees*, 310–25. Berlin: Springer-Verlag.

Menzel, R. 1990. Learning, memory, and "cognition" in honey bees. In R. P. Kesner and D. S. Olton, eds., *Neurobiology of comparative cognition*, 237–92. Hillsdale, N.J.: Lawrence Erlbaum Associates.

Menzel, R., L. Chittka, S. Eichmüller, K. Geiger, D. Peitsch, and P. Knoll. 1990. Dominance of celestial cues over landmarks disproves map-like orientation in honey bees. *Zeitschrift für Naturforschung* C 45:723–26.

Mittelstaedt, H. 1985. Analytical cybernetics of spider navigation. In F. G. Barth, ed., *Neurobiology of arachnids*, 298–316. Berlin, Heidelberg, New York: Springer-Verlag.

Morris, R. G. M. 1981. Spatial localization does not require the presence of local cues. *Learning and Motivation* 12:239–61.

Müller, M., and R. Wehner. 1988. Path integration in desert ants, *Cataglyphis fortis*. *Proceedings of the National Academy of Sciences U.S.A.* 85:5287–90.

O'Keefe, J., and L. Nadel. 1978. *The hippocampus as a cognitive map*, Oxford: Clarendon Press.

Opfinger, E. 1931. Über die Orientierung der Biene an der Futterquelle. *Zeitschrift für vergleichende Physiologie* 15:431–87.

Poucet, B. 1993. Spatial cognitive maps in animals: New hypotheses on their structure and neural mechanisms. *Psychological Review* 100:163–82.

Rau, P. 1929. Experimental studies in the homing of carpenter and mining bees. *Journal of Comparative Psychology* 9:35–70.

Real, L. A. 1990. Predator switching and the interpretation of animal choice behavior: The case for constrained optimization. In R. N. Hughes, ed., *Behavioural mechanisms of food selection*, 1–21. Berlin, Heidelberg: Springer-Verlag.

Real, L. A. 1991. Animal choice behavior and the evolution of cognitive architecture. *Science* 253:980–86.

Rescorla, R. A. 1988. Pavlovian conditioning: It's not what you think it is. *American Psychologist* 43:151–60.

Ristau, C. A., ed. 1991. *Cognitive ethology*. Hillsdale, N.J.: Lawrence Erlbaum Associates.

Robinson, G. E., and F. C. Dyer. 1993. Plasticity of spatial memory in honey bees: Reorientation following colony fission. *Animal Behaviour* 46:311–20.

Roitblat, H. L. 1982. The meaning of representation in animal memory. *Behavioral and Brain Sciences* 5:353–406.

Roitblat, H. L. 1987. *Introduction to comparative cognition*. New York: W. H. Freeman.

Roitblat, H. L., and L. von Fersen. 1992. Comparative cognition: Representations and processes in learning and memory. *Annual Review of Psychology* 43:671–710.

Rosenheim, J. A. 1987. Host location and exploitation by the cleptoparasitic wasp. *Argochrysis armilla:* The role of learning (Hymenoptera: Chrysidae). *Behavioral Ecology and Sociobiology* 21:401–6.

Rumelhart, D. E., and J. L. McClelland. 1986. *Parallel distributed processing.* Vol. 1. Cambridge, Mass.: Bradford/MIT Press.

Schmidt-Hempel, P. 1984. The importance of handling time for the flight directionality in bees. *Behavioral Ecology and Sociobiology* 15:303–9.

Schöne, H. 1984. *Spatial orientation: The spatial control of behavior in animals and man.* Princeton: Princeton University Press.

Sherry, D. F. 1987. Learning and adaptation in food-storing birds. In R. C. Bolles and M. D. Beecher, eds., *Evolution and learning,* 79–95. Hillsdale, N.J.: Lawrence Erlbaum Associates.

Sherry, D. F., and D. L. Schachter. 1987. The evolution of multiple memory systems. *Psychological Review* 94:439–54.

Shettleworth, S. J. 1990. Spatial memory in food-storing birds. *Philosophical Transactions of the Royal Society of London* B 329:143–51.

Shettleworth, S. J., J. R. Krebs, D. W. Stephens, and J. Gibbon. 1988. Tracking a fluctuating environment: A study of sampling. *Animal Behaviour* 36:87–105.

Skinner, B. F. 1971. *Beyond freedom and dignity.* New York: Knopf.

Srinivasan, M. V., S. W. Zhang, and B. Rolfe. 1993. Is pattern vision in insects mediated by "cortical" processing? *Nature* 362:539–40.

Stephens, D. W., and J. R. Krebs. 1986. *Foraging theory.* Princeton: Princeton University Press.

Tinbergen, N. 1951. *The study of instinct.* Oxford: Oxford University Press.

Tinbergen, N., and W. Kruyt. 1938. Über die Orientierung des Bienenwolfes (*Philanthus triangulum* Fabr.) III. Die Bevorzugung bestimmter Wegmarken. *Zeitschrift für vergleichende Physiologie* 25:292–334.

Tolman, E. C. 1948. Cognitive maps in rats and men. *Psychological Review* 55:189–208.

Uttal, W. R. 1982. Internal representations and indeterminacy: A skeptical view. *Behavioral and Brain Sciences* 5:392–93.

Waage, J. K. 1979. Foraging for patchily-distributed hosts by the parasitoid, *Nemeritis canescens. Journal of Animal Ecology* 48:353–71.

Wasserman, E. A. 1993. Comparative cognition: Beginning the second century of the study of animal intelligence. *Psychological Bulletin* 113:211–28.

Wehner, R. 1981. Spatial vision in arthropods. In H. Autrum, ed., *Handbook of Sensory Physiology,* vol. VII/6C, 287–616. Berlin, Heidelberg, New York: Springer-Verlag.

Wehner, R. 1982. Himmelsnavigation bei Insekten: Neurophysiologie und Verhalten. *Neujahrsblatt der Naturforschenden Gesellschaft in Zürich* 184:1–132.

Wehner, R. 1983. Celestial and terrestrial navigation: Human strategies—insect strategies. In F. Huber and H. Markl, eds., *Neuroethology and behavioral physiology,* 366–381. Berlin: Springer-Verlag.

Wehner, R. 1987. "Matched filters"—neural models of the external world. *Journal of Comparative Physiology.* A 161:511–31.

Wehner, R. 1991. Visuelle Navigation: Kleinstgehirn-Strategien. *Verhandlungen der Deutschen Zoologischen Gesellschaft* 84:89–104.

Wehner, R., and I. Flatt. 1977. Visual fixation in freely flying bees. *Zeitschrift für Naturforschung* C 32:469–71.

Wehner, R., and B. Lanfranconi. 1981. What do the ants know about the rotation of the sky? *Nature* 293:731–33.

Wehner, R., and R. Menzel. 1990. Do insects have cognitive maps? *Annual Review of Neuroscience* 13:403–14.

Wehner, R., and M. Müller. 1985. Does interocular transfer occur in visual navigation by ants? *Nature* 315:228–29.

Wehner, R., and F. Räber. 1979. Visual spatial memory in desert ants, *Cataglyphis bicolor* (Hymenoptera: Formicidae). *Experientia* 35:1569–71.

Wehner, R., and M. V. Srinivasan. 1981. Searching behavior of desert ants. *Journal of Comparative Physiology* 142:315–38.

Wehner, R., and S. Wehner. 1986. Path integration in desert ants: Approaching a long-standing puzzle in insect navigation. *Monitore Zoologico Italiano (NS)* 20:309–31.

Wehner, R., and S. Wehner. 1990. Insect navigation: Use of maps or Ariadne's thread? *Ethology, Ecology, and Evolution* 2:27–48.

Wehner, R., R. D. Harkness, and P. Schmid-Hempel. 1983. *Foraging strategies in individually searching ants*. Cataglyphis bicolor (*Hymenoptera: Formicidae*). Stuttgart, New York: G. Fischer Verlag.

Wehner, R., S. Bleuler, C. Nievergelt, and D. Shah. 1990. Bees navigate by using vectors and routes rather than maps. *Naturwissenschaften* 77:479–82.

Yoerg, S. I. 1992. Ecological frames of mind: the role of cognition in behavioral ecology. *Quarterly Review of Biology* 66:287–301.

Yoerg, S. I., and A. C. Kamil. 1991. Integrating cognitive ethology with cognitive psychology. In C. A. Ristau, ed., *Cognitive ethology*, 273–89. Hillsdale, N.J.: Lawrence Erlbaum Associates.

FIVE

Information Processing and the Evolutionary Ecology of Cognitive Architecture

LESLIE A. REAL

BEHAVIORAL ECOLOGY has focused primarily on the functional (i.e., evolutionary) significance of behavior in an attempt to explain the origin and maintenance of observed patterns of resource use, mate choice, and reproductive tactics. Early models in behavioral ecology often characterized environments as static and attributed to organisms remarkable cognitive abilities. Organisms "knew" the global properties of all pertinent resources that could be reaped from an unchanging environment free of competitors, predators, pathogens, and disturbances. Such benign omniscience hardly represents the reality of animals under natural circumstances. Instead, organisms must learn to exploit resources that are often unfamiliar, change over time, and may be subject to depletion through competitive interactions. They must remember the location and qualities of the resources they learn about, attend to conflicting needs and sensory inputs, engage in social interactions, and balance all of these considerations using a cognitive machinery that may be limited by perceptual capabilities and biases. Much of contemporary theory and experimentation in behavioral ecology is directed at incorporating these complications into traditional approaches. The earliest models in behavioral ecology can be used as a baseline for evaluating the importance of these complicating factors in accounting for and predicting patterns of animal behavior.

The cognitive machinery involved in information processing may be treated either as an adaptation or as a constraint. One may ask, for example, how limits to the recall of information (i.e., memory) may influence behavioral choices and, on the other hand, one may ask how memory systems may have evolved to store information concerning appropriate aspects of the environment. Every aspect of animal cognition may act simultaneously as both an adapted component of the organism's phenotype and as a potential constraint to the processing of information.

Current research programs that incorporate cognitive complexity have largely focused on aspects of animal learning and memory (e.g., Kacelnik and Krebs 1985; Sherry and Schacter 1987; Stephens 1987; Shettleworth et al. 1988; Krebs and Kacelnik 1991; Kamil, chap. 2; Krebs and Inman, chap. 3). While learning and memory are of critical importance in shaping animal behavior, they are only two of a host of cognitive functions that must be understood if we are to have a general predictive theory of the organism's behavioral response to the environment. In this chapter I

This chapter was previously published in a slightly different form in *American Naturalist* 140, Supplement (November 1992), S 108–145. © 1992 by The University of Chicago.

will focus on cognitive aspects of decision making and the processing of information by the individual organism as it encounters and perceives an uncertain environment, as well as on the implications of individual information processing across levels of biological organization. I will (1) outline an evolutionary approach to cognition and cognitive architecture, (2) discuss cognitive aspects of decision making in stochastic environments, (3) argue for the bumblebee as a model organism for experimental investigations into the evolution of cognitive structures, (4) relate economic and ecological studies of risk-sensitive foraging to traditional conditioning approaches in experimental psychology, (5) point out the necessity to measure utility and probability effects in assessments of risk taking, (6) consider potential algorithms for characterizing an individual bumblebee's expectations of energetic reward, (7) outline a research strategy for uncovering the neuronal mechanisms by which decision making can be implemented, and last, (8) consider the significance of bumblebee choice behavior in the evolution of pollinator-plant interactions. The uniting theme across each section is the overwhelming importance of information processing in the organism's performance of ecologically important tasks.

COGNITION AND COGNITIVE ARCHITECTURE

A cognitive approach to the analysis of animal behavior is characterized by the recognition of three fundamental processes: encoding, computation, and representation (Gardner 1985; Roitblat 1987, Gallistel 1989, 1990). Information from the environment is translated through the senses into an encoded form that can be manipulated and/or stored through various mental operations. These mental operations, characterized by different computational schemes, generate different representations of the environment; the specific representations depend on both the specific set of informational inputs and the nature of the computational algorithm applied to these inputs. The concept of representation remains the most controversial aspect of the cognitive approach, and there is considerable confusion as to what may legitimately be called a representation (Roitblat 1982). Throughout this chapter I will be using *representation* in a sense consistent with that used in neural modeling (Rumelhart and McClelland 1986; Wasserman 1989; Levine 1991). The components leading to a given representation can be viewed as parts or stages in a single dynamic system mechanistically tied to the organism's nervous system. The encoding of information corresponds to initial sensory inputs translated into electrical signals. The computational rules correspond to the transmission of electrical signals along neural pathways that are modifiable through a variety of mechanisms, including learning. A representation then corresponds to the equilibrium configuration that emerges from the transient dynamics.

The Necker cube (fig. 5.1) can be used to illustrate the relations between these different components. In one spatial orientation, corner B is perceived as being in front of corner A (state 1). In another spatial orientation, corner A is perceived as being in front of corner B (state 2). Individuals can switch back and forth between these two spatial orientations, and some individuals may show a preference for one of the two possible images. It is clear that which particular orientation is currently visualized does not depend solely on the informational input. Rather, it is the result of configuring the inputs through a set of mental operations. The particular spatial orientation corresponds to a particular representation, and in this example there are two possible representational states. The configuring of information leading to these

Necker Cube

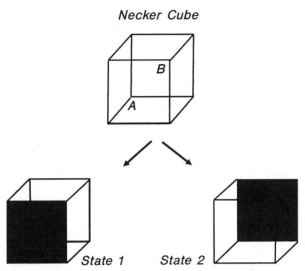

Fig. 5.1 The Necker cube illustrates the dependent relationship between the representation of information and internal mental operations. The cube can take on either of two spatial orientations: corner A in front of corner B, or vice versa. However, the visualization of a particular spatial orientation is not determined solely through sensory inputs, since they remain the same for the two possible states.

two representations has been modeled as a simple neural network with two stable nodes (Rumelhart 1989); the details of connecting neural models to representations and decision making will be discussed later. For now it is sufficient to note that my use of the word *representation* does not imply some special awareness or conscious attention. Representations are information-bearing internal mental states relating the perceived attributes of the environment after the important informational inputs have been processed. In this way representations are in accord with and complement more traditional associationist theories of behavior (Pearce 1987).

The principal thesis of this chapter is that the design of an animal's cognitive machinery—its "cognitive architecture"—is subject to natural selection and evolution as is any other aspect of the organism's phenotype. That there should be an evolutionary origin and differentiation in mental function has been noted since Darwin (Richards 1987). However, there has been considerable recent interest in further exploring the implications of evolution and natural selection in interpreting an organism's cognitive abilities and limitations (e.g., Staddon 1983; Hookway 1984; Shettleworth 1983, 1988; Gallistel 1990; Wuketits 1990; Rozin and Schull 1988; Cheney and Seyfarth 1990; Real 1991, 1993; Dukas and Real 1993a). While many studies document varying degrees of adaptive specialization in cognitive function (e.g., Shettleworth 1983; Wilson et al. 1985; Sherry and Schacter 1987; Balda and Turek 1984; Krebs et al. 1990; Dukas and Real 1991), few studies explore the adaptive nature of specific computational rules as aspects of the organism's cognitive architecture.

Different computational algorithms can generate selection differentials, at least in principle, either through affecting computational efficiency and accuracy or by producing different representations of the environment. As an example of algorithmic efficiency in cognitive function, consider the following problem: Sum the numbers between 1 and 100. The way most of us might attack this problem is to simply sequentially sum numbers—1 plus 2 equals 3, plus 3 equals 6, plus 4 equals 10, and so on—until arriving at 5,050, which is the answer. Alternatively, one could note that 1 plus 100 equals 101, 2 plus 99 equals 101, 3 plus 98 equals 101, and that there are 50 such pairs of sums. Thus the answer is 50 times 101, which equals 5,050. Both methods lead to the same answer, but the second method is clearly more efficient and quicker than the first. If we try to carry out this calculation in our heads, then the second method is probably more accurate as well. This example illustrates how different computational schemes may lead to different degrees of efficiency in mental operations. Of course, it is not clear how increased efficiency in this example leads to any selective advantage. However, presumably selection could operate to increase the efficiency of essential computations involving behavior of immediate consequence, such as food gathering or mate selection.

Different computational rules can also give rise to very different representations of the environment, which may be crucial for leading to appropriate actions and decisions. For example, the central tendency in a series of events that may form the basis for an animal's expectation of rewards can be represented by a variety of computations. Consider the central tendency of a population of flowers in which nectar rewards conform to a uniform distribution between 1 and 20 μl. The three most common measures of central tendency—the arithmetic mean ($\overline{X}_A$), the geometric mean ($\overline{X}_G$), and the harmonic mean ($\overline{X}_H$)—generate three different measures and could potentially lead to very different choices and decisions:

$$\overline{X}_A = \frac{1}{20} \sum_{X=1}^{20} = 10.5 \tag{5.1}$$

$$\overline{X}_G = 20 \sqrt[]{\prod_{x=1}^{20} X} = 8.3 \tag{5.2}$$

$$\overline{X}_H = \left[\frac{1}{20} \sum_{x=1}^{20} \frac{1}{X} \right]^{-1} = 5.6 \tag{5.3}$$

Which of the three particular representations of central tendency best corresponds to observed patterns of behavior may depend upon the specific task. For example, the arithmetic mean of rates of net energetic gain may characterize floral choice in bumblebees (Real et al. 1990; Real 1991), while the harmonic mean of intertrial delays may be employed by starlings when making foraging decisions in operant chambers (Reboreda and Kacelnik 1991). As can be seen from the equations for central tendencies, in some cases currencies may be related. For example, minimizing delay (X) in the harmonic mean is equivalent to rate maximization in the arithmetic mean, since $1/X$ is a rate. Which of the possible representations of central tendency is actually used may relate to general arguments for the applicability of each type of averaging method. For example, the geometric mean may best characterize systems with compounding effects, such as growth. Arithmetic averaging may best be employed in situations in which rewards are spatially distributed (see Hastings and Caswell 1979 for a general discussion of the uses of the arithmetic versus the geometric mean in biological systems). Thus, natural selection may lead to one or another method of computational representation depending on the particular task or problem faced by the organism.

PROBLEM SOLVING AND DECISION MAKING UNDER UNCERTAINTY

One common type of problem faced by all organisms is the choice of which resources to exploit when the outcomes of particular choices are not completely predictable. Uncertainty in the environment may be associated with variation in the quantity and quality of resources (as well as the rate and timing of their availability) and with the physiological condition of the animal. How organisms allocate their time and energy toward the acquisition of critical resources in stochastic environments remains a central concern in behavioral ecology and human decision science.

Unquestionably, the dominant model of human decision making and choice behavior is the expected utility approach, introduced by Bernoulli (1738) and axiomatized by von Neumann and Morganstern in 1944. Bernoulli was concerned with explaining the following sorts of gambling behaviors observed in humans. Imagine a fair coin toss in which you receive $10 if the coin comes up heads but must surrender $10 if the coin comes up tails. Since the mathematical expectation of the game is $0 (i.e., you have an equal chance of winning or losing $10), you should be indifferent as to whether you play or not. Most people opt not to play, and hence are avoiding the risk of a loss. Most individuals choose the sure outcome of $0 rather than the variable outcome with comparable expected value. As the range of potential losses and gains increases (say, $1,000 won or lost), more people decide not to take the gamble. Why do we avoid fair gambles of this kind?

Bernoulli suggested that what matters to the gambler is not the absolute amount of money won or lost, but the "utility" of money. The utility function translates some variable, in this case money, into perceived value. If the utility (perceived value) of money shows diminishing returns (concave-down), then the expected utility of variable gambles will be lower than the expected utility of sure gambles with a comparable arithmetic mean (fig. 5.2A). Bernoulli suggested that the utility of money was logarithmic and hence would account for players' choices not to engage in a fair coin toss. On the other hand, if utility shows increasing returns (concave-up), then the expected utility of variable gambles will be higher than the expected utility of sure gambles with an equal mean (fig. 5.2B). Concave-down utility functions lead to *risk-averse behavior*, while concave-up utility functions lead to *risk-seeking* behavior.

The organism's response to environmental variability depends on the specific variables under investigation and how these variables translate into utility (perceived value, fitness, or fitness component). A general method for expressing the anticipated response to variability for any class of variables and for an unspecified utility function is to approximate the expected utility associated with different variables using Taylor's series (Real 1980a, 1980b; Real and Caraco 1986; Stephens and Krebs 1986; Ellner and Real 1989). Consider some unspecified utility function, $U(X)$, of some specific random variable, X, measuring some aspect of the environment over which the organism must make choices. The expected utility approximated at the random variable's arithmetic mean value within the environment is given by:

$$E\, U(X)|\mu_X \simeq U(\mu_X) + \frac{1}{2}U''(\mu_X)\sigma_x^2 \qquad (5.4)$$

where μ_X and σ_x^2 correspond to the arithmetic mean and variance of X, respectively. Uncertainty about the random variable is expressed by the variance, and the organism's response to that uncertainty at the mean value for X is revealed by the curva-

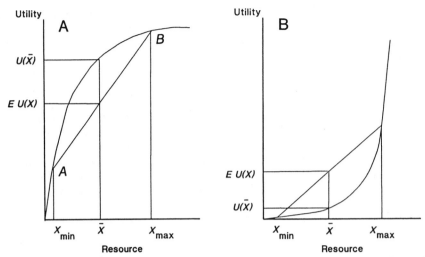

Fig. 5.2 (A) A representation of the utility associated with some resource variable X when utility shows diminishing returns (concave-down). The intersections of the chords (AB and CD) with the resource variable mean determines the expected utility generated by different gambles. The chord AB represents low-variance gambles, while the chord CD corresponds to high-variance gambles. When utility is concave-down, the expected utility decreases with increasing variance. Expected utility maximizers then prefer low-risk options; in other words, they demonstrate risk-aversion. (B) When utility is an increasing and accelerating function of resource X, then the expected utility is greater for options with higher variances. Expected utility maximizers should prefer higher-risk options; in other words, demonstrate risk-prone behavior.

ture of the utility function at that mean value, $U''(\mu_X)$. When there is no variance in X, then choices are governed by different mean values. The option with the highest arithmetic mean generates the highest expected utility. When there is variability and the utility function is concave-down ($U'' < 0$), the organism should be risk-averse. When there is variability and the utility function is concave-up ($U'' > 0$), then the organism should be risk-seeking.

There are several features of this approach to decision making under uncertainty that I wish to emphasize. First, I have purposefully not specified the nature of the random variable X over which decisions are to be made. Expected utility models can be applied to decision making over any number of alternative variables. In my earliest presentation of this approach (Real 1980a, 1980b), I gave examples in which X corresponded to fractional allocations of a cohort by mothers to different offspring genotypes or to the two sexes, visitation rates to different floral types by pollinators, allocation of oviposition behavior among host types, and differential patterns of reproductive allocation in plants. Expected utility models have been used to examine the evolution of life history phenomena in plants where the random variable of interest is the genotypic expression of plant phenotypic traits as they contribute nonlinearly to fitness components (Lacey et al. 1983; Real and Ellner 1992). Within the domain of foraging ecology, however, the random variable has generally been restricted to characterizations of rates of energetic returns from different resources (Caraco et al. 1980; Krebs and Kacelnik 1991), though the method is equally applicable to other currencies, such as amounts of energetic reward (Real 1981; Real et al. 1982; Wunderle et al. 1987) or temporal delays in the delivery of rewards. I will say more about currencies below when I discuss the psychology of choice.

A second feature of the general expected utility model is the lack of specificity in the utility function. Utility can correspond to any nonlinearity associated with the translation of some measurable attribute of the environment into some meaningful biological property or function. For example, utility might correspond to a nonlinear translation of floral number into seed set. Or it might correspond to the relation between energetic rate of return and survivorship. Or it might correspond to size of prey and nutrient loading into egg mass. The specific nature of utility depends entirely on the problem under study. The generality of this approach, in which both the random variable X and its translation into utility are quite problem-specific—has led to our suggestion that the variance-discounting approach (equation 5.4) be viewed as a method rather than as a specific model (Ellner and Real 1989). This methodological approach, however, is very different from the concept of utility as applied in economics. Utility in economics is intended to merely reflect preferences over different options. It is an entirely descriptive bookkeeping device for indexing choices. As such, there is no functional explanation for the patterns of preferences that are observed. On the other hand, biological formulations seek to uncover some functional relationship that is encapsulated in the utility function, though the specific relation may change from problem to problem.

The variance-discounting method embraces a wide range of risk responses to resource and environmental variability. The organism's risk sensitivity is reflected in the coefficient of risk, $A = 1/2\ U''\ (\mu_X)$. As can be seen, the magnitude and sign of the risk response can (and most likely will) change as the arithmetic mean of X changes. Some organisms may alter their risk-taking behavior as a function of the arithmetic mean rate of energetic gain relative to their daily energy requirements (Caraco et al. 1980; Caraco et al. 1990; Cartar and Dill 1990). When μ_X is below requirement, then the organism tends toward risk-prone behavior. When μ_X is above requirement, then the organism tends toward risk-averse behavior. Stephens (1981) and Stephens and Charnov (1982) have explored the logic behind switching in risk-taking as a function of energy balance, and have produced an alternative model (the z-score) that captures the essential features of this energy balance rule. Real and Caraco (1986) and Ellner and Real (1989) have pointed out that there are some theoretical abnormalities inherent in the z-score model and that switching in risk-taking behavior as a function of energetic state can be incorporated easily into a cubic model for utility (fig. 5.3). If μ_X is below the critical level (X_R), then the utility function is concave-up and generates risk-seeking behavior. When μ_X is above X_R, then the organism turns risk-averse. If we understand the temporal dynamics of both the organism's energetic state and μ_X, then we can produce a dynamic model of temporal patterns of risk-taking behavior during the foraging cycle. Such dynamic models may effectively combine the traditionally static approach to decision making under uncertainty with the dynamic state-dependent approach to risk-sensitive foraging advocated by McNamara and Houston (1992).

The expected utility approach may be most usefully employed when the random variable of interest translates into organismal perception through identifiable physiological processes. For example, risk-sensitive foraging can be related to temporal patterns of delivery of rewards, and animals generally tend to choose options that minimize average delays (Fantino and Abarca 1985; Reboreda and Kacelnik 1991). Temporal patterns of delivery can be encoded through temporal patterns of neuronal activity, and there may be some nonlinearities associated with the temporal states of the nervous system and the perception of time. A more concrete example might relate nonlinear translations of amounts of reward through different excitation levels

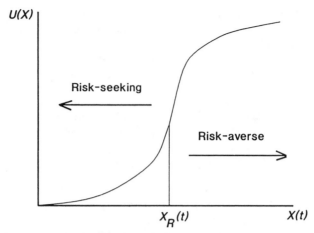

Fig. 5.3   An all-purpose utility function that is both concave-up and concave-down over different regions of resource level. The infection point $X_R(t)$ represents the critical resource requirement that must be met through foraging at some time t. For average resource levels below the critical requirement, organisms are anticipated to be risk-prone; above the requirement, risk averse.

in specific neurons. For example, nectarivorous insects probably show nonlinear and diminishing levels of excitation of the stretch receptors in the gut with increasing meal size (Taylor and Prochazka 1981). By establishing the connection between specific physiological processes and perception of rewards, we can begin to build functional theories of choice grounded in measurable proximate mechanisms.

THE BUMBLEBEE AS A MODEL SYSTEM

My discussion of the evolutionary design of cognitive function has been a bit abstract, so it may prove worthwhile to examine our experiments using bumblebees as a concrete example of decision making under uncertainty. Bumblebees are particularly well suited for evolutionary studies of decision making. Worker bees engage in the psychologically interesting task of choosing a subset of resources to exploit from a wider set of potential resources. However, this task is not complicated by other needs such as mating, territorial defense, or predator avoidance. Choices and their consequences are tied to evolutionary fitness since energetic input to the colony is a direct determinant of the colony's reproductive success and future genetic representation.

Our experiments with bees typically restrict individual workers to foraging on an artificial array of floral types in which the distribution of nectar rewards can be manipulated (Real 1981; Real et al. 1982; Ott et al. 1985; Real 1991). In all of our experiments we manipulate the amount and probability of reward in each floral type (the random variable X corresponds to nectar volume in microliters, μl). If the amount of reward translates nonlinearly into some measure of utility, and if bees are maximizing the expected utility with respect to amount of reward, then individual bees should be sensitive to the variance in reward as well as the arithmetic mean reward characterizing each of the floral types (following equation 5.4).

For example, in one experimental series the artificial patch of flowers consisted

Visits to Blue

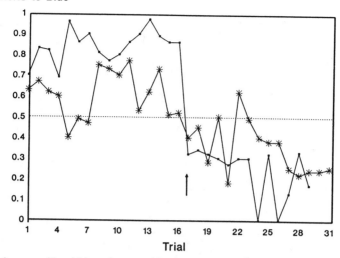

Trig. 5.4 Preferences of bumblebees (measured by the proportion of visits to blue flowers) for constant over variably rewarding floral types in an artificial patch of blue and yellow flowers. Individual bumblebees visited approximately 40 flowers during any single foraging sequence (a "trial"). The solid line with filled squares corresponds to experimental trials in which the constant floral type contained 2 μl of nectar in every flower, and the variable type contained no nectar in two-thirds of its flowers and 6 μl per flower in the remaining one-third. The solid line with asterisks corresponds to trials in which the constant floral type contained 2 μl in every flower, but the variable floral type contained 5 μl in one-third and 0.5 μl in two-thirds of the flowers. For trials 1–16 the constant floral type was blue. For trials 17–31 the constant floral type was switched (arrow) to yellow. If bees formed preferences on the basis of expectation alone, then visitation to blue flowers should always equal 0.5.

of 100 yellow and 100 blue flowers randomly mixed in space and distributed over a 1.2-m² patch. The blue flowers each contained 2 μl of nectar (honey diluted with water; 30% sucrose equivalents by weight). Yellow flowers, on the other hand, contained 6 μl of nectar in one-third of the flowers and no nectar in the remaining two-thirds. If bees are risk-neutral with respect to variation in the amount of reward, then they should show no preference for either color type, since the arithmetic mean for blue and yellow is the same (2 μl). However, bees show a significant preference for the constant blue over the variable yellow (fig. 5.4), and when color condition is switched, so that yellow is constant and blue is variable, individual bees show a preference for yellow over blue (Real 1981). Thus, bees appear to be risk-averse with respect to amount of reward, and behave in a manner consistent with expected utility maximization.

A further implication of the expected utility hypothesis is that uncertainty can be compensated for by increasing expectations. This implication is explicitly rendered through equation 5.4 by determining those combinations of arithmetic mean reward and variance in reward that generate the same expected utility (i.e., the "mean-variance indifference set"). Indifference curves of this sort can be empirically constructed by using the method of behavioral titration. For example, in one of our experimental arrays (Real et al. 1982), yellow flowers always contained 0.5 μl of nectar, while the distribution of nectar in blue was manipulated to give different combinations of arithmetic mean and variance in reward. Those combinations of

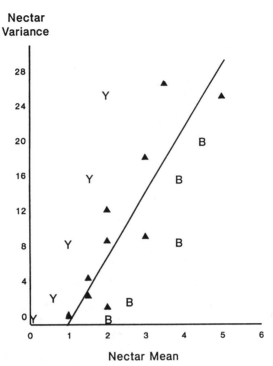

Fig. 5.5   A plot of the reward means and variances in the blue floral type that generated an ordinal blue preference (B), indifference (solid triangles), or yellow preference (Y) for eight bees when nectar content of yellow flowers was held constant at 0.5 μl/flower, and that of blue was variable. Individual bees' responses are pooled since bees show similar ordinal preferences, measured as proportion of visits to a given floral type out of a foraging sequence of approximately 40 visits. Both blue and yellow flowers were randomly distributed over the artificial floral patch. A linear regression through the combinations of means and variances in blue that generated indifferent foraging reveals a significant positive relationship (variance = −6.78 + 7.11 mean, SE of the slope = 1.38, $p < .01$). Thus increasing levels of uncertainty in reward can be compensated for by increasing expectation in reward.

mean and variance in blue that generated no significant preference for either the blue or the yellow flowers then represented the mean-variance indifference set (fig. 5.5). As can be seen, there is a significant positive correlation in the indifference set suggestive of the kind of trade-off anticipated if bees are expected utility maximizers. The exact nature of this trade-off (i.e., intercept, slope, curvature) can be influenced by the ecological conditions of the experiment (e.g., spatial arrangement of flowers), and often risk sensitivity can be quite variable in magnitude of response (Real et al. 1982; see also Caraco and Lima 1986 for indifference curve analysis of risk-sensitive foraging in birds).

The observed pattern of risk aversion with respect to amount of reward in bumblebees implies that the individual bee's utility must be an increasing but decelerating function of reward size. Harder and Real (1987) suggested that the nonlinear relation between reward size and rate of net energetic gain could constitute a utility function for individual bees. The rate of net energy uptake (E) per flower for an individual foraging bee is represented by

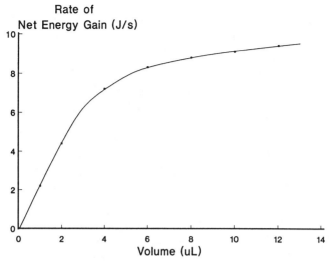

Fig. 5.6 A plot of the model "utility" function (equation 5.5), based on the empirically established biomechanics of nectar extraction, which shows rate of net energy gain (J/sec) in individual bumble-bees as a function of nectar volume (μl) in visited flowers. When bees maximize the expected rate of net energy gain, the nonlinear relationship accounts for the bees' sensitivity to variance in nectar rewards per flower.

$$E = \frac{e\rho SV - W[K_p(T_a + V/I)K_fT_f]}{T_f + T_a + (V/I)} \tag{5.5}$$

where e is the energy content of 1 mg of sucrose (15.48 J); ρ is the nectar density (mg/μl); S is the nectar concentration (% sucrose equivalents); V is the nectar volume (μl); W is the bee's mass (g); $T_p = T_a + V/I$ is duration of visit (sec); T_f is flight time between flowers (sec); K_p and K_f are energetic costs of probing and flying (J/g/sec); T_a is time required for entering and leaving a flower (sec); and I is ingestion rate (μl/sec).

For the specific physiological parameters of an average worker of our test species *Bombus pennsylvanicus* and for the nectar and flower characteristics used in the artificial patch experiments, a plot of equation (5.5) indicates that the rate of net energy gain is an increasing, decelerating function of nectar volume (fig. 5.6). We further showed that this physiologically based model was a fairly accurate predictor of the bumblebees' choice behavior over the different nectar distributions used in the earlier experiments (Harder and Real 1987; Real et al. 1990). Consequently, we have a first principles account of risk aversion in bumblebees that does not require constructing a utility function through choice behavior and in which the biological utility function clearly relates to animal fitness.

In our experiments we have manipulated the amount of reward associated with different floral types and have consistently shown risk aversion. The preference for a constant over a variable amount of nectar can be accounted for by simple rate maximization, in which rates of return are established through the utility function defined by equation (5.5). Cartar and Dill (1990) have analyzed risk-sensitive foraging in bumblebees in a manner different from ours: they take rate of net energetic gain as the random variable of interest (X = J/sec) and proceed to manipulate the energetic state of the colony to test the energy balance rule. In their experiments, the choice is between visiting flowers that give a constant rate of energy gain versus flowers with a variable rate, but both types have the same arithmetic mean rate of gain. When colonies were in negative energy balance (the colony's honey pots were drained), individual bees turned risk-seeking; when colonies were in positive energy balance (their honey pots were full), individual foragers were risk-averse in their choices. This pattern of behavior is consistent with the energy balance rule, and the Cartar-Dill experiment underscores the importance of identifying both the currency that is manipulated (e.g., temporal delays, amounts of reward, or rates of reinforcement) and the evolutionary/ecological context of behavior.

The Cartar-Dill experiment also raises important issues concerning the mechanisms of neuronal or sensory integration of information that may underlie choice. Currencies that rely on temporal delay or reward size as random variables may be easily encoded and translated into patterns of nervous system activity. Currencies that rely on rates as random variables may combine sensory inputs from temporal pattern and amount of reward. Thus some currencies may represent information processing at higher levels of nervous system integration that may be limited by lower-level processing of the separate input variables (i.e., time, magnitude). How specific sensory inputs are integrated into a currency appropriate for decision making in the natural environment remains a major unanswered question.

THE PSYCHOLOGY OF CHOICE AND RISK SENSITIVITY

The approach that I have outlined is primarily an economic and cognitive analysis of animal choice behavior rather than an operant conditioning analysis of choice. The infusion of psychological concepts and constraints into economic models of choice is progressing (e.g., Kahneman and Tversky 1979), and there has been considerable cross-fertilization between economics and psychology (e.g., Staddon 1983). Nonetheless, a synthesis of these two approaches that can be readily applied to ecological problems has failed to emerge. One obstacle to a synthesis is the lack of clear mapping between the structure of choice situations found under natural conditions and the variety of reinforcement schedules and paradigms frequently employed in the operant conditioning literature (Bolles 1985; Kamil, chap. 2). Often, the techniques used to carefully ascertain the mechanisms that account for behavior in the laboratory are not those that will usefully reveal ecologically important aspects of behavior. Nonetheless, some lessons from instrumental learning relate to the experiments I have performed, and I wish to highlight those lessons. What I will argue is the need to unite economic and cognitive theories and concepts (such as expected utility and subjective probability perception) with traditional concepts from operant psychology (such as delay reduction and the value of immediate reinforcement).

Research on associative learning in choice situations has focused primarily on three factors: magnitude of reinforcement (e.g., amount of reward), delay to reinforcement, and the probabilities of amount and delay (Mackintosh 1974; Staddon and Ettinger 1989). By far, the second of these factors, delay to reinforcement, has received the greatest attention. The literature on this subject is enormous. My discussion is restricted to those aspects of operant conditioning that relate to risk-sensitive foraging and particularly probability assessment.

The amount of reward is clearly important in choice situations, and it is the variable that I have systematically manipulated in our bumblebee experiments. Under conditions in which there is no delay to reinforcement, animals appear to choose options that generate the highest magnitude of reward (Davenport 1962; Hill and Spear 1962; Catania 1963; Clayton 1964). When there is variation in amounts of reward with equal arithmetic mean amounts, animals prefer the less variable option—that is, they are risk-averse with respect to amount of reward (Reboreda and Kacelnik 1991). These results are in keeping with our results on bumblebee choice behavior, in which the amount of reward is determined by the quantity of nectar found in individual flowers. Note that the amount of reward a bee receives will be positively correlated with the amount of time spent at an individual flower.

Some authors have suggested that there is a confounding of the effects of time and the effects of amount in risk-sensitive foraging experiments (Krebs and Kacelnik 1991). For example, in Caraco's experiments, birds were given choices between two feeding stations with identical arithmetic mean rewards (number of seeds) but different variances in reward (Caraco et al. 1980). To control for overall energy intake rate and to maintain birds at similar hunger levels, a subject that happened to receive a large reward at the variable station was subjected to an increased delay to the next presentation. Birds preferred the constant over the variable option. This response, however, could result from choice governed by minimizing the average delay to reinforcement as well as minimizing the variance in amount of reward or reward rate (Fantino and Abarca 1985; Krebs and Kacelnik 1991). A risk-averse response would also be consistent with explanations for violations of self-control (Logue 1988) since large rewards are correlated with longer average delays.

In our experiments there is no enforced delay correlated with the size of reward that foragers happen to receive. The only correlation is between the size of reward and the time required for handling and acquiring the reward (exploitation time). It has been traditional within operant analyses to treat exploitation time as separate from delay in reinforcement. In fact, the classic experiments that examine the effects of amount of reward on learning manipulate reward by increasing the amount of time animals are allowed to feed. For example, the amount of reward available to a pigeon in Catania's (1963) classic experiments was controlled by the amount of time it was allowed to feed from a hopper. In our experimental arrangement, we also consider time for exploitation an aspect of the magnitude of reinforcement. Since there is no enforced delay associated with reward size—that is, bees that receive large rewards at a flower experience the same average delay to reinforcer availability as bees receiving small rewards—the effects of delay can be determined independently from the effects of amount. Thus, while there is an amount–time to exploitation correlation, there is no amount–delay to reinforcement correlation. As I will show, delay should best be considered a consequence of the probabilistic nature of repeated gambles, rather than a consequence of different magnitudes of reward, as has been suggested by Rachlin et al. (1986, 1991).

Considerable attention of late has been devoted to assessing the mechanisms by which bees form associations between floral stimuli and magnitude of reinforcement. Bitterman and his colleagues (Couvillon and Bitterman 1984, 1987; Buchanan and Bitterman 1988, 1989; Couvillon et al. 1991) have undertaken an extensive analysis of how honeybees learn the association between different floral stimuli (simple and compound) and amounts of reward. Their work has primarily focused on testing the linear operator model of learning (Rescorla and Wagner 1972) and on establishing whether strength of association between reward and stimulus (color and/or odor) can account for choice and asymptotic rates of visitation to different floral types. They conclude that strength of association can account for learning and that the amount of reward affects the asymptote of the acquisition curve (Couvillon et al. 1991). The major difference between Bitterman's experiments and ours is that we are looking at the allocation of behavior *after* learning has occurred. We always subject individuals to several pretrial foraging sessions for any given distribution of nectar rewards before we measure the bees' preferences. This period of pretrial learning is sufficient to train bees to their asymptotic levels of visitation to each floral type (Dukas and Real 1991). We are measuring the asymptotic allocation of behavior and determining how trade-offs in reward magnitude and probability affect the bees' asymptotic preferences. Bitterman's analyses are complementary to ours in that he ascertains the mechanism by which bees form associations between stimuli and rewards, while we then look at the global allocation of choice behavior, assuming that bees have already formed the appropriate associations.

Choices involving different amounts of reward can become quite complicated when there are different delays to reinforcement. The general conclusion from research involving interval scheduling is that animals tend to reduce both the absolute delay to reinforcement and the average delay, though this effect can be offset by increasing amount of reward (Logan 1965). For example, pigeons trained on concurrent fixed-interval or variable-interval schedules tend to select the alternative correlated with the shorter average delay before reinforcement (Killeen 1970). As Mackintosh (1974, 187) states, "differences in delay of reinforcement may be assumed to influence choice behavior by affecting the association between the consequences of each choice and subsequent reinforcement."

When amounts of reward are held constant and options differ in the variability in delay, then foragers tend to be risk-seeking with respect to delay; that is, they choose variable-interval over fixed-interval schedules of comparable average delay, presumably due to the discounting of future rewards (Mazur 1984, 1987; Mazur et al. 1985). Killeen's incentive theory and Gibbon's scalar expectancy theory can also account for the preference for immediate over delayed reinforcement (Gibbon 1977; Killeen 1985). Both delay reduction and scalar expectancy theories have been applied to optimal foraging problems with some success (Fantino and Abarca 1985; Kacelnik et al. 1991).

Delay to reinforcer availability can enter into our experimental arrangement through the probabilistic nature of repeated gambles. To illustrate, consider my first experiments in which blue flowers each contained 2 μl of nectar, and yellow flowers contained no nectar in two-thirds and 6 μl of nectar in one-third. On average it will take three visits to receive a reward from yellow, but only one visit to receive a reward from blue. If amount of reward is irrelevant, then the risk-averse response that we observed may reflect the average shorter delay to reinforcement associated with blue flowers. Note, however, that delay here is not associated with time between landing on a given flower and delivery of a reward, nor is delay a forced period of inactivity that results from receiving a large reward; rather, it is a consequence of the probability of hitting particular strings of rewarding and nonrewarding flowers. Rachlin et al. (1986, 1991) have shown that there is a formal relation between temporal delay and probability distributions in repeated gambles that is a simple modification of the discounting procedure suggested by Mazur (1984, 1987; Mazur et al. 1985). The probability-time discounting procedure can be applied to our experiments on risk aversion, but, as I will show, cannot give a complete explanation for the bee's pattern of response.

While in my first experiment, the average time to reinforcement differed between blue and yellow flowers, consider an alternative experiment I performed (Real 1981) in which blue flowers contained 2 μl in every flower and yellow contained 0.5 μl in two-thirds and 5.0 μl in the remaining one-third. In this experiment bees preferred the constant blue floral type and were therefore risk-averse. Expected time to reward is identical for both floral types (i.e., one visit). The simplest way of discounting the reward is to divide the size of the reward by its position in the foraging sequence. For example, the blue flowers would generate a sequence of values 2/1, 2/2, 2/3, and so on. The discounted value of the foraging bout would then correspond to $\Sigma_n\, 2/n$. The discounted value of yellow flowers would be the expected reward at each flower visit discounted by its position in the sequence, that is:

$$\sum_n \frac{\frac{2}{3}(0.5) + \frac{1}{3}(5)}{n} = \sum_n \frac{2}{n}.$$

An easy way to see the equality of the two types of flowers is to restrict the calculation to the first three visits. For yellow there are three possible sequences with equal likelihood: (0.5, 0.5, 5), (0.5, 5, 0.5), or (5, 0.5, 0.5). The expected discounted value would then be

$$E_y V(X) = \frac{1}{3}\left[\left(\frac{0.5}{1} + \frac{0.5}{2} + \frac{5}{3}\right) + \left(\frac{0.5}{1} + \frac{5}{2} + \frac{0.5}{3}\right) + \left(\frac{5}{1} + \frac{0.5}{2} + \frac{0.5}{3}\right)\right] = \frac{11}{3}.$$

The expected discounted value of blue for the first three visits is

$$E_b V(X) = \frac{1}{3}\left[3\left(\frac{2}{1} + \frac{2}{2} + \frac{2}{3}\right)\right] = \frac{2}{1} + \frac{2}{2} + \frac{2}{3} = \frac{11}{3}.$$

The temporal discounting of rewards makes no difference in the assignment of value to each of these floral types. The basic reasoning behind this lack of difference may be put as follows: Flowers are randomly distributed in space, so there is no difference in the variance of time to acquisition across floral types. Since reward sizes are randomly distributed across flowers within a type, the temporal delays and amounts average out. *Therefore, time discounting cannot explain risk aversion in the bumblebee experiments.* What does explain risk aversion?

The expected size of reward may not be the appropriate currency for choice; instead, one must look to the expected utility of reward size $\overline{U(X)}$, measured as net rate of energy gain (in joules/sec). Since the utility of reward is nonlinear, the average utilities of the two floral types are not the same. For blue, $\overline{U(X)} = U(2) \approx 5.3$ J/s, while for yellow, $\overline{U(X)} = \frac{1}{3}U(0.5) + \frac{2}{3}U(5) \approx 3.1$ J/s. Thus, bumblebee choice in our experiments seems to be attributable to nonlinear utility effects rather than to temporal patterns of delay.

To ascertain whether bumblebees' choices are sensitive to variation in temporal delay, another set of experiments is needed in which variance in time to reward is introduced. Perhaps the simplest experiment would be to arrange one flower patch (blue) so that flowers are uniform in spatial distribution (e.g., a hexagonal array) with equal flight distances from flower to flower (constant time). In another patch (yellow), flowers would be distributed in a Poisson fashion with predetermined mean and variance in distance between flowers (variable time). In the experiment, mean distance (mean time to reward) in the two patches would be identical and all flowers would give the same reward. Consequently, the difference between patches would be entirely due to variance in the time to acquisition of reward. If bumblebees are influenced by variance in temporal delays, as predicted by scalar timing theory (Gibbon et al. 1988), then they should prefer variable delay over a fixed delay equal to the mean of the variable set. That is, bees should be risk-prone with respect to variance in time, and prefer the yellow flowers over the blue.

MEASURING UTILITY AND PROBABILITY EFFECTS

The formation of preferences and choices under the expected utility hypothesis is influenced by both the nature of the utility function and the organism's perception of probabilities. Attitudes toward risk taking can be the result of either utility or probability perception, as was noted by Arrow in 1971:

> It may be argued that the gambler is one who believes the odds are more favorable to him than they really are: according to his *subjective* probabilities, the bet is favorable to him, but there is, for one reason or another, a divergence between the subjective and objective probabilities. Then gambling can be consistent with risk-aversion, when the risks are understood subjectively. (Arrow 1971, 91).

Given the possible influence of probability perception on choice behavior, we must devise methods and experiments that simultaneously assess and partition the relative contributions of utility and probability in the formation of preferences. There are two common methods for partitioning probability and utility effects in economic models of choice. The first method employs the economic theory of state preference and manipulates reward amounts over a fixed probability set (Yaari 1965, 1969; Rosett 1971; Machina 1983; Real 1987). A second method constructs the so-called "Marschak-Machina triangle" and manipulates probabilities over a fixed set of rewards.

Method 1. This procedure for assessing subjective probability and utility assumes that individuals follow an expected utility model for choice (Real 1987). Let a flower type be represented by two reward states, x_1 and x_2, with objective probabilities p_1 and $p_2 = (1 - p_1)$. Let $U(x_1, x_2)$ represent the utility of using flower type X. Then the expected utility is given by

$$E\,U(x_1,x_2) = p_1U(x_1) + (1 - p_1)\,U(x_2) \tag{5.6}$$

For a fixed probability set $\{p_1, 1 - p_1\}$, we can determine those values of reward states $\{x_1, x_2\}$ that generate the same expected utility by setting equation 5.6 equal to a constant and taking the derivative with respect to x_1 to obtain

$$p_1U'(x_1) + (1 - p_1)\,U'(x_2)\frac{dx_2}{dx_1} = 0.$$

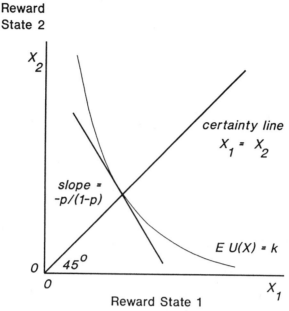

Reward
State 2

certainty line
$X_1 = X_2$

slope =
$-p/(1-p)$

$E\ U(X) = k$

0 45°

0

Reward State 1

X_1

Fig. 5.7 An indifference curve E $U(x_1, x_2) = k$ in reward state space for a concave-down utility function defined over the set of reward states $\{x_1, x_2\}$ with fixed probabilities $\{p_1, 1 - p_1\}$. The subjective probability is operationally defined by the slope of the indifference curve at the certainty point $-p_1/(1 - p_1)$.

Upon rearrangement we can see that the rate of substitution of X_2 for X_1 along a curve generating equal expected utility—that is, a "utility indifference curve"—is given by

$$\frac{dx_2}{dx_1} = - \frac{p_1}{(1 - p_1)} \frac{U'(x_1)}{U'(x_2)} \qquad (5.7)$$

This represents the conditional substitution of reward states with fixed probabilities that generate the same benefit to the bee.

A plot of this indifference curve in the (x_1, x_2) coordinate system gives a curve that is concave from the origin and downwardly sloping if $U(x_1, x_2)$ shows diminishing returns (fig. 5.7). At the 45° line $x_1 = x_2$, so $U'(x_1)/U'(x_2) = 1$. Therefore, at the 45° line, $dx_2/dx_1 = - p_1/(1 - p_1)$. This condition is what can be used to operationally derive subjective probability.

Method 2. The state preference theory approach outlined above represents the "classic" economic approach to probability assessment. In this approach reward levels are varied over a fixed probability distribution. An alternative, more recent approach, the "Marschak-Machina probability triangle," varies probabilities over fixed reward levels (Machina 1983, 1987a, 1987b; Kagel et al. 1990). Theoretically, estimates of probability bias should be similar regardless of whether one varies reward levels with fixed probabilities or varies probabilities with fixed reward levels.

A *Risk-Averse Forager*

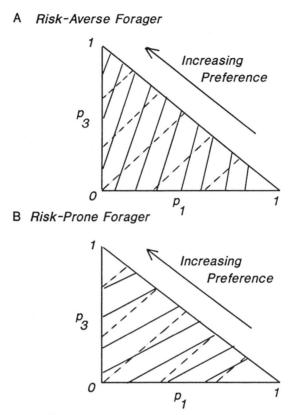

B *Risk-Prone Forager*

Fig. 5.8   Representations of Marschak-Machina probability triangles for (A) risk-averse and (B) risk-prone foragers. In this method reward states are fixed and probabilities are manipulated to generate indifference relations. Linear utility functions (risk-neutral foragers) generate the dashed-line indifference curves. Risk-averse foragers should have indifference curves with greater slope than risk-neutral foragers. Risk-prone foragers generate indifference curves with lower slopes. In both cases, the slope of the indifference curve is constant over the triangle.

Traditional expected utility theory assumes that decisions are linear in the probabilities; that is, that decisions satisfy the "independence axiom" (Shoemaker 1982). Graphically, the property of linearity in the probabilities can be illustrated by considering the set of all reward prospects over the fixed levels $x_1 < x_2 < x_3$, which can be represented by the set of all probability triples of the form $p = (p_1, p_2, p_3)$, where $p_i = \mathrm{prob}(x_i)$ and $\Sigma p_i = 1$. Since $p_2 = 1 - p_1 - p_3$, all probability combinations of reward states can be represented by points in the unit triangle in the (p_1, p_3) plane (fig. 5.8). For example, the lower left corner would correspond to the reward level x_2 occurring with probability equal to one, $p_2 = 1$. Northwest movement within the triangle results in increasing preference since upward movement in the triangle shifts probabilities from the less rewarding state x_2 to the more rewarding state x_3, and left movement shifts probability from x_1 to x_2.

The most important feature of this representation of choice outcomes is that the organism's indifference curves in the (p_1, p_3) triangle are given by solutions to the linear equation

$$E\,U(X) = \sum_{i=1}^{3} U(x_i)p_i = U(x_1)p_1 + U(x_2)(1 - p_1 - p_3) + U(x_3)p_3 = \text{constant} \qquad (5.8)$$

and will consist of parallel straight lines with slope $[U(x_2) - U(x_1)]/[U(x_3) - U(x_2)]$. If we experimentally construct only one indifference curve in the unit triangle, then, according to the linearity assumption, all choices over the rest of the triangle should be predictable based on the slope of the indifference curve. As I will discuss below, violations of this prediction can be used to assess nonlinearities in subjective probability formation and subjective probability bias. First, however, I will show how the Marschak-Machina triangle relates to assessing risk-sensitive foraging.

For the risk-neutral forager, $U(x_i) = x_i$, and, consequently, the indifference curves given by equation (5.8) reduce to

$$E\,(X) = \sum_{i-1}^{3} x_i p_i = x_1 p_1 + x_2(1 - p_1 - p_3) + x_3 p_3 = \text{constant} \qquad (5.9)$$

and correspond to the dashed lines in figure 5.8 with slope $(x_2 - x_1)/(x_3 - x_2)$.

For a risk-averse forager, $U'' < 0$, $U(X)$ is concave, and the slope of these indifference curves will be steeper than the risk-neutral indifference curves (fig. 5.8A), since

$$[U(x_2) - U(x_1)]/(x_2 - x_1) > [U(x_3) - U(x_2)]/(x_3 - x_2) \qquad (5.10)$$

whenever $x_1 < x_2 < x_3$ (Machina 1987a, 1987b). Conversely, for the risk-seeking forager, $U'' > 0$, $U(X)$ is convex, and the slope of these indifference curves will be shallower than those for the risk-neutral forager (fig. 5.8B).

By experimentally constructing indifference relationships over the unit probability triangle we can measure the curvature of the organism's utility function (i.e., determine the utility contribution to risk-sensitive foraging). Similarly, determination of subjective probability bias using the Marschak-Machina triangle relies on assessing whether choices are linear in the probabilities and determining the slopes of indifference relations over different reward and probability sets.

One of the earliest and best-known examples of systematic violations of linearity of probabilities in human subjects is the Allais paradox (Allais 1953). The paradox is revealed through choices over pairs of options of the form:

Choice 1: A: reward x_2 with probability p

reward x_3 with probability $1 - p$

or

B: reward x_1 with probability q

reward x_3 with probability $1 - q$

Choice 2: C: reward x_2 with probability rp

reward x_3 with probability $1 - rp$

or

D: reward x_1 with probability rq

reward x_3 with probability $1 - rq$

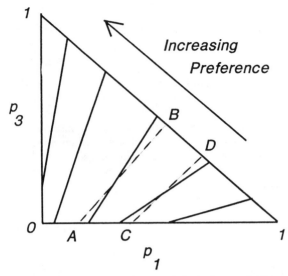

Fig. 5.9   Fanning out of indifference curves in the Marschak-Machina probability triangle can account for certain forms of paradoxical preferences. This figure illustrates potential resolution of the Allais paradox where option A is preferred to option B but option D is preferred to option C even though the sets of options maintain the same relative position with respect to utility (the "common-ratio effect").

where $p > q$, $0 < x_1 < x_2 < x_3$, and $0 < r < 1$. The paradox formed this way is called the *common-ratio effect* and derives from the equality of $\text{prob}(x_2)/\text{prob}(x_1)$ in A versus B and in C versus D. If the axioms of expected utility theory are obeyed, then an individual who prefers A over B must also prefer C over D, and vice versa. However, human subjects show a systematic tendency to prefer A over B and D over C if the outcomes x_i are gains (as in the reward example), and to prefer B over A and C over D when the outcomes are losses.

One explanation for this violation suggests that indifference curves in the probability triangle fan out from the origin (fig. 5.9) as a result of nonlinear subjective probability bias (Machina 1987a, 1987b; Kagel et al. 1990). Fanning out of the indifference curves can explain the Allais paradox. The dashed lines in figure 5.9 represent linear combinations of the prospects A and B or C and D. Solid lines are the indifference curves. Since the dashed lines are parallel by construction, fanning out can account for the apparent violation since prospect A will be preferred to prospect B, but prospect D will be preferred to C in the illustration.

While this new approach and the attendant theoretical developments have been exciting, there are very few experimental tests assessing the structure of indifference relations using the Marschak-Machina triangle. There is to my knowledge only one series of experiments using laboratory rats that explores these effects in nonhuman animals (Battalio et al. 1985; Kagel et al. 1990). Preliminary results indicate that rats commit Allais-type violations in choice behavior. However, the results are rather weak, since they do not construct entire indifference curves. Nonetheless, test animals seem to exhibit some fanning out of their indifference curves, and the magnitude of fanning out can be used to operationally measure subjective probability bias.

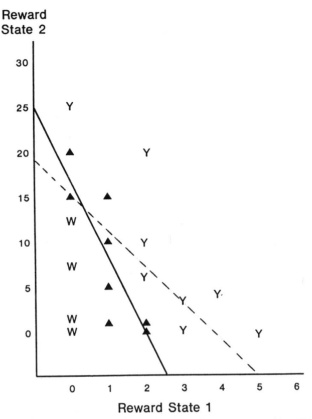

Reward
State 2

Reward State 1

Fig. 5.10   A state preference plot of combinations of reward states in variable yellow flowers that generated either ordinal preference for yellow (Y), indifference (solid triangles), or preference for the constant white floral type (W) in five individual bees. Individual bees visited approximately 40 flowers during each of the foraging sequences and showed similar ordinal preferences when foraging on identical reward distributions. Reward states in the yellow flowers occurred with fixed probabilities $[p(x_1) = 0.2$ and $p(x_2) = 0.8]$. A linear regression through the indifference points has a slope of -8.25 $(x_2 = 15.88 - 8.25x_1$, SE of the slope $= 2.30$, $p < .01)$. The expected slope based on the objective probabilities is -4.0. Consequently, bees appear to overestimate the likelihood of common events and underestimate the likelihood of rare events.

Tests of Subjective Probability Bias in Bumblebees

Using an enclosed colony of bumblebees, *Bombus bimaculatus*, I constructed a state preference indifference curve (method 1). We constrained bees to foraging on an artificial patch with 200 artificial white and yellow flowers of equal abundance. Methods were the same as in Real (1981), and Real et al. (1982). White flowers always contained 4 μl of nectar, while yellow flowers had two possible reward states occurring with fixed probabilities of $\text{prob}(x_1) = 0.8$ and $\text{prob}(x_2) = 0.2$. Indifference was determined by adjusting the two reward states in the yellow flowers until bees foraged equally from the constant white and the variable yellow flowers. Figure 5.10 represents the combined responses of five bees that foraged regularly from the colony (Real 1991). However, most of the foraging for the colony was carried out by only three individuals.

The first important point to note about the constructed indifference curve is the lack of significant nonlinearity, which simplifies the determination of the subjective probability. If the bees followed the objective probabilities, then the slope of the regression through the indifference points would equal $-0.8/0.2 = -4$. Instead, the slope equals -8.25. A slope of this value indicates that $prob(x_1) = 0.89$ and $prob(x_2) = 0.11$. Thus, subjective probabilities are biased in a manner that overestimates common events and underestimates rare events.

I have repeated these experiments using the responses of individual worker bees rather than responses averaged across individuals. In these experiments I used 50 artificial blue and 50 artificial yellow flowers. The reward states had probabilities identical to those in our earlier experiment, $prob(x_1) = 0.8$ and $prob(x_2) = 0.2$, only now individual responses could be monitored explicitly. We have been able to complete tests on six individuals as of this date. In all cases the individual bee misjudges the objective probabilities and underestimates the rare event (L. A. Real, unpubl.).

My results so far contrast with the conclusions of Kahneman and Tversky (1982), who assert that rare events are subjectively overestimated. One explanation for this inconsistency is that humans and bees differ markedly in how they gather and process information about potential rewards. These differences may reflect differences in learning and memory constraints as well as differences in habitat structure (Real 1991).

The perception of rarity by bumblebees may be analyzed using some data collected previously on switching behavior (Real 1990a). To examine switching behavior, I varied the ratio of floral types (blue and yellow) present at any given trial; the total number of flowers in the patch always equaled 200. In all trials, blue and yellow flowers contained the same variable distribution of nectar reward. Individual workers foraged from the artificial patch, and the percentage of visits to yellow flowers was monitored as a function of their relative abundance. The derived curve represents the responses of several different foragers to different ratios of blue and yellow flowers when the energetic qualities of blue and yellow flowers are constant across ratios. The experiment clearly demonstrates switching behavior, and yellow flowers are underexploited when they are relatively rare (fig. 5.11). Once again we observe that rare events are misperceived and underrepresented in the foraging behavior of the bees.

Both the subjective probability experiments and the switching experiments can be viewed as pattern recognition problems in which the representational pattern is frequency of occurrence. How is the subjective bias in probability tied to the bees' sampling of flowers, and can this information-gathering phase be linked to probability formation?

Frequency estimates in any sampling process tend to converge asymptotically with the objective probabilities as the distribution is repeatedly sampled. If the sampling process is truncated, there can still remain a difference between the subjective probability estimate and the objective probability. If bees are framing their decisions on the basis of small samples or assessing floral properties based on short sequences, then estimation bias can result.

We have calculated expected net energy gain for different floral reward distributions based on different sample sizes, and have compared the predictions based on sample size with observed choice behavior (Real et al. 1990). Our analysis suggests that bees frame their decisions on the basis of only a few flower visits, though apparently more than three or four sequentially exploited flowers (Dukas and

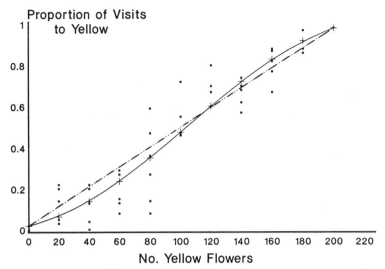

Fig. 5.11 Switching response of individual worker bumblebees to changes in relative floral abundance when both blue and yellow floral types varied in nectar quantities (one-third with 6 µl and two-thirds with no nectar). The best fit cubic regression is $Y = .149 - 4.23 \times 10^{-3}X + 1.11 \times 10^{-4}X^2 - 3.62 \times 10^{-7}X^3$ ($R^2 = .87$, df $= 45$, $p < .001$).

Real 1993b). If our analysis is correct, then this can explain the observed high frequency of bias in probability estimation observed in my experiments. Since bees are sampling from a highly skewed distribution of rewards, truncation of the sampling process or calculations of probability based on a short sequence of flower visits will tend to produce subjective bias favoring the skew in most experiments—in this case, the high- versus the low-probability event (Real 1991).

Disproportionate use of resources that vary in relative frequency has been documented in operant conditioning experiments, and generally goes under the names of "undermatching" and "overmatching" (Baum 1974; Davison and McCarthy 1988). The mathematical formulation of the generalized matching law that embraces nonlinear responses to relative frequency is identical to the mathematical equation that I have proposed for translations of objective probabilities into subjective probabilities (Real 1987). Both predatory switching and under- and overmatching have been mechanistically tied to some forms of signal detection, and signal detection theory may prove very useful in forming an information processing foundation for this phenomenological pattern of behavior (Nevin et al. 1982; Davison and McCarthy 1988; Getty and Krebs 1985).

The high frequency of experiments showing bias in frequency estimates may then be the result of constraints due to sampling or signal processing, which in turn may be structured by memory or learning constraints.

COMPUTATIONAL ALGORITHMS AND REPRESENTATION

The biological account of risk aversion that I have suggested assumes that information about floral rewards is processed in a particular fashion. There are numerous ways that a particular string of information can be partitioned or "chunked" while maintaining the order sequence, but different patterns of parsing the sequence can give rise to very different representations and meanings (Terrace 1987, 1991; Terrace and Chen 1991; Swartz et al. 1991). For example, consider the sentence

On the east side street, boys play ball.

The meaning of this sentence changes rather dramatically if we shift the comma to before "street":

On the east side, street boys play ball.

even though there is no change in the sequence of words. In an analogous fashion, how bumblebees parse and chunk the information from particular strings of flower visits may profoundly influence their representation of the energetic environment.

For each flower in the sequence of n visits let R_i represent the reward from the ith flower and T_i the time required to get the reward. The Harder-Real model assumes that bees maximize the expected net energy gain from calculations based on attributes of single flowers; that is, the average of the set of R_i/T_i:

$$E_1 = \text{avg}\left\{\frac{R_1}{T_1}, \frac{R_2}{T_2}, \ldots, \frac{R_n}{T_n}\right\} \tag{5.11}$$

Such an averaging scheme amounts to calculating the expectation of the ratio of two random variables, $E(R/T)$, which leads to the maximization of short-term energy gain (Turelli et al. 1982). Several authors have argued, however, that organisms should maximize long-term gain rather than short-term gain (Possingham et al. 1990).

Maximizing long-term gain amounts to calculating the ratio of expectations of the random variables, $E(R)/E(T)$. Actually, there is a range of possibilities between completely short-term and completely long-term gain corresponding to different ways of processing information from a string of flower visits. For example, rewards and times could be pooled across successive pairs of flowers in the string:

$$E_2 = \text{avg}\left\{\frac{R_1 + R_2}{T_1 + T_2}, \frac{R_3 + R_4}{T_3 + T_4}, \ldots, \frac{R_{n-1} + R_n}{T_{n-1} + T_n}\right\} \tag{5.12}$$

or across triplets of flowers (E_3), quadruplets of flowers (E_4), and so on. In the theoretical limit $E_n \to \infty$ for which $E_\infty = E(R)/E(T)$, which corresponds to the long-term average rate of energetic gain (Turelli et al. 1982).

For each distribution of floral rewards used in our 1982 experiments, and using the physiologically based expressions for R and T from the Harder-Real model (equation 5.5), we calculated the average rate of net energy gain E_k as a function of the frame size k over which samples are pooled. The frame size determines the fine structure of the processing of a particular string of information. In all of our experiments, the correlation between average rate of energetic gain and observed preference was highest for short-term optimization (i.e., frame length $k = 1$ or 2), and lowest for long-term optimization (Real et al. 1990). These results strongly suggest that bumblebees frame their decisions on the basis of individual flowers or on pairs of flowers, resulting in short-term energy maximization.

There are at least three scenarios under which calculations based on small frame lengths will prove advantageous.

First, if bees are limited in their memory capacity and are therefore constrained to base decisions on small sample sizes, then calculations based on E(R/T) may be more accurate than calculations based on E(R)/E(T). Short-term calculations are a more robust estimation of performance when sampling is truncated. This may occur especially when the reward (R) and the time to acquire the reward (T) are correlated (Real et al. 1990). A positive correlation between R and T is guaranteed in floral systems simply because it takes more time to ingest greater quantities of nectar.

Second, short-term optimization may be adaptive when there is a high degree of spatial autocorrelation in the distribution of floral rewards. In most field situations there is intense local competition among pollinators for floral resources. When "hot" and "cold" spots in fields of flowers are created through pollinator activity, then such activity will generate a high degree of spatial autocorrelation in nectar rewards. If information about individual flowers is pooled, then the spatial structure of reward distributions will be lost, and foraging over the entire field will be less efficient. In spatially autocorrelated environments ("rugged landscapes") averaging obscures the true nature of the environement, and it may pay not to pool information.

A third explanation combines both memory constraints and the hierarchical processing of information. In his seminal paper on human memory, Miller (1956) suggested that there were limits to the recall of specific information. If seven objects of a specific class are presented to subjects, recall of the set of objects is fairly accurate. If many more objects are presented, then recall deteriorates. We obviously can recall more than seven objects, so how do we store large amounts of information in a manner that will allow for recall? Miller suggested that information is hierarchically nested, so that classes of information are pooled together. For example, if a subject were presented with seven pieces of fruit and seven hand tools, recall would be fairly efficient, since "fruit" and "tools" can be used as class identifiers.

Information on floral qualities could be hierarchically nested if rewards and times at individual flowers are pooled into simple quotients (rates). More information may be processed and recalled if the strings are hierarchically structured. Pairing reward and time as a rate by flower seems a very reasonable hierarchical pattern for nesting the information.

The memory constraint and hierarchical processing schemes depend on some truncation of sampling or limits to recall in the nervous system. The experimental evidence on how bumblebees form subjective estimates of probabilities supports the view that bees engage in truncated sampling that leads to bias in subjective probability formation.

The short-term averaging algorithm that seems to best characterize the bumblebee's choice behavior should not be interpreted to mean that bees possess only a short-term memory window or that bees remember only the last flower visited. The algorithm that we have proposed assumes that information is assembled on an individual-flower basis, but makes no statement about how many individual flowers are involved in computing the average. Under some circumstances, limited memory, leading to truncated sampling or recall of previously visited flowers, may lead to an advantage in using short-term versus long-term averaging algorithms. However, at this time, the relation between short-term averaging and memory constraints remains an object of empirical study. Throughout this discussion I have concentrated on the memory patterns associated with immediate foraging sequences, and therefore have restricted the discussions to short-term memory constraints. More work is needed on the relation between short-term memory and long-term memory in bumblebees, comparable to the kinds of studies carried out by Menzel (1990) on honeybees, and on how long-term memory affects choice behavior.

LOOKING DOWN: NEURONAL IMPLEMENTATION OF CHOICE

At this point the development of questions on decision making and choice under uncertainty can follow two tracks. One track leads down, and explores in more detail the psychological and neurobiological foundations of complex decision making; the other track leads up, and examines the ecological and evolutionary consequences of observed patterns of behavior on, for example, pollinator-plant interactions and community organization. The last two sections of this chapter hint at ventures along each of these paths.

One cannot help but wonder how bees are capable of performing the kinds of complicated tasks that they seem to perform. Obviously, the machinery for prcoessing the information that goes into decision making resides in the nervous system, and a mechanistic account of decision making in bees must ultimately seek out the neuronal processes implementing specific tasks. There has been some work on aspects of nervous system organization and behavior (primarily learning and memory) in the honeybee (Erber 1983; Menzel and Mercer 1987; Menzel 1990). However, almost nothing is known about nervous system organization and *decision making* in bees, other insects, or simpler invertebrates.

Exploration into the neural basis of decision making in bees will rely at first almost exclusively on constructing model nervous systems. The emerging paradigm of parallel processing allows for the construction of model nervous systems that can potentially solve quite complex problems (McClelland and Rummelhart 1986; Wasserman 1989; Gabriel and Moore 1990; Levine 1991). Risk-sensitive foraging can either be formulated as a special problem in nonlinear programming (e.g., quadratic programming) (Real 1980a) or as a problem in stochastic dynamic programming (McNamara and Houston 1992). Many new methods are being developed for dealing with nonlinear problem solving using neural network structures, and we can apply these new techniques to foraging behavior in bees.

The value of neural networks resides in their ability to model nervelike functions. In either the quadratic programming or the stochastic dynamic programming approach to risk-sensitive foraging, one assumes that the organism can calculate a mean and variance. As a first step, the mean and variance in rewards offered by flowers can be implemented in a simple nervous system model by the scheme outlined in figure 5.12.

This system comprises a three-layer neural net. There are n inputs that correspond to rewards obtained from flowers. The information on nectar volumes, for example, might come from the bee's stretch receptors. Each input neuron is excitatory and is connected to two second-layer neurons: one that calculates the arithmetic mean and another that tracks the initial input value. The second-layer neuron that calculates the mean acts as an inhibitory neuron on each of the tracking neurons. The second-layer tracking neurons are connected to the third layer via a truncated transfer function that squares the outputs. Outputs from the second layer all feed into a neuron that measures the variance and another neuron that measures the arithmetic mean. These two values can be internally stored, and act as inputs to the larger nonlinear optimization problem.

The scheme presented above is obviously very simplistic. I present it as a heuristic device for illustrating how one can begin to construct models of the neural networks that may be involved in the kinds of decision processes we are attempting to understand. There are obvious unrealistic features in the above scheme. More realistic neural network models must include the following features.

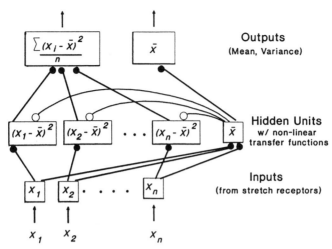

Fig. 5.12 A three-layer feed-forward neural net that calculates a mean and variance based on n simultaneous inputs. Solid circles correspond to excitatory connections; open circles correspond to inhibitory connections. Output units are the result of applying a squaring transfer function to the differences between inputs and the inhibitor effects of an averaging hidden unit.

First, in the above scheme, inputs from floral rewards are assumed to be simultaneous rather than sequential. Since, however, floral inputs are sequential, neuronal integration must involve inputs that have undergone some degree of excitatory decay. Temporally dependent neural networks that mimic sequential processes have been developed for some aspects of learning (Barto et al. 1990; Sutton and Barto 1990), but much more theoretical work is needed in this area.

Second, inputs from flower rewards are not only sequential, but come from many different species of plants. Consequently, there must be a categorization scheme built into the network that allows for separate calculation of means and variances for each floral species in the habitat. Neural networks that categorize information are not uncommon, but linking such category-forming processes to optimization problems is still a challenge.

Third, once means and variances can be established based on sequential inputs from several sources, they must be utilized to solve the nonlinear programming problem that maximizes the difference between the mean and variance for a given exploitation pattern. Problems of this sort will involve adding sampling procedures in which the temporal input of new information is dependent on past information and mediated by current behavior. Bees learn to visit one or another floral species based on whether current estimates of differences between means and variances are increasing.

Finally, a bee does not know in advance what types of flowers it will encounter on a foraging bout. Indeed, as in our experimental work, it may find types of artificial flowers that it has never seen before. Yet the bee is capable of associating expected nectar rewards even with these unfamiliar flowers. To do so, bees must first be able to *learn* the characteristics (e.g., color, shape) that distinguish one flower type from another. Learning in neural network models is accomplished by modifying the *weights* of connections between neurons according to a learning rule. A large variety of learning rules have been developed, and we must choose appropriate, biologically plausible ones, such as variants of unsupervised Hebbian rules (Wasserman 1989; Hanson and Olson 1990; Levine 1991). Neural network modeling of bumblebee behavior is an especially appealing research topic since the model can comprise an entire closed-loop system, from perception through action. These modifications represent first steps toward a theoretical foundation for foraging behavior under uncertainty grounded in nervous-system-like operations.

LOOKING UP: IMPLICATIONS FOR POLLINATOR-PLANT INTERACTIONS

The laboratory studies we have carried out indicate significant changes in individual bumblebee foraging behavior as a consequence of different distributions of nectar in artificial floral species. Do patterns of nectar distribution in natural populations of flowering plants influence patterns of floral visitation, and can these influences affect reproductive success in individual plants? We have examined this question in a set of studies on natural populations of the mountain laurel (*Kalmia latifolia*), a common evergreen shrub growing in heath communities in mountainous regions of eastern North America (Real and Rathcke 1991). Thirty-two individual shrubs were monitored for average 24-hour nectar production in individual flowers, average per flower visitation rate by pollinators, and one measure of reproductive success (percent fruit set). Significant positive correlations existed between arithmetic mean nectar production and visitation rate (fig. 13A) as well as between visitation rate and percent fruit set (fig. 13B). Pollinators appear to be differentially attracted to individual shrubs that are characteristically high nectar producers, and this differential attraction can have significant effects on plant fitness.

In some plant systems, visitation rates appear to correlate with nectar production rates, but only affect male components of fitness (pollen removal rates) (Pleasants and Chaplin 1983). Nevertheless, increasing visitation rates will increase the probability of high-quality pollen expressing itself through intergametic selection. In such cases, female fitness may still be affected through the production of superior offspring when there is a correlation between gametophytic fitness and sporophytic fitness (Mulcahy 1983).

CONCLUSION

The theoretical and experimental analysis of choice behavior is the ideal starting place for constructing biological theories of decision making. In principle, every complex decision can be decomposed into a series of dichotomous choices. The decision (choice) to exploit a given resource (or not to exploit it) is of central concern in most aspects of behavioral and community ecology. Choice behavior can be extended to include choices over a variety of resource types, such as habitat selection, prey selection, and mate selection, all of which can be viewed and analyzed as dichotomous choices. For example, the habitat and mate selection problems can be viewed as decisions whether to stop searching or to continue searching (Real 1990b).

Dichotomous series of choices are likely to be analyzable at the neural level, providing proximate mechanistic explanations. At the same time, choices have obvious implications for ecological organization and evolutionary processes. Unfortunately, the three major fields that have been concerned with aspects of choice—psychology, economics, and ecology—have largely developed their theories

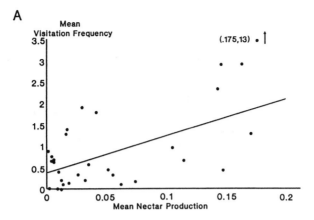

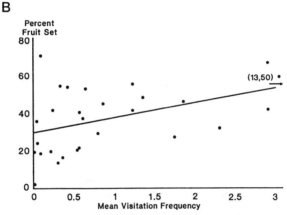

Fig. 5.13 (*A*) Correlation between individual plant mean 24-hour nectar production and corresponding pollinator visitation rate. Individual shrubs that are characteristically high nectar producers show significantly higher visitation rates (Kendall's rank correlation $\tau = 0.30$, $p < .01$). (*B*) Individual plants receiving higher visitation rate also experience higher pollination success, measured as percent fruit set ($\tau = 0.33$, $p < .01$).

independently of each other. It is my belief that the uniting principle behind all of these fields is the basic observation that all organisms are information processors that may have undergone various degrees of evolutionary specialization for processing information in specific ways. It is this principle that is basic to an evolutionary theory of cognition.

Both the public and academic imagination have been captivated by an emerging "cognitive revolution." New books are appearing almost monthly that extol the virtues of the cognitive view, and many have argued that we need an evolutionary perspective in order to understand the nature of the human mind and of human intelligence (Hookway 1984; Cherniak 1986; Wuketits 1990; Donald 1991; Bickerton 1990; Dennett 1991). Ironically, much of the traditional approach used in behavioral ecology is implicitly cognitive; yet cognitive psychologists are only beginning to appreciate the last 15 years of research in ethology and behavioral ecology (Gallistel 1990; Roitblat 1987; Kesner and Olton 1990; Kamil, chap. 2). The opportunity for mutual benefit is great. Behavioral ecologists will benefit from incorporating explicit treatment of cognitive processing of information into their studies and by appreciating the limits imposed by mental and neurobiological machinery. Cognitive scientists will gain a clearer understanding of the nature of adaptation and evolutionary response. Together the two approaches allow for a synthesis across levels of organization—from communities of neurons to communities of species—that may be the most evolutionary aspect of the cognitive revolution.

ACKNOWLEDGMENTS

I would like to thank R. Cartar, R. Dukas, D. Eckermann, S. Ellner, A. Kacelnik, E. Marschall, and B. Roche for their comments on the manuscript and/or for the many helpful and exciting discussions we have had on these ideas. This research has been supported through National Science Foundation Grants BSR 8500203, BNS 8719292, BNS 9096209, and IBN 9305931.

REFERENCES

Allais, M. 1953. Le comportement de l'homme rational devant le risque, Critique des postulates et axiomes de l'École Américaine. *Econometrica* 21:503–46. [Translated in M. Allais and O. Hagen, eds., 1979, *Expected utility hypotheses and the Allais paradox* (Dordrecht: Reidel).]

Arrow, K. 1971. *Essays in the theory of risk bearing*. Amsterdam: North Holland.

Balda, R. P., and R. J. Turek. 1984. The cache-memory system as an example of memory capabilities in Clark's nutcracker. In H. L. Roitblat, T. G. Bever, and H. S. Terrace, eds., *Animal cognition*, 513–32. Hillsdale, N.J.: Lawrence Erlbaum Associates.

Barto, A. G., R. S. Sutton, and C. J. C. H. Watkins. 1990. Learning and sequential decision-making. In M. Gabriel and J. Moore, eds., *Learning and computational neuroscience: Foundations of adaptive networks*, 539–602. Cambridge, Mass.: MIT Press.

Battalio, R. C., J. H. Kagel, and D. N. MacDonald. 1985. Animals' choices over uncertain outcomes: Some initial experimental results. *American Economic Review* 75:597–613.

Baum, W. M. 1974. On two types of deviation from the matching law: Bias and undermatching. *Journal of the Experimental Analysis of Behavior* 22:231–42.

Bernoulli, D. 1738. Specimen theoriae novae de mensura sortis. Commentarii Academiae Scientiarum Imperialis Petropolitanae, Tomas V. [Translated by L. Sommer, *Econometrica* 22:23–36 (1954).]

Bickerton, D. 1990. *Language and species*. Chicago: University of Chicago Press.

Bolles, R. C. 1985. The slaying of Goliath: What happened to reinforcement theory. In T. D. Johnston and A. T. Pietrewicz, eds., *Issues in the ecological study of learning*. Hillsdale, N.J.: Lawrence Erlbaum Associates.

Buchanan, G. M., and M. E. Bitterman. 1988. Learning in honeybees as a function of amount and frequency of reward. *Animal Learning and Behavior* 16:247–55.

Buchanan, G. M., and M. E. Bitterman. 1989. Learning in honeybees as a function of amount of reward: tests of the equal-asymptote assumption. *Animal Learning and Behavior* 17:475–80.

Caraco, T., and S. Lima. 1986. Survivorship, energy budgets and foraging risk. In M. Commons, S. J. Shettleworth, and A. Kacelnik, eds., *Quantitative analysis of behavior*, vol. 6, 1–22. Cambridge, Mass.: Harvard University Press.

Caraco, T., S. Martindale, and T. S. Whitham. 1980. An empirical demonstration of risk-sensitive foraging preferences. *Animal Behaviour* 28:820–30.

Caraco, T., W. U. Blackenhorn, G. M. Gregory, J. A. Newman, G. M. Recer, and S. M. Zwicker. 1990. Risk-sensitivity: Ambient temperature affects foraging choice. *Animal Behaviour* 39:338–45.

Cartar, R., and L. Dill. 1990. Why are bumblebees risk-sensitive foragers? *Behavioral Ecology and Sociobiology* 26:121–27.

Catania, A. C. 1963. Concurrent performances. *Journal of the Experimental Analysis of Behavior* 6:253–63.

Cheney, D. L., and R. M. Seyfarth. 1990. *How monkeys see the world.* Chicago: University of Chicago Press.

Cherniak, C. 1986. *Minimal rationality.* Cambridge, Mass.: MIT Press.

Clayton, K. N. 1964. T-maze learning as a joint function of the reward magnitudes for the alternatives. *Journal of Comparative Physiological Psychology* 58:333–38.

Couvillon, P. A., and M. E. Bitterman. 1984. The overlearning-extinction effect and successive negative contrast in honeybees (*Apis mellifera*). *Journal of Comparative Psychology* 98:100–109.

Couvillon, P. A., and M. E. Bitterman. 1987. Discrimination of color-odor compounds by honeybees: Tests of a continuity model. *Animal Learning and Behavior* 15:218–27.

Couvillon, P. A., Y. Lee, and M. E. Bitterman. 1991. Learning in honeybees as a function of amount of reward: Rejection of the equal asymptote assumption. *Animal Learning and Behavior* 19:381–87.

Davenport, J. W. 1962. The interaction of magnitude and delay of reinforcement in spatial discrimination. *Journal of Comparative Physiological Psychology* 55:267–73.

Davison, M., and D. McCarthy. 1988. *The matching law.* Hillsdale, N.J.: Lawrence Erlbaum Associates.

Dennet, D. 1991. *Consciousness explained.* Boston: Little, Brown.

Donald, M. 1991. *Origins of the modern mind.* Cambridge, Mass.: Harvard University Press.

Dukas, R., and L. A. Real. 1991. Learning foraging tasks by bees: A comparison between social and solitary species. *Animal Behaviour* 42:269–76.

Dukas, R., and L. A. Real. 1993a. Cognition in bees: From stimulus reception to behavioral modification. In D. Papaj and A. Lewis, eds., *Insect learning: Ecological and evolutionary perspectives,* 343–78. London: Chapman and Hall.

Dukas, R., and L. A. Real. 1993b. Effects of recent experience on foraging decisons by bumble bees. *Oecologia* 94:244–46.

Ellner, S., and L. A. Real. 1989. Optimal foraging models for stochastic environments: Are we missing the point? *Comments on Theoretical Biology* 1:129–58.

Erber, J. 1983. The search for neural correlates of learning in the honeybee. In F. Huber and H. Markl, eds., *Neuroethology and behavioral physiology,* 216–30. Heidelberg: Springer-Verlag.

Fantino, E., and N. Abarca. 1985. Choice, optimal foraging, and the delay-reduction hypothesis. *Behavioral and Brain Sciences* 8:315–30.

Gabriel, M., and J. Moore, eds. 1990. *Learning and computational neuroscience: Foundations of adaptive networks.* Cambridge, Mass.: MIT Press.

Gallistel, C. R. 1989. Animal cognition: The representation of space, time and number. *Annual Review of Psychology* 40:155–89.

Gallistel, C. R. 1990. *The organization of learning.* Cambridge, Mass.: MIT Press.

Gardner, H. 1985. *The mind's new science: A history of the cognitive revolution.* New York: Basic Books.

Getty, T., and J. R. Krebs. 1985. Lagging partial preferences for cryptic prey: A signal detection analysis of great tit foraging. *American Naturalist* 125:239–56.

Gibbon, J. 1977. Scalar expectancy theory and Weber's law in animal timing. *Psychological Reviews* 84:279–325.

Gibbon, J., R. M. Church, S. Fairhurst, and A. Kacelnik. 1988. Scalar expectancy theory and choice between delayed rewards. *Psychological Review* 95:102–14.

Hanson, S. J., and C. R. Olson. 1990. *Connectionist modeling and brain function: The developing interface.* Cambridge, Mass.: MIT Press.

Harder, L., and L. A. Real. 1987. Why are bumble bees risk-averse? *Ecology* 68:1104–8.

Hastings, A., and H. Caswell. 1979. Role of environmental variability in the evolution of life history strategies. *Proceedings of the National Academy of Sciences U.S.A.* 76:4700–4703.

Hill, W. F., and N. E. Spear. 1962. Choice between magnitudes of reward in a T-maze. *Journal of Comparative Physiological Psychology* 56:723–26.

Hookway, C., ed. 1984. *Minds, machines, and evolution*. Cambridge: Cambridge University Press.

Kacelnik, A., and J. R. Krebs. 1985. Learning to exploit patchily distributed food. In R. M. Sibley and R. H. Smith, eds., *Behavioural ecology, 25th symposium of the British Ecological Society*, 189–205. Oxford: Blackwell Scientific Publications.

Kacelnik, A., D. Bonner, and J. Gibbon. 1991. Timing mechanisms in optimal foraging: Some applications of scalar expectancy theory. In R. N. Hughes, ed., *Behavioural mechanisms in food selection*, 61–82. Heidelberg: Springer-Verlag.

Kagel, J. H., D. N. MacDonald, and R. C. Battalio. 1990. Tests of "fanning out" of indifference curves from animal and human experiments. *American Economics Review* 80:912–21.

Kahneman, D., and A. Tversky. 1979. Prospect theory: An analysis of decisions under risk. *Econometrica* 47:263–91.

Kahneman, D., and A. Tversky. 1982. The psychology of preference. *Scientific American* 246: 160–73.

Kesner, R. P., and D. S. Olton, eds. 1990. *Neurobiology of comparative cognition*. Hillsdale, N.J.: Lawrence Erlbaum Associates.

Killeen, P. 1970. Preference for fixed-interval schedules of reinforcement. *Journal of the Experimental Analysis of Behavior* 14:127–31.

Killeen, P. 1985. Incentive theory: IV. Magnitude of reward. *Journal of the Experimental Analysis of Behavior* 43:407–17.

Krebs, J. R., and A. Kacelnik. 1991. Decision-making. In J. R. Krebs and N. D. Davies (eds) *Behavioural Ecology*, 3d ed., 105–36. Oxford: Blackwell Scientific Publications.

Krebs, J. R., S. D. Healy, and S. J. Shettleworth. 1990. Spatial memory of Paridae: Comparison of a storing and a non-storing species, the coal tit, *Parus ater*, and the great tit, *P. major*. *Animal Behaviour* 39:1127–37.

Lacey, E. P., L. A. Real, J. Antonovics, and D. G. Heckel. 1983. Variance models in the study of life-histories. *American Naturalist* 122:114–31.

Levine, D. S. 1991. *Introduction to neural and cognitive modeling*. Hillsdale, N.J.: Lawrence Erlbaum Associates.

Logan, F. A. 1965. Decision-making by rats: Delay versus amount of reward. *Journal of Comparative Physiological Psychology* 59:1–12.

Logue, A. W. 1988. Research on self-control: An integrated framework: *Behavioral and Brain Sciences* 11:665–79.

Machina, M. J. 1983. *The economic theory of individual behavior toward risk: Theory, evidence, and new directions*. Technical Report no. 433. Institute for Mathematical Studies in the Social Sciences, Stanford University.

Machina, M. J. 1987a. Choices under uncertainty: Problems solved and unsolved. *Economic Perspectives* 1:121–54.

Machina, M. J. 1987b. Decision-making in the presence of risk. *Science* 236:537–43.

Mackintosh, N. J. 1974. *The psychology of animal learning*. New York: Academic Press.

McNamara, J. M., and A. I. Houston. 1992. Risk-sensitive foraging: A review of the theory. *Bulletin of Mathematical Biology* 54:355–78.

Mazur, J. E. 1984. Tests of an equivalence rule for fixed and variable reinforcer delays. *Journal of Experimental Psychology: Animal Behavior Processes* 10:426–36.

Mazur, J. E. 1987. An adjusting procedure for studying delayed reinforcement. In M. Commons, J. E. Mazur, J. A. Nevin, and H. Rachlin, eds., *Quantitative analysis of behavior*, vol. 5, 55–76. Hillsdale, N.J.: Lawrence Erlbaum Associates.

Mazur, J. E., M. Snyderman, and D. Coe. 1985. Influences of delay and rate of reinforcement on discrete-trial choice. *Journal of Experimental Psychology: Animal Behavior Processes* 11:565–75.

Menzel, R. 1990. Learning, memory, and "cognition" in honey bees. In R. P. Kesner and D. S. Olton, eds., *Neurobiology of comparative cognition*, 237–92. Hillsdale, N.J.: Lawrence Erlbaum Associates.

Menzel, R., and A. Mercer, eds. 1987. *Neurobiology and behavior of honeybees*. Heidelberg: Springer-Verlag.

Miller, G. A. 1956. The magical nunber seven, plus or minus two: Some limits on our capacity for processing information. *Psychological Review* 63:81–97.

Mulcahy, D. 1983. Models of pollen tube competition in *Geranium maculatum*. In L. A. Real, ed., *Pollination biology*, 152–62. New York: Academic Press.

Nevin, J. A., P. Jenkins, S. Whittaker, and P. Yarensky. 1982. Reinforcement contingencies and signal detection. *Journal of the Experimental Analysis of Behavior* 37:65–79.

Ott, J. R., L. A. Real, and E. Silverfine. 1985. The effect of nectar variance on bumblebee patterns of movement and potential gene flow. *Oikos* 45:333–40.

Pearce, J. M. 1987. *An introduction to animal cognition*. Hillsdale, N.J.: Lawrence Erlbaum Associates.

Pleasants, J. M., and S. J. Chaplin. 1983. Nectar production rates of *Aesclepias quadrifolia:* Causes and consequences of individual variation. *Oecologia* 59:232–38.

Possingham, H. P., A. I. Houston, and J. M. McNamara. 1990. Rick-averse foraging in bees: Comments on the model of Harder and Real. *Ecology* 71:1622–24.

Rachlin, H., A. W. Loque, J. Gibbon, and M. Frankel. 1986. Cognition and behavior in studies of choice. *Psychological Review* 93:33–45.

Rachlin, H., A. Raineri, and D. Cross. 1991. Subjective probability and delay. *Journal of the Experimental Analysis of Behavior* 55:233–44.

Real, L. A. 1980a. Fitness, uncertainty, and the role of diversification in evolution and behavior. *American Naturalist* 115:623–38.

Real, L. A. 1980b. On uncertainty and the law of diminishing returns in evolution and behavior. In J. E. R. Staddon, ed., *Limits to action*, 37–64. New York: Academic Press.

Real, L. A. 1981. Uncertainty and pollinator-plant interactions: The foraging behavior of bees and wasps on artificial flowers. *Ecology* 62:20–26.

Real, L. A. 1987. Objective benefit versus subjective perception in the theory of risk-sensitive foraging. *American Naturalist* 130:399–411.

Real, L. A. 1990a. Predator switching and the interpretation of animal choice behavior: The case for constrained optimization. In R. N. Hughes, ed., *Behavioral mechanisms in diet selection*, 1–21. Heidelberg: Springer-Verlag.

Real, L. A. 1990b. Search theory and mate choice. I. Models for single-sex discrimination. *American Naturalist* 136:376–404.

Real, L. A. 1991. Animal choice behavior and the evolution of cognitive architecture. *Science* 253:980–86.

Real, L. A. 1993. Toward a cognitive ecology. *Trends in ecology and evolution* 8:413–17.

Real, L. A., and R. Caraco. 1986. Risk and foraging in stochastic environments: Theory and evidence. *Annual Review of Ecology and Systematics* 17:371–90.

Real, L. A., and S. Ellner. 1992. Life-history evolution in stochastic environments: A graphical mean-variance approach. *Ecology* 73:1227–36.

Real, L. A., and B. J. Rathcke. 1991. Individual variation in nectar production and its effects on fitness in *Kalmia latifolia*. *Ecology* 72:149–55.

Real, L. A., J. R. Ott, and E. Silverfine. 1982. On the trade-off between the mean and variance in foraging: Effects of spatial distribution and color preference. *Ecology* 63:1617–23.

Real, L. A., S. Ellner, and L. D. Harder. 1990. Short-term energy maximization and risk-aversion in bumblebees: Comments on Possingham et al. *Ecology* 71:1625–28.

Reboreda, J. C., and A. Kacelnik. 1991. Risk sensitivity in starlings: Variability in food amount and food delay. *Behavioral Ecology* 2:301–8.

Rescorla, R. A., and A. R. Wagner. 1972. A theory of Pavlovian conditioning: Variations on the effectiveness of reinforcement and non-reinforcement. In A. H. Black and W. F. Prokasy, eds., *Classical conditioning*, vol. 2, *Current research and theory*, 64–99. New York: Appleton Century Crofts.

Richards, R. J. 1987. *Darwin and the emergence of evolutionary theories of mind and behavior*. Chicago: University of Chicago Press.

Roitblat, H. L. 1982. The meaning of representation in animal memory. *Behavioral and Brain Sciences* 5:353–406.

Roitblat, H. L. 1987. *Introduction to comparative cognition*. New York: W. H. Freeman and Co.

Rosett, R. N. 1971. Weak experimental verification of the expected utility hypothesis. *Review of Economic Studies* 38:481–92.

Rozin, P., and J. Schull. 1988. The adaptive-evolutionary point of view in experimental psychology. In R. Atkinson, R. J. Herrnstein, G. Lindsey, and R. D. Luce, eds., *Handbook of experimental psychology: Motivation*, 1–40. New York: John Wiley.

Rumelhart, D. E. 1989. The architecture of mind: A connectionist approach. In M. I. Posner, ed., *Foundations of cognitive science*, 133–60. Cambridge, Mass.: MIT Press.

Rumelhart, D. E., and J. L. McClelland. 1986. *Parallel distributed processing: Explorations in the microstructure of cognition*. Vols. 1 and 2. Cambridge, Mass.: MIT Press.

Sherry, D. F., and D. L. Schacter. 1987. The evolution of multiple memory systems. *Psychological Review* 94:439–54.

Shettleworth, S. J. 1983. Function and mechanism in learning. *Advances in Analysis of Behavior* 3:1–37.

Shettleworth, S. J. 1988. Foraging as operant behavior and operant behavior as foraging: What have we learned? In G. Bower, ed., *The psychology of learning and motivation: Advances in research and theory*, vol. 22. New York: Academic Press.

Shettleworth, S. J., J. R. Krebs, D. W. Stephens, and J. Gibbon. 1988. Tracking a fluctuating environment: A study in sampling. *Animal Behaviour* 36:87–105.

Shoemaker, P. 1982. The expected utility model: Its variants, purposes, evidence, and limitations. *Journal of Economic Literature* 20:529–63.

Staddon, J. E. R., ed. 1983. *Limits to action*. New York: Academic Press.

Staddon, J. E. R., and R. H. Ettinger. 1989. *Learning*. New York: Harcourt Brace Jovanovitch.

Stephens, D. W. 1981. The logic of risk-sensitive foraging preferences. *Animal Behaviour* 29:628–29.

Stephens, D. W. 1987. On economically tracking a fluctuating environment. *Theoretical Population Biology* 32:15–25.

Stephens, D. W., and E. L. Charnov. 1982. Optimal foraging: Some simple stochastic models. *Behavioural Ecology and Sociobiology* 10:251–63.

Stephens, D. W., and J. R. Krebs. 1986. *Foraging theory*. Princeton: Princeton University Press.

Sutton, R. S., and A. G. Barto. 1900. Time-derivative models of Pavlovian reinforcement. In M. Gabriel and J. Moore, eds., *Learning and computational neuroscience: Foundations of adaptive networks*, 497–538. Cambridge, Mass.: MIT Press.

Swartz, K. B., S. Chen, and H. S. Terrace. 1991. Serial learning by rhesus monkeys: I. Acquisition and retention of multiple four-item lists. *Journal of Experimental Psychology: Animal Behavior Processes* 17:396–410.

Taylor, A., and A. Prochazka, eds. 1981. *Muscle-receptors and movement*. New York: Macmillan.

Terrace, H. S. 1987. Chunking by a pigeon in a serial learning task. *Nature* 325:149–51.

Terrace, H. S. 1991. Chunking during serial learning by a pigeon: I. Basic evidence. *Journal of Experimental Psychology: Animal Behavior Processes* 17:81–93.

Terrace, H. S., and S. Chen. 1991. Chunking during serial learning by a pigeon: III. What are the necessary conditions for establishing a chunk? *Journal of Experimental Psychology: Animal Behavior Processes* 17:107–18.

Turelli, M., J. H. Gillespie, and T. W. Schoener. 1982. The fallacy of the fallacy of the averages in ecological optimization theory. *American Naturalist* 119:879–84.

von Neumann, J., and O. Morganstern. 1944. *Theory of Games and Economic Behavior*. Princeton: Princeton University Press.

Wasserman, P. D. 1989. *Neural computing*. New York: Van Nostrand Reinhold.

Wilson, B., N. J. Mackintosh, and R. A. Boakes. 1985. Transfer of relational rules in matching and oddity learning in pigeons and corvids. *Quarterly Journal of Experimental Psychology, B*, 37:313–32.

Wuketits, F. M. 1990. *Evolutionary epistemology*. Albany, N.Y.: State University of New York Press.

Wunderle, J. M., M. Santo Castro, and N. Fetcher. 1987. Risk-averse foraging by bananaquits on negative energy budgets. *Behavioural Ecology and Sociobiology* 21:249–55.

Yaari, M. 1965. Convexity and the theory of choice under risk. *Quarterly Journal of Economics* 79:278–90.

Yaari, M. 1969. Some remarks on measures of risk aversion and their uses. *Journal of Economic Theory* 1:315–29.

SIX

Optimizing Learning and Its Effect on Evolutionary Change in Behavior

Daniel R. Papaj

AT LEAST since the time of Lamarck, there has been interest in discovering how learning (or more generally, plasticity) might influence evolutionary change. Among the more successful efforts were those of Baldwin (1896), Morgan (1896), and Osborn (1896), who collectively erected a theory of "organic selection" in which phenotypic "accommodation" ("plasticity" in the modern vernacular) facilitated the evolution of congenitally expressed traits through natural selection. In their view, plasticity permitted a population to persist in a new environment long enough for latent genetic variation to be acted upon by selection and for congenital responses (i.e., instincts) to arise. The key elements of this process as presented by Morgan were as follows:

1. A population is placed in a new environment.
2. Selection acts on existing plasticity (e.g., learning) to permit the population to persist in the new environment.
3. Latent genetic variation in congenital responses is expressed (or new genetic variation in those responses arises).
4. Selection gives rise to the appropriate congenital responses.

Since the turn of the century, the issue has arisen now and again (Wright 1931; Waddington 1953; Ewer 1956; Haldane and Spurway 1956; Tierney 1986); currently it is being considered most earnestly by students of artificial intelligence (AI) and robotics (Hinton and Nowlan 1987; Nolfi et al. 1990; Cecconi and Parisi 1991).

In the first part of this chapter, I critically examine the alternative notions that learning, on one hand, facilitates evolutionary change and, on the other, inhibits it. I review models, including my own (Papaj 1993a), that illustrate how learning—and particularly learning of an adaptive sort—can influence evolutionary change in behavior. In the second part, I use results of these models to generate a hypothesis (Papaj 1993b) that optimal behavior in extant taxa is the manifestation of optimizing mechanisms that evolved in ancestral taxa and does not reflect the recent action of natural selection.

HOW LEARNING GUIDES EVOLUTIONARY CHANGE

Learning Is Adaptive

Thorpe (1956) defined learning as "that process which manifests itself by adaptive changes in behavior as a result of experience." The definition was endorsed by ethologists as prominent as Konrad Lorenz (cf. Lorenz 1965) and, at the same time, criticized by a number of anthropologists and behavioral biologists (Hailman 1985; Goodall 1986; Papaj and Prokopy 1989). These critics argued that declaring a priori that learning was adaptive undermined efforts to evaluate the evolution of learning.

The objection is compelling if the term "adaptive" is taken to refer to behavioral change "that serves a definable function and has evolved under the action of natural selection" (Alcock 1989). This is the sense in which the term is most often used by contemporary biologists. However, Lorenz and possibly even Thorpe used the term "adaptive" in another sense. Lorenz meant only that individual behavior after learning had higher survival value in the current environment than did behavior before learning (cf. Lorenz 1965). In other words, animals generally learn to do what it is sensible to do. Animals generally learn, for example, to orient toward—not away from—stimuli that are rewarding and away from—not toward—stimuli that are punishing. Learning, especially associative learning, almost always seems to adapt the animal to its current environment. This sense for the term "adaptive" is one used often by psychologists (cf. Staddon 1983).

Adaptive Properties of Learning and the Evolution of Instinct

Saying that learning generally increases the fitness value of current and future behavior is not the same as saying that learning evolved under natural selection. The conditions under which learning will be favored by selection have been addressed by a number of authors (Johnston 1982; Stephens 1987, 1993). The difference between "adaptive" as used by Lorenz and "adaptive" as used by these authors can be summarized as follows: the latter are interested primarily in how learning evolves as a consequence of natural selection, the former primarily in how behavior evolves (in the more general meaning of that word) as a consequence of experience. The root cause of this semantic confusion seems to be the tendency for effects of experience to mimic effects of natural selection: animals often appear to alter their behavior over their lifetimes in the manner in which we presume it would be altered by natural selection over evolutionary time.

For the remainder of the chapter, I will distinguish between adaptiveness that arises over an individual's lifetime and that which arises over an evolutionary timescale. The distinction is a critical one because, as I argue below, adaptive evolutionary change in behavior may be influenced greatly by the degree to which behavior changes in adaptive ways over an individual's lifetime. While the following argument is set forth in terms of an animal foraging for essential resources (specifically, host selection behavior in insects, a focus of my own research), it should apply equally well to other kinds of behavior.

How Learning Inhibits Evolutionary Change: A Simulation Model

Suppose a population of herbivorous insects initially exploits a variety of host plant species, each of which is unpredictable in abundance or quality over time or space. Selection has presumably favored insects that have a weak congenital predisposition to find and use each potential species as well as an ability, through learning, to adopt the optimal preference for the species that yields the highest fitness. Suppose the environment suddenly changes such that one and only one host species is available. Assuming that learning is costly in predictable environments (Johnston 1982; Papaj and Prokopy 1989; Alcock 1993), there will be selection for the immediate expression of fully functional behavior, that is, for instinctive (congenitally expressed) behavior of a functional sort. The following simulation model shows that

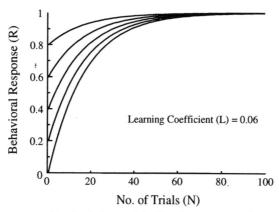

$$R = I + \left[(1 - I)\left(1 - e^{-NL}\right) \right]$$

R = Behavioral Response

I = Instinct Coefficient

L = Learning Coefficient

N = Number of Trials

Fig. 6.1 Algorithm assumed in simulation model. Also shown are learning curves generated by algorithm for various values of *I* with $L = 0.06$.

the extent to which such congenital responses evolve depends critically on how quickly insects learn.

The basic algorithm used in these simulations is a modification of one constructed by Jaenike (fig. 6.1; Jaenike and Papaj 1992). A behavioral response (R) has an instinctual component (denoted by I, the instinct coefficient) and a learned component. The learned component is a function of I, but also of a learning coefficient (L) and the number of experiences (N) that modify the response. This algorithm generates a learning curve similar to those commonly observed in animal learning studies (fig. 6.1). All individuals with L greater than 0 eventually adopt a response of value equal to 1. Because the model assumes that this value is optimal in terms of fitness, this learning is an "optimizing" learning.

At the outset of a simulation run, either I or L or both may vary genetically. Where either coefficient is genetically variable, heritability is always set at a value of 0.50 (i.e., half of the variation in either trait has a genetic basis). Population sizes are set at 1,000. In a typical simulation, selection acts generation by generation on mean response and molds any available genetic variability in instinct and learning coefficients. The fitness function on which selection is based is an inverted parabola truncated at 0 (fig. 6.2). All individuals gain the same amount of experience over their lifetimes (specifically, 100 trials) and contribute progeny in direct proportion to the fitness associated with their particular behavioral response profile.

In the first set of simulations, I asked whether learning influenced the evolution of a wholly congenital response to a host species in a predictable environment where

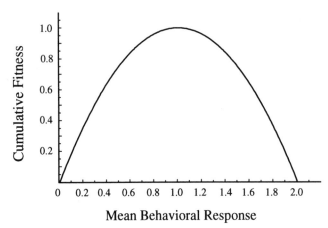

Fig. 6.2   Fitness function assumed in simulations of instinct evolution: Fitness = $F_{max} - [c(1 - R)^2]$ where F_{max} = maximum fitness = 1, c is a flatness parameter equal to 1, and R is the behavioral response.

such a response would be favored. In this simulation, only the congenital response was genetically variable. The ability to learn, if any, was identical for all individuals. The results are shown in figure 6.3A. In the absence of learning, congenital responses (expressed in terms of I, the instinct coefficient) evolve in a negatively accelerating fashion until individuals express the optimal response congenitally. In the presence of learning, congenital responses evolve more slowly. How much more slowly they evolve depends on how fast individuals learn. Fast learning (denoted by high values of L, the learning coefficient) slows the rate of evolution of instinctive behavior more than slow learning. When learning is relatively rapid ($L = 0.05$), evolution of a congenital response is virtually arrested. Though rapid (figure 6.3B), learning at this point is far from single-trial learning. In fact, rates of learning of this magnitude are considerably lower than those reported for insects (Menzel et al. 1974; Lewis 1986; Dukas and Real 1991).

Learning suppresses the evolution of congenital responses by permitting any individual, regardless of its genetic constitution, to adopt the optimal behavior in the current environment (in this case, an optimal response to the only available plant species). Learning makes all individuals more similar in overall fitness than if they did not learn (fig. 6.4). When individuals learn rapidly, even relatively large genetic differences in congenital response are associated with negligible differences in overall fitness. Under selection, evolutionary change is correspondingly negligible.[1] This potential for learning to forestall its own demise may explain why we often find learned responses where we expect instinctive ones. For example, we often find evidence of learning in feeding or egg-laying behavior by specialist insects even though learning would seem unnecessary and even disadvantageous for such insects (Papaj 1986; Papaj and Prokopy 1989). In fact, arguments below suggest that specialist insects might learn *better* than generalist ones.

1. In this way, learning can preserve genetic variation for long periods of time. In fact, one might test the notion that learning arrests evolutionary change by looking for positive associations across species between learning ability and genetic variation in congenital responses (J. Jaenike, pers. comm.).

A. *Evolution of instincts when animals learn*

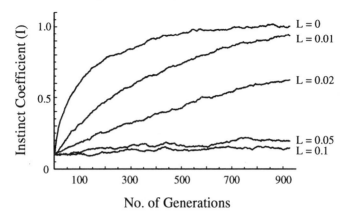

No. of Generations

B. *Learning curves assumed in simulations*

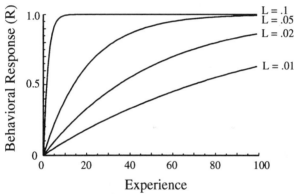

Experience

Fig. 6.3 Results of simulation in which only *I* is genetically variable. *L* is varied in successive simulation runs between 0 and 0.1. (*A*) Evolutionary trajectory of mean congenital response, denoted by instinct coefficient, *I*. (*B*) Corresponding learning curves associated with values of *L* assumed in simulation.

Thus far we have assumed that only congenital responses are heritable. However, it is reasonable to assume that learning is heritable as well (cf. Brandes and Menzel 1990). In that case, the response to selection in a predictable environment might conceivably take either of two forms. As before, selection might favor a stronger congenital ability to find and use the available host species (fig. 6.5A). In addition, selection might favor individuals that learn faster to find and use that species (fig. 6.5B).[2]

These alternative responses to selection are not mutually exclusive. Given genetic variation in both learning ability and congenital response, selection in predict-

---

2. Elsewhere (Papaj 1993a), I address the possibility that instincts arise when individuals evolve to learn so fast that virtually no experience is needed.

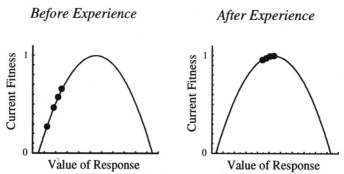

Before Experience *After Experience*

Fig. 6.4 Diagram showing how optimizing learning moves individuals to optimal responses in terms of current (and future) fitness. Individuals are represented as points on a current fitness function. Experience causes individuals to converge on the optimal response, regardless of initial congenital differences.

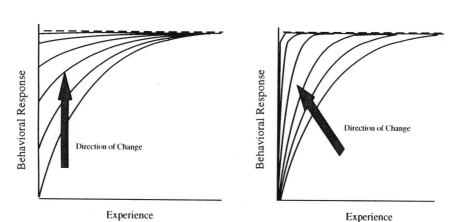

A. *Evolution of congenital response*   B. *Evolution of faster learning*

Fig. 6.5 Alternative routes to evolution of instinct. (*A*) Evolution of congenital response. (*B*) Evolution of faster learning. In each figure, the dashed line indicates the value of the optimal behavioral response in a predictable environment.

able environments might simultaneously favor both congenital responses and faster learning. We might ask if one or the other effect predominates. The answer is yes. In another simulation, I assumed that all individuals learned and that there was genetic variation in both congenital responses and learning ability; all other conditions were as described above. The results, shown in figure 6.6A, indicate that both congenital responses and faster learning evolve rapidly at first. However, learning soon evolves to a rate at which evolution of the congenital response is essentially arrested. At this point, the mean congenital response in the population is rather weak. Learning, by contrast, is relatively rapid (fig. 6.6B).[3]

---

3. Learning at this point is reminiscent of a common form of learning called alpha-conditioning, in which a preexisting response is heightened by experience (see Marler and Terrace 1984 for discussions of alpha-conditioning).

A. *Evolution of Instinct and Learning Coefficients*

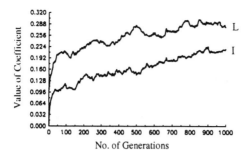

B. *Evolution of Behavioral Response Function*

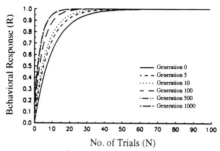

Fig. 6.6   Results of simulation in which both *I* and *L* are genetically variable. *L* in generation 0 was equal to ca. 0.06. (*A*) Evolutionary trajectory of mean values of *I* and *L*. (*B*) Change in mean behavioral response function, composed of congenital responses and learning curve. Selection causes only a slight change in mean congenital response, while causing a conspicuous change in the mean rate at which individuals learn.

In the previous simulation, an ability to learn was an initial or ancestral condition, as supposed by early naturalists. In a final simulation (Papaj 1993a), I assumed that individuals originally had virtually no learning ability and only a very weak congenital response, but that there was genetic variability in both traits. In this situation, I found exactly the same patterns depicted in figure 6.6. Once again, both congenital response (*I*) and learning ability (*L*) initially increased in value, and once again, learning quickly evolved to a rate at which further evolution of congenital responses was arrested. Again, the congenital response at this point was far less than 1, and again, learning at this point was rather rapid.

The reason that evolution of fast learning dominates that of congenital responses is that *L* has an exponential effect on the value of *R* while *I* has only an additive one. A change of a given magnitude in the value of *L* generally thus causes a much greater change in *R* than does a change of the same magnitude in *I*. Even though both coefficients are equally heritable, initial changes in *L* have larger effects on behavior and fitness than similar changes in *I*, and so learning ability appears to dominate congenital response. Whether this is biologically realistic—whether changes at loci affecting learning have larger effects on behavior than changes at loci affecting congenital responses—is unknown. In this instance, information about mechanism, specifically about how genes specify traits, would strengthen inferences about the evolution of behavior.

The Effect of Number of Experiences

While genetic in nature, the model lacks genetic realism in some respects. For example, individuals reproduce asexually, yet genetic variation in particular traits (if any) is maintained at a constant level even in the face of strong selection. This amounts to assuming an extraordinarily high mutation rate that compensates perfectly each generation for the loss of genetic variation due to selection (although this process is not considered explicitly in the model). Recently, a student, Kees Swaans, and I built a more realistic model in which diploid individuals reproduce sexually. Both congenital responses and learning ability are assumed to have a polygenic basis that is explicitly represented in simulated meiosis. Selection is again stabilizing, but the fitness function is Gaussian in form. In addition, an algorithm based on the single linear operator model of learning (a model that is widely accepted in the learning literature: Atkinson et al. 1965; Boyd and Richerson 1985) is substituted for the one used above. As in the previous model, generations are discrete and population size is held constant at 1,000.

Despite these changes, our principal results were consistent with those outlined above and were robust over a range of selection intensities. The model yielded other results worth noting. For example, we varied the intensity of selection and discovered that, under intense selection, learning permitted genetic variation in congenital responses to persist where it otherwise would have been depleted by drift before optimal values for the trait were achieved. This result implies that, under some selective regimes, learning can actually permit congenital responses to evolve to values closer to the optimum than they would in the absence of learning.

We also manipulated the number of experiences (N) occurring over an individual's lifetime. As expected, we found that, if N is small enough, congenital responses evolve as fast as does learning ability (fig. 6.7). Of greater interest were the findings that learning ability evolves fastest for intermediate values of N and that, as N increases, the degree to which congenital responses evolve falls to almost negligible levels but then begins again to increase. Key to these results is the assumption that the learning curves of all genotypes share a common asymptote (i.e., every genotype eventually adopts the optimum behavior and differs from any other only with respect to how fast that optimum is attained). At large N, all individuals find the optimum and remain there for a substantial portion of their lives. Differences in relative fitness due to differences among individuals in the rate at which the optimum is reached are thus diminished. Learning ability evolves more and more slowly, and learning inhibits less and less the evolution of congenital responses. As a consequence, congenital responses evolve slightly more rapidly at large N than at intermediate N.

How Learning Facilitates Evolutionary Change: A Neural Network Model

The tendency for learning to inhibit genetic change in congenitally expressed behavior was also obtained in a model of genetic transmission of behavior put forth by Boyd and Richerson (1985). In their words, "as the organism's skill at moving towards the optimum increases, selection for a good a priori guess about the environment decreases" (121). The notion that learning might inhibit the evolution of congenital responses, though, is seemingly at odds with the conclusions of Morgan (1896), Baldwin (1896), and Osborn (1896) that plasticity facilitates their evolution.

The crux of the conflict between the results of the above simulations and the scheme of Morgan and colleagues lies in assumptions about initial conditions. In the simulations, at least some genotypes in the population have nonzero fitness

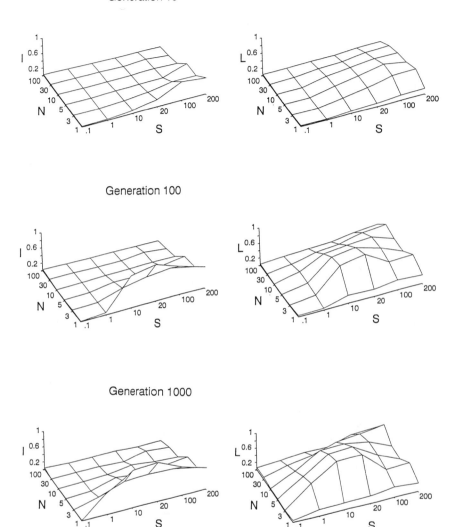

Fig. 6.7 Effects of number of experiences (*N*) and selection intensity (*S*) on the evolution of congenital response (*I*) and learning ability (*L*) in a genetic model incorporating meiosis. Over low to moderate selection intensities, *L* rises with *N* and then falls. Conversely, over the same intensities, *I* falls, then rises slightly with *N*.

even in the absence of learning. In Morgan's formulation in particular, individuals that do not learn (or do not learn well enough) have no fitness at all in the new environment. Genetic variation in congenital response in the population is absent (or latent), having been removed (or suppressed) under past selective regimes. If at least some individuals cannot learn the appropriate responses, the entire population goes extinct. Learning in the situation envisaged by Morgan permits the population time enough to generate (through mutation, recombination and migration) individuals that would have nonzero fitness even in the absence of learning.

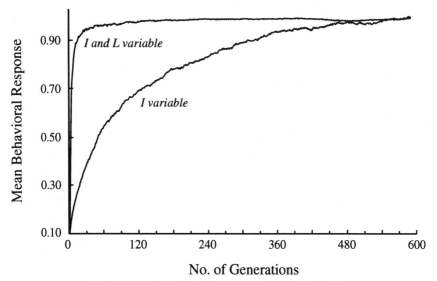

No. of Generations

Fig. 6.8   Effect of genetic variation in L on evolution of R.

The premise that individual learning might prevent populations from going extinct is a reasonable one. However, the simulations above suggest that whether congenital responses will evolve in the face of learning depends on how well animals learn initially and how readily selection alters learning ability. Where learning itself is heritable, congenital responses may be suppressed even in a predictable environment and even where animals are not initially able to learn the appropriate response.

While learning does not enable change in congenital responses directly, it can enable change in the overall mean behavioral response. Provided learning is heritable, it increases the rate at which an animal's overall mean response evolves. Fig. 6.8 depicts the evolution of mean behavioral responses under conditions in which L is or is not heritable. The rate at which the mean behavioral response evolves is clearly greater when there is potential for evolutionary change in learning ability.

What we do in assuming genetic variation in L (as well as I) is to increase effectively the number of loci specifying R at which evolutionary change can occur. It is thus not surprising that R evolves at a faster rate when learning is heritable. A similar outcome is obtained for more interesting reasons in a model developed by Nolfi et al. (1990). In this model, an organism (actually an *animat*) is asked to find food located at random positions on a rectangular grid. The animat's moves are decided by a neural network whose decisions depend on the type and strength of connections between nodes in the network (hence the term "connectionist model" applied to this and similar models).

The neural networks used by Nolfi et al. in simulations have three layers of nodes: an input layer, an output layer, and a single intermediate "hidden layer" (fig. 6.9A). In the input layer, two nodes are dedicated to sensory input (one encoding the angle of the nearest food item and one encoding the distance to the nearest food item). The other two nodes encode which of four possible actions (stop, right, left, forward) was last executed by the network. The two nodes in the output layer encode the next action of the network.

A. Hard-Wired Network

B. Flexible Network

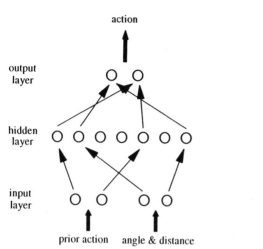
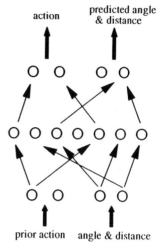

Fig. 6.9   Neural networks employed in the Nolfi et al. (1990) models of learning and evolution. (*A*) Hardwired network. (*B*) Flexible network. (Adapted from Nolfi et al. 1990.)

Each node in a lower layer has a connection of a given weight (which can vary between 0 and 1) with each node in the next higher level. At the outset of a simulation, weights are assigned at random and the sensory nodes in the input layer are activated. As in all connectionist models, nodes are activated (or fire) on the basis of an activation rule relating input to output. The activation rule in the Nolfi et al. model is sigmoidal in form (i.e., output increases sigmoidally with input). As is typical in a connectionist model, activation always moves up the network (never down).

In one simulation, networks were permitted to evolve under a kind of truncation selection in which the top 20% of networks in the population (in terms of foraging effectiveness) reproduced in an asexual fashion. Weights on the connections were equivalent to continuously varying alleles at loci; mutation at those loci was simulated by taking 5 of the 40 weights and tweaking their values in a random fashion over a specified range. The resulting evolutionary trajectory in foraging effectiveness is shown in figure 6.10: over successive generations, the animat population got better and better at finding food in a negatively accelerating fashion.

A second simulation was run with a type of network similar in all respects to the first except that it learned (fig. 6.9B). In order to teach the network, two nodes that define predicted sensory input after a move is made were added to the output layer. As an animat with such a network moved through the grid in search of food, actual sensory input was compared with predicted input and errors calculated. Adjustments to weights were made in response to errors (the principal changes were made to connections between hidden-layer nodes and prediction nodes). Because these adjustments propagate backward through the network, this technique for teaching a neural network is called "back propagation of error" (Rumelhart et al. 1986). Back propagation of error constitutes a popular means in the AI literature by which networks are modified by experience.

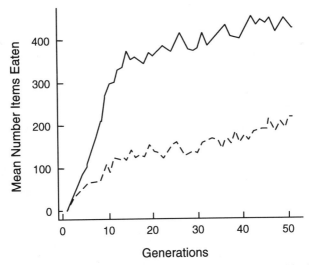

Fig. 6.10   Evolution of neural networks in the Nolfi et al. model. Evolution of flexible networks is indicated by solid line; evolution of the hardwired networks, by the dashed line. (Adapted from Nolfi et al. 1990.)

Figure 6.10 shows the results of simulated selection on such a learning network. The effectiveness with which the animat finds food in the grid evolves at a faster rate and reaches a higher asymptote when the network learns than when it does not. Nolfi et al.'s results are explained in part by evolution of better and faster learning and in part by evolution of congenital responses (fig. 6.11). In this sense, their results are consistent with my own simulations. The tendency for changes in congenital responses (i.e., the starting weights assigned to connections in the network) to become arrested long before peak response is achieved is also consistent with my model.

However, the way learning guides both the initial rate of change in foraging effectiveness and its asymptote owes part of its explanation to the unique way in which error is propagated and behavior altered. Over a wide range of initial weights on connections in the network, back propagation of error usually, but not always, leads to a global minimum of error (a global optimum in behavior). Nolfi et al. argue that the exceptions—those instances in which the network is trapped in local error minima (local fitness maxima)—are of critical importance to the way learning guides evolutionary change. Their argument is summarized briefly below.

Consider an N-dimensional space in which each dimension represents a connection in the network (in this case, 56 dimensions). On that space lie all possible weights (from 0 to 1) of these connections. Add to that space a dimension $N + 1$, which defines the error (or fitness, the inverse of error) associated with each possible combination of weights. The $N + 1$ dimensional space represents an error (fitness) landscape. Moreover, this fitness surface is not a single peak, but a landscape of peaks of different heights. Learning, because it is an adaptive learning, moves the animat around the landscape in essentially the manner in which natural selection will move the animat. Since selection takes place *after* learning, adaptive learning permits selection to "see" which animats have a good local landscape.

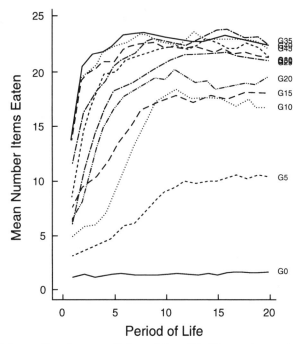

Fig. 6.11   Selected mean "learning curves" of population of flexible neural networks over evolutionary time. Notation to right of curves (G0 . . . G50) reflects number of generations since beginning of simulation. (Adapted from Nolfi et al. 1990.)

Consider, for example, two points (i.e., two genotypes) on the landscape that start out at the same fitness value but which are located on different hills with different peak elevations. Assume that learning moves each genotype up its respective peak. The genotype that ends up at the higher elevation will be represented in greater proportion in the next generation. That genotype may end up at a higher elevation either because its initial weights conferred upon it a better learning ability (so that it moves faster up the hill) or because its initial weights conferred upon it the opportunity to evolve to a higher maximum fitness. In the latter case, learning has essentially acted to screen genotypes with respect to their potential for future evolutionary change.

This argument for how learning facilitates adaptive evolution is valid so long as learning and natural selection lead to the same outcome. So long as the action of experience within generations works as a metaphor for selection across generations, learning will tend to facilitate adaptive evolutionary change. One instance in which we might expect a different outcome is when the problem that the animat is asked to solve changes across generations. Suppose, for example, the animat is asked to solve an optimal diet choice problem. Suppose that an exclusive preference for diet item A is optimal in generation 1, an exclusive preference for diet item B is optimal in generation 2, a preference for item A is optimal in generation 3, and so on. In that situation learning might facilitate evolutionary change, but of a maladaptive sort. In generation 1, genotypes that learn faster to prefer diet item A will be favored; in generation 2, the mean genotype may well be less adapted to an environment in which diet item B is optimal than if the animat did not learn. This example is extreme, but illustrates how dependent is the conclusion that learning facilitates adaptive evolutionary change (currently a widely accepted conclusion in the literature; cf. Hinton and Nowlan 1987; Maynard Smith 1987) on a particular set of environmental conditions.

Costs of Learning

Both the phenomenological and the neural network models make an assumption that is critical to the observed tendency for learning to facilitate evolutionary change even in predictable environments, namely, that learning has no cost except in time wasted in adoption of the optimal response. In the neural network model, for example, the two additional nodes found in the learning network have no costs in terms of reproduction. Similarly, computations made using the back propagation algorithm (computations on which changes in behavior are entirely based) cost nothing. The connectionist models are particularly likely to mislead us in this regard, as these computations might appear to take place outside the neural network itself. Realistically, one might expect some nontrivial amount of neural tissue to be devoted to this task. In either case, neural costs associated with learning would be expected to reduce learning's tendency to facilitate evolutionary change.

LEARNING AND THE EVOLUTION OF OPTIMAL BEHAVIOR

Optimal Behavior and the Recency of Genetic Change

I argued above that (1) learning of an adaptive form will evolve even for simple tasks in predictable environments, and that (2) once evolved, such learning can retard adaptive evolutionary change in behavior by mimicking the effects of selection. The capacity of learning to put adaptive evolutionary change in check prompts the following question: Is optimal behavior in extant species a consequence of the incessant and recent action of natural selection, or is it the result of adapting, or even optimizing, mechanisms evolved in ancestral species?

Although sound reasons for believing that natural selection should not produce optimal behavior have been advanced (Gould and Lewontin 1979; Pierce and Ollason 1987), behavior often seems to be optimal or nearly optimal in a wide range of contexts. Certainly many students of behavior share this perspective (cf. Boyd and Richerson 1985; Krebs and Davies 1993). This widespread pattern of adaptive or optimal behavior is often taken (implicitly or explicitly) as an indication of the ongoing and prevalent action of selection on behavior (cf. Alcock 1993). However, such patterns have an alternative explanation (Papaj 1993b). It is possible that extant species possess adapting or even optimizing behavioral mechanisms that arose early on in the history of a particular group, and that these mechanisms have been commandeered in more or less their original form to solve various problems as environments and organisms have changed. The expression of optimal or nearly optimal behavior in new species or by old species in new environments may have required little or no genetic change.

Assume that animals possess a single optimizing mechanism called O. O permits an animal to adopt optimal or nearly optimal behavior in the environment in which it currently finds itself; it is, in short, a problem-solving mechanism. O represents a system of neurological and hormonal processes that permit an animal to express behavior appropriate to a given environment. While these processes obviously include learning and motivation, behavioral plasticity does not require these processes. Even behavior fixed with respect to experience or deprivation can be plastic in the sense that different and possibly optimal behavior patterns may be expressed by a given genotype in different environments (Lorenz 1981).

On one hand, one might suppose that distinct optimizing mechanisms $O_1 \ldots O_n$ have arisen independently in recently evolved species. As taxa diversified and entered new ecological niches, optimizing mechanisms arose or, if already present, were refined extensively by selection. In this case, optimal or nearly optimal behavior in any given species reflects the recent action of natural selection.

On the other hand, O may have arisen early on in a particular group's history. As taxa diversified and species entered new ecological niches, O may have been implemented with essentially no genetic change to solve problems associated with those niches. Under this scenario, optimal behavior differs in form among species simply because the problems that the species are asked to solve differ themselves in form. The optimizing mechanism O is the same in all species. In this instance, optimal behavior in modern species does not reflect the recent action of optimizing selection. The ability to express optimal behavior evolved under selection early on and was simply retained by descendant species. Were we to infer that natural selection acted consistently across species to optimize behavior, we would be guilty of pseudoreplication (Rosenheim 1993). In this case, the behavior of all species in a particular group constitutes in essence just a single data point.

If the latter scenario is the more correct one, it would challenge the perspective that broad patterns of optimal behavior observed in extant species are evidence of the prevalent, enduring action of optimizing selection. In the extreme, optimal behavior in current circumstances would instead be nothing more than the manifestation of a mechanism that evolved under selection long ago. This alternative perspective is appealing in part because it allows the opinion that behavior is often optimal to be reconciled with the view of field ecologists and ecological geneticists that optimizing selection in nature is rare (Endler 1986; Travis 1989).

Nowhere in these arguments am I appealing to examples of nonadaptive behavior or invoking reasons to think that behavior will not be perfectly or nearly perfectly adapted (although such arguments are persuasive ones; cf. Gould and Lewontin 1979; Pierce and Ollason 1987). I am actually adopting an adaptationist perspective in postulating the origin and maintenance of an optimizing mechanism, O, designed to maximize individual fitness. However, I am also posing the possibility that O arose relatively long ago and that different modern species employ essentially the same O in ways that always seem to maximize fitness.

If true, this would mean that we as behavioral ecologists are frequently fleshing out the adapting properties of O without saying a great deal about how behavior has changed under selection. We may be engaged more in the study of a special "proximate" cognitive mechanism of behavioral flexibility (i.e., the workings of O) than of "ultimate" evolutionary causes of behavior. This last point is especially germane to the theme of this volume: not only might it pay to give more attention to mechanism, but many evolutionary ecologists may have been studying mechanisms all along. In that respect, the distinction between proximate and ultimate causes of behavior may be a hazy one.

Implications and Evidence

As I have noted elsewhere (Papaj 1993b), the alternative scenarios for the evolution of optimal behavior outlined above are not mutually exclusive, and the truth presumably falls somewhere between them. Below I list and evaluate two elements of the second scenario.

O or Elements of O Are Relatively Ancient. If optimal behavior in extant species has been achieved with little or no genetic change, it follows that behavioral plasticity of an optimizing form predates the taxon or taxa whose optimal behavior is under scrutiny. If we were, in particular, to suggest that optimal patterns of behavior in diverse phyla share a common underlying mechanism, then that mechanism must be very ancient.

One element of O is learning. Despite a long history of interest in the subject (Morgan 1896; Bitterman 1965, 1975, 1988), data bearing on phylogenetic aspects of learning are sparse. Nevertheless, there is some indication that two forms of learning, habituation and associative learning, are relatively ancient (Papaj 1993b). Habituation appears almost as soon as organisms acquire nerve nets (i.e., in the Cnidaria). The pattern for associative learning is more open to debate. It may or may not occur in the Cnidaria; it was, however, almost certainly present in ancestors of the acoelomates. We must be wary of inferences about homologies in learning among taxa until we know for certain that the same mechanism accounts for a given form of learning in different taxa (Smith 1993). Still, evidence to date suggests that learning arose almost as soon as did nervous tissue (even before, depending on one's definition of learning; cf. Staddon 1983) and that even associative learning was present in relatively primitive groups. Tierney (1986) has gone so far as to speculate that plasticity is an inherent property of nervous tissue and, hence, an inherent property of behavior. In short, at least one element of O appears to have been in place for a very long time.

O Can Be Used to Solve Novel Problems. Also implied in the idea that optimal behavior has been achieved in recent groups with little or no genetic change is an ability for existing O's to solve novel problems. As taxa diversify, or as old species are thrust into new environments, new problems arise. The second scenario implies that animals use extant optimizing mechanisms to solve these new problems; otherwise the mechanisms would be altered by selection. Do animals apply existing cognitive abilities to solve novel problems?

Although data are once again scarce, I believe that the answer to this question is yes. A classic example of an animal learning to solve a novel problem is that of blue tits in Europe, which learned to remove the caps from milk bottles and extract the cream within (Fisher and Hinde 1948). While this kind of novel problem solving apparently involved social learning, there are numerous other examples among the vertebrates that do not. Anyone who has been to the circus or attended a marine mammal show or watched a squirrel at a bird feeder has observed animals solving all sorts of problems that are, to varying degrees, novel ones. The capacity for parrots and other animals (Pepperberg 1987) to learn human sounds and syntax may reflect an ability to solve novel problems related to communication.

There are corresponding examples among the insects, although data for insects are even more meager than for vertebrates. Honeybees thrust by humans into alien habitats appear to have solved, probably through learning and perhaps without genetic change, problems associated with finding and extracting nectar from novel flower species. In some cases (e.g., alfalfa in North America), these flowers appear to be different in key respects from the ones to which honeybees have been exposed over evolutionary time. We might expect the same of native bees presented with novel flower species in the form of crop plants and ornamentals.

Similarly, herbivorous insects have solved problems associated with exploitation of plants in alien habitats to which the insects have been introduced or plants (usually crops or ornamentals) thrust into the insects' native habitat. Occasionally these plants are in systematic terms very different from native hosts and might thus constitute a novel problem for the herbivores. It is conceivable that optimal solutions to these problems have sometimes been obtained in part through behavioral plasticity and with perhaps less genetic change in behavioral traits than currently thought (Courtney et al. 1989; Jaenike and Papaj 1992). Much work needs to be done before we can conclude that this is so. However, it is worth noting that practitioners of integrated pest management are increasingly aware that behavioral flexibility may permit pest insects to circumvent our best efforts to reduce their fitness and control their numbers (Roitberg and Angerilli 1986).

Examples of such problem solving abound in the recent spate of reports on parasitic wasps that learn the odor of microhabitats in which their hosts live. The use of exotic flavors such as chocolate and vanilla in some of these experiments (Lewis and Tumlinson 1988; Lewis and Takasu 1990) may amuse us, but it also makes a point: these insects can learn to make associations between odors and host presence that they very likely never had cause to make over their evolutionary history. It is easy to suppose that, were a host species to expand its niche breadth or shift altogether to a novel microhabitat, its parasite might follow along with little or no genetic change in host-finding behavior. Novel problem solving obviously has interesting implications for coevolution between host and parasite or predator and prey.

A final example from the insects relates to kin recognition in the social Hymenoptera. The ability to recognize kin is thought to have facilitated the evolution of nepotistic behavior. In the social Hymenoptera, such recognition is frequently mediated by discrimination among odors, and such discrimination is usually learned at least in part. Hölldobler and Michener (1980) proposed that the ability of advanced eusocial species to learn to discriminate kin from nonkin was a consequence of a preexisting ability in solitary ancestors to learn odors associated with recognition of nest site entrances. An ability to use learning to solve novel problems may thus have facilitated the evolution of nepotism in eusocial insects.

Our own species provides the best examples of how existing problem-solving mechanisms have been applied by animals to solve novel problems. In humans, cultural change has greatly outstripped biological evolution. Nobody would argue that human behavior has evolved under natural selection in lockstep with cultural change, and yet human behavior remains principally adaptive in form (Boyd and Richerson 1985). Apparently, learning (insight and social learning in particular) helps us human beings today to solve all sorts of culture-based problems that never confronted us over the course of our biological evolution. Associative learning even helps us to solve effectively problems of a kind that natural selection could not otherwise solve; that is, problems that arise over the course of a single generation and never appear again. Does it require so much of a leap of faith to suppose that associative learning has been similarly commandeered (though perhaps in less conspicuous ways) in animals other than human beings? Behavioral ecologists place great faith in general solutions to specific optimization problems that can be applied to new problems as they occur. Would we really be surprised if nature endowed animals early on in their phylogenetic history with general problem-solving mechanisms that were assigned to novel problems materializing later in their history?

LEARNING, ADAPTATION, AND MECHANISM

In closing, I would like to relate the major conclusions of this chapter to the theme of this volume, namely, the role of mechanism in evolutionary ecology.

In the first part of the chapter, simulation models are used to illustrate how learning can both arrest and facilitate evolutionary change. The models are based on algorithms that generate learning curves resembling those obtained for real animals (cf. Dukas and Real 1991, who use a similar function to fit learning curves to data for bumblebees). These algorithms describe the phenomenon of learning, but certainly not its mechanism. The connectionist model of Nolfi and colleagues incorporates a neural mechanism and leads to similar results. How close networks of this type are to describing cognitive mechanisms, though, is a source of continuing controversy (rev. Zeidenberg 1990). Implicit in this controversy is skepticism that the conclusions of connectionist models will be robust to improvements in "neural realism." Such skepticism provides reason enough to involve neural mechanisms in the study of the evolution of behavior.

In addition to neural realism, we need to consider further the genetic realism of our models, and so we need information on the genetic mechanisms underlying behavior. My models assume, for example, that genetic variation in congenital responses is wholly independent of genetic variation in learning ability. Studies in quantitative genetics could do much to assess the plausibility of such assumptions.

In the second part of the chapter, I suggest that behavioral ecologists may be more often studying the nuts and bolts of an optimizing mechanism than the outcome of recent selection on behavior. I am not maintaining that optimal behavior in extant taxa is never a consequence of recent selection. Rather, I regard the proposition that optimal behavior in extant taxa may have been achieved with a minimum of genetic change and in the absence of strong selection as a null hypothesis. Since genuine progress in science occurs roughly at the rate at which null hypotheses are rejected, the possibility that optimizing mechanisms have been co-opted for purposes very different from those for which they were selected should be considered seriously by evolutionary ecologists.

Such consideration should prompt a search for a better understanding of this mechanism termed O. Current explorations of animal cognition (Real 1991; Real, chap. 5) are legitimate steps in this direction. However, more effort is needed. We need to investigate further the ability of animals to solve novel problems with existing cognitive abilities. We need to determine to what extent animals solve the range of existing problems using the same cognitive mechanisms. The arguments above imply, for example, that animals will come to use similar or even the same cognitive mechanisms to solve very different problems. With respect to learning, this inference is supported by studies of *Drosophila* mutants (Hall 1985), as well as by selection experiments on honeybees (Brandes and Menzel 1990) in which a genetic change in learning on one task is usually paralleled by a change in learning on other (sometimes very different) tasks. This inference is also consistent with recent efforts to define neural networks that compute means and variances (Real, unpubl.). It is easy to imagine that a network that evolves to compute sample statistics in, say, a diet choice task could be co-opted for computation of similar statistics in, say, a mate choice task.

Finally, the possibility that behavior is generally anticipatory—that is, that behavioral mechanisms have evolved to anticipate new situations and to adapt to them with relatively little genetic change—should force us to integrate the study of mechanism with phylogenetic approaches. Knowledge of both mechanism and history is key to defining the extent to which current patterns of optimal behavior are the consequence of optimizing mechanisms that evolved under past selective regimes. The current proliferation of molecular and theoretical techniques for phylogeny reconstruction (Maddison 1990; Brooks and McLennan 1991; Rosenheim 1993) should be of great help in this regard.

ACKNOWLEDGMENTS

I thank Ann Hedrick, John Jaenike, Bill Mitchell, Peter Smallwood, and Bill Wcislo for useful discussions.

REFERENCES

Alcock, J. 1989. *Animal behavior: An evolutionary approach.* 4th ed. Sunderland, Mass.: Sinauer Associates.

Alcock, J. 1993. *Animal behavior: An evolutionary approach.* 5th ed. Sunderland, Mass.: Sinauer Associates.

Atkinson, R. C., G. H. Bower, and E. J. Crothers. 1965. *An introduction to mathematical learning theory.* New York: Wiley.

Baldwin, J. M. 1896. A new factor in evolution. *American Naturalist* 30:441–553.

Bitterman, M. E. 1965. Phyletic differences in learning. *American Psychologist* 20:396–410.

Bitterman, M. E. 1975. The comparative analysis of learning. *Science* 188:699–709.

Bitterman, M. E. 1988. Vertebrate-invertebrate comparisons. In H. J. Jerison and I. Jerison, eds., *Intelligence and evolutionary biology,* 251–75. New York: Springer-Verlag.

Boyd, R., and P. J. Richerson. 1985. *Culture and the evolutionary process.* Chicago: University of Chicago Press.

Brandes, C., and R. Menzel. 1990. Common mechanisms in proboscis extension conditioning and visual learning revealed by genetic selection in honeybees (*Apis mellifera capensis*). *Journal of Comparative Physiology* A 166:545–52.

Brooks, D. R., and D. A. McLennan. 1991. *Phylogeny, ecology and behavior: A research programme in comparative biology.* Chicago: University of Chicago Press.

Cecconi, F., and D. Parisi. 1991. Evolving organisms that can reach for objects. In J. A. Meyer and S. W. Meyer, eds., 391–99. *From animals to animats,* Cambridge, Mass.: MIT Press.

Courtney, S. P., G. K. Chen, and A. Gardner. 1989. A general model for individual host selection. *Oikos* 55:55–65.

Dukas, R., and L. A. Real. 1991. Learning foraging tasks by bees: A comparison between social and solitary species. *Animal Behaviour* 42:269–76.

Endler, J. A. 1986. *Natural selection in the wild.* Princeton: Princeton University Press.

Ewer, R. F. 1956. Imprinting in animal behaviour. *Nature* 177:227–28.

Fisher, R. A., and R. A. Hinde. 1948. The opening of milk bottles by birds. *British Birds* 42:347–57.

Goodall, J. 1986. *The chimpanzees of Gombe: Patterns of behavior.* Cambridge, Mass.: Belknap Press of Harvard University Press.

Gould, S. J., and R. C. Lewontin. 1979. The spandrels of San Marcos and the Panglossian paradigm: A critique of the adaptionist programme. *Proceedings of the Royal Society of London series B* 205:581–98.

Hailman, J. P. 1985. Historical notes on the biology of learning. In T. D. Johnston and A. T. Pietrewicz, eds., *Issues in the ecological study of learning*, 27–57. Hillsdale, N.J.: Lawrence Erlbaum Associates.

Haldane, J. B. S., and H. Spurway. 1956. Imprinting and the evolution of instincts. *Nature* 178:85–86.

Hall, J. C. 1985. Genetic analysis of behavior in insects. In G. A. Kerkut and L. I. Gilbert, eds., *Comprehensive insect physiology, biochemistry and pharmacology,* vol. 9, 287–373. Oxford: Pergamon Press.

Hinton, H., and S. J. Nowlan. 1987. How learning guides evolution. *Complex Systems* 1:495.

Hölldobler, B., and C. D. Michener. 1980. Mechanisms of identification and discrimination in social Hymenoptera. In H. Markl, ed., *Evolution of social behavior: Hypotheses and empirical tests,* 35–58. Dahlem Conferences 1980. Weinheim: Verlag Chemie.

Jaenike, J., and D. R. Papaj. 1992. Learning and patterns of host use by insects. In M. Isman and B. D. Roitberg, eds., *Insect chemical ecology: An evolutionary approach,* 245–64. New York: Chapman and Hall.

Johnston, T. D. 1982. Selective costs and benefits in the evolution of learning. *Advances in the Study of Behavior* 12:65–106.

Krebs, J. R., and N. B. Davies. 1993. *Introduction to behavioural ecology.* 3d ed., Oxford: Blackwell Scientific Publications.

Lewis, A. C. 1986. Memory constraints and flower choice in *Pieris rapae. Science* 232:863–65.

Lewis, W. J., and K. Takasu. 1990. Use of learned odours by a parasitic wasp in accordance with host and food needs. *Nature* 348:635–36.

Lewis, W. J., and J. H. Tumlinson. 1988. Host detection by chemically mediated associative learning in a parasitic wasp. *Nature* 331:257–59.

Lorenz, K. Z. 1965. *Evolution and modification of behavior.* Chicago: University of Chicago Press.

Lorenz, K. Z. 1981. *The foundations of ethology.* New York: Springer-Verlag.

Maddison, W. P. 1990. A method for testing the correlated evolution of two binary characters: Are gains or losses concentrated on certain branches of a phylogenetic tree? *Evolution* 44:539–57.

Marler, P., and H. S. Terrace. 1984. *The Biology of Learning.* Berlin: Springer-Verlag.

Maynard Smith, J. 1987. When learning guides evolution. *Nature* 329:761–62.

Menzel, R., J. Erber, and T. Masuhr. 1974. Learning and memory in the honeybee. In L. Barton-Browne, ed., *Experimental analysis of insect behavior,* 195–217. Berlin: Springer-Verlag.

Morgan, L. 1896. *Habit and instinct.* London: Arnold.

Nolfi, S., J. L. Elman, and D. Parisi. 1990. *Learning and evolution in neural networks.* CRL Technical Report 9019. Center for Research in Language, University of California.

Osborn, H. F. 1896. Ontogenic and phylogenic variation. *Science* 4:786–89.

Papaj, D. R. 1986. Interpopulation differences in host preference and the evolution of learning in the butterfly, *Battus philenor. Evolution* 40:518–30.

Papaj, D. R. 1993a. Automatic behavior and the evolution of instincts: Lessons from learning in parasitoids. In D. R. Papaj and A. C. Lewis, eds., *Insect learning: Ecological and evolutionary perspectives,* 243–72. New York: Chapman and Hall.

Papaj, D. R. 1993b. Learning as an adapting process. In D. R. Papaj and A. C. Lewis, eds., *Insect learning: Ecological and evolutionary perspectives,* 374–86. New York: Chapman and Hall.

Papaj, D. R., and R. J. Prokopy. 1989. Ecological and evolutionary aspects of learning in phytophagous insects. *Annual Review of Entomology* 34:315–50.

Pepperberg, I. M. 1987. Interspecies communication: A tool for assessing conceptual abilities in the African Grey parrot (*Psittacus erithacus*). In G. Greenberg and E. Tobach, eds., *Language, cognition, and consciousness: Integrative levels,* 31–56. Hillsdale, N.J.: Lawrence Erlbaum Associates.

Pierce, G. J., and J. G. Ollason. 1987. Eight reasons why optimal foraging theory is a complete waste of time. *Oikos* 49:111–18.

Real, L. A. 1991. Animal choice behavior and the evolution of cognitive architecture. *Science* 253:980–86.

Roitberg, B. D., and Angerilli, N. P. D. 1986. Management of temperate-zone deciduous fruit pests: Applied behavioural ecology. *Agricultural Zoological Review* 1:137–65.

Rosenheim, J. A. 1993. Comparative and experimental approaches to understanding insect learning. In D. R. Papaj and A. C. Lewis, eds., *Insect learning: Ecological and evolutionary perspectives*, 273–307. New York: Chapman and Hall.

Rumelhart, D. E., G. E. Hinton, and R. J. Williams. 1986. Learning representations by back-propagating errors. *Nature* 323:533–36.

Smith, B. 1993. Merging mechanism and adaptation: An ethological approach to learning and generalization. In D. R. Papaj and A. C. Lewis, eds., *Insect learning: Ecological and evolutionary perspectives*, 126–57. New York: Chapman and Hall.

Staddon, J. E. R. 1983. *Adaptive behavior and learning*. Cambridge: Cambridge University Press.

Stephens, D. W. 1987. On economically tracking a variable environment. *Theoretical Population Biology* 32:15–25.

Stephens, D. W. 1993. Learning and behavioral ecology: Incomplete information and environmental unpredictability. In D. R. Papaj and A. C. Lewis, eds., *Insect learning: Ecological and evolutionary perspectives*, 195–218. New York: Chapman and Hall.

Thorpe, W. H. 1956. *Learning and instinct in animals*. London: Methuen.

Tierney, A. J. 1986. The evolution of learned and innate behavior: Contributions from genetics and neurobiology in a theory of behavioral evolution. *Animal Learning and Behavior* 14: 339–48.

Travis, J. 1989. The role of optimizing selection in natural populations. *Annual Review of Ecology and Systematics* 20:279–96.

Waddington, C. H. 1953. Genetic assimilation of an acquired character. *Evolution* 7:118–26.

Wright, S. 1931. Evolution in Mendelian populations. *Genetics* 16:97–159.

Zeidenberg, M. 1990. *Neural networks in artificial intelligence*. New York: Ellis Horwood.

Part II

COMMUNICATION

SEVEN

Errors, Exaggeration, and Deception in Animal Communication

R. HAVEN WILEY

MY CONTENTION in this chapter is that animals make mistakes and that this simple but neglected circumstance has deep implications for the evolution of animal communication. The possibility of error is explicitly incorporated into mathematical theories of communication and choice, in particular, information theory, signal detection theory, and decision theory. Although the basic concepts of these theories are neither new nor complex, they have yet to be integrated into evolutionary theories of animal communication. An objective of this chapter is to show, by expanding themes introduced earlier (Wiley 1983), that these basic concepts provide explanations for the evolution of some fundamental features of communication. In particular, they can provide sufficient conditions for the evolution of exaggerated displays and for the evolutionary stability of deception. This approach also indicates that some of the parameters needed to understand the evolution of communication have been overlooked.

THREE EXAMPLES OF SITUATIONS IN WHICH ERROR OCCURS

To begin, consider three situations in which animals might make mistakes in responding to signals. In the present context, a mistake or error is an evolutionarily inappropriate response, one that reduces chances for the spread of genes associated with the response. Some definitions of terms used in the study of communication warrant attention in a later section. For the moment, an evolutionary perspective provides an operational definition of error. The following examples illustrate the possibilities for error during communication at long range, in dense aggregations, and in the presence of deception.

Long-Range Communication with Song

First, as an example of communication at long range, consider a male territorial passerine bird listening for conspecifics' songs in a forest. Appropriate responses to those songs might include approaching any conspecific male that sings within the listener's territory or countersinging with a neighbor. Song at close range and at full power might almost always evoke an appropriate response; listeners, in other words, would make few, if any, errors. Yet even slight hesitation by a temporarily distracted listener or confusion of songs with superficially similar sounds in the environment could make the listener's response less than optimal. At a distance, the chances for error increase. The signal heard by the listener is attenuated and distorted, sometimes to the extent that detection fails completely. Possibilities for error increase further if the listener's task is more complicated than simply detecting conspecifics; for instance, it might include recognizing different neighbors' songs, localizing the source, discriminating different song types, or associating songs with memories of previous locations or interactions. In addition, there is the possibility that singers might deceive or manipulate listening rivals by minimizing possibilities for recognition or localization (Krebs 1977; Morton 1986). Then of course the possibilities for error increase still more.

Current information suggests that communication between territorial birds by means of song requires complex discriminations. Territorial male passerines discriminate between conspecific songs and others, between neighbors' and strangers' songs, and between songs of individual neighbors in normal and abnormal locations (Falls and Brooks 1975; Wiley and Wiley 1977; Becker 1982; Falls 1982; Godard 1991, 1992; Stoddard et al. 1991). When individuals have repertoires of distinct song types, their use of these song types differs, at least in some species, with location within their territories or with probabilities of subsequent actions (Lein 1978; Smith et al. 1978; Schroeder and Wiley 1983; Temerin 1986; Dabelsteen and Pedersen 1990). In a few cases, experimental playbacks of songs have shown that males respond differently to these variants (Järvi et al. 1980; Schroeder and Wiley 1983; Dabelsteen and Pedersen 1990). Females also recognize conspecific and even individual males' songs (King and West 1983; Searcy et al. 1981; Searcy 1990; Wiley et al. 1991). Woodland birds can also judge the distance to a singer by attenuation or degradation of songs (Richards 1981b; MacGregor and Krebs 1984; MacGregor 1991). A listening bird's response thus depends on a set of discriminations based on the features of a song. Communication with song often occurs over tens to hundreds of meters and in the presence of other species with some similar features in their songs, conditions that make perfect detection and recognition of songs unlikely.

Experimental playbacks of recorded songs in the field elicit variable responses, often frustratingly so. In analyzing these responses, investigators have always focused on the group means across subjects, rather than on variation in responses by individual subjects to replicate presentations. The variation, however, emphasizes that inconsistency in responses persists even to a standardized signal in standardized situations. An understanding of error in communication requires an analysis of this variation.

Communication in Mating Aggregations of Frogs

As a second example of the possibility for error in communication, consider frogs that mate in aggregations. This situation raises the possibility of error in females' choice of mates. In an experiment designed to determine the effects of background sounds on acoustic communication, Gerhardt and Klump (1988a) tested the ability of female green treefrogs (*Hyla cinerea*) to orient toward a male's calls when they were masked by the sounds of a large chorus. The female's problem was to detect and to locate an intermittent call in the presence of continuous background sounds of similar frequency. Only when the intensity of a male's call equaled or exceeded that of the background sound did most females succeed. As a result of this limitation, a female presumably detects only some three to five males at a time in a large chorus.

In mixed choruses, there is also the possibility of interference from calls of different species. Frogs active at the same season and in the same habitats sometimes avoid overlap in calling (Littlejohn and Martin 1969; Zelick and Narins 1982;

Schwartz and Wells 1983a, 1983b; Schwartz 1987) and differ distinctly in the spectral or temporal patterning of their calls (Littlejohn 1977). Nevertheless, the partitioning is sometimes incomplete. For example, *Hyla ebraccata* often calls within choruses of *H. microcephala* in Panama; the spectrum of *ebraccata*'s calls broadly overlaps the lower peak in the spectrum of its congener's calls. To reduce masking, *ebraccata* often remain silent during cyclic peaks of calling by the louder and denser *microcephala* (Schwartz and Wells 1983a, 1983b). In *H. microcephala* and *H. versicolor*, playback experiments suggest that both males and females are more likely to respond appropriately to nonoverlapping calls (Schwartz 1987). The possibility of inappropriate responses, including failure to respond, thus seems plausible in the presence of interfering calls. The possibility for errors might depend on the distance of communication, although Gerhardt and Klump (1988b) showed that female barking treefrogs (*H. gratiosa*) preferred sounds of choruses that contained conspecific males over those that did not at a distance of 160 m. Experiments with green treefrogs suggest that low-frequency components in the males' calls might attract females from a greater distance than the high-frequency components (Gerhardt 1987).

Females approaching these mixed choruses in some cases must make fine discriminations. *H. chrysoscelis* and *H. versicolor* differ primarily in the temperature-dependent pulse rates of their calls. Female *versicolor* prefer lower pulse rates than do female *chrysoscelis* at any temperature. They can discriminate pulse rates differing by much less than a factor of two. In addition, when pulse rates differ by at least a factor of two, females' preferences are stable even when the sound pressure level of the nonpreferred call exceeds that of the preferred one by 18 dB (Gerhardt 1982). Even though *chrysoscelis* is diploid and *versicolor* tetraploid, there appears to be some hybridization where the two species occur together (Gerhardt pers. comm.), so errors in choice of mates apparently do occur. Errors in mating might occur even more often than indicated by hybridization.

Hybridization between sympatric *H. cinerea* and *H. gratiosa* has been carefully documented (Gerhardt et al. 1980). In addition, *H. squirella* also shares breeding sites with these two species. When given no choice, females of each species occasionally approach playbacks of one of the other species' calls (Oldham and Gerhardt 1975), even though some features, including the peak frequency in the lower of two spectral bands in the calls of each species, have little or no overlap. In two-choice tests, female *cinerea* prefer synthetic calls with a low-frequency peak (LFP) of 900 Hz over those with an LFP lower than or higher than 800–1,000 Hz and thus discriminate against sounds with LFPs typical of *gratiosa* and *squirella*, respectively. Nevertheless, some female *cinerea* chose the alternative calls. Furthermore, discrimination is less clear when females have a simultaneous choice of four calls (Gerhardt 1982, 1987). In general, reducing the difference in frequency between alternative calls and increasing the number of simultaneous choices both reduce a female's selectivity (fig. 7.1). Furthermore, as Gerhardt (1982) indicates, females can hardly rely on experience in making these discriminations, in the absence of extended associations of parents and offspring or of mates. Despite these possibilities for errors by females, hybrids are rarely found (Schlefer et al. 1986). However, when heterospecific matings produce less viable offspring, the frequency of hybrids underestimates the number of mismatings.

Intraspecific Deception

A third source of error in animal communication is deception. Deception occurs when a signaler, in specified circumstances, gains from a receiver's response to a

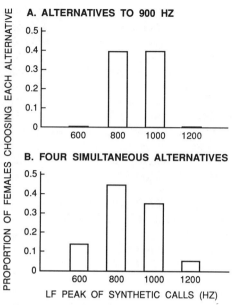

Fig. 7.1 Female green tree frogs are less discriminating in four-choice than in two-choice experiments. (A) When females are given a choice between two speakers, one playing a naturalistic synthetic call with a low-frequency peak (LFP) at 900 Hz and another playing a call at one of four other frequencies, they never respond to calls with LFPs at 600 or 1,200 Hz ($n \geq 8$ females for each bar). (B) When presented with four choices of synthetic calls, females sometimes find their way to the speakers presenting calls with LFPs at 600 or 1,200 Hz ($n = 60$ females). In both experiments sounds were presented to females at 75 dB. (Adapted from Gerhardt 1982.)

signal while the receiver loses. This situation arises when a signaler takes advantage of a receiver's rules for decoding signals (Wiley 1983). Gains and losses in this context are changes in the signaler's and receiver's expected survival and reproduction that in turn affect the spread of genes associated with signaling and responding. The possibility of this sort of interaction between signalers and receivers was first emphasized by Dawkins and Krebs (1978) in their discussion of manipulation in communication.

Many recently documented cases of intraspecific deception fall into two categories: mimicry of females by males to gain surreptitious matings, and use of alarm calls to gain temporary advantages in competition for food or territories. An example of the former is the occurrence of small, female-like males in several species of fish and crustaceans. In small populations of the bluehead wrasse (*Thalassoma bifasciatum*), a Caribbean reef fish, most females mate at preferred sites monopolized by large, terminal-phase males, which have completed protogynous sex reversal. Small males form spawning groups or attempt to accompany females spawning with large males (Warner and Hoffman 1980; Warner 1984, 1987). Either way they presumably fertilize a small proportion of any one spawning female's eggs. Large males aggressively exclude small males from their territories. The small males' success in parasitizing large males depends on their close resemblance to females, which makes discrimination difficult for terminal-phase males. In large populations, terminal-phase males cannot monopolize the preferred mating sites, and thus females more often spawn with small males. As expected from their relative success in mating, small males constitute a greater proportion of large populations. Because individuals do not change reefs after settling, small males facultatively allocate their resources to either growth or spawning in accordance with the size of the reef they inhabit and their consequent chance for reproduction at an early age (Warner 1984).

In other cases, individuals can change tactics within their lifetimes. In the scorpionfly *Hylobittacus apicalis,* males can alternate between hunting for prey to use as nuptial offerings to attract females—behavior that risks capture in spiders' webs—and posing as females to steal other males' offerings (Thornhill 1979). A male with a nuptial offering sometimes recognizes a deceptive male in time to protect its offering, but on other occasions presents it to the deceptive male and thus loses it. In both wrasses and scorpionflies, males with an inherent advantage in mating confront the problem of discriminating between the similar appearances of females and deceptive males. Errors, either responding to a deceptive male as if it were a female or to a female as if it were a deceptive male, result in lower success in mating.

Examples of deceptive use of alarm calls include the great tits (*Parus major*) studied by Møller (1988). Tits at an artificial feeding station often flee when they hear the distinctive alarm call usually uttered by another tit after spotting a hawk. Occasionally a tit prevented from feeding by a dominant opponent utters an alarm call and thus gains access to the feeding site after the opponent departs. Tits use this ploy only when a feeding site is occupied by a dominant opponent. In another example, individual antshrikes (*Thamnomanes schistogynus*) in mixed-species flocks in Amazonian forests produce a conspicuous call both when a raptor is spotted and when a nearby individual of some other species is about to seize a large insect. In the latter case, during the moment of hesitation by the nearby individual when it hears the possible alarm, the antshrike often catches the prey (Munn 1986a, 1986b). Spectrograms of calls produced in the two contexts do not differ. In these cases, a signal virtually indistinguishable from one used in other contexts serves to evoke a response inappropriate for the receiver. From the receiver's point of view, failure to discriminate any subtle differences between the normal and the deceptive signals or their contexts results in an error.

FUNDAMENTAL PROPERTIES OF A COMMUNICATION SYSTEM

Before proceeding, it is worthwhile to review some fundamental features of communication and some basic definitions. A framework for studying errors in communication develops naturally from these basic concepts.

Definitions of deception or of errors in communication that rely entirely on evolutionary criteria will not suit everyone. Cognitive ethologists have tended to insist that true deception requires intention, an awareness by the signaler of its influence on the receiver's state of mind (Woodruff and Premack 1979; Ristau 1991; Ristau and Robbins 1982; Cheney and Seyfarth 1990). The issue of intentions also surfaces in definitions of communication itself or, at least, of the distinctive human form of communication by means of language (Grice 1969; Austin 1973; Dennett 1987, 1988). One motivation for stipulating intentions as a criterion for communication has been to separate communication from behavior that is unconscious, conditioned, innate, or even accidental.

Any attempt to include intentions in the necessary conditions for communication faces the difficulty of specifying generally acceptable definitions for mental states in other individuals. It is not yet clear that it is possible to devise such definitions for mental states like intentions (see, for instance, Putnam 1967; Bennett 1976; Harré 1984; Harris 1984; Dennett 1988; Cheney and Seyfarth 1990). In any case, the present evolutionary analysis does not require them. Nor does the present analysis require specification of the proximate behavioral or neural mechanisms of communication, but it assumes that they exist and can evolve.

Another frequent stipulation in definitions of communication, at least in the past, has been mutually advantageous interaction between sender and receiver. This condition seemed reasonable because it excludes such intuitively noncommunicative interactions as predation. However, it also excludes deception, as defined above, and indeed any possibility for error. It thus excludes some of the most interesting possibilities for communication.

Communication can be defined in a way that avoids these difficulties as *any alteration in a receiver produced by a signaler by means of a signal*. A signal is any pattern of energy or matter produced by one individual (the signaler) and altering some property of another (the receiver) *without providing the power to produce the entire response*. The last phrase excludes all interactions between individuals in which the actor overpowers the recipient; for instance, it makes a distinction between communication and predation. It sets aside all questions about the mechanisms involved, including intentions or other mental states, however they might be defined by behavioral or neural events. Indeed, these basic definitions apply to machines as well as animals.

In this approach, the components of any system of communication consists of a signaler, a signal, and a receiver. In this triplet of components, first introduced by Shannon and Weaver (1949) in their analysis of rates of information transfer, the signaler and receiver are unspecified mechanisms with outputs that are associated by means of a signal produced by one and detected by the other. This scheme differs crucially from the earlier triplet of signal, receiver, and referent. This latter triplet, diagrammed for the first time by Ogden and Richards (1923), emphasizes the relationship between a signal and its referent, or as Saussure (1959 [1915]) put it, between a "sound-image" and an associated "concept." This relationship, mediated by the receiver and often called interpretation or representation, raises all the issues about the operational definitions of mental states mentioned above.

Any scientific investigation of communication actually requires a fourth essential component, a nonparticipant observer, first introduced by Shannon and Weaver (1949) and emphasized by Cherry (1957). Of course, few observers can completely avoid all participation or interference in their subjects' interactions, so perhaps "minimally participant," or "scientific," observer is a more accurate term. It behooves a scientific observer to assess all possible influences on the subjects of study. Such an observer should be able to intercept the signal, often at different points in its propagation between signaler and receiver, and to measure behavioral and physiological events and states in both signaler and receiver.

In this view of communication, a receiver is an unspecified mechanism for associating signals and responses. The stipulation that the signal cannot produce the "*entire* response" is important, because of course the signal must provide enough energy to effect *some* response in the receiver's sensors; yet some, usually most, of the energy to produce the response must come from the receiver itself. A receiver must thus include, in some form, both transducers and amplifiers (or sensors and effectors). As a behavioral, neural, and, ultimately, physical mechanism, it must also follow specifiable rules; alterations in the receiver as a result of a signal must be a function of the signal and the current state of the receiver (Wiley 1983). An alteration in a receiver might include overt changes in effectors, such as a movement of limbs, or vocal cords, that could provide a signal for another receiver. It might also go unnoticed by an observer and thus remain covert. The alterations in the receiver of a signal might involve great complexity.

These possibilities allow us to decompose any act of communication into components each one of which admits further analysis into smaller, but conceptually similar, components. Thus one analysis might focus on a signal, receiver, and response that consisted of the mating calls of a particular species of frog, conspecific females, and movement toward a speaker. Another might focus on artificially synthesized sounds, auditory receptors in the basilar papillae of a frog, and impulses in the eighth cranial nerve; another on sounds, perhaps unspecifiable neural circuits, and evoked potentials in the tectum. Responses might include complex patterns of neural activity without unique motor consequences and not measurable with currently available equipment.

These examples emphasize that many responses are apparent only to a scientific observer with special apparatus and that some remain covert even to observers with the best available equipment. Note also that any signal-receiver-response system consists of a nested hierarchy of conceptually similar systems and that a hierarchy of these systems constitutes the full description of any act of communication. An essential feature of this view of communication is the emphasis on associations between signals and responses. A receiver at any level of analysis, viewed as a whole, is a mechanism that associates signals with responses.

The simplest way to think of such a system is to consider two alternative (mutually exclusive) signals associated by a receiver with two alternative responses. Thus incoming signals are of two types, to each of which the receiver makes one of two kinds of responses. More complex situations are elaborations of this simple one. For instance, to study *detection of a single type of signal*, we can divide stimulation impinging on the receiver into two categories, signal present or signal absent. To study *recognition of alternative signals*, we can compare two signals, such as conspecific and heterospecific songs or familiar and unfamiliar ones. We can divide responses into two categories, such as approaching or not, vocalizing for longer or shorter periods, or, in the case of humans, saying "yes" or "no."

A scientific observer of this simplest system can identify four possible stimulus-response pairs, each of two signals crossed with each of two responses. The observer must accurately differentiate the two categories of signals and the two categories of responses. It is this situation that signal detection theory analyzes, although the mathematical elaboration of the theory requires additional assumptions: that the two stimulus states differ along a single dimension (intensity, for instance); and that both include random perturbations with Gaussian distributions and equal variances along this dimension (Green and Swets 1966; Egan 1975). These assumptions of unidimensionality and homoscedasticity are not essential for some basic conclusions about communication, however.

In cases in which one of the two categories of response is appropriate for one of the two categories of signal, the four signal-response pairs have clear interpretations (table 7.1). When the appropriate signal is present, responses are either *correct detections* (CD) or *missed detections* (MD), and, when the appropriate signal is absent, responses are either *false alarms* (FA) or *correct rejections* (CR). These four possibilities arise whenever two signal categories are not perfectly discriminated by the receiver. Unpredictable variation in the properties of signals, from whatever source, can produce overlap between two signal categories, as sensed by a receiver, so the receiver

Table 7.1   Four Possibilities when a Signal Is Not
Perfectly Discriminated by a Receiver

| | RESPONSE APPROPRIATE FOR SIGNAL | |
SIGNAL	Present	Absent
Present	Correct detection	Missed detection
Absent	False alarm	Correct rejection

cannot perfectly distinguish them. Note that a scientific observer, by means of a special vantage or equipment, might be able to distinguish two signal categories that a receiving animal could not. Whenever a receiver cannot perfectly distinguish two signal categories, it cannot simultaneously reduce both its probability of false alarm and its probability of missed detection to zero.

The fundamental consequence of this situation is that *a receiver cannot simultaneously maximize its probability of correct detection and minimize its probability of false alarm.* This conclusion does not depend on any assumptions concerning the distribution of perturbations or equality of variances. It applies whenever the categories of signals (including signal and no signal) preclude perfect discrimination by the receiver's sensors for any reason.

ERROR AND NOISE IN ANIMAL COMMUNICATION

Any conclusion about "error" as an evolutionarily inappropriate response by a receiver requires estimates of the spread of genes associated with responses, as indicated, for instance, by the receiver's (and its kin's) probabilities of survival and reproduction. It is useful to distinguish error, in this sense, from noise.

The term "noise," in everyday usage, comes close to meaning irrelevant masking energy. Quite a different view emerges from Shannon and Weaver's (1949, 20–21, 66–70) discussion of noise as it applies to information theory. They define error as equivocation, $H_y(x)$, the average uncertainty (or entropy) in the input from the source (x) when the output from the receiver (y) is known:

$$(1/n) \, \Sigma_y \, \Sigma_x \, p_{x|y} \log_2 p_{x|y},$$

where $p_{x|y}$ is the probability of the xth signal category provided a response of the yth response category has occurred and n is the number of possible responses. In terms appropriate for behavioral interactions, equivocation is a scientific observer's average uncertainty about which signal (x) has occurred when the observer knows only the response of a receiver (y). Transmission of information between signaler and receiver occurs at the maximal rate when every signal category produces a unique response category. The rate of transmission, R, then equals the uncertainty in either the signals or the responses considered alone: $R_{max} = H(x) = H(x) = H(y)$. To obtain the actual rate of transmission of information, an observer must subtract the measure of equivocation from the uncertainty in the signals alone,

$$R = H(x) - H_y(x).$$

Note that if a signal is completely predictable, so that $H(x) = 0$ and thus $H_y(x) = 0$, then $R = 0$. Only when signals occur with some uncertainty can they transfer information. In addition, if for any reason the signaler's behavior has no association with the receiver's, so that $H_y(x) = H(x)$, then again $R = 0$. Thus R measures the amount of information that signals actually convey to a receiver. A mathematically equivalent statement,

$$R = H(y) - H_x(y),$$

expresses the rate of transmission as the difference between uncertainty in the responses and the average conditional uncertainty of responses when the signal is known. In the first equation, $H_y(x)$ represents the amount of information needed to correct the received message. In the second equation, $H_x(y)$ represents the portion of the uncertainty in the receiver's responses that is "due to noise," as Shannon and Weaver put it.

Therefore, for these authors, transmission is *noiseless* when there is no uncertainty about the response once the signal is known. This perspective merges with our conventional view that noise is masking energy, so that responses are functions of both signals and noise. On the other hand, according to Shannon and Weaver, transmission is *error-free*, or needs no correction, when there is no uncertainty about the signal when the response is known.

In applying these ideas to animals, some problems arise. As engineers, Shannon and Weaver view communication as a reconstruction by the receiver of symbols exactly as they originate from the source—a dot for a dot, a dash for a dash, or a letter for a letter. In animal communication, however, the action of the receiver is often different from the action of the signaler. We cannot judge errors by comparing the form of the output with that of the input. Instead, it makes more sense to define errors in terms of the evolutionary appropriateness of the response to a signal in a particular situation. Just as Shannon and Weaver indicate in their discussion of error, the inappropriateness of a response must be determined by an observer with knowledge of the signaler, signal, receiver, and context.

The engineer's objective of reconstructing the signal from the response assumes that the signaler and receiver share the same interests in communication. In considering the evolution of animal communication, this assumption does not necessarily apply. However, the engineer's definitions of noise and error do apply. From the *signaler's* perspective, a signal should evoke a particular response from an intended receiver in a particular situation; any uncertainty about the response following a signal is noise from the signaler's viewpoint. Thus, for a signaler, communication is noise-free when $H_x(y) = 0$. From a *receiver's* perspective, a response should be made to a particular kind of signal in a particular situation; any uncertainty about the preceding signal raises the possibility of an inappropriate response, an error for the receiver. Thus, for a receiver, communication is error-free when $H_y(x) = 0$. These two conditional uncertainties are usually nonzero and not necessarily equal.

For a signaler, noise results from anything that reduces the probability that a signal will evoke a particular response from the intended receiver. For a male frog in a chorus, noise could result from other males' calls, environmental attenuation and degradation of its own calls, or reluctance of females. The last constitutes, from the signaler's perspective, neural noise in the receiver, analogous to amplifier noise in an electronic receiver. Such neural noise might result from limitations on attention or memory or from a high threshold for response. For a male frog, communication is just as "noisy" regardless of the reasons for a female's failure to respond.

For a receiver, noise results from anything that contributes to error—in other words, anything that reduces the probability of an appropriate response to a signal. For a female frog, noise could result from heterospecific or suboptimal conspecific males with calls similar to those of optimal mates or from environmental attenuation and degradation of males' calls.

Note that Shannon and Weaver's use of the term "noise" corresponds to the signaler's perspective, whereas ordinary usage in general corresponds to the receiver's. In using this term, extra care is clearly needed. A common element in these two perspectives is the problem of environmental attenuation and degradation.

ENVIRONMENTAL ATTENUATION AND DEGRADATION

All signals eventually become inseparable from background energy in the environment at some distance from the source. This distance, the effective range of a signal, depends on the signal's intensity at the source, its attenuation and degradation during propagation through the atmosphere, and the level of irrelevant energy in the background. These influences are best understood for acoustic signals in terrestrial environments (Piercy and Embleton 1977; Wiley and Richards 1982; Gerhardt 1983; Michelsen and Larsen 1983). Effective ranges, or active spaces, have been determined for a variety of such signals: songs of red-winged blackbirds (*Agelaius phoeniceus*) (Brenowitz 1982); calls of blue monkeys (*Cercopithecus mitis*) (Brown and Waser 1984); and calls of a variety of frogs (Loftus-Hills and Littlejohn 1971) and insects (Römer and Bailey 1986).

An obvious way to increase the effective range of acoustic signals is to use frequencies not masked by background sounds. For instance, birds in a tropical forest in Panama appear to use frequencies lower than those of diurnal insects, mostly cicadas, in the same habitat (Ryan and Brenowitz 1985). Another way to extend the range of signals is to use bands of frequencies less susceptible to attenuation. For instance, in European marshes, *Acrocephalus* warblers emphasize frequencies in their songs that propagate best from the typical locations of their perches (Jilka and Leisler 1974).

The physical explanations for differences in the attenuation of sound are usually complex. Attenuation results from spherical spreading and, in addition, from atmospheric absorption, scattering, and interactions with the ground (Wiley and Richards 1982). The last is probably the least familiar of these effects to biologists. Propagation of sound between a source and a receiver that are both within centimeters of a porous surface with low acoustic impedance, like soil covered with a thin layer of vegetation, results in pronounced attenuation of frequencies above about 500 Hz (Embleton et al. 1976). This effect presumably explains why so many small insects and vertebrates avoid calling or singing from the ground when potential receivers are also on the ground (Paul and Walker 1979; Wiley and Richards 1982; Michelsen and Larsen 1983).

Transmission farther above the ground can result instead in pronounced attenuation of low frequencies, as a result of interference between the direct wave reaching the listener and the phase-shifted wave reflected from the ground (Piercy and Embleton 1977; Wiley and Richards 1982). This attenuation shifts to lower frequencies as elevation above the ground increases. Most small birds do not produce frequencies low enough to be affected by this form of attenuation when singing more than about 1 meter above the ground. Primates in the canopies of tropical forests use frequencies much lower than those of birds in their long-range calls but still always above the band subject to attenuation by reflection from the ground (Waser and Brown 1984). The destructive interference between reflected and direct waves in these habitats results from the large phase shift as sound reflects from the porous soil, a surface of low acoustic impedance. Over a nonporous surface with high acoustic impedance, like water, low frequencies are not attenuated by interference in this way. As a consequence, birds in marshes use lower frequencies than those in grasslands (Cosens and Falls 1984; Wiley 1991).

In addition to reflection from the ground, atmospheric absorption and scattering from foliage attenuate sound in natural environments. Both of these forms of attenuation increase with frequency, and attenuation as a result of scattering from foliage increases more steeply with frequency in forests than in open areas (Morton 1975; Marten and Marler 1977). As a consequence, the frequencies used by birds for territorial songs in forests might differ from those used in open habitats. The expected differences, however, are not immediately clear. It is important to note that, above the band of frequencies affected by reflection from the ground, low frequencies attenuate least in all habitats. Thus, for songs with maximal range (or minimal effort for a given range), birds should always use dominant frequencies as low as possible for their body sizes. In forests, the attenuation of higher frequencies might result in a lower upper limit of acceptable frequencies. In fact, when body mass is controlled, birds in forests in eastern North America have lower maximal frequencies on average than those in open habitats. In contrast, with body mass controlled, there are no significant differences among major habitats in dominant frequencies (fig. 7.2; Wiley 1991). The narrower band of acceptable frequencies in forests thus influences the maximal, rather than the dominant, frequencies in songs.

Attenuation is not the only process that makes signals less distinctive to a receiver. Degradation of the temporal structure of signals can mask features that allow detection or recognition. Reverberation is a particularly important source of temporal degradation of acoustic signals. Among birds that communicate over distances of tens to hundreds of meters, the reverberations from trees in a forest obscure temporal structure within any frequency band. As a consequence, forest-inhabiting birds of North America tend to avoid rapid repetitions of frequencies in their long-range songs (fig. 7.3; Wiley 1991a). At long ranges in open habitats, sounds acquire irregular amplitude fluctuations as a result of refraction from moving cells of air that differ in velocity or temperature from the surrounding air (Richards and Wiley 1980). In these circumstances, birds produce songs of great temporal complexity, with rapid trills and other rapid repetitions of frequencies, often resulting in a tinkling quality (Wiley 1991a).

Similar considerations apply to communication with other sensory modalities. Vision in water, for instance, is limited by the pronounced frequency-dependent attenuation of light. Light attenuates within meters in water, rather than kilometers as in air. In either medium, objects that are distant in relation to the rate of attenuation lose contrast and take on the color and brightness of the general background. Underwater, a bright object near the surface is usually illuminated by downwelling light but is often viewed against the dimmer, horizontally scattered light called spacelight. The white tips on the wide fins of the epipelagic shark *Carcharhinus*

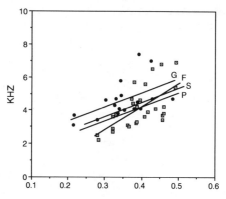

A. DOMINANT FREQUENCY

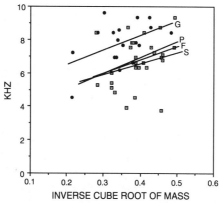

B. MAXIMAL FREQUENCY

INVERSE CUBE ROOT OF MASS

Fig. 7.2   Territorial passerine birds in eastern North America have songs that match the transmission properties of their habitats. (A) Within each of four broad categories of habitats (F, broad-leaved or mixed forest; P, parkland with scattered trees; S, shrubland; G, grassland), the dominant frequencies in songs of different species vary primarily with body size. (B) Maximal frequencies in songs, however, also differ significantly among habitats, presumably because lower attenuation of high frequencies in open grassland increases their effective range. Each graph includes the regressions of frequency on the inverse cube of body mass for species in the four categories of habitat; values for each species in forest (open squares) and grassland (solid circles) illustrate the variation within habitats (N = 29, 35, 21, and 18 for habitats F, P, S, and G, respectively). (See Wiley 1991a for statistical analysis.)

longimanus, for instance, resemble a school of fish when the shark's body is lost in the background spacelight. Presumably the shark thus lures its prey within striking range (Myrberg 1991).

Because short wavelengths both scatter best and attenuate least, optical signals at even modest distances or depths underwater become nearly monochromatic. At close range near the surface, yellow provides contrast with the dim blue spacelight and attenuates less than red. As a consequence, detection is improved by visual signals and receptor pigments offset toward longer wavelengths than the transmission maximum for water. Indeed, many fish in shallow water have receptor pigments with absorption maxima offset in this way (Lythgoe 1979, 1984). At a certain

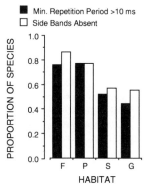

Fig. 7.3   Songs of birds differ among habitats in inclusion of rapid amplitude modulation (indicated by sidebands) or short repetition intervals at any one frequency. In forested habitats, reverberation from foliage masks rapid recurrence of elements at the same frequency. F, broad-leaved or mixed forest; P, parkland with scattered trees; S, shrubland; G, grassland. (From Wiley 1991a, © 1991 by The University of Chicago.)

distance or depth, however, yellow becomes gray and red black. Dark/light contrast is the only effective way to encode signals in these conditions. The striking red and white patterns of some fishes that live at moderate depths or have crepuscular habits appear effectively black and gray. It may be easier for such species to acquire red pigments from the diet than to synthesize melanin, so that red pigments are the most efficient for making "black" patterns.

This discussion of environmental attenuation and degradation has focused on its consequences for simple detection of signals. These consequences also affect discrimination between similar signals. Signals that are distinct at their sources might become barely discriminable after attenuation and degradation.

COMPROMISES FOR SIGNALERS AND RECEIVERS

How can we understand the evolution of signaling, on the one hand, and responding to signals, on the other? Each involves its own compromises, as in each case it is not normally possible to maximize all objectives simultaneously. A brief overview of these compromises provides some orientation before the following section attempts a thorough examination.

The evolution of signaling should tend to maximize the probability of correct responses from the intended receivers. This trend, however, is subject to two constraints: efficiency in the production of signals, and risks from unintended receivers. The efficiency of a signal is appropriately expressed as a ratio: the probability of response by the intended receiver divided by the signaler's effort or risk (Wiley 1983). An efficient signal thus reliably produces a response in the intended receiver with a minimum of effort by or risk to the signaler. Unintended receivers are those that might intercept a signal and respond in a way disadvantageous to the signaler. For instance, a predator or rival might intercept a signal intended for a mate or offspring, to the signaler's disadvantage. A signaler must balance the advantages of responses to a signal by intended receivers against the effort and risks of producing a signal and the consequences of its interception by unintended receivers. Signals should thus evolve *to maximize the efficiency of obtaining responses by intended receivers and to minimize the probability of interception.*

Receivers also face balancing advantages and disadvantages. Receivers must discriminate those signals to which a response is advantageous from those to which a response is disadvantageous. For each of the four basic associations of signals and responses (correct detections, missed detections, false alarms, and correct rejections), there is a net advantage (positive or negative) for the receiver. The mechanisms for discriminating signals should thus evolve *to maximize the sum of the net advantages of each possible outcome times its probability.* This sum is called the *expected utility* of the receiver's responses.

RECEIVERS' ADAPTATIONS

For a receiver, the object is to distinguish signals from irrelevant patterns of energy (or to distinguish different categories of signals) and then to associate each signal with the appropriate response. Because signals attenuate, degrade, and mix with irrelevant energy during propagation, some patterns of stimulation produced by a particular signal are not distinguishable by the receiver from irrelevant energy or from patterns produced by other signals.

The possibility of confusion is easily illustrated in terms of a simple signal discriminated from irrelevant masking energy by a single parameter, such as intensity. The masking energy often fluctuates randomly in intensity; the addition of a signal of constant intensity displaces this distribution toward higher, but often not distinctly higher, intensities (fig. 7.4). A receiver might respond or not depending on whether or not the intensity exceeds some threshold. This threshold then represents a simple criterion for response. In this case, the receiver can alter the probabilities of correct detection and false alarm (and consequently the probabilities of missed detection and correct rejection) by adjusting its criterion. Yet, as emphasized above, it cannot simultaneously maximize its probability of correct detection (P_{CD}) and minimize its probability of false alarm (P_{FA}). The possible compromises are represented

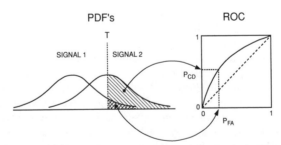

Fig. 7.4   Derivation of the receiver operating characteristic (ROC) from the probability density functions (PDFs) for two signals (or signal plus irrelevant energy versus irrelevant energy alone) along a single stimulus dimension. For any criterion selected by the receiver (T), the probability of a correct detection P_{CD}) equals the integral of one PDF from the threshold to infinity (shading with negative slope), and the probability of a false alarm (P_{FA}) equals the integral of the other PDF from the threshold to infinity (shading with positive slope). This pair of probabilities produces one point on the ROC, which is generated in its entirety by moving the threshold continuously from positive to negative infinity. The resulting ROC is a convex line lying above the positive diagonal in the unit square.

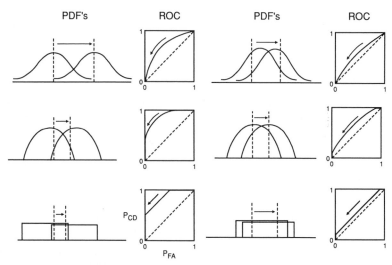

Fig. 7.5   Receiver operating characteristics vary in shape as PDFs vary. Less overlap of PDFs (greater differences in means in relation to variances) results in a more convex ROC (compare plots on the left with corresponding plots on the right), one that reaches closer to ideal performance ($P_{CD} = 1.0$, $P_{FA} = 0$, the upper left corner). Deviations from normality in the PDFs result in variations in the shapes of the ROCs without altering their general properties.

by the locus of P_{CD} as a function of P_{FA}, as a receiver's criterion varies continuously; this function is called the *receiver operating characteristic* (ROC, fig. 7.4).

The ROC for discriminations of this sort always lies on or above the positive diagonal and thus forms a convex set (Egan 1975; Macmillan and Creelman 1991). If the distributions of masking energy and of signal plus masking energy have the same shapes, then the ROC is symmetrical about the negative diagonal (fig. 7.5). Different variances or higher moments of the distributions produce an asymmetrical ROC. Furthermore, for all $P_{FA} > 0$, P_{CD} increases monotonically with P_{FA}. Note that the greater the inherent detectability of the signal (the larger the difference in the means of the two distributions in relation to their variances), the more convex is the ROC. Conversely, an almost undetectable signal results in a nearly flat ROC, close to the positive diagonal.

This simple scenario easily generalizes to more complex situations. In particular, it applies to recognition of two signals with values that overlap along some dimension. Even when a signal and irrelevant masking energy, or two different signals, are distinctly separable at close range, all signals become less recognizable as they attenuate or as their temporal properties degrade.

This discussion has so far assumed that a response occurs whenever the receiver registers a value exceeding its criterion. This situation suggests a threshold in responses—in other words, an open-ended, all-or-nothing pattern of response as some parameter of a stimulus varies. In contrast, most open-ended patterns of response fit a more continuously increasing function of the features of a stimulus. Such patterns, often discussed in terms of a supernormal stimulus, occur widely among animals (Staddon 1975; Cohen 1984; Rowland 1989; Ryan 1990; Ryan, chap. 8). Increasing response as a continuous function of some parameter of a stimulus, however, does not necessarily indicate an underlying continuity in responsiveness.

Continuity could also result from a threshold for detection of a stimulus in noisy conditions. In this case signals with low values of the parameter would exceed the receiver's threshold less often than those with high values. As a consequence, despite an underlying threshold, any response to the rate of stimulation would change continuously as the stimulus varied.

An example of a threshold for response is provided by the treefrog *Hyla chrysoscelis*, whose mating calls differ primarily in pulse rate from those of its cryptic sibling species *H. versicolor*. In choices between two calls, female *chrysoscelis* prefer those with pulse rates above a temperature-dependent threshold, which corresponds to the pulse rate in calls of conspecific males at the same temperature (Gerhardt 1982). This threshold tends to inhibit responses to the wrong species, as *chrysoscelis* males at any temperature have higher pulse rates than do *versicolor* males.

Some patterns of response suggest tuned rather than open-ended preferences. In these cases, an optimal stimulus or range of values for a stimulus evokes a response. Female *H. versicolor*, unlike females of their sibling species, have preferences tuned to pulse rates similar to those of conspecific males at the same temperature (Gerhardt and Doherty 1988). The female green treefrogs mentioned earlier provide another example of tuned preferences (Gerhardt 1987).

Adjustments in tuning have much the same consequences as do adjustments in thresholds. If two different signals (or signals and masking noise) overlap along some dimension, then decisions about the tuning of responsiveness, like those about simple thresholds, affect the probabilities of correct detection and false alarm. Any tuning curve, behavioral or neuronal, is specified by its best value, its bandwidth, and its symmetry. When the properties of received signals vary, there is inevitably a trade-off between narrow tuning to reduce false alarms and broad tuning to reduce missed detections, analogous to the selectivity-sensitivity trade-off for electronic filters. Furthermore, continuous variation in any parameter of a tuning curve generates a receiver operating characteristic with properties like those discussed above (fig. 7.6). In particular, P_{CD} is a convex function of P_{FA}, monotonically increasing for $P_{FA} > 0$. In addition, a more convex ROC results when two signals are inherently more discriminable (differ more in their means in relation to their variances).

A receiver adopts a criterion for responding to a signal by adjusting the parameters of a threshold or a tuning curve. The mechanism for this adjustment might be evolutionary, developmental, or physiological. This decision then affects the probabilities of the four possible outcomes for a receiver (correct and missed detections, false alarms, and correct rejections). How should this choice be made? Basic decision theory suggests that a criterion should maximize the expected utility for the receiver

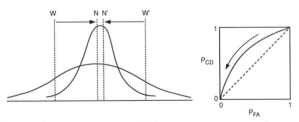

Fig. 7.6   The principles that apply to thresholds for response also apply to filters or tuned responses. As the width of the filter narrows from W–W' to N–N', the probabilities of correct detection (P_{CD}) and false alarm (P_{FA}) fall along a convex line, just as in the case of variation in a simple threshold.

(see Sperling 1983 for a clear exposition of expected utilities and their applications to receiver operating characteristics). As a result of its imperfect performance in responding to a particular signal (or in responding differently to two signals), a receiver's expected utility, $E(U)$, depends on (1) the *rates of occurrence* of different signals, (2) the *vector of payoffs* for the four possible outcomes, and (3) the *criterion* for a response, which sets the probabilities for the four outcomes.

Consider two signals (or a signal with background energy) that have different appropriate responses and occur with probabilities α and $1 - \alpha$, respectively. The probability density distributions of the two signals along some dimension separating them, together with the receiver's choice of a criterion for response, determine the probabilities of correct detection (P_{CD}), missed detection ($1 - P_{CD}$), false alarm (P_{FA}) and correct rejection ($1 - P_{FA}$). Each of these four outcomes has an associated payoff: h, m, a, and j, respectively. Each is a positive or negative net advantage. Thus the expected utility equals the sum, over all four outcomes, of the probability of each outcome times its payoff:

$$E(U) = \alpha h P_{CD} + \alpha m(1 - P_{CD}) + (1 - \alpha)a P_{FA} + (1 - \alpha)j(1 - P_{FA})$$
$$= \alpha(h - m)P_{CD} + (1 - \alpha)(a - j)P_{FA} + \alpha m + (1 - \alpha)j$$

The locus of values that yields constant expected utility is often called an *indifference curve*. To obtain such indifference curves in this case, we let $E(U)$ equal a constant, U, and rearrange the preceding equation, as follows:

$$P_{CD} = \frac{(1 - \alpha)(j - a)}{\alpha(h - m)} P_{FA} + \alpha(j - m) - j + U.$$

Each of these lines, like the ROC, represents the probability of correct detection (P_{CD}) expressed as a function of the probability of false alarm (P_{FA}). The slope is positive provided the payoff for a correct detection is greater than that for a missed detection ($h > m$) and the payoff for a correct rejection is greater than that for a false alarm ($j > a$). Both of these conditions are presumably met whenever it is advantageous for the receiver to respond to the signal.

Note that the slope remains constant but the intercept increases with increasing expected utility (U, above). We can therefore plot a family of parallel indifference lines on the same axes as the ROC (fig. 7.7A). The maximum realizable expected utility corresponds to the intersection of the ROC and the indifference line tangent to it. Therefore, for any ROC (determined by the probability density distributions of the features of signals and their alternatives) and for any expected utilities (determined by the probabilities of signals and alternatives and by the payoff vector for the four possible outcomes), maximum expected utility is attained by choosing a criterion for response that achieves P_{FA}^* and P_{CD}^* (the point of tangency in fig. 7.7).

The slope (S) of the indifference lines,

$$S = \frac{(1 - \alpha)(j - a)}{\alpha(h - m)},$$

determines the point of tangency on the ROC and thus both the optimal criterion for response and the optimal combination of P_{FA}^* and P_{CD}^* (fig. 7.7B). For large S (steep indifference lines), the optimal P_{FA}^* and P_{CD}^* are relatively small; for small S, the optimal P_{FA}^* and P_{CD}^* are relatively large.

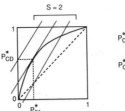
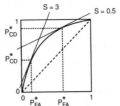

Fig. 7.7 (A) Indifference lines plotted on an ROC reveal the combination of P_{CD} and P_{FA} that maximizes the expected utility and thus specifies (see fig. 7.4) the optimal criterion for response. (B) For any given ROC, indifference lines with high slopes (S) result in lower optimal combinations of P_{CD} and P_{FA}. As a consequence, receivers with high S should evolve adaptive gullibility or adaptive fastidiousness, criteria that result in many missed detections.

Consider two cases that present different problems of signal detection: detecting a subtle cheater (case 1); and searching for subtly discriminable objects (case 2), such as cryptic prey or an optimal mate. These cases serve to illustrate an important point: a receiver maximizing its expected utility should often not detect all deceivers nor respond to all optimal stimuli.

Case 1 applies to males that try to detect females but are parasitized by males that resemble females and occasionally steal matings. In this case, let the payoff for failing to detect a cheater equal 0 ($m = 0$), because the subject has now lost its advantage in obtaining a mate. Then the payoff for detecting a cheater equals the probability of finding a mate at some time in the future ($h > 0$); the payoff for correctly recognizing a female (rejecting the possibility of a cheater) equals the probability of successful mating once a female has been found ($j > h$); and that for a false alarm, in which the subject attacks a female mistaken for a cheating male, equals the probability of finding another female some time in the future ($a = h$ or perhaps 0). Under these conditions,

$$S = \frac{(1 - \alpha)(j - h)}{\alpha h}.$$

If the probability of finding a mate after detecting a cheater (h) is small compared with the probability of mating with a correctly identified female (j), then the slope (S) of the indifference lines is steep. It would pay for males to adopt a criterion for response to cheaters with low, but nonzero, P_{CD}^* and consequently low P_{FA}^*. In other words, males in this case should not evolve to detect cheaters with complete accuracy. Instead, they should evolve some susceptibility to deception, a condition appropriately called *adaptive gullibility*.

Two other conditions also have this consequence: small α, and a flat ROC. Evolution should favor susceptibility to deception (low P_{CD}^*) when cheaters are infrequent in comparison with actual females (large $[1 - \alpha]/\alpha$) or when cheaters and actual females are virtually indistinguishable (flat ROC, fig. 7.8). In the latter case, even a moderately steep slope for the indifference lines can favor complete gullibility, with $P_{CD}^* = P_{FA}^* = 0$.

Case 2 applies to most forms of searching behavior; for instance, a female attempting to choose an optimal mate from among many similar ones or a predator seeking cryptic prey. In these cases, a false alarm incurs the costs of selecting a suboptimal mate or prey, and a correct detection, the benefits of selecting an optimal

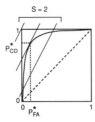
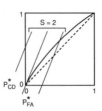

Fig. 7.8   For any given slope of indifference lines, a more convex ROC increases the probability of correct detections in relation to false alarms. (*A*) A signal that has low overlap with irrelevant stimulation and thus a highly convex ROC has high inherent detectability. (*B*) Steep indifference lines in combination with a nearly flat ROC can favor a criterion that excludes any response. Such signals are best ignored.

one ($h < a$). A missed detection (or a suitable mate or prey) or a correct rejection (of a suboptimal object) entails additional search, with its associated costs of time lost and risks encountered. We can think of these costs as devaluing the expected utility at the next opportunity for a response (U) by a factor λ. Under these conditions,

$$S = \frac{(1 - \alpha)(\lambda U - a)}{\alpha\,(h - \lambda U)},$$

where $h > \lambda U$ and $1 > \lambda > 0$. Here the steepness of the slope, S, increases with λ, which varies inversely with the costs of additional search. So low search costs favor a criterion that yields low P^*_{CD} and low P^*_{FA}. In other words, searchers should evolve to pass many suitable objects of their choice (low P^*_{CD}), a tendency appropriately called *adaptive fastidiousness*. Real (1990) reached a similar conclusion in discussing the theory of sequential search.

Again two other conditions favor low P^*_{CD} and P^*_{FA}: small α, and a flat ROC. For any constant cost of search, fastidiousness is favored when optimal objects of search are infrequent (large $[1 - a]/a$) or when optimal objects have low inherent detectability (flat ROC, fig. 7.8).

Cases 1 and 2 represent two views of situations that generate indifference lines with steep slopes. In both cases, maximizing expected utility leads to low P^*_{CD} and low P^*_{FA}. Case 1 emphasizes the advantages of gullibility in response to suboptimal signals; case 2 emphasizes the advantages of fastidiousness in response to optimal signals. Gullibility and fastidiousness are two aspects of the same problem.

SIGNALERS' ADAPTATIONS

Signalers face different problems depending on whether or not it is in the receiver's interest to respond. In cases of manipulation or deception, signalers receive an advantage from a response to a signal that is disadvantageous to the receiver. In these cases, the signaler must maximize the probabilities of a false alarm by a receiver. This objective is attained by producing a signal that mimics another signal (the model) that evokes the desired response in appropriate circumstances for the receiver. Mimicking a female to take advantage of a more successful male's attraction of females, simulating the presence of several rivals rather than just one (a Beau Geste effect: Krebs 1977, but see Yasukawa 1981), and mimicking an alarm call to trigger withdrawal of an opponent all fall into this category. Maximizing a response from the receiver in these cases, as demonstrated in the discussion of case 1 above, depends in general on close mimicry of the model, to achieve low inherent detectability by the receiver and thus a flat ROC, and on infrequent use, to assure low α.

When it is advantageous for the intended receiver to respond to a signal, then signalers should evolve to maximize the probability of correct detections by the receiver and to minimize disadvantageous interception by unintended receivers. The best way to increase a receiver's correct detections depends on its search costs, as described above for case 2. When the receiver's search costs (λ) are low enough that the indifference lines have slope $S - 1$, then frequent signaling (high α) and high inherent detectability (convex ROC) favor increased correct detections by the receiver. On the other hand, when the receiver's search costs are high enough that $S < 1$, then increased detectability could instead favor lower probabilities of correct detection.

The inherent detectability of signals becomes a problem especially in situations with many similar signals or with high attenuation and degradation of signals. In such situations, there are four basic ways to increase the inherent detectability of signals: (1) increasing the intensity of signals, (2) increasing their contrast with other patterns in the environment, (3) increasing the spatial or temporal stereotypy (redundancy) of signals, and (4) increasing the possibilities for the receiver to act as a matched filter (Wiley 1983, Harper 1991). For example, calling male frogs, to increase the chances of a female's response, could (1) evolve increased intensity of calling, (2) shift to frequencies or times with less masking sound, (3) repeat the call more often, and (4) use a stereotyped frequency (or frequency-temperature relationship) in stereotyped temporal patterns and at stereotyped locations. In general, frogs exemplify these adaptations.

The first two of these adaptations increase the signal's contrast with irrelevant energy. Increases in intensity, whether louder sounds, larger or brighter visual stimuli, or greater quantities of a pheromone, beyond some point require changes in morphology or synthetic pathways. These changes in turn require compromises with other demands on the signaler. Shifts to new parameters not masked by irrelevant energy in the environment require a correlated shift on the part of the receiver. In the initial stages, the advantage of the new signal to a receiver as a result of increased detectability would have to balance any disadvantage of additional search time for a rare signal. On the other hand, if a receiver at first had a more broadly tuned receptor, rather than one with the best stimulus shifted, then this variant receiver and the variant signaler might both realize immediate advantages.

Increased redundancy in signals has a well-known influence on the probability of correct detection (Shannon and Weaver 1949; Wiley 1983). Redundancy can include both spatial and temporal relationships among the components of a signal. In either case, predictable relationships among the components of a signal allow a receiver to reconstruct the correct signal from an imperfectly received one. The receiver thus has more than one chance for correct detection of the signal. Redundancy in animals' displays for communication is thus closely related to the classic ethological concept of ritualization (Cullen 1966; Wiley 1983).

Stereotyped temporal and spatial relationships among the components of a signal have the disadvantage of reducing the amount of information encoded in signals in unit time (or area). Consequently, these signals find their application in situations in which the probability of error in responses is high but the advantage of diverse or variable signals is low (Wiley 1983).

Receivers can detect signals most reliably when they are maximally prepared for them. For instance, the ability of human observers to detect signals in masking noise is improved by giving them only one, rather than several possible, signals to detect (reviewed by Wiley 1983). When a signal must be learned, those with easily remembered parameters enhance detection (Guilford and Dawkins 1991). In addition, human observers perform better when alerted about the time interval and the location in which a signal might occur. For instance, a light before or just after the possible occurrence of a sound in masking noise makes detection of the sound more reliable. In this case, the light serves as an easily detected alerting signal that specifies the time at which a less easily detected signal might occur. Such an alerting signal would also improve recognition, which involves classification of signals in addition to simple detection (Richards 1981a; Wiley and Richards 1982; Wiley 1983).

These improvements in the receiver's performance as a result of its knowledge of the possible occurrence or exact parameters of a signal could have two explanations. For electronic systems, such prior knowledge could be used to develop a filter optimally matched to the expected signal. Animals might conceivably alter the filtering characteristics of their sensors as well. Sensitization to repeated stimuli suggests such an effect. The improvement might also result from greater attention focused on the task at hand. Indeed, attention, in psychology, seems analogous to matched filtering, in engineering. Provided it is advantageous for receivers to detect or to recognize a signal, signaling should evolve to provide opportunities for attention or matched filtering, and receivers in turn should evolve to take advantage of these opportunities.

Unintended receivers can constrain the evolution of efficient signals. In some cases, special properties of signals can limit their detection to intended receivers. Use of high-frequency sounds, for instance, limits potential receivers to those in the immediate vicinity of the signaler and thus reduces possibilities for detection by predators or rivals. This consideration might explain their use in some alarm calls, juvenile begging calls, and precopulatory calls of birds. Low intensity also results in a relatively small active space, but high frequencies have the advantage of allowing high intensity close to the source in combination with a small active space. A particularly striking case of "private" communication occurs in a deep-sea fish, in which the photophores produce red light that has a high attenuation rate in water but is exactly matched by the species' receptor pigments (O'Day and Fernandez 1974).

IMPLICATIONS OF ERRORS IN COMMUNICATION

To explore possible adaptations for reducing errors in communication, the remaining sections consider the evolution of (1) behavioral boundaries between species, (2) spectacular displays, and (3) honesty and deception. These topics shed some light on the evolution of communication in the cases described in the opening section: mate choice by female frogs; interactions of territorial birds by singing; and deceptive use of alarm calls. The discussion emphasizes how little we know about errors in communication despite their central place in explaining the evolution of receivers' and signalers' behavior. A recent controversy concerning the nature of species provides a start.

Evolution of Behavioral Boundaries between Species

Do populations evolve adaptations for recognizing suitable mates or for rejecting unsuitable ones? Recent discussions of species limits have disagreed over whether species are more consistently characterized by recognition of conspecifics or by isolation from heterospecifics (Patterson and Macnamara 1984; Endler 1989). In other words, does the evolution of mate choice result in acceptance of conspecifics or rejection of heterospecifics? This distinction is obscured when a receiver's task is viewed as a problem in signal detection. As we have seen, probabilities of correct detection (response to a suitable mate) and false alarm (response to an unsuitable mate) do not usually admit independent adjustment. Consequently, probabilities of correct detection and correct rejection (rejecting an unsuitable mate) cannot usually be maximized simultaneously. To permit such simultaneous maximization, conspecific and heterospecific signals would have to be completely distinct to the receiver's sensors despite attenuation and degradation.

If hybridization is disadvantageous, selection should favor reduction of false alarms in mate choice. This situation thus fits case 2 above, with $a < 0$. With steep indifference lines, high probabilities of correct detection combined with low probabilities of false alarms can only occur with a highly convex receiver operating characteristic, a result of signals with high inherent distinctiveness. Consequently, signals might evolve greater distinctiveness in order to reduce either the frequency of dysgenic hybridization or the time required to find a conspecific mate. Scenarios for the divergence of mating signals thus fall between two extremes: either (1) the two populations initially hybridize after contact but subsequently diverge in their signals for mate choice as a result of selection against hybridization; or (2) the populations do not hybridize at the outset as a result of differences in the final (often short-range) signals for mate choice but subsequently diverge in their initial (long-range) signals as a result of selection for reduced time spent searching for a mate. In each case, selection for divergence would have to come from advantages of increased inherent discriminability of conspecific signals. Of course, populations in the first situation might evolve into the second situation if final (short-range) signals diverged before initial (long-range) signals did. Note that once initial signals are distinct, there is little or no selection for divergence of final signals for mating.

Few, if any, studies of hybrid zones have provided evidence specifically for the first scenario, often called reinforcement of reproductive isolation in sympatry (Butlin 1989). Several zones of contact between species of frogs provide evidence for greater divergence of signals in sympatry than in allopatry (Littlejohn 1965, 1981; Fouquette 1975; Ralin 1977), although whether each of these cases represents the outcome of the first scenario above (reinforcement) or the second (often called reproductive character displacement) remains uncertain.

Divergence of signals might not occur if changes in the properties of signals brought disadvantages in propagation. If so, these disadvantages in attracting conspecifics would cancel any advantages of attracting proportionately fewer heterospecifics. Reinforcement might also occur by changes in females' preferences without any change in signal properties. To increase selectivity for conspecific signals (decrease false alarms in proportion to correct detections), receivers could raise their thresholds (or narrow their tuning) for responses. These changes would also be likely to increase search time (a result of increased missed detections), so that advantages of selectivity would balance disadvantages of additional search. These considerations make it clear that any investigation of the evolution of behavioral boundaries between species must consider the net advantages of changes in signals and receivers' criteria in relation to the probabilities and consequences of false alarm, correct rejection, missed detection, and correct detection.

Mimicry of mating signals by potential predators drastically alters the optimal probabilities of correct detection and false alarm for the receiver. In the best-known case of this possibility, male *Photinus* fireflies responding to conspecific females' flash patterns must avoid the similar flash patterns of predatory *Photuris* females of several species (Lloyd 1981, 1985). In this case, the low inherent discriminability of conspecific from predatory signals results in a relatively flat receiver operating characteristic, and the high cost of false alarms in relation to the benefits of correct detections, as in case 2 above, produces steep indifference lines. Thus this situation should favor receivers with adaptive fastidiousness: low probabilities of correct detection as a result of high thresholds (or sharp tuning) for response. The males' cautious approaches to flashing females suggest such a situation. Female *Photuris* lure males of several species of *Photinus;* apparently the difficulty of mimicking several species provides opportunities for cautious *Photinus* males to detect most such traps. Female *Photuris* in turn appear to leave themselves open to conspecific males that mimic the flashes of their prey and thus locate them before they are ready to mate. It would be fascinating to have a full analysis of signal detection by these fireflies, including estimation of the expected utilities of response criteria.

Evolution of Spectacular Signals

Elaborations of displays that result in increased intensity, contrast with irrelevant patterns of energy, and temporal and spatial redundancy all serve to improve an intended receiver's performance by increasing the inherent detectability of the signal. Such signals thus offer advantages to signalers when attenuation and degradation are high or when many different signals require discrimination, so that receivers would otherwise miss many signals. We can call signals that have evolved in this way spectacular displays; their defining characteristic is high inherent detectability (or discriminability from other signals).

Previous explanations for the evolution of elaborate displays fall into two general categories: (1) runaway sexual selection for exaggerated features of signals as a result of arbitrary female preferences (those with no consequences for the female's survival or reproductive success or for her progeny's success) (Fisher 1930; Lande 1981; Kirkpatrick 1987); and (2) selection for costly signals to ensure honesty (Zahavi 1975; Kodric-Brown and Brown 1984; Andersson 1982a, 1982b; Grafen 1990a, 1990b; Maynard Smith 1991). Exaggerated or costly displays are, however, not necessarily spectacular, in the sense of being inherently highly detectable. Basic considerations of signal detection suggest, however, that inherent detectability is a crucial consideration.

For females, the consequences of choosing the wrong mate can include less than optimal genes for her progeny or less assistance in raising her offspring. If the cost of a false alarm (choosing a suboptimal mate) is high (net advantage low) and the cost of a missed detection (additional search for an optimal mate) is low (net advantage high), this situation resembles case 2 above. Steep indifference lines and a shallow receiver operating characteristic, as a result of signals with low inherent discriminability, favor the evolution of adaptive fastidiousness. Because the low probabilities of response in this case create "noise" for a signaler, signaling by optimal males should evolve toward more spectacular displays.

It is important to emphasize again that fastidiousness of a receiver in itself creates "noisy" communication from the viewpoint of the signaler (see above). For a signaler, the situation is exactly the same whether receivers fail to respond to signals because of environmental attenuation or because of "neural attenuation," a result of high thresholds or narrow filters for response. In either case, more spectacular signals—ones with greater intensity, redundancy, and distinctiveness—are advantageous (Wiley 1983).

From the female's perspective, the difficulty of detecting optimal mates presumably increases when she has little time to make her choice, when she has little experience with potential mates, or when she responds to attenuated signals from potential mates at long distances (Wiley 1983). These three situations, respectively, reduce the redundancy from prolonged communication with potential mates, make mates of different quality inherently less discriminable, or require discrimination of attenuated and degraded signals. Females should encounter greater problems in optimal mate choice when their previous experience with these signals is limited, for instance, when young are reared by females alone, when sexes do not associate regularly except during mating, and when mate choice occurs quickly. These situations favor a shift in females' preferences toward signals with greater inherent detectability (or discriminability). Thus when a response is made after a short time, with little experience, or at long range, a female's preference for spectacular signals has advantages in reducing false alarms and missed detections. Such a preference is adaptive rather than arbitrary.

The process of runaway sexual selection, as a result of genetic correlation between signal properties and female preferences, can occur in theory with completely arbitrary female preferences. Nevertheless, the conditions for sexual selection depend strongly on the net advantages of female preferences (Heisler 1984; Pomiankowski 1987, 1988). From the preceding analysis of female preferences as a problem in signal detection, it seems likely that these preferences are rarely, if ever, completely arbitrary. A full analysis of these advantages must incorporate both the probabilities of false alarms, missed detections, and correct detections and their consequences (suboptimal mating, additional search, and optimal mating, respectively) for a female's reproductive success.

Any shift in a female's preferences for mates seems likely to alter the expected utility of her criterion for response, as derived above. Direct selection for female preferences, such as selection for an increase in the expected utility of the criterion for detecting optimal mates, makes the evolution of adaptive preferences more likely, because it lowers the threshold required for runaway evolution (Heisler 1984). Thus the universality of the problems of signal detection tends to favor the evolution by sexual selection of exaggerated signals that are also inherently detectable or discriminable—in other words, spectacular displays.

The distinctive acceleration of sexual selection requires genetic correlation as a result of assortative mating between signalers and receivers. This acceleration, despite Zahavi's (1991) recent arguments, can thus only apply to signals that affect recognition of mates. Otherwise, signals and receivers' criteria no doubt still coevolve, in the sense that the advantages of any signal depend on the criteria of receivers in the population and, conversely, the advantages of any criterion depend on the available signals, but the acceleration as a result of sexual selection is absent.

Limitations on the evolution of spectacular displays include the disadvantages to signalers from interception of their signals by unintended receivers, often predators, parasites, or rivals. Bright coloration increases predation in fishes (MacPhail 1969; Endler 1980, 1983); calling attracts parasites to crickets (Cade 1979) and predators to katydids and frogs (Bellwood and Morris 1987; Ryan et al. 1981). These disadvantages of spectacular displays, however, are often partially mitigated by mechanisms for concealment when not in use. Many lekking birds with spectacular plumage can conceal conspicuous features when not displaying (Wiley 1991b). Túngara frogs (*Physalaemus pustulosus*) utter the wideband component of their calls, the "chuck," primarily when calling in groups; this component attracts both females and predaceous bats (Ryan et al. 1981; Rand and Ryan 1981; Ryan 1983). Thus, by adding "chucks" when calling in groups, males increase their mating success in competition with other males and minimize the consequent risks of predation.

Spectacular signals also have disadvantages in rapid transmission of information, as a consequence of their high spatial and temporal redundancy. Thus situations that require high rates of transmission of information, such as negotiations during close-range interactions of mates or opponents, should favor diverse and variable signals rather than stereotyped ones. This contrast is evident in several comparisons of close-range and long-range communication in birds and primates (Marler 1973; Wiley 1973). The evolution of spectacular displays is thus constrained not only by risks of predation or parasitism but also by limitations on rates of transmission of information.

Evolution of Honesty and Deception

Sexual selection, when female preferences are adaptive, is a special case of the evolution of honesty in signaling. Much as in the case of sexual selection, the evolution of honesty in general must often result from the evolution of receivers' preferences for inherently reliable signals. Simple signals, easily bluffed or mimicked, tend to become unreliable. For honest signalers, it then pays to escalate the intensity or persistence of display until imitators can no longer match them (Andersson 1982a, 1982b; Kodric-Brown and Brown 1984). For receivers, it is advantageous to respond selectively to these escalated signals or to any others that preclude mimicry or bluffing. In the case of signals indicating overall strength or vigor, costly displays or those directly revealing vigor should prevent imitation by less vigorous signalers. Either way, the result is honest signals.

When communication is viewed in terms of signal detection, the conditions for the evolution of honest, costly signals nearly match those for the evolution of spectacular, exaggerated displays. Both sorts of signals can result from receivers' objectives of increasing correct detections and decreasing false alarms. These objectives can be achieved in any one, or a combination, of at least three ways: receivers can (1) adopt more stringent criteria, (2) switch to inherently more reliable signals, or (3) probe the honesty of signalers at irregular intervals (Wiley 1983). Each possibility raises additional issues.

If receivers evolve more stringent criteria for response, honest signalers should evolve progressively more exaggerated displays. Exaggeration, spectacular display, and high cost might thus often evolve together. When a spectacular, exaggerated signal requires overall vigor, its high cost results in high inherent detectability for a receiver, because deceiving signalers cannot match the costly signal. Note that it is not enough for such a signal to have high cost or exaggeration; it must also be spectacular, so that a receiver can easily discriminate it from less costly displays. Otherwise, there is no increase in the receiver's expected utility from increasing its criterion for response.

If a receiver changes its criterion to reduce false alarms (from bluffs or mimicry), it also usually increases its missed detections. Consider a newly arrived migrant bird looking for a territorial vacancy. If it adopts a lax criterion for judging whether an area is already occupied by a serious defender, it risks false alarms, cases in which it fails to challenge minimal advertisement by a weak opponent or by an opponent trying to claim an oversized territory, and thus incurs additional costs of search or risks failing to find any territory. On the other hand, if it adopts a stringent criterion, it risks missed detections, cases in which opponents fight vigorously when challenged. Thus a more stringent criterion (higher threshold or narrower tuning) encounters both the advantage of fewer false alarms (vacancies passed up by mistake) and the disadvantage of more missed detections (challenges that are lost causes). The balance of advantages and disadvantages then determines whether or not it pays to adopt more stringent criteria for response to a signal and hence whether or not exaggeration of the signal occurs.

A second alternative for receivers, instead of a more stringent criterion for response to an unreliable signal, is a switch to a signal with less bluffable parameters, ones that inherently permit discrimination of honest and deceptive signalers. The intensity of the carotenoid pigments of male guppies (*Poecilia reticulata*), as an indicator of feeding efficiency provides a possible example of such a parameter (Endler 1980; Kodric-Brown 1989). In this case honest signals need not be exaggerated to assure high detectability.

A third alternative for receivers subject to deception is to probe signalers by calling their bluff. If bluffers or mimics, once discovered, incurred high losses, then these disadvantages of bluffing could promote honesty in signaling. On the other hand, as any gambler knows, calling a strong opponent's bluff has its own costs. The consequences for exaggeration of signals seem complex. If probing occurred frequently, honesty would tend to spread at the expense of bluffing. On the other hand, probing would tend to spread only if bluffing occurred frequently. These reciprocal influences could lead to evolutionarily stable mixed strategies of probing and gullible receivers, on the one hand, and honest and bluffing signalers, on the other. In general, probing could stall the evolution of exaggerated signals in response to bluffing and mimicry by assuring high inherent detectability of simple signals.

This discussion emphasizes that, as a result of the interaction of selection on signalers and receivers, signals normally evolve toward honesty. The problem in the evolution of deception is thus the nature of limitations on the evolution of honesty. Since deception relies on errors by receivers, selection should always favor receivers that minimize deception by increasing discrimination, shifting to more reliable signals, or probing. However, receivers face trade-offs that can limit their adaptations to deception.

The theory of signal detection makes these trade-offs clear (Wiley 1983). As explained above, a receiver's performance can be understood only in terms of the inherent detectability of signals (and consequently the receiver operating characteristic) and the criterion for response. The expected utility of any criterion in turn depends on the probabilities and net advantages of each of the four possible outcomes of a receiver's decisions to respond or not. Dawkins and Guilford (1991) have also emphasized the importance of receivers' costs in explaining the evolution of deception. These costs are incorporated in the expected utility of a receiver's criterion for response. As shown above, maximizing this expected utility can lead to evolutionarily stable deception.

Such adaptive gullibility can, for example, explain the evolutionary stability of deception by males that mimic females in order to steal matings. For males intrinsically more likely to attract females, the costs of false alarms (attacking an actual female mistaken for a deceptive male) can make it advantageous to accept some deception (missed detections of female mimics). Thus, depending on the payoff vector and probabilities of true and false signals, the receiver's expected utility is maximized by accepting some level of deception rather than by changing its criterion for response in order to minimize deception. In this case, deception becomes an evolutionarily stable feature of communication.

In the case of deceptive alarm calls, the situation is similar. Here the costs of missed detections (failing to respond to a true alarm) make it advantageous for a receiver to accept some deception (false alarms). Again, the receiver maximizes its expected utility by a criterion that results in some deception, and thus deception becomes an evolutionarily stable feature of communication.

EVOLUTIONARILY STABLE DECEPTION

In a series of elegant expositions, Grafen (1990a, 1990b, 1991) presented a strong case for the evolution of honesty in signaling. In his model of communication, the evolutionarily stable strategy for receivers is to respond only to reliable signals. Consequently, the evolutionarily stable strategy for signalers is to produce such signals. He concludes trenchantly, "Receivers must get what they want in a stable signalling system" (Grafen 1990a, 526). The model of communication analyzed by Grafen involves advertisement and assessment. A signaler's objective is to persuade the receiver that its quality is as high as possible, and the receiver's objective is to evaluate the signaler's quality.

In this model, the deduction that honesty in signaling is evolutionarily stable requires only a few assumptions. Most important is the assumption of "continuity in everything" (Grafen 1990b, 476): the signaler's fitness is a continuous function of its actual signaling level, its perceived signaling level, and its quality. Reliable or honest signals require costs for signaling, and in particular, greater costs for signalers with lower quality. Provided there is continuity in everything, it does not pay to try deception: for any increase in signaling, above the signaler's evolutionarily stable level, the gain from enhanced perception by the receiver is more than compensated by the increased cost. Because every signaler has its own equilibrium related to its quality, honesty in signaling is the only evolutionarily stable strategy.

In his discussion of the limitations of this model, Grafen (1990a: sections 5 and 6) considered two possibilities that might restrict honesty in communication. One of these possibilities, the most obvious element of the present approach missing in Grafen's models, is error by receivers. His brief discussion of this issue (Grafen 1990a, 528) makes it clear, however, that imperfect perception by receivers, short of outright blindness, has little influence on his deductions. The assumption of "continuity in everything" assures that an error-prone receiver's perceptions are continuously related to expectations of the signaler's quality. Receivers' errors alone are not sufficient to limit honesty in signaling.

The situation changes, however, if we abandon the assumption of continuity. As Grafen notes, when he considers this second possible restriction on honesty, there might exist an alternative set of signalers or modes of signaling for which "signals are much cheaper for a given quality [of the signaler]" (Grafen 1990a, 533). This possibility of cheap imitations of honest signals, in combination with error by receivers, produces the conditions for evolutionarily stable deception.

Such deceptive signals cannot occur too frequently in comparison with honest signals, otherwise receivers would do better to ignore the signals. A receiver unable to discriminate between deceptive and honest signals must rely on the average gain when deciding to respond (Wiley 1983, Grafen 1990a, 534–35). The present treatment of communication has clarified this condition. To maximize its fitness, a receiver must adjust its criterion for response to maximize its expected utility, a quantity that depends on the frequencies of honest and deceptive signals and on the consequences for fitness of responding or failing to respond to each.

Grafen's analysis is thus in substantial agreement with the present approach. As a general rule, the interaction between signaler and receiver leads to honesty in communication. Evolutionarily stable deception is an exception to the rule. Grafen's analysis clarifies the importance for the evolution of stable deception of some discontinuity in the relations between signals or costs and signalers' states. The present analysis emphasizes the importance of imperfect discrimination by receivers. This condition in turn requires an evolutionarily stable limitation on discrimination. The application of decision theory to communication makes this limitation clear: maximizing the receiver's expected utility in responding to a signal can limit the advantages of discrimination by a receiver and, provided there is not "continuity in everything," can lead to evolutionarily stable error and deception.

CONCLUSION

Errors by receivers, in the sense of evolutionarily inappropriate responses to signals, are likely to occur during communication at long range, in dense aggregations, or in the presence of deception. Such errors have major implications for the evolution of communication. For signals that cannot be completely distinguished by a receiver, there are four possible results of any decision to respond or not to respond to a signal: correct detection, missed detection, false alarm, and correct rejection. In general, the probabilities of these results cannot be independently adjusted by a receiver; in particular, criteria for response that increase the probability of correct detection also inevitably increase the probability of false alarm. The inherent discriminability of the signal, together with the probability and net advantage of each outcome, determines the expected utility of any criterion for response. Maximizing the expected utility of a receiver's criterion can lead to *adaptive gullibility* (evolutionarily stable susceptibility to deception) or to *adaptive fastidiousness* (low responsiveness to signals). In the latter case, signalers must contend with increased uncertainty in the responses to a signal. Signalers in such situations can improve signaling efficiency by increasing the inherent discriminability of signals, often by exaggeration. Female choice of mates, as an example of adaptive fastidiousness, can result in such exaggeration of signals. In general, explanations for the evolution of any receivers' performance, as in the cases of female choice or evolutionarily stable susceptibility to deception, require evaluation of the expected utility of the receivers' criterion as well as the probability and inherent detectability of the signal.

ACKNOWLEDGMENTS

I thank members of the Animal Behavior Seminar and Ecolunch at Chapel Hill for thorough review of the ideas expressed here. Many helpful suggestions were also received from Carl Gerhardt, Michael Ryan, Lori Wollerman, Jean Boal, and Helmut Mueller.

REFERENCES

Andersson, M. 1982a. Evolution of condition-dependent ornaments and mating preferences: Sexual selection on viability differences. *Evolution* 40:804–16.

Andersson, M. 1982b. Sexual selection, natural selection and quality advertisement. *Biological Journal of the Linnaean Society* 17:375–93.

Austin, J. L. 1973. *How to do things with words*. 2d ed. Cambridge, Mass.: Harvard University Press.

Becker, P. H. 1982. The coding of species-specific characteristics in bird sounds. In D. E. Kroodsma and E. H. Miller, eds., *Acoustic communication in birds*, vol. 1, 213–52. New York: Academic Press.

Bellwood, J. J., and G. K. Morris. 1987. Bat predation and its influence on calling behavior in neotropical katydids. *Science* 238:64–67.

Bennett, J. 1976. *Linguistic behaviour*. Cambridge: Cambridge University Press.

Brenowitz, E. A. 1982. The active space of red-winged blackbird song. *Journal of Comparative Physiology* 147:511–22.

Brown, C. H., and P. M. Waser. 1984. Hearing and communication in blue monkeys (*Cercopithecus mitis*). *Animal Behaviour* 32:66–75.

Butlin, R. 1989. Reinforcement of premating isolation. In D. Otte and J. A. Endler, eds., *Speciation and its consequences*, 158–79. Sunderland, Mass.: Sinauer Associates.

Cade, W. 1979. The evolution of alternative male reproductive strategies in field crickets. In M. Blum and N. Blum, eds., *Sexual selection and reproductive competition in insects*, 343–80. New York: Academic Press.

Cheney, D. L., and R. M. Seyfarth. 1990. *How monkeys see the world*. Chicago: University of Chicago Press.

Cherry, C. 1957. *On human communication*. Cambridge, Mass.: MIT Press.

Cohen, J. 1984. Sexual selection and the psychophysics of mate choice. *Zeitschrift für Tierpsychologie* 64:1–8.

Cosens, S. E., and J. B. Falls. 1984. A comparison of sound propagation and song frequency in temperate marsh and grassland habitats. *Behavioral Ecology and Sociobiology* 15:161–70.

Cullen, J. M. 1966. Reduction of ambiguity through ritualization. *Philosophical Transactions of the Royal Society of London* B 251:363–74.

Dabelsteen, T., and S. B. Pedersen. 1990. Song and information about aggressive responses of blackbirds, *Turdus merula*: Evidence from interactive playback experiments with territory owners. *Animal Behaviour* 40:1158–68.

Dawkins, M. S., and T. Guilford. 1991. The corruption of honest signalling. *Animal Behaviour* 41:865–74.

Dawkins, R., and J. R. Krebs. 1978. Animal signals: Information or manipulation? In J. R. Krebs and N. B. Davies, eds., *Behavioural ecology*, 2d ed., 282–309. Oxford: Blackwell Scientific Publications.

Dennett, D. C. 1987. *The intentional stance*. Cambridge, Mass.: MIT Press.

Dennett, D. C. 1988. The intentional stance in theory and practice. In R. W. Byrne and A. Whiten, eds., *Machiavellian intelligence*, 180–202. Oxford: Oxford University Press.

Egan, J. P. 1975. *Signal detection theory and ROC analysis*. New York: Academic Press.

Embleton, T. F. W., J. E. Piercy, and N. Olson. 1976. Outdoor sound propagation over ground of finite impedance. *Journal of the Acoustical Society of America* 59:267–77.

Endler, J. 1980. Natural selection on color patterns of *Poecilia reticulata*. *Evolution* 31:76–91.

Endler, J. A. 1989. Conceptual and other problems in speciation. In D. Otte and J. A. Endler, eds., *Speciation and its consequences*, 625–48. Sunderland, Mass.: Sinauer Associates.

Falls, J. B. 1982. Individual recognition by sounds in birds. In D. E. Kroodsma and E. H. Miller, eds., *Acoustic communication in birds*, vol. 2, 237–78. New York: Academic Press.

Falls, J. B., and R. Brooks. 1975. Individual recognition by song in white-throated sparrows. II. Effects of location. *Canadian Journal of Zoology* 59:2380–85.

Fisher, R. A. 1930. *The genetical theory of natural selection*. London: Oxford University Press.

Fouquette, M. J. 1975. Speciation in chorus frogs. I. Reproductive character displacement in the *Pseudacris nigrita* complex. *Systematic Zoology* 24:16–23.

Gerhardt, H. C. 1982. Sound pattern recognition in North American treefrogs (Anura: Hylidae): Implications for mate choice. *American Zoologist* 22:581–95.

Gerhardt, H. C. 1983. Communication and the environment. In T. R. Halliday and P. J. B. Slater, eds., *Animal behaviour*, vol. 2, *Communication*, 82–113. Oxford: Blackwell Scientific Publications.

Gerhardt, H. C. 1987. Evolutionary and neurobiological implications of selective phonotaxis in the green treefrog, *Hyla cinerea*. *Animal Behaviour* 35:1479–89.

Gerhardt, H. C., and J. A. Doherty. 1988. Acoustic communication in the gray treefrog, *Hyla versicolor*: Evolutionary and neurobiological implications. *Journal of Comparative Physiology* A 162:261–78.

Gerhardt, H. C., and G. M. Klump. 1988a. Masking of acoustic signals by the chorus background noise in the green treefrog: A limitation on mate choice. *Animal Behaviour* 36: 1247–49.

Gerhardt, H. C., and G. M. Klump. 1988b. Phonotactic responses and selectivity of barking treefrogs (*Hyla gratiosa*) to chorus sounds. *Journal of Comparative Physiology* A 163:795–802.

Gerhardt, H. C., S. I. Guttman, and A. A. Karlin. 1980. Natural hybrids between *Hyla cinerea* and *Hyla gratiosa*: Morphology, vocalization and electrophoretic analysis. *Copeia* 1980: 577–84.

Godard, R. 1991. Long-term memory of individual neighbors in a migratory songbird. *Nature* 350:228–29.

Godard, R. 1993. Tit for tat among neighboring hooded warblers. *Behavioral Ecology and Sociobiology* 33:45–50.

Grafen, A. 1990a. Biological signals as handicaps. *Journal of Theoretical Biology* 144:517–46.

Grafen, A. 1990b. Sexual selection unhandicapped by the Fisher process. *Journal of Theoretical Biology* 144:473–516.

Grafen, A. 1991. Modelling in behavioural ecology. In J. R. Krebs and N. B. Davies, eds., *Behavioural ecology*, 3d ed., 5–31. Oxford: Blackwell Scientific Publications.

Green, D. M., and J. A. Swets. 1966. *Signal detection theory and psychophysics*. New York: Krieger.

Grice, H. P. 1969. Utterer's meanings and intentions. *Philosophical Reviews* 78:147–77.

Guilford, T., and M. S. Dawkins. 1991. Receiver psychology and the evolution of animal signals. *Animal Behaviour* 42:1–14.

Harper, D. G. C. 1991. Communication. In J. R. Krebs and N. Davies, eds., *Behavioural ecology*, 3d ed., 374–98. Oxford: Blackwell Scientific Publications.

Harré, R. 1984. Vocabularies and theories. In R. Harré and V. Reynolds, eds., *The meaning of primate signals*, 90–106. Cambridge: Cambridge University Press.

Harris, R. 1984. Must monkeys mean? In R. Harré and V. Reynolds, eds., *The meaning of primate signals*, 116–37. Cambridge: Cambridge University Press.

Heisler, I. L. 1984. A quantitative genetic model for the origin of mating preferences. *Evolution* 38:1283–95.

Järvi, T., T. Radesäter, and S. Jacobsson. 1980. The song of the willow warbler *Phylloscopus trochilus* with special reference to singing behaviour in agonistic situations. *Ornis Scandinavica* 11:236–42.

Jilka, A., and B. Leisler. 1974. Die Einpassung dreier Rohrsängerarten (*Acrocephalus schoenobaenus*. *A. scirpaceus*, *A. arundinaceus*) in ihre Lebensräume in bezug auf das Frequenzspektrum ihrer Reviergesänge. *Journal für Ornithologie* 115:192–212.

King, A. P., and M. J. West. 1983. Female perception of cowbird song: A closed developmental program. *Developmental Psychobiology* 16:335–42.

Kirkpatrick, M. 1987. Sexual selection and the evolution of female choice. *Evolution* 36:1–12.

Kodric-Brown, A. 1989. Dietary carotenoids and male mating success in the guppy: an environmental component to female choice. *Behavioral Ecology and Sociobiology* 25:393–401.

Kodric-Brown, A., and L. H. Brown. 1984. Truth in advertising: The kinds of traits favored by sexual selection. *American Naturalist* 124:309–23.

Krebs, J. R. 1977. The significance of song repertoires: The Beau Geste hypothesis. *Animal Behaviour* 25:475–78.

Lande, R. 1981. Models of speciation by sexual selection on polygenic traits. *Proceedings of the National Academy of Sciences U.S.A.* 78:3721–25.

Lein, M. R. 1978. Song variation in a population of chestnut-sided warblers (*Dendroica pensylvanica*): Its nature and suggested significance. *Canadian Journal of Zoology* 56:1266–83.

Littlejohn, M. J. 1965. Premating isolation in the *Hyla ewingi* complex (Anura: Hylidae). *Evolution* 19:234–43.

Littlejohn, M. J. 1977. Long-range acoustic communication in anurans: An integrated and evolutionary approach. In D. H. Taylor and S. I. Guttman, eds., *The reproductive biology of amphibians*, 263–94. New York: Plenum Press.

Littlejohn, M. J. 1981. Reproductive isolation: A critical review. In W. R. Atchley and D. S. Woodruff, eds., *Evolution and speciation*, 298–334. Cambridge: Cambridge University Press.

Littlejohn, M. J., and A. A. Martin. 1969. Acoustic interaction between two species of leptodactylid frogs. *Animal Behaviour* 17:785–91.

Lloyd, J. E. 1981. Mimicry in the sexual signals of fireflies. *Scientific American* 245(July):138–45.

Lloyd, J. E. 1985. On deception, a way of all flesh, and firefly signalling and systematics. *Oxford Surveys in Evolutionary Biology* 1:48–84.

Loftus-Hills, J. J., and M. J. Littlejohn. 1971. Mating-call sound intensities of anuran amphibians. *Journal of the Acoustical Society of America* 49:1327–29.

Lythgoe, J. N. 1979. *The ecology of vision*. Oxford: Clarendon Press.

Lythgoe, J. N. 1984. Visual pigments and environmental light. *Vision Research* 24:1539–50.

MacGregor, P. K. 1991. The singer and the song: On the receiving end of bird song. *Biological Reviews* 66:57–81.

MacGregor, P. K., and J. R. Krebs. 1984. Sound degradation as a distance cue in great tit (*Parus major*) song. *Behavioral Ecology and Sociobiology* 16:49–56.

Macmillan, N. A., and C. D. Creelman. 1991. *Detection theory*. Cambridge: Cambridge University Press.

MacPhail, J. D. 1969. Predation and the evolution of a stickleback (*Gasterosteus*). *Journal of the Fisheries Research Board of Canada* 26:3183–3208.

Marler, P. 1973. A comparison of vocalizations of red-tailed monkeys and blue monkeys, *Cercopithecus ascanius* and *C. mitis*, in Uganda. *Zeitschrift für Tierpsychologie* 33:223–47.

Marten, K., and P. Marler. 1977. Sound transmission and its significance for animal vocalization. I. Temperate habitats. *Behavioral Ecology and Sociobiology* 2:271–90.

Maynard Smith, J. 1991. Honest signalling—the Philip Sidney game. *Animal Behaviour* 42:1034–35.

Michelsen, A., and O. N. Larsen. 1983. Strategies for acoustic communication in complex environments. In F. Huber and H. Markl, eds., *Neuroethology and behavioral physiology*, 322–32. Berlin: Springer.

Møller, A. P. 1988. False alarm calls as a means of resource usurpation in the great tit *Parus major. Ethology* 79:25–30.

Morton, E. S. 1975. Ecological sources of selection on avian sounds. *American Naturalist* 109:17–34.

Morton, E. S. 1986. Predictions from the ranging hypothesis for the evolution of long-distance signals in birds. *Behaviour* 83:66–86.

Munn, C. A. 1986a. Birds that cry "wolf." *Nature* 319:1433–35.

Munn, C. A. 1986b. The deceptive use of alarm calls by sentinel species in mixed-species flocks of neotropical birds. In R. Mitchell and N. Thompson, eds., *Deception: Perspectives on human and nonhuman deceit*, 169–75. Albany: State University of New York Press.

Myrberg, A. A., Jr. 1991. Distinctive markings of sharks: Ethological considerations of visual function. *Journal of Experimental Zoology*, Supplement 5:156–66.

O'Day, W. T., and H. R. Fernandez. 1974. *Aristomias scintillans* (Malacosteidae): A deep sea fish with visual pigments adapted to its own bioluminescence. *Vision Research* 14:545–50.

Ogden, C. K., and I. A. Richards. 1923. *The meaning of meaning*. London: Routledge and Kegan Paul.

Oldham, R. S., and H. C. Gerhardt. 1975. Behavioral isolation of the treefrogs *Hyla cinerea* and *Hyla gratiosa. Copeia* 1975:223–31.

Patterson, H. E. H., and M. Macnamara. 1984. The recognition concept of species. *South African Journal of Science* 80:312–18.

Paul, R. C., and T. J. Walker. 1979. Arboreal singing in a burrowing cricket, *Anurogryllus arboreus. Journal of Comparative Physiology* 132:217–23.

Piercy, J. E., and T. F. W. Embleton. 1977. Review of noise propagation in the atmosphere. *Journal of the Acoustical Society of America* 61:1403–18.

Pomiankowski, A. N. 1987. The costs of choice in sexual selection. *Journal of Theoretical Biology* 128:195–218.

Pomiankowski, A. N. 1988. The evolution of female mate preferences for male genetic quality. *Oxford Surveys in Evolutionary Biology* 5:136–84.

Putnam, H. 1967. The mental life of some machines. In H. N. Castaneda, ed., *Intentionality, minds and perception,* 177–200. Detroit: Wayne State University Press.

Ralin, D. B. 1977. Evolutionary aspects of mating call variation in a diploid-tetraploid species complex of treefrogs (Anura). *Evolution* 31:721–36.

Rand, A. S., and M. J. Ryan. 1981. The adaptive significance of a complex vocal repertoire in a neotropical frog. *Zeitschrift für Tierpsychologie* 57:209–14.

Real, L. 1990. Search theory and mate choice. I. Models of single-sex discrimination. *American Naturalist* 136:376–405.

Richards, D. G. 1981a. Alerting and message components in the songs of rufous-sided towhees. *Behaviour* 76:223–49.

Richards, D. G. 1981b. Estimation of distance of singing conspecifics by the Carolina wren. *Auk* 98:127–33.

Richards, D. G., and R. H. Wiley. 1980. Reverberations and amplitude fluctuations in the propagation of sound in a forest: Implications for animal communication. *American Naturalist* 115:381–99.

Ristau, C. 1991. Aspects of the cognitive ethology of an injury-feigning bird, the piping plover. In C. A. Ristau, ed., *Cognitive ethology,* 91–126. Hillsdale, N.J.: Lawrence Erlbaum Associates.

Ristau, C. A., and D. Robbins. 1982. Cognitive aspects of ape language experiments. In D. R. Griffin, ed., *Animal mind—human mind,* 299–331. Berlin: Springer.

Römer, H., and W. J. Bailey. 1986. Insect hearing in the field. II. Male spacing behaviour and correlated acoustic cues in the bushcricket *Mygalopsis marki. Journal of Comparative Physiology* 159:627–38.

Rowland, W. J. 1989. Mate choice and the supernormality effect in female sticklebacks (*Gasterosteus aculeatus*). *Behavioral Ecology and Sociobiology* 24:433–38.

Ryan, M. J. 1983. Sexual selection and communication in a neotropical frog, *Physalaemus pustulosus. Evolution* 37:261–72.

Ryan, M. J. 1990. Sexual selection, sensory systems, and sensory exploitation. *Oxford Surveys in Evolutionary Biology* 7:157–95.

Ryan, M. J., and E. A. Brenowitz. 1985. The role of body size, phylogeny, and ambient noise in the evolution of bird song. *American Naturalist* 126:87–100.

Ryan, M. J., M. D. Tuttle, and L. K. Taft. 1981. The costs and benefits of frog chorusing behavior. *Behavioral Ecology and Sociobiology* 8:273–78.

Saussure, F. de. 1959. *Course in general linguistics.* New York: Philosophical Library. Original edition in French, Paris, 1915.

Schlefer, E. K., M. A. Romano, S. I. Guttman, and S. B. Ruth. 1986. Effects of twenty years of hybridization in a disturbed habitat on *Hyla cinerea* and *Hyla gratiosa. Journal of Herpetology* 20:210–21.

Schroeder, D. M., and R. H. Wiley. 1983. Communication with repertoires of song themes in tufted titmice. *Animal Behaviour* 31:1128–38.

Schwartz, J. J. 1987. The function of call alternation in anuran amphibians: A test of three hypotheses. *Evolution* 41:461–71.

Schwartz, J. J., and K. D. Wells. 1983a. An experimental study of acoustic interference between two species of neotropical frogs. *Animal Behaviour* 31:181–90.

Schwartz, J. J., and K. D. Wells. 1983b. The influence of background noise on the behavior of a neotropical frog, *Hyla ebraccata. Herpetologica* 39:121–29.

Searcy, W. A. 1990. Species recognition of song by female red-winged blackbirds. *Animal Behaviour* 40:1119–27.

Searcy, W. A., P. Marler, and S. S. Peters. 1981. Species song discrimination in adult female song and swamp sparrows. *Animal Behaviour* 29:997–1003.

Shannon, C. E., and W. Weaver. 1949. *The mathematical theory of communication*. Urbana: University of Illinois Press.

Smith, W. J., J. Pawlukiewicz, and S. T. Smith. 1978. Kinds of activities correlated with singing patterns in the yellow-throated vireo. *Animal Behaviour* 26:862–84.

Sperling, G. 1983. A unified theory of attention and signal detection. In R. Parasuraman and D. R. Davies, eds., *Varieties of attention*, 103–81. Orlando: Academic Press.

Staddon, J. E. R. 1975. A note on the evolutionary significance of "supernormal" stimuli. *American Naturalist* 109:541–45.

Stoddard, P. K., M. D. Beecher, C. L. Horning, and S. E. Campbell. 1991. Recognition of individual neighbors by song in the song sparrow, a species with song repertoires. *Behavioral Ecology and Sociobiology* 29:211–15.

Temerin, H. 1986. Signing behaviour in relation to polyterritorial polygyny in the wood warbler (*Phylloscopus sibilatrix*). *Animal Behaviour* 34:146–52.

Thornhill, R. 1979. Adaptive female-mimicking behavior in a scorpionfly. *Science* 205:412–14.

Warner, R. R. 1984. Deferred reproduction as a response to sexual selection: A test of the life-historical consequences in a coral reef fish. *Evolution* 38:148–62.

Warner, R. R. 1987. Female choice of sites versus mates in a coral reef fish, *Thalassoma bifasciatum*. *Animal Behaviour* 35:1470–78.

Warner, R. R., and S. G. Hoffman. 1980. Local population size as a determinant of mating system and sexual composition in two tropical marine fishes (*Thalassoma* spp.). *Evolution* 34:508–18.

Waser, P. M., and C. H. Brown. 1984. Is there a "sound window" for primate communication? *Behavioral Ecology and Sociobiology* 15:73–76.

Wiley, R. H. 1973. The strut display of male sage grouse: A "fixed" action pattern. *Behaviour* 47:129–52.

Wiley, R. H. 1983. The evolution of communication: Information and manipulation. In T. R. Halliday and P. J. B. Slater, eds., *Animal behaviour*, vol. 2, *Communication*, 156–89. Oxford: Blackwell Scientific Publications.

Wiley, R. H. 1991a. Associations of song properties with habitats for territorial oscine birds of eastern North America. *American Naturalist* 138:973–93.

Wiley, R. H. 1991b. Lekking in birds and mammals: Behavioral and evolutionary issues. *Advances in the Study of Behavior* 20:201–91.

Wiley, R. H., and D. G. Richards. 1982. Adaptations for acoustic communication in birds: Sound propagation and signal detection. In D. E. Kroodsma and E. H. Miller, eds., *Acoustic communication*, vol. 1, 131–81. New York: Academic Press.

Wiley, R. H., and M. S. Wiley. 1977. Recognition of neighbors' duets by stripe-backed wrens. *Campylorhynchus nuchalis*. *Behaviour* 62:10–34.

Wiley, R. H., B. J. Hatchwell, and N. B. Davies. 1991. Recognition of individual males' songs by female dunnocks: A mechanism increasing the number of copulatory partners and reproductive success. *Ethology* 88:145–53.

Woodruff, G., and D. Premack. 1979. Intentional communication in the chimpanzee: The development of deception. *Cognition* 7:333–62.

Yasukawa, K. 1981. Song repertoires in the red-winged blackbird (*Agelaius phoeniceus*): A test of the Beau Geste hypothesis. *Animal Behaviour* 29:114–25.

Zahavi, A. 1975. Mate selection—a selection for a handicap. *Journal of Theoretical Biology* 53:205–14.

Zahavi, A. 1991. On the definition of sexual selection, Fisher's model, and the evolution of waste and of signals in general. *Animal Behaviour* 42:501–3.

Zelick, R. D., and P. M. Narins. 1982. Analysis of acoustically evoked call suppression behavior in a neotropical treefrog. *Animal Behaviour* 30:728–33.

EIGHT

Mechanisms Underlying Sexual Selection

MICHAEL J. RYAN

WHETHER TO THE SCIENTIST or to the layperson, the bizarre or extreme is more intriguing than the normal. This can be evidenced by the public popularity of tabloids, freak shows, and beauty contests; perhaps this is also why sexual selection has fascinated so many for so long.

Every major group of animals offers examples of extreme morphologies and behaviors that have been attributed to sexual selection: the bright colors of guppies, tails of peacocks, penises of primates, and songs of birds, frogs, and insects are just a few. There is a general consensus that these secondary sexual characteristics, usually of males, can evolve under the influence of sexual selection even if they reduce the survivorship of the bearer. If, on average, a trait increases the individual's ability to acquire mates in numbers sufficient to offset the fitness cost due to reduced survivorship, then the trait will evolve to a compromise between the different optima dictated by sexual selection and natural selection (Darwin 1871; Fisher 1958). Traits favored by sexual selection could enhance mate acquisition ability in different ways: they could increase the male's ability to gain direct access to females, or they could make him more attractive to females. What we do not always understand fully is why females prefer males with these death-defying traits (Bradbury and Andersson 1987; Kirkpatrick and Ryan 1991). Sometimes it is obvious: in many cases females select mates that enhance their immediate reproductive success. Some males might have better territories, be better parents, fertilize more eggs, or have fewer parasites. In each instance the female's mating preference is under direct selection because it immediately affects her fitness; these preferences are easy to understand because they are obviously adaptive. It is in situations in which males have nothing to offer but their sperm that it is difficult to understand why females choose.

Choice of mates take place during courtship, and courtship is an act of communication. The male is the signaler, the female the receiver, and it is the interaction of the male's signal and the female's sensory system that influences the female's choice of mates. We will address this interface of signal and sensory system in an attempt to gain some additional insights into the biology of sexual selection by female choice. None of the ideas about mechanisms of female choice presented here are new. Some are from ethology (e.g., Tinbergen 1951; Morris 1956; Nottebohm 1972; Barlow 1977). But more recently in behavioral and evolutionary ecology there has been a serious effort toward elucidating mechanisms of sexual selection and discerning how they can contribute to our understanding of the behavior and evolution of these systems (West-Eberhard 1979, 1984; Burley 1985; Ryan 1985, 1990, 1991; Kirkpatrick 1987; Endler 1989, 1992; Ryan and Rand 1990; Ryan et al. 1990a; Christy and Salmon 1991; Guilford and Dawkins 1991; Kirkpatrick and Ryan 1991; Enquist and Arak 1993; Weary et al. 1993). The results of this effort will be reviewed here.

APPROACHES TO THE STUDY OF SEXUAL SELECTION
BY FEMALE CHOICE

There have been two major, and often disparate, approaches to studying female mating preferences. One approach, which invokes a common paradigm in behavioral and evolutionary ecology, attempts to document how female preferences contribute to variance in male mating success and, by extrapolation, generate selection on male traits. Ideally (Searcy and Andersson 1986), such studies show that mating success favors males with some trait variants over others, that females have the opportunity to choose mates in nature, and that variation in these traits per se influences the female's preference.

Another approach is more typical of ethology and neuroethology. These studies systematically examine female preferences along the multiple dimensions in which a trait might vary in order to elucidate the precise constellation of signal properties that release female preference. Although many of these studies were conducted for the purpose of understanding sensory processing, it is clear that they can also contribute to our understanding of sexual selection (e.g., Gerhardt 1988, 1991, 1992).

WHY STUDY MECHANISMS OF FEMALE CHOICE?

For too long physiologists and ecologists have tended to work in isolation from each other. The dichotomy of proximate versus ultimate sometimes seems not only to justify but also to demand ignorance of the alternative approach. The history of the study of animal behavior clearly reflects such a schism. Prior to the advent of sociobiology, much of the field concentrated on mechanisms underlying behavior, as evidenced by ethological concepts such as fixed action patterns, innate behavior, behavioral drive, and feature detectors. Physiology was then a major part of animal behavior. But with the excitement generated by kin selection and other issues in sociobiology, much of the field has shifted toward addressing questions of the current function and adaptive significance of behaviors, often with little or no consideration of how these behaviors are controlled. Population genetics also began to have an important influence (e.g., Marler 1985). Fortunately, many researchers in behavioral and evolutionary ecology now recognize the importance of understanding mechanisms of behavior, and the schism seems to be yielding to studies characterized by an integration of proximate and ultimate paradigms (e.g., see contributions in: Krebs and Davies 1991; volume 139, Supplement [March], *The American Naturalist* 1992; *Philosophical Transactions of the Royal Society of London*, series B [April] 1993). Such an approach has recently shown great utility in studies of sexual selection by female choice.

There are several reasons why behavioral and evolutionary ecologists working on sexual selection should appreciate the need to understand the sensory mechanisms underlying female choice. First, there is great interest in the evolution of female preferences; this is the most controversial area of sexual selection (Kirkpatrick and Ryan 1991). A preference, however, is merely the behavioral manifestation of underlying neural properties. To completely understand how a preference evolves, it is necessary to know what changes take place in the sensory mechanisms governing this behavior.

A second reason for defining the mechanisms of choice, independent of whether choice currently generates selection, is that we thereby obtain a better understanding of both the past role and the future potential of sexual selection. For example, the túngara frog, *Physalaemus pustulosus*, has a call with two components, a whine, which is necessary and sufficient for species recognition, and a chuck, which, when added to the call, makes the call more attractive to both females and frog-eating bats. It seems likely that the chuck evolved under the influence of sexual selection. There does not appear to be any variation in males' ability to produce chucks, however, and in a large chorus all males appear equally likely to add chucks to their calls. Therefore, currently there is no selection for adding this call component, since selection can act only on variation. But phonotaxis experiments demonstrating females' preference for calls with chucks implicate this preference as the selection force responsible for the evolution of chucks, and also suggest that this preference would generate selection if some males were to lose the ability to produce chucks. A similar example comes from Andersson's (1982) study of widowbirds. Males of this species have extremely long tails. In nature, the variation among males in tail length is relatively small, and there is no correlation between a male's tail length and his mating success; thus there is no selection on tail length. But when Andersson manipulated tail length, this trait then contributed to variance in male mating success. Both of these studies show that concentrating only on how current preferences generate selection on extant male traits can limit our understanding of both mechanisms and evolutionary history.

Not only is it valuable to explore preferences regardless of extant variation in the trait, but it can also be informative to examine preferences for traits that do not exist. This suggestion is motivated by ethological studies of supernormal stimuli (Tinbergen 1951). A complete understanding of the mechanisms of female mate choice can help us understand why certain male traits have or have not been favored by sexual selection. Furthermore, this information can help us understand why female preferences might have evolved. As will be discussed extensively below, some theories of the evolution of mating preferences predict that preferences and traits coevolve. If it can be shown that a preference exists prior to the evolution of a trait, this demonstrates that coevolution of this particular preference and this particular trait did not occur and allows rejection of some hypotheses, such as good genes and runaway sexual selection, for the evolution of female mating preferences (Ryan 1990). This does not exclude the possibility that good genes, runaway, or other forces were responsible for establishing the preference, which later in evolution other, different traits then exploited (Ryan 1990). I will address this issue further below in my discussion of specificity of preferences.

A final reason for exploring mechanisms of female mate choice is that the sensory systems used in choice have other important tasks that must be carried out for the survival and reproduction of the species (Kirkpatrick 1987; Proctor 1991). For example, as will be discussed below, only by understanding the physiology of the crab's eye is it apparent that female preferences for certain structures used in courtship probably derive from ecological selection for avoiding predators (Christy and Salmon 1991).

PREFERENCES FOR AVERAGE AND EXTREME MALE TRAITS

Much of the work on female mating preferences has focused on species recognition. Most definitions of species require assortative mating among species, and it has been amply demonstrated that females commonly exhibit mating preferences for their own conspecific males over heterospecifics. An implied corollary of species recognition is preference for the average trait, and there is no doubt that in many cases females prefer traits that approximate the population or species mean. It would hardly be surprising if preference for average traits were the more usual case. For some researchers, preference for average traits implies no sexual selection. This deduction comes from a tendency to dichotomize species recognition and sexual selection as mutually exclusive processes, implying that stabilizing selection is the result of species recognition and directional selection results from sexual selection. However, sexual selection by female choice will occur whenever female mating preferences influence the variance in male mating success, regardless of whether the effect on the distribution of male phenotypes is stabilizing or directional (Ryan and Rand 1993c).

At the sensory level, species-specific preferences predict a close congruence between the response properties of the sensory system and the properties of the courtship signals (see fig. 8.1). This has been demonstrated, for example, in acoustic systems of insects (Huber 1990), anurans (Fuzessery 1988; Walkowiak 1988), and birds (Konishi 1989); pheromone systems of moths (Grant et al. 1989); and electric-discharge systems of electric fishes (Hopkins 1983). Although species recognition is ubiquitous, this fact by itself does not address why such communication systems have evolved. There is no doubt that selection favors the ability of females to recognize potential mates, but this fact does not discriminate among various hypotheses, such as reproductive character displacement and incidental divergence in allopatry, for why this congruence occurs (but see Coyne and Orr 1989).

The striking feature of sexual selection is that it can generate selection for extreme traits. Therefore, many studies of female preferences in intraspecific mate choice have examined the response of females to traits at either extreme of the population range or to average traits as compared with extreme traits. Recently, Ryan and Keddy-Hector (1992) surveyed studies of intraspecific female mating preferences. The question they posed was quite simple: in cases in which females prefer traits that deviate from the population mean, is there any consistency in the direction in which the trait deviates? There were two noteworthy findings. First, there have been a large number of studies, especially in the last decade, demonstrating female preferences for extreme traits; they cited more than 150 studies involving preferences based on acoustic and visual cues. Second, there is a simple, striking generalization that applies to most of the studies: females prefer traits of greater quantity. In acoustic systems, females prefer calls and songs that are more intense, complex, and longer and are delivered at a greater repetition rate than that average for the species or population. The same generalization applies to courtship in the visual domain: females prefer larger, brighter displays that are delivered at a greater rate than average.

We can consider what the second generalization tells us about preferences at the proximate and ultimate level. Traits of greater quantity probably have greater signal value; they are more likely to be perceived by females, and when perceived, they will directly elicit greater sensory stimulation. Therefore, sexual selection favors traits that are more stimulatory to females, although these signals must also indicate an appropriate mate. At the sensory level this could arise from several mechanisms. First, and most obviously, signals that are delivered repeatedly will elicit greater stimulation. Second, signals that vary, such as a large song repertoire in birds, might elicit greater stimulation because they release the sensory system from habituation that could be induced by repeating the same signal. Third, the response properties of the sensory system could be biased away from the population mean; for example, auditory neurons could be tuned to lower call frequencies than those exhibited by most males.

Unfortunately, such a striking generalization about directional preferences tells us almost nothing about why these preferences evolved. There are many possibilities, but three hypotheses have received the most attention (Kirkpatrick and Ryan 1991). "Good genes" hypotheses state that females prefer males with traits indicating a heritable genetic basis for greater physical fitness or vigor. The genes that influence vigor evolve directly under natural selection, and the indicator trait and female preference evolve under indirect selection because they become genetically correlated with the good genes. As Zahavi (1975) has noted, indicator traits must be costly or they could easily be bluffed. For example, only the healthiest males could afford to produce displays that required great energy expenditures or exposed the male to greater predation risk. Thus in a general sense, good genes hypotheses predict the trend prevalent in the review by Ryan and Keddy-Hector: females prefer signals that are more costly.

Another hypothesis for the evolution of female preferences is Fisher's hypothesis of "runaway" sexual selection. In its simplest form, a female preference and male trait become genetically correlated; this correlation can arise either from direct selection on females or from random processes. The trait evolves directly because it is favored by the preference, and the preference evolves under indirect selection, as a correlated response to evolution of the trait. There is no inherent directionality to runaway sexual selection; it need not necessarily lead to the elaboration of a trait, and it does not predict the trend that females prefer traits of greater quantity. However, Lande (1981) and Kirkpatrick (1982) have shown that certain biases, for example, a bias of the sensory system to perceive certain kinds of signals, could easily skew the direction of the runaway process. Thus although not a necessary prediction, preference for traits of greater quantity is easily accommodated by the runaway hypothesis.

A third possibility is that female preferences evolve under direct selection, and not as a correlated response to evolution of the trait. As mentioned previously, this might be the usual case when resources are at stake. Females might prefer males that have better territories, are better parents, fertilize more eggs, or have fewer parasites because these factors have an immediate influence on the number of offspring successfully sired. It is possible that these males would also be healthier, regardless of genetic differences among males, and thus could afford to produce more costly signals. If so, preferences for these signals would be favored because of their direct effect on female reproductive success.

Direct selection can be important in a more subtle manner. We must remember that the sensory systems used by females to perceive courtship signals are also used for other tasks crucial to survival and reproduction. Eyes did not initially evolve in female birds to identify conspecific plumage, nor did ears evolve to allow them to identify song. Direct selection outside of the context of mate choice could have important effects on properties of the sensory system that, in turn, could have consequential effects on female mating preferences. For example, in many sensory systems there appears to be strong directional selection to maximize signal detection. The minimum movement of the basilar membrane in the mammalian ear is constrained by random Brownian motion; if the ear were any more sensitive, it would hear random movement of molecules in the environment (Hudspeth 1989). Similarly, some pheromone and visual systems respond to single molecules of chemicals or photons of light. Thus there will be strong selection for a signal to be detectable or to be conspicuous relative to its background. Thus, as Nottebohm (1972) suggests, much of signal evolution might be in response to signal value rather than iconic value; the form of the signal might be dictated more by its stimulatory effect on the sensory system than by any meaning it has. Guilford and Dawkins (1991) suggest a similar dichotomy in their discussion of "tactical" versus "strategic" aspects of signals, and emphasize the role of receiver psychology in favoring certain signal structures (i.e., tactical aspects).

In summary, studies have documented species-specific preferences, female preferences for average traits, and congruence between species-specific properties of signals and receivers. Also, there are many studies that demonstrate female preferences for traits that deviate from the mean. The direction of deviation is usually consistent, and suggests that females prefer both traits of greater quantity and traits that elicit greater sensory stimulation. This fact in itself does not tell us why preferences have evolved; that issue will be addressed below.

SENSORY BASES OF SEXUAL SELECTION

As mentioned above, neuroethological studies of species recognition have shown that response properties of sensory systems are biased toward those properties of courtship signals indicating species identity (fig. 8.1). For example, in crickets, a set of local brain neurons responds only to the syllable repetition rate of the conspecific song (Huber 1990). In the song control area of the brain in birds, the HVc, there are auditory neurons that respond best to that bird's own individual song rather than to other songs (Margoliash 1983); interestingly, when this area is lesioned, females still respond behaviorally to song but no longer discriminate between conspecific and heterospecific song (Brenowitz 1991). Electric fishes use their electric discharge organs for prey localization, but these also function in communication between the sexes. The electroreceptor organ, the knollenorgan, is most sensitive to the peak in the power spectrum of the electric discharge (Hopkins 1983). A close congruence between signal and receiver is also evident in pheromone systems. In moths, females produce long-range pheromones that attract males, and males then release a short-range pheromone that females use to evaluate the male as a potential mate. The receptor organs on the female's antennae are more sensitive to extracts of the male's pheromone than to other substances (Grant et al. 1989).

Species-specific matching between signal and receptor has also been demonstrated quite extensively in anurans. Since I will use several examples from studies of frogs, I will briefly review some of the details of how female frogs recognize the calls of conspecific males (see Fritzsch et al. 1988 for an extensive review of this area; see also Ryan 1991).

There are more than 2,500 species of frogs, and in almost all of them, males produce advertisement calls that function in attracting females and repelling other males. All of these calls have some species-specific properties and thus differ from one another. Although many of these differences are in temporal properties, such as pulse repetition rate, we will consider only how information in the spectral domain is processed.

Frogs have two organs in the inner ear that are sensitive to sound: the amphibian papilla (AP) and the basilar papilla (BP). The hair cells of the AP are innervated by fibers from the auditory nerve, and each of these fibers has a distinct frequency

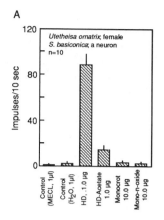

A

Impulses/10 sec

Utetheisa ornatrix, female
S. basiconica, a neuron
n=10

Control (MECL, 1µl)
Control (H₂O, 1µl)
HD, .1.0 µg
HD-Acetate 1.0 µg
Monocrot 10.0 µg
Mono-n-oxide 10.0 µg

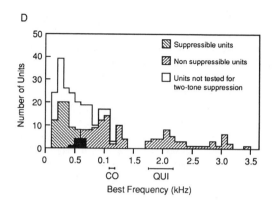

D

Number of Units

▨ Suppressible units
▧ Non suppressible units
☐ Units not tested for two-tone suppression

Best Frequency (kHz)

CO   QUI

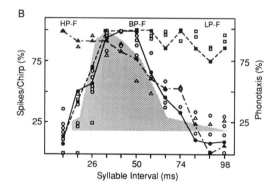

B

Spikes/Chirp (%)

Phonotaxis (%)

HP-F   BP-F   LP-F

26   50   74   98

Syllable Interval (ms)

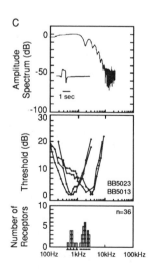

C

Amplitude Spectrum (dB)

1 sec

Threshold (dB)

BB5023
BB5013

Number of Receptors

n=36

100Hz   1kHz   10kHz   100kHz

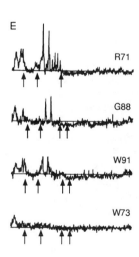

E

R71

G88

W91

W73

to which it is most sensitive, usually referred to as its best excitatory frequency (BEF). For most species, these frequencies are always below 1,200 Hz. At the population level, the AP is usually characterized by two peaks of frequency sensitivity. The auditory nerve fibers emanating from the BP all have similar BEFs, and this organ is always tuned to higher frequencies than the AP, usually above 1,500 Hz. The spectral distribution of energy in the species' advertisement call usually matches the tuning properties of its peripheral auditory system (Zakon and Wilczynski 1988). If the call has two major concentrations of energy, these will be matched to the most sensitive frequencies of both the AP and the BP. Some calls have only a single concentration of energy, and this dominant frequency will match the most sensitive frequencies of either the AP or the BP. There is a good deal known about the processing of spectral information in the brain (Fuzessery 1988; Walkowiak 1988), but clearly the tuning of the peripheral auditory system relative to the spectral properties of the call makes an important contribution to call recognition.

Within each species, the congruence between auditory tuning and call frequencies is very close. In at least some species, however, there is a slight mismatch, and it appears that this mismatch can bias female preference for call frequencies away from the mean and thus potentially generate directional selection on males' calls.

The advertisement call of the cricket frog (*Acris crepitans*) is a short click that is rapidly repeated in call groups. The dominant frequency of the call is usually 3–4 kHz; therefore, auditory processing in the periphery is probably restricted to the BP. There is significant variation in most call properties, including the dominant frequency of the call, among populations of cricket frogs (Ryan and Wilczynski 1988, 1991; Ryan et al. 1992). The tuning properties of the female's BP also differ among populations, and usually females are tuned more closely to the calls of their own males than to calls from other populations (Ryan et al. 1992; Wilczynski et al. 1992). But unexpectedly, in each population studied, the average BEF of the female is always lower in frequency than the average male's call (fig. 8.2A).

If the relationship between tuning and call frequency guides mating preferences, females should prefer calls that are lower than average in frequency. This seems to be the case. Female cricket frogs from three populations were given a choice between two calls. Both calls had the same temporal properties, but one had the dominant

---

Fig. 8.1 Examples of the congruence of neural properties and mate recognition signals in moths, crickets, fish, frogs, and birds. (*A*) Histogram showing the mean (± SE) responses from neurons on the antennae of female moths (*Utetheisa ornatrix*) to stimulation by the main component of the male pheromone (HD) relative to other chemicals. (Adapted from Grant et al. 1989.) (*B*) The tuning, as spikes per chirp, of various classes of neurons (HP-F, high pass filter; BP-F, band pass filter; LP-F, low pass filter) and the strength of phonotaxis as a function of the syllable interval for the call of the cricket *Gryllus bimaculatus*. The peak phonotaxis response is to the species-specific syllable interval, and the graph shows how the combination of the tuning of the three classes of neurons matches the phonotaxis function. (Adapted from Huber 1990.) (*C*) Top, the power spectrum of a single electric discharge from the electric fish *Brienomyrus brachyistius*. Middle, representative tuning curves of the receptor organ, the knollenorgan. Bottom, distributions of knollenorgan best excitatory frequencies. (Adapted from Hopkins 1983.) (*D*) The distribution of best excitatory frequencies from the peripheral auditory system of females of the frog *Eleutherodactylus coqui*. The frequency ranges of the species' two call components, the "co" and "qui," are indicated. (Adapted from Zakon and Wilczynski 1988.) (*E*) Response of neuronal clusters from a forebrain nucleus, HVc, of a white-crowned sparrow (*Zonotrichia leucophrys*) to its own song (R71) and three other songs. The arrows represent the onset of phrases in the songs. The response to its own song is greater than it is to other conspecific songs. (Adapted from Margoliash 1985.)

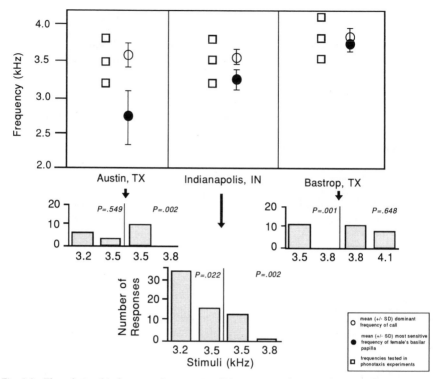

Fig. 8.2 The relationship between the average call frequency and average best excitatory frequency of the female's basilar papilla for three populations of cricket frogs, *Acris crepitans*. Also shown is the number of phonotactic responses by female cricket frogs in response to calls with a dominant frequency characterizing that of calls from their own population (middle two bars of each graph) versus calls with higher or lower frequencies. (Adapted from Ryan et al. 1992.)

frequency of the local call and the alternative had the dominant frequency characterizing another population's call, which was either 300 Hz higher or 300 Hz lower than the local call. Females exhibited significant preferences in four of the six experiments (Ryan et al. 1992; fig. 8.2B). Sometimes they preferred the local call and sometimes the foreign call, but if there was a preference, it was always for the lower-frequency call. In the two experiments in which there was not a significant preference, the majority of the females were attracted to the lower-frequency call. This preference for lower-frequency calls also can act on within-population variation. Wagner (cited in Ryan et al. 1992) presented females with a choice between calls that were either 200 Hz above or 200 Hz below the average call, and females preferred the lower-frequency call.

The effect of body size on the tuning properties of the auditory system also can affect patterns of mate choice. Both among (Zakon and Wilczynski 1988) and within (Keddy-Hector et al. 1992) species, the tuning of the BP is correlated with body size; larger frogs have BPs tuned to lower frequencies. Ryan et al. (1992) attempted to determine whether variation in the body size of female cricket frogs influenced female call preferences. The only experiment in which sample sizes were sufficient for statistical analysis was one in which females from Indiana were given a choice between calls of 3.2 and 3.5 kHz. Although there was a significant preference for

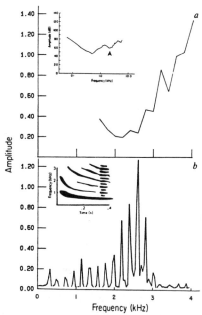

Fig. 8.3 (*A*) The mean audiogram of the basilar papilla of the túngara frog (*Physalaemus pustulosus*) derived from five individuals. The audiogram is truncated below 1.5 kHz to remove the influences of the amphibian papilla units. The insert is a full audiogram from a single frog; the best excitatory frequency of the basilar papilla is indicated by an arrow. (*B*) Fourier spectrum of a single chuck. The insert is a sonogram of a whine followed by one chuck. (Adapted from Ryan et al. 1990a.)

the foreign, lower-frequency call, a substantial number of females were attracted to the higher-frequency call. The average body size of these females was significantly smaller than that of the females that preferred the lower-frequency calls. Thus within a population there might be significant variation in female mating preferences, and this variation might derive from allometric relationships of body size and neural properties of the sensory system.

The túngara frog (*Physalaemus pustulosus*) offers another example of the relationship between sensory tuning and sexual selection. As mentioned above, these frogs produce a call with two components: a whine and a chuck. The whine can be produced alone (the simple call), or can be followed by up to six chucks (the complex call; fig. 8.3). The whine is both necessary and sufficient for species recognition, and all the close relatives of this species produce whinelike calls. The chuck by itself elicits neither male vocalizations nor female phonotaxis, but when added to the whine it increases vocal responses from other males and makes the call more attractive to females. Only one other species, *P. petersi*, the sister species of *P. pustulosus*, is known to produce a chuck, and this has only been documented in a single population.

The whine appears to be processed primarily by the AP. The dominant frequency of this call component closely matches the most sensitive frequency of the AP (Ryan et al. 1990a), and only the fundamental frequency, which sweeps from 900 to 400 Hz, is used in call recognition; the higher harmonics are superfluous (Rand et al. 1992). The chuck has a broad frequency range that encompasses the frequencies to which both the AP and BP are most sensitive. Synthetic chucks that have either only the lower seven (215–1,505 Hz) or only the upper seven (1,720–3,010 Hz) harmonics are as effective as the normal chuck in increasing the attractiveness of the simple call, as long as each of the chuck variants has the same amount of energy as the normal chuck (Ryan and Rand 1990). Although females prefer a whine plus chuck to the simple call, they also prefer whines with half-chucks to simple calls, even though these traits do not exist. White noise too is as effective as the normal chuck when added to the whine (Rand et al. 1992).

These results suggest that we need to be a bit more circumspect when defining a preference. Female *P. pustulosus* have a preference for certain kinds of acoustic stimuli, and a chuck is one of several alternatives that can elicit this preference. These experiments also demonstrate that the sensory mechanisms underlying female preference can be flexible. In this case the whine-only call, which stimulates primarily the AP, can be made more attractive by adding a half-chuck, which stimulates either the AP or the BP, or a normal chuck and white noise, which stimulate both the AP and BP.

Although a chuck encompasses a wide frequency range, most of the energy is concentrated in frequencies to which the BP is sensitive: on average, 90% of the total energy is above 1,500 Hz. Furthermore, the dominant frequency of the chuck is tuned near the most sensitive frequency of the BP. This suggests that the chuck is processed by the BP, and that stimulation of the AP does not contribute to female preference for normal complex calls. To test this hypothesis, females were given a choice between two calls. Each had a whine, one of which was followed by a low-frequency half-chuck and the other by a high-frequency half-chuck. In each case the half-chuck had the same amount of energy it would in nature: 10% and 90% of the normal full chuck, respectively. Females did not show a preference between the whine + low-frequency half-chuck and the whine only (12 versus 9, one-tailed exact binomial probability, $P = .33$), but did show a strong tendency to prefer the whine + high-frequency half-chuck to the whine only (14 versus 6, $P = .06$; Wilczynski et al., in press). This demonstrates that for the average chuck, in which 90% of the energy is in the higher half of the frequency range, BP stimulation is necessary and probably is sufficient to explain the enhanced attractiveness of the call derived from adding the chuck. Stimulation of the AP adds little if anything to perception of the chuck.

Closer examination of the relationship between tuning of the BP and chuck dominant frequency also reveals a pattern of mismatch similar to that just discussed in the cricket frog. The average tuning of the BP in *P. pustulosus* is about 2,100 Hz, while the average dominant frequency of the chuck is 2,500 Hz (see fig. 8.3). Ryan (1985) had already shown that larger males are more likely to be chosen as mates, that larger males produce chucks with lower frequencies, and that females prefer synthetic chucks with lower frequencies. The neurophysiological study suggested the reason for these results: females are tuned below the average dominant frequency of most chucks. To test this hypothesis, a computer model of the filter properties of the BP was constructed. Then the amount of energy from a chuck that would pass through the filter was determined—this result estimated the amount of stimulation the BP would receive from various chucks. Lower-frequency chucks elicited greater neural stimulation (Ryan et al. 1990a). The hypothesis predicts that the preference for lower-frequency chucks should reside in the BP. This prediction was recently tested using calls in which a whine was followed by a tone with the temporal properties of the chuck but which consisted of only a single frequency of 2,100 Hz or 3,000 Hz. The results were equivocal. Although there was a strong trend toward preference for the lower-frequency tone, the trend was not statistically significant (19 versus 11, one-tailed exact binomial probability, $P = .10$; Wilczynski et al., in press).

An earlier ethological study of the fritillary butterfly by Magnus (1958) also probes the underlying sensory mechanisms that guide preference for extreme traits. In these butterflies males choose females, and that choice is influenced by wing-flapping rate. Females normally flap their wings at a rate of 8 to 10 Hz. Using a model that simulated the flapping of females, Magnus showed that males were attracted preferentially to higher flapping rates within the normal range. He also showed that this preference extended well into the supernormal range—up to 140 Hz. In a separate set of experiments Magnus determined that the flicker-fusion rate of the butterfly's eye was 140 Hz. This result suggests that the behavioral preference for high wingbeat rates was directly related to the increased rate of visual stimulation; at 140 Hz, when the rates of flapping and visual stimulation became uncoupled, the preference was extinguished.

Rowland (1989a, 1989b) also shows how male mate choice of females can be mediated by the amount of visual stimulation. Male sticklebacks prefer females with distended abdomens, supposedly because this indicates a more fecund female. However, using dummies, Rowland showed that the form of distention was unimportant to males. In fact, males preferred dummies with abnormal distentions that resembled females laden with high levels of parasites rather than with large numbers of eggs. His conclusion was that mate choice was regulated by the total amount of retinal stimulation.

One of the more complicated and better-known behaviors in the animal kingdom is the song repertoire of oscine birds. Hartshorne's (1956) monotony hypothesis suggests that selection favors the addition of different elements to the repertoire to avoid the narcoticizing effects of habituation (see also Nottebohm 1972). His hypothesis was applied to interactions among males, but the same principle might apply to the song's effect on females. Although there are exceptions, in many cases studies have shown that female birds are more likely to exhibit courtship solicitation displays to speakers broadcasting larger song repertoires (Catchpole 1987). Recently, Searcy (1992) investigated the effect of repertoire size on female preferences in grackles. Some females received a series of identical song elements, while other females received the same amount of song, but the song switched between several different types of song elements. As with other studies, more courtship solicitation displays were exhibited to the more complex repertoires. However, Searcy also showed that the female's response decayed with the repetition of the same element, but immediately increased when the song switched to a new element (fig. 8.4). This result suggests that the enhanced stimulatory effect of repertoires might result from a female's release from habituation when a new song element is perceived. Another interesting aspect of this study is that, despite this female preference, male grackles do not produce repertoires; each male produces only one song element. We will return to this intriguing point later.

In some animals species recognition is a learned phenomenon. Mechanisms of learning and recognition can also result in biased female mating preferences among conspecific males. Weary et al. (1993) recently developed suggestions by O'Donald (1980) and Staddon (1975) that the psychological phenomenon known as "peak shift displacement" could result in females exerting directional selection on male traits. Peak shift was first discussed within the paradigm of operant conditioning, and occurs when an animal is exposed to two stimuli along a stimulus gradient in which

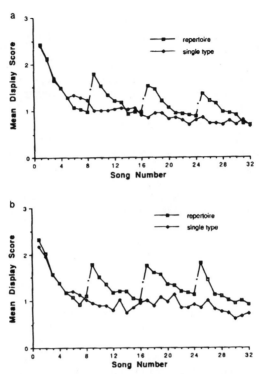

Fig. 8.4   The response of female common grackles to (*a*) the first set of 32 songs of either four song types (repertoire) or a single song type and (*b*) the second set of 32 songs of either four types or a single type. * within a break of a continuous line indicates a change in response significant at the .05 level (two-tailed Wilcoxon matched-pairs test). (Adapted from Searcy 1992.)

there is a reward associated with the positive stimulus and no reward associated with the negative stimulus. After training, it is found that the greatest response is not to the original positive stimulus but to one that is more extreme than the positive stimulus in the direction away from the negative stimulus. As an example, they cite the classic study of Hanson (1959) in which pigeons were presented with two light stimuli, and given food following light at 550 nm but denied food following light at 590 nm. Generalization during testing produced a peak response not at 550 nm but at 530 nm. Weary et al. suggest that similar situations might occur in nature when individuals learn to discriminate between the sexes. For example, if males have slightly longer tails than females, then peak shift could result in males with longer-than-average tails being more attractive to females. This phenomenon also might explain preferences exhibited by quail for traits that are more extreme than the traits to which they were imprinted (ten Cate and Bateson 1989).

   Another recent foray into the interface between learning and sexual selection is the clever use of neural networks to explore mate recognition by Enquist and Arak (1993). Although the details of the model can be complicated, their essence can be briefly summarized and the reader referred to the original paper for most details.

   The simulations using the neural networks revolve around a simple model of the visual system containing the analogue of a retina, which receives input from the environment and then outputs onto an array of "hidden cells." These hidden cells,

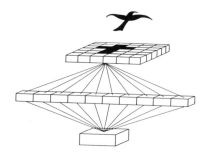

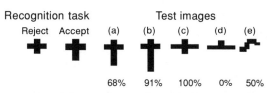

Fig. 8.5　*Top:* A representation of the neural network model of Enquist and Arak (1993). The image of the bird projects to a set of detectors that function analogously to a retina. This set of cells projects onto a set of "hidden" cells, which in turn project onto a single "feature detector." All cells in the retina project to all of the hidden cells, but only one such set of connections is shown for simplicity's sake. The response properties of the network can be varied by changing the weighting of the connections between calls at different levels. *Bottom:* Stimuli that the network was trained to discriminate, and test images to which the network's response was tested after training. Also shown are the percentages of instances in which the test image yielded a response that was greater than that in response to the "conspecific" image. (Adapted from Enquist and Arak 1993.)

in turn, have outputs onto a feature detector, which sums its inputs from above and indicates that the relevant stimulus has or has not been perceived (fig. 8.5A). The network is trained to recognize a "conspecific" image during simulations in which the weightings of the inputs are "mutated" and the network that best recognizes the image is retained. There are two surprising results in how these networks behave. First, once the networks are trained to recognize the conspecific image, other images with which it has had no experience, including some images that are quite dissimilar to the conspecific stimulus, are also recognized as appropriate. Second, some of these images that have supernormal aspects relative to the conspecific stimulus are not only recognized but are more attractive (fig. 8.5B). Enquist and Arak suggest this might be another example of peak shift displacement.

Both of these studies investigating recognition show that selection could favor the learning of one task, recognizing the correct sex or species, but that this could have unintended consequences that generate directional sexual selection on male traits. The fact that these studies address two very different kinds of mechanisms suggests that such unintended responses or preexisting preferences might be general properties of how recognition comes about.

In summary, it appears that when comparing mate recognition systems in different species, there is close congruence within each species between properties of the signal and the receiver. There are a number of cases within a species, however, in which the match between signal and receiver is not perfect, but in which responses of the receiver are best elicited by signals that deviate from the mean. These two concepts are not mutually exclusive and are analogous to among- and within-group variation in an analysis of variance. The relative magnitude of the among-group variation influences the partitioning of the explained variance, but it does not diminish the biological significance of the within-group variation. The fact that females prefer conspecific to heterospecific signals neither excludes nor predicts an effect of intraspecific variation on female preferences.

DIRECT SELECTION ON MATING PREFERENCES OUTSIDE
THE CONTEXT OF MATE CHOICE

When we review the evidence for intraspecific mate choice, it is apparent that in many cases females prefer males exhibiting traits of greater quantity. Also, in some cases this preference is governed by sensory systems in which the rate or amount of neural stimulation is directly related to the strength of the preference. None of this information tells us why females have evolved such preferences, or why they have evolved sensory biases that result in such preferences. In this section we will explore the hypothesis that in some cases selection on sensory systems in other contexts besides mate choice might influence the evolution of female mating preferences.

Females use their eyes, ears, and other sense organs when choosing a mate, but they also use these organs to perform other tasks crucial for survival and reproduction. Endler (1992) has discussed the variety of forces that can influence signal evolution in his hypothesis of sensory drive, suggesting that sensory systems and sensory conditions can drive evolution in particular directions (fig. 8.6). Therefore, natural selection forces having nothing to do with mating should have some influence on how these sense organs evolve biases. The degree to which males evolve traits to match these sensory biases is referred to as sensory exploitation (Ryan 1990;

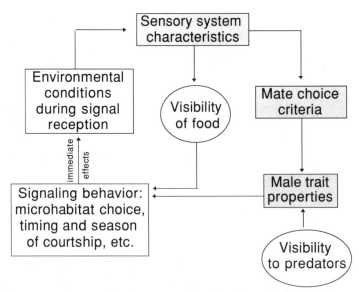

Fig. 8.6   The major components of Endler's sensory drive hypothesis. The arrows indicate evolutionary influences, with the exception of that labeled "immediate effects." (Adapted from Endler 1992.)

also see Nottebohm 1972; Barlow 1977; West-Eberhard 1979, 1984; Burley 1985; Ryan 1985, 1991; Kirkpatrick 1987; Endler 1989, 1992; Ryan and Rand 1990; Ryan et al. 1990a; Christy and Salmon 1991; Guilford and Dawkins 1991; Kirkpatrick and Ryan 1991; Enquist and Arak 1993; Weary et al. 1993). As Endler (1992) indicated, sensory exploitation can be considered a model within his concept of sensory drive. Several examples follow.

In many species of fruit flies the vibration of the male's wings causes vibrational disturbances that stimulate the female. Bennet-Clark and Ewing (1967) showed that the love song of *Drosophilia melanogaster* stimulates receptors that females use for monitoring their own wingbeat pattern. All arctiid moths have ears that are able to perceive ultrasonics, and these receptors are thought to have evolved to allow the detection of bats. Some arctiids produce ultrasonics that function in courtship; apparently these signals are perceived by organs that originally evolved under the force of predation (Conner 1987). West-Eberhard (1984) reports that in some bees males use pheromones that mimic floral scents, and suggests that female response to these scents evolved in response to selection forces involved in foraging, not in mate searching. Similarly, Fleischman (1992) points out that anoline lizards, all of which are insectivorous, are exquisitely adapted for perceiving the small, quick movements of insects, and that this aspect of the visual system then selects for males who make short, jerky motions characteristic of the push-up display used in courtship.

There are also some good examples detailing how sexual selection currently acting on male traits might have resulted from selection on female sensory systems in other contexts. In fiddler crabs of the genus *Uca* the ommatidia are arranged in a pattern that enhances the perception of objects projecting vertically from the horizon. *Uca beebei* constructs vertical pillars in front of its burrows. These pillars aid crabs of both sexes in orienting to burrows when escaping from predators, but also increase male mating success since females are more likely to enter burrows with pillars when they are searching for mates. Thus Christy has argued that the female's sensory properties that result in preference for burrows with pillars—that is, the specific arrangement of the ommatidia—evolved under selection for predator detection, not mate choice (Christy and Salmon 1991).

Prey detection appears to have played an important role in mate attraction in the water mite *Neumania papillator* (Proctor 1991). These mites feed on copepods, and males attract females by mimicking the wave vibration pattern of this prey (fig. 8.7). A female approaches a vibrating male as if he were prey. After she grabs the male, she eventually realizes that he is not prey and releases him; he then begins to court the female. Females deprived of food are more likely to respond to males' prey-mimicking vibrations than are satiated females.

There seems to be little doubt that female preferences could evolve as a direct result of selection for species recognition, as well as under a runaway process or under selection to choose males with good genes or superior resources (Kirkpatrick and Ryan 1991). However, it is now becoming clear that sensory systems used in choice have other tasks and thus are subject to other forms of selection. This realization forces us to reexamine how female preference might evolve. The two hypotheses of female preference evolution that have been most popular are the good genes hypothesis and Fisher's theory of runaway sexual selection (Bradbury and Andersson 1987; Kirkpatrick and Ryan 1991). In both of these hypotheses female preferences coevolve with already existing male traits. However, if preferences arise as pleiotropic effects of selection in other contexts, this suggests that males might then evolve traits that exploit, manipulate, or better match these biases in the female's sensory

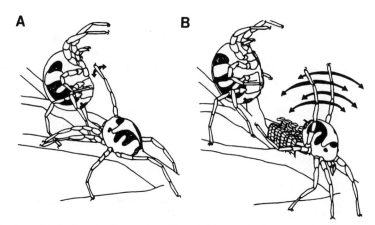

Fig. 8.7   Courtship in the mite *Neumania papillator*. (*A*) The male vibrates his foreleg in front of the female, who is in net stance. (*B*) The female has turned slightly in response to the trembling, and the male fans his fourth legs over the spermatophores he has just deposited. Sausage-shaped objects on top of spermatophore stalks are the sperm packets. (Adapted from Proctor 1991.)

system. This hypothesis of sensory exploitation, fortunately, makes some testable predictions that allow it to be discriminated from the hypotheses based on coevolution.

PHYLOGENETIC PATTERNS AND THE EVOLUTION OF PREFERENCES

If males evolve traits to match already existing biases in the female's sensory system, this should result in phylogenetic patterns of traits and preferences not possible under hypotheses that predict coevolution. Specifically, the sensory exploitation hypothesis predicts the existence of sister taxa in which one taxon exhibits both a trait that evolved under sexual selection in that taxon and a preference for that trait, whereas the other taxon lacks the trait but exhibits the preference. If the trait also is absent in the outgroup species, this suggests that only one taxon possesses the trait because it evolved after the divergence of the two sister taxa, but the two taxa share the preference because the preference evolved in a common ancestor before the two taxa diverged.

The principle of parsimony is crucial in interpreting such a phylogenetic pattern as support for the sensory exploitation hypothesis because it assumes that the observed pattern of preference and trait has arisen with the smallest possible number of evolutionary changes. Other alternatives are possible. For example, the trait and preference could have arisen simultaneously, and the trait subsequently lost in one taxon but not the other. This is not the most parsimonious explanation, but it still might be correct. Unfortunately, in examining historical patterns of evolution we can use parsimony as an operating principle, but can never be sure it is right. It must be recognized that any single example that supports the sensory exploitation hypothesis rests on this principle. Only when many such examples are available should we consider this hypothesis as a viable alternative to other explanations as to how the correlation of traits and preferences comes about.

Studies of mate preference in frogs of the *Physalaemus pustulosus* species group

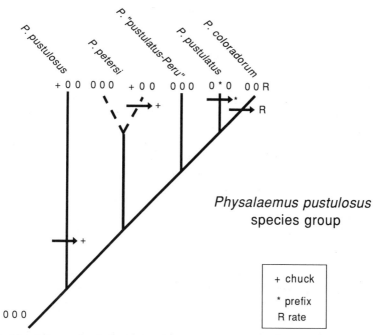

Fig. 8.8 The phylogenetic relations within the *Physalaemus pustulosus* species group. The phylogeny is based on data from morphology, allozymes, and sequence of a portion of the mitochondrial DNA genome (D. M. Hillis et al., unpubl.; see Ryan and Rand 1993b for details). Call characters were not used in reconstructing phylogenetic relationships. *P.* "pustulatus-Peru" is an undescribed species. Two populations of *P. petersi* are discriminated. The one in the Rio Napo area of Amazonian Ecuador does not produce chucks, while some males in Tambopato, Peru, do. Call characters are superimposed on the phylogeny. The most parsimonious character state transitions are shown by arrows. "0" indicates that the character is absent. The species used as the outgoup are: *P. enesefae, P. ephippifer,* and an undescribed species from Roraima, Brazil.

provide several indications that sensory exploitation has operated (Ryan and Rand 1993b). As discussed earlier, in *P. pustulosus* the BP is tuned near to the dominant frequency of the chuck, and although the chuck encompasses a frequency range to which both the AP and BP are most sensitive, 90% of its energy is within the frequency range of the BP. Furthermore, when females are presented with whines followed by either low or high half-chucks with the same amount of energy as found in nature, only the high half-chuck is effective.

The importance of the BP in processing the chuck gives us insights into how this trait evolved. *P. pustulosus* belongs to a species group that includes four other species: *petersi, coloradorum, pustulatus,* and an as yet undescribed species, "pustulatus-Peru." All of these species, as well as those we think are the three closest relatives of the *P. pustulosus* species group (*enesefae, ephippifer,* and an undescribed species, "roraima") produce whinelike advertisement calls. All of these calls have most of their energy below 1,000 Hz and thus probably use only the AP in call processing (Ryan and Drewes 1990). Only *P. petersi* has a call component with frequencies higher than 1,000 Hz, because in one of the several populations studied it too produces a chuck, and thus might also use the BP in auditory perception (fig. 8.8). But since all of the other species lack call components with high frequencies, it appears that these species do not use the BP in communication.

Whether or not the frogs produce high-frequency call components, they still have a BP, and it is tuned similarly in all these closely related species. This suggests that when the chuck first evolved it was more attractive to females because it was able to tap into a communication channel that had not been used in the recent evolutionary history of this group of frogs. Since the tuning is the same for all the species, it also suggests that there was not necessarily any evolution of the BP in response to evolution of the chuck. This hypothesis was tested by presenting female *P. coloradorum* with a choice between the whinelike advertisement call of their own males and the same call to which three chucks from a *P. pustulosus* call had been added. There was a significant preference for the call with chucks (Ryan and Rand 1993a, 1993b). Since both *P. pustulosus* and *P. coloradorum* exhibit the preference for chucks, the most parsimonious explanation is that this trait evolved in a common ancestor of the two species before the chuck evolved (fig. 8.8). If this interpretation is correct, then hypotheses suggesting that trait and preference coevolved cannot explain the preexisting preference for chucks; instead, it suggests that males that produced chucks were favored by the already existing preference.

There are two other examples suggesting sensory exploitation in these frogs. *P. pustulatus* appears to have uniquely derived an extreme amplitude–modulated prefix to its call, and *P. coloradorum* is the only species among those examined to produce calls in doublets or triplets in response to conspecific male vocalizations (fig. 8.8). Although both of these traits are lacking in *P. pustulosus*, these females prefer their conspecific call with the AM prefix of *P. pustulatus* added over their normal conspecific call, and they prefer their own call produced in doublets to the normal calling rate of conspecific males (Ryan and Rand 1993b).

Sensory exploitation might also explain the evolution of preferences for lower-frequency chucks. As discussed previously, Ryan et al. (1990a) argued that the slight mismatch between the tuning of the BP and the average chuck's dominant frequency is responsible for the preference for lower-frequency chucks and the greater mating success of larger males. The tuning of the BP does not differ statistically among members of the species group. Therefore, the sensory properties that result in low-frequency preference are shared by all the taxa and thus existed prior to the evolution of the chuck itself. But, as mentioned previously, a prediction of this hypothesis is that the low-frequency preference is due only to the tuning of the BP. When female *P. pustulosus* were given a choice between higher and lower tones within the BP's range, there was a strong trend suggesting preference, but not a statistically significant preference, for the lower frequency. Thus support for this hypothesis regarding preference for low-frequency chucks is merely suggestive, although there is stronger support for preexisting preferences for the chuck of *P. pustulosus*, the amplitude-modulated prefix of *P. pustulatus*, and the calling rate of *P. coloradorum*.

As discussed previously, Proctor (1991) showed that in one species of mite, males mimic the surface water vibration of potential prey, and that this enhances their mating success because they attract females that are using surface water vibrations to hunt for prey. Proctor (1992) also has reconstructed the evolutionary history of both the predatory behavior and the male courtship behavior. The phylogenetic results suggest that the male behavior evolved to exploit the female hunting behavior—strong evidence in support of the sensory exploitation hypothesis.

Similar results with swordtails and platys of the genus *Xiphophorus* also support the hypothesis of sensory exploitation (Basolo 1990a). The swordtails represent a monophyletic clade within the genus (Rosen 1979); the sword evolved within this clade and is absent in the platys. In the swordtail *X. helleri*, females prefer males with longer swords (Basolo 1990b). Basolo (1990a) artificially attached swords to male platys, *X. maculatus*, and females of that species preferred males with these artificial swords. Again, this suggests that the preference for swords evolved prior to the sword itself. (Recently, Meyer et al. (1994) suggested an alternative hypothesis for relations among xiphophorine fishes, which complicates this interpretation.)

Another study of swordtails offers an interesting possibility of sensory exploitation. *X. nigrensis* males exhibit a large range of body sizes, which are determined by a single allele on the Y chromosome (Kallman 1989). Larger males court while small males sneak after females. Large males have greater mating success in nature, and in the laboratory females prefer larger males to smaller males (Ryan et al. 1990b). The closely related *X. pygmaeus* has only small males, none of which court. Female *X. pygmaeus* prefer the larger, courting *X. nigrensis* males to their own smaller, noncourting conspecific males (Ryan and Wagner 1987). In this case we do not know if large size and courtship were lost in *X. pygmaeus* or if this is the primitive condition and those traits were later gained in the other species; Kallman (1989) suggests the latter. Nevertheless, there is a preference for larger and/or courting males in *X. pygmaeus* even though such males do not exist. This finding suggests that if males evolved large size they could exploit this preexisting preference—and in fact this seems to have happened. Recently, M. J. Ryan et al. (unpubl.) found two populations of *X. pygmaeus* with large males. Interestingly, these males do not show courtship behavior. However, when females from populations without larger males were given a choice between the normal, small males and the larger males, they preferred the latter. This is an example in which a preference for a trait might have evolved in the context of mate choice, perhaps due to good genes or runaway sexual selection, but the trait was subsequently lost and the preference remained. In such a situation males could still exploit what, in this case, would now be a preexisting preference. Sensory exploitation only requires that the preference exist when the trait evolves or reevolves; the preference need not necessarily exist because of direct selection outside of the context of mate choice.

There are other cases in which females prefer traits that exist in closely related species but are absent in their own males. A close relative of *Uca beebei*, the crab that erects pillars by its burrows, is *U. stenodactylus*. This species lacks pillars, but when its burrows were artificially augmented with pillars, they were more likely to be visited by females (Christy and Salmon 1991). Similarly, Searcy (1992) showed that females of both red-winged blackbirds, which possess song repertoires, and grackles, which lack repertoires, prefer larger repertoires. In both of these cases the phylogenetic data needed to hypothesize the sequence of trait/preference evolution are not available. But, as with the swordtail *X. pygmaeus*, these are examples of preferences that could be exploited if males were to evolve or reevolve the preferred trait.

Even when sensory exploitation can be argued from phylogenetic data, it must be remembered that this does not exclude the role of other forces in the evolution of the preference. The preexisting preference could have initially been established in the context of mate choice, as might be true with *X. pygmaeus*. Also, the preference could become further elaborated after the trait evolved. Searcy (1992) suggests this is true by showing that among some species of birds, the strength of the preference for repertoires is positively correlated with repertoire size. (The same pattern does not hold when comparing the strength of the preference for chucks between *P. pustulosus* and *P. coloradorum*, but might be due to small sample sizes [Ryan and Rand 1993b].) However, support in some species of sensory exploitation does suggest that in others the correlated presence of both a preference and a trait need not be the result of coevolution, and thus of good genes or runaway sexual selection.

SPECIFICITY OF PREFERENCES

At first consideration, the concept of preexisting female preferences seems odd. For example, as far as we know, only twice in the history of anurans has a frog evolved the ability to add a chuck to its call. It seems unlikely that females would have previously evolved a sensory bias toward this unique trait. However, the concept of preexisting preferences seems less odd, in fact almost likely, if we reevaluate the way we define preferences. Consider the following studies.

Female túngara frogs prefer calls with chucks, a very specific acoustic structure that spans a frequency range of about 200–3000 Hz. However, as mentioned previously, female preferences are equally strong for calls to which a low or high half-chuck is added as long as each variant has the same amount of energy as the full chuck (Ryan and Rand 1990). Also, a stimulus in which the chuck is replaced by white noise having the same amplitude envelope as the chuck is as effective as the full chuck (Rand et al. 1992), as are some, but not all, pure tone stimuli (Wilczynski et al., in press). Therefore, it appears that female mating preferences could have exerted strong selection on a variety of signal variants if they were to evolve. Perhaps the evolution of the full chuck tells us more about morphological constraints on males than it does about the specificity of preferences exhibited by females.

In another example, Burley (cited in Trivers 1985) showed that the mating success of male zebra finches could be enhanced if they were adorned with red leg bands. She also manipulated the feathers of the males' heads, fitting them with elaborate hats. Although head feathers are not elaborated in this species, females preferred these oddly adorned males. We also need to recall the behavior of the neural networks trained by Enquist and Arak (1993; see fig. 8.5). Some of the stimuli recognized were obviously exaggerated forms of the training stimuli (e.g., similar forms but with longer tails), but others bore no resemblance at all to the training stimuli. Their simulation is analogous to selection for a species-specific trait, with an incidental consequence of preexisting preferences for other traits.

There might be a variety of preferences that exist but will never be detected by researchers because of the population-level approach generally used in behavioral and evolutionary ecology. By defining a preference only by the extant male trait that elicits the preference, we greatly limit our ability to understand the potential of sexual selection by female choice to quickly act on new variation as it arises. Clearly, some preferences possess a latitude that is merely glimpsed by studying preferences only in the context of extant traits; this is especially true if, as some have suggested, there might be selection on male traits for novelty per se (West-Eberhard 1979). An instructive exercise might be to map in multivariate space the preference function of females, and to extend this function past the normal variation in male traits. This would constitute a combinatorial approach by which we could consider why certain male traits have not evolved if preferences for them exist, and why some preferences have extreme latitude relative to the extant variation in male traits.

This broader view of preferences also provides insights into evolutionary processes. Consider the preferences for swords in swordtails and platys (Basolo 1990a, 1990b). In many species of poeciliid fishes, females prefer larger males (e.g., Ryan and Wagner 1987). Is it possible that both platys and swordtails have inherited the same preference for larger males, and that the sword is one trait that has evolved to exploit that preference? Deeper bodies, longer bodies, and larger dorsal fins could be other traits also favored by this same preference (see also Rowland 1989a, 1989b). It is possible that this preference for large size evolved under a good genes or runaway scenario, but this does not at all contradict Basolo's conclusion about the history of trait/preference evolution, or her rejection of these hypotheses for the evolution of the sword per se. This point has not always been understood (Reeve and Sherman 1993). This is not to suggest that the preference cannot be modified. As discussed previously, Searcy (1992) presents data suggesting a role for sensory exploitation and then further elaboration of repertoire size in songbirds. Also, if a preference for exploitive traits resulted in strong selection against females, we would predict that natural selection would influence certain aspects of that preference.

CONCLUSIONS

Demonstrating current sexual selection by female choice necessitates showing that male traits do, in fact, influence female preference, and that the preference influences variance in male reproductive success. Without such, there is no evidence that sexual selection is acting (Searcy and Andersson 1986). The introduction of quantitative analysis of selection differentials and gradients into the study of animal behavior has added a rigor to speculations about the adaptiveness of behavior, a rigor sorely lacking in earlier ethological studies. However, there are several important aspects of evolution that are not addressed by merely quantifying current selection, and these omissions appear especially obvious in studies of sexual selection.

Female mating preferences are behavioral manifestations of neural properties. As has been shown in this chapter, these neural properties can be studied, and an understanding of these mechanisms can give us a better appreciation of how preferences evolve. Also, if we recognize the importance of the sensory systems, it becomes readily apparent that these systems are under selection outside of the context of mate choice, and thus that there might be important pleiotropic effects on female mating preferences.

Another omission in studies of current selection is their failure to address the past. Often, the relevant question is why or how a trait or preference evolved. Notice the tense. This question cannot be framed in the present tense, and it thus requires methods used for historical analysis. We have reviewed two examples of the importance of history. First, some studies show that extant variation in a male trait does not influence female preferences, but when the variance is manipulated, female preferences are influenced. This suggests at least the potential for a historical effect of preferences on the evolution of traits, a suggestion that could not be derived from a study of current selection. Second, in some cases it appears that certain biases in the female's sensory system might exist prior to the male traits now favored by female choice. If so, these results suggest other, perhaps more pertinent, hypotheses for the joint evolution of traits and preferences.

Evolution is a multivariate process that can be studied at several levels of integration using a variety of techniques. If nothing else, this chapter should suggest that such an integrative approach is likely to provide further insights into sexual selection.

REFERENCES

Andersson, M. 1982. Female choice selects for extreme tail lengths in widowbirds. *Nature* 299:818–20.

Barlow, G. W. 1977. Model action patterns. In T. A. Sebeok, ed., *How animals communicate,* 98–134. Bloomington: Indiana University Press.

Basolo, A. L. 1990a. Female preference predates the evolution of the sword in swordtail fish. *Science* 250:808–10.

Basolo, A. L. 1990b. Female preferences for male sword length in the swordtail *Xiphophorus helleri* (Pisces: Poeciliidae). *Animal Behaviour* 40:322–38.

Bennet-Clark, H. C., and A. W. Ewing. 1967. Stimuli provided by courtship of male *Drosophila melanogaster. Nature* 215:669–71.

Bradbury, J. W., and M. B. Andersson, eds. 1987. *Sexual selection: Testing the alternatives.* Chichester: John Wiley and Sons.

Brenowitz, E. A. 1991. Altered perception of species-specific song by female birds after lesions of a forebrain nucleus. *Science* 251:303–5.

Burley, N. 1985. The organization of behavior and the evolution of sexually selected traits. *Ornithological Monographs* 37:22–44.

Catchpole, C. K. 1987. Bird song, sexual selection and female choice. *Trends in Ecology and Evolution* 2:94–97.

Christy, J. H., and M. Salmon. 1991. Comparative studies of reproductive behaviors in mantis shrimp and fiddler crabs. *American Zoologist* 31:329–37.

Conner, W. E. 1987. Ultrasound: Its role in the courtship of the arctiid moth, *Cycnia tenera. Experentia* 43:1029–31.

Coyne, J. A., and H. A. Orr. 1989. Patterns of speciation in *Drosophila. Evolution* 43:362–81.

Darwin, C. 1871. *The descent of man and selection in relation to sex.* Random House, New York. (Reprint of the original.)

Endler, J. A. 1989. Conceptual and other problems in speciation. In D. Otte and J. A. Endler, eds., *Speciation and its consequences,* 625–48. Sunderland, Mass.: Sinauer Associates.

Endler, J. A. 1992. Signals, signal conditions, and the direction of evolution. *American Naturalist* 139:S125–53.

Enquist, M., and A. Arak. 1993. Selection of exaggerated traits by aesthetic female preferences. *Nature* 361:446–48.

Fisher, R. A. 1958. *The genetical theory of natural selection,* 2d rev. ed. New York: Dover.

Fleischman, L. J. 1992. The influence of the sensory system and the environment on motion patterns in the visual displays of anoline lizards and other vertebrates. *American Naturalist* 139:S36–S60.

Fritzsch, B., M. Ryan, W. Wilczynski, T. Hetherington, and W. Walkowiak, eds. 1988. *The evolution of the amphibian auditory system.* New York: John Wiley and Sons.

Fuzessery, Z. M. 1988. Frequency tuning in the anuran central auditory system. In B. Fritzsch, M. Ryan, W. Wilczynski, T. Hetherington, and W. Walkowiak, eds., *The evolution of the amphibian auditory system,* 253–73. New York: John Wiley and Sons.

Gerhardt, H. C. 1988. Acoustic properties used in call recognition by frogs and toads. In B. Fritzsch, M. Ryan, W. Wilczynski, T. Hetherington, and W. Walkowiak, eds., *The evolution of the amphibian auditory system,* 455–83. New York: John Wiley and Sons.

Gerhardt, H. C. 1991. Female mate choice in treefrogs: Static and dynamic acoustic criteria. *Animal Behaviour* 42:614–35.

Gerhardt, H. C. 1992. Multiple messages in acoustic signals. *The Neurosciences* 4:391–400.

Grant, A. J., R. J. O'Connell, and T. Eisner. 1989. Pheromone-mediated sexual selection in the moth *Utetheisa ornatrix:* Olfactory receptor neurons responsive to a male-produced pheromone. *Journal of Insect Behavior* 2:371–85.

Guilford, T., and M. S. Dawkins. 1991. Receiver psychology and the evolution of animal signals. *Animal Behaviour* 42:1–14.

Hanson, H. M. 1959. Effects of discrimination training on stimulus generalization. *Journal of Experimental Psychology* 58:321–34.

Hartshorne, C. 1956. The monotony threshold in singing birds. *Auk* 95:758–60.

Hopkins, C. D. 1983. Neuroethology of species recognition in electroreception. In J.-P. Ewert, R. R. Capranica, and D. J. Ingle, eds., *Advances in vertebrate neuroethology*, 871–81. New York: Plenum Publishing Co.

Huber, F. 1990. Nerve cells and insect behavior—studies on crickets. *American Zoologist* 30: 609–27.

Hudspeth, J. A. 1989. How the ear's works work. *Nature* 341:397–404.

Kallman, K. D. 1989. Genetic control of size at maturity in *Xiphophorus*. In G. K. Meffe, and F. F. Snelson, eds., *Ecology and evolution of live-bearing fishes* (Poeciliidae), 163–200. Englewood Cliffs, N.J.: Prentice Hall.

Keddy-Hector, A. C., W. Wilczynski, and M. J. Ryan. 1992. Call patterns and basilar papilla tuning in cricket frogs. II. Intrapopulation variation and allometry. *Brain, Behavior and Evolution* 39:328–46.

Kirkpatrick, M. 1982. Sexual selection and the evolution of female choice. *Evolution* 36:1–12.

Kirkpatrick, M. 1987. The evolution of female preferences in polygynous animals. In J. W. Bradbury and M. B. Andersson, eds., *Sexual selection: Testing the alternatives*, 67–82. Chichester: John Wiley and Sons.

Kirkpatrick, M., and M. J. Ryan. 1991. The evolution of mating preferences and the paradox of the lek. *Nature* 350:33–38.

Konishi, M. 1989. Birdsong for neurobiologists. *The Neuron* 3:541–49.

Krebs, J. R., and N. B. Davies. 1991. *Behavioral ecology: An evolutionary approach*. 3d ed. Oxford: Blackwell Scientific Publications.

Lande, R. 1981. Models of speciation by sexual selection on polygenic traits. *Proceedings of the National Academy of Sciences U.S.A.* 78:3721–25.

Magnus, D. 1958. Experimentelle Untesuchungen zur Bionomie und Ethologie des aisermantels *Argynnis paphia* Girard (Lep. Nymph.). *Zeitschrift für Tierpsychologie* 15:397–426.

Margoliash, D. 1983. Acoustic parameters underlying the response of song-specific neurons in the white-crowned sparrow. *Journal of Neurosciences* 3:1039–57.

Margoliash, D. 1985. Auditory representation of autogenous song in the song system of white-crowned sparrows. *Proceedings of the National Academy of Sciences U.S.A.* 82:5997–6000.

Marler, P. 1985. Forward. In M. J. Ryan, *The túngara frog: A study in sexual selection and communication*. Chicago: University of Chicago Press.

Meyer, A., J. M. Morrissey, M. Schartl. 1994. Recurrent origin of a sexually selected trait in *Xiphophorus* fishes inferred from a molecular phylogeny. *Nature* 368:539–42.

Morris, D. 1956. The function and causation of courtship ceremonies. Fondation Singer-Polignac: Colloque Intertan. sur L'Instinct. June 1954.

Nottebohm, F. 1972. The origins of vocal learning. *American Naturalist* 106:116–40.

O'Donald, P. 1980. *Genetic models of sexual selection*. Cambridge: Cambridge University Press.

Proctor, H. C. 1991. Courtship in the water mite *Neumania papillator:* Males capitalize on female adaptations for predation. *Animal Behaviour* 42:589–98.

Proctor, H. C. 1992. Sensory exploitation and the evolution of male mating behaviour: A cladistic test using water mites (Acari: Parasitengona). *Animal Behaviour* 44:745–52.

Rand, A. S., M. J. Ryan, and W. Wilczynski. 1992. Signal redundancy and receiver permissiveness in acoustic mate recognition by the túngara frog, *Physalaemus pustulosus*. *American Zoologist* 32:81–90.

Reeve, H. K., and P. W. Sherman. 1993. Adaptation and the goals of evolutionary research. *Quarterly Review of Biology* 68:1–32.

Rosen, D. E. 1979. Fishes from the uplands and intermontane basins of Guatemala: Revisionary studies and comparative geography. *Bulletin of the American Museum of Natural History* 162:263–375.

Rowland, W. J. 1989a. The ethological basis of mate choice in male threespine sticklebacks, *Gasterosteus aculeatus*. *Animal Behaviour* 38:112–20.

Rowland, W. J. 1989b. Mate choice and the supernormality effect in female sticklebacks *Gasterosteus aculeatus*. *Behavioral Ecology and Sociobiology* 24:433–38.

Ryan, M. J. 1985. *The túngara frog, a study in sexual selection and communication.* Chicago: University of Chicago Press.

Ryan, M. J. 1990. Sexual selection, sensory systems, and sensory exploitation. *Oxford Surveys in Evolutionary Biology* 7:158–95.

Ryan, M. J. 1991. Sexual selection and communication in frogs. *Trends in Ecology and Evolution* 6:351–54.

Ryan, M. J., and R. Drewes. 1990. Vocal morphology of the *Physalaemus pustulosus* species group (Family Leptodactylidae): Morphological response to sexual selection for complex calls. *Biological Journal of the Linnean Socity* 40:37–52.

Ryan, M. J., and A. C. Keddy-Hector. 1992. Direction patterns of female mate choice and the role of sensory biases. *American Naturalist* 139:S4–S35.

Ryan, M. J., and A. S. Rand. 1990. The sensory basis of sexual selection for complex calls in the túngara frog, *Physalaemus pustulosus* (Sexual selection for sensory exploitation). *Evolution* 44:305–14.

Ryan, M. J., and A. S. Rand. 1993a. Phylogenetic patterns of behavioral mate recognition systems in the *Physalaemus pustulosus* species group (Anura: Leptodactylidae): The role of ancestral and derived characters and sensory exploitation. In D. Edwards and D. Lees, eds., *Evolutionary patterns and processes*, 251–67. London: The Linnean Society of London.

Ryan, M. J., and A. S. Rand. 1993b. Sexual selection and signal evolution: The ghost of biases past. *Philosophical Transactions of the Royal Society of London*, series B, 340:187–95.

Ryan, M. J., and A. S. Rand. 1993c. Species recognition and sexual selection as a unitary problem in animal communication. *Evolution* 47:647–57.

Ryan, M. J., and W. E. Wagner, Jr. 1987. Asymmetries in mating preferences between species: Female swordtails prefer heterospecific mates. *Science* 236:595–97.

Ryan, M. J., and W. Wilczynski. 1988. Coevolution of sender and receiver: Effect on local mate preference in cricket frogs. *Science* 240:1786–88.

Ryan, M. J., and W Wilczynski. 1991. Evolution of intraspecific variation in the advertisement call of a cricket frog (*Acris crepitans*, Hylidae). *Biological Journal of the Linnean Society* 44: 249–71.

Ryan, M. J., J. H. Fox, W. Wilczynski, and A. S. Rand. 1990a. Sexual selection for sensory exploitation in the frog *Physalaemus pustulosus*. *Nature* 343:66–67.

Ryan, M. J., D. K. Hews, and W. E. Wagner Jr. 1990b. Sexual selection on alleles that determine body size in the swordtail *Xiphophorus nigrensis*. *Behavioral Ecology and Sociobiology* 26: 231–37.

Ryan, M. J., S. A. Perrill, and W. Wilczynski. 1992. Auditory tuning and call frequency predict population-based mating preferences in the cricket frog, *Acris crepitans*. *American Naturalist* 139:1370–83.

Searcy, W. A. 1992. Song repertoire and mate choice in birds. *American Zoologist* 32:71–80.

Searcy, W. A., and M. Andersson. 1986. Sexual selection and the evolution of song. *Annual Review of Ecology and Systematics* 17:507–33.

Staddon, J. E. R. 1975. A note on evolutionary significance of supernormal stimuli. *American Naturalist* 109:541–45.

ten Cate, C., and P. Bateson. 1989. Sexual selection: The evolution of conspicuous characteristics in birds by means of imprinting. *Evolution* 42:1355–58.

Tinbergen, N. 1951. *The study of instinct*. Oxford: Oxford University Press.

Trivers, R. L. 1985. *Social Evolution*. Menlo Park, Calif.: Benjamin/Cummings.

Walkowiak, W. 1988. Central temporal coding. In B. Fritzsch, M. Ryan, W. Wilczynski, T. Hetherington, and W. Walkowiak, eds., *The evolution of the amphibian auditory system*, 275–294. New York: John Wiley and Sons.

Weary, D. M., T. C. Guilford, and R. G. Weisman. 1993. A product of discrimination learning may lead to female preferences for elaborate males. *Evolution* 47:333–36.

West-Eberhard, M. J. 1979. Sexual selection, social competition, and evolution. *Proceedings of the American Philosophical Society* 123:222–34.

West-Eberhard, M. J. 1984. Sexual selection, competitive communication and species-specific signals in insects. In T. Lewis, ed., *Insect communication*, 283–324. London: Academic Press.

Wilczynski, W., A. C. Keddy-Hector, and M. J. Ryan. 1992. Call patterns and basilar papilla tuning in cricket frogs. I. Differences among populations and between sexes. *Brain, Behavior and Evolution* 39:229–37.

Wilczynski, W., A. S. Rand, and M. J. Ryan. In press. The processing of spectral cues by the call analysis system of the túngara frog. *Animal Behaviour*.

Zahavi, A. 1975. Mate selection: A selection for a handicap. *Journal of Theoretical Biology* 53: 205–14.

Zakon, H., and W. Wilczynski. 1988. The physiology of the anuran eighth nerve. In B. Fritzsch, M. Ryan, W. Wilczynski, T. Hetherington, and W. Walkowiak, eds., *The evolution of the amphibian auditory system*, 125–55. New York: John Wiley and Sons.

Part III

NEURAL, DEVELOPMENTAL,
and GENETIC PROCESSES

Critical Events in the Development of Bird Song: What Can Neurobiology Contribute to the Study of the Evolution of Behavior?

ARTHUR P. ARNOLD

BIOLOGICAL EXPLANATIONS come in two forms. If one were to answer the question, "Why do birds sing?," one could either concentrate on the physiological, environmental, and developmental events within the bird's own lifetime that lead to singing behavior (proximate mechanisms), or discuss events occurring before the bird was conceived that favored the evolution of that behavior (ultimate mechanisms). Unfortunately, all too often behavioral biologists also fall into two groups, those who study mechanisms of behavior and those who study evolution. Perhaps the reason for this division of behavioral biology is that each approach requires a different set of skills and background, and there are few that can master both fields enough to bridge the disciplines. Yet, a knowledge of proximate mechanisms can illuminate ultimate mechanisms, and vice versa. An understanding of species diversity in behavior is aided by an understanding of the functional development of the neural circuits that control that behavior.

In this chapter, I discuss the physiological and morphological development of the neural circuits controlling song in passerine birds (order Passeriformes). These circuits are highly sexually dimorphic and mediate the development of learned song. Although this discussion is mostly a consideration of proximate mechanisms underlying the development of this behavior, we can nevertheless occasionally discern interesting differences among species that constrain the way we think about the evolution of such differences. Thus, neurobiological investigations can suggest what kinds of changes in the genome are required to explain the evolution of different developmental programs for sexual differentiation and song learning.

DESCRIPTION OF SONG AND THE SONG SYSTEM

Song is the most complex of vocalizations of passerine birds, and is part of a coordinated sequence of behaviors that occur during reproduction. A male zebra finch (*Poephila guttata*) sings a quiet courtship song to a female as part of the courtship ritual. In other species, song is also used for territorial advertisement or in aggressive encounters with other males. In all of these contexts, song is regulated by steroid hormones. In many cases castration decreases the frequency of singing behavior, and treatment with testosterone or its metabolites increases singing (Arnold 1975a; Harding et al. 1983). Presumably, the sensitivity of song to sex steroid hormones

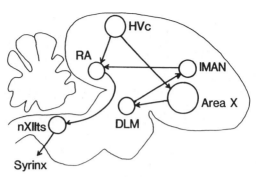

Fig. 9.1 A schematic drawing of the two main portions of the neural circuit for song in oscine passerine birds. The main descending motor circuit consists of HVC, which projects to RA, which in turn projects to the nXIIts, which innervates the vocal organ, the syrinx. The accessory circuit runs from HVC to area X, which projects to DLM, which projects to lMAN. lMAN projects back to the main efferent path at RA.

has evolved to insure that singing behavior occurs at the time when the testes are secreting testosterone, which is required for production of mature sperm.

Many of the brain regions controlling song are anatomically discrete and stain differently from surrounding regions using a variety of histological stains. Two major parts of the neural song circuit have been recognized: the main descending motor circuit, and an accessory or recurrent loop (fig. 9.1). The main motor circuit comprises the higher vocal center (HVC), which projects to the robust nucleus of the archistriatum (RA), which in turn projects to the tracheosyringeal portion of the hypoglossal motoneurons (nXIIts) that innervate the muscles of the vocal organ, the syrinx (Nottebohm et al. 1976; Nottebohm et al. 1982). The strong descending projections from HVC to RA to nXIIts themselves imply motor control. This inference is supported by lesion studies that confirm that damage to these nuclei disrupts vocalizations (Nottebohm et al. 1976), and by electrophysiological experiments that detect electrical activity in this pathway preceding vocalization (McCasland and Konishi 1981; McCasland 1987). The accessory loop comprises the circuit from HVC to area X of the lobus parolfactorius, which projects to the medial dorsolateral nucleus of the thalamus (DLM), which projects to the lateral magnocellular nucleus of the anterior neostriatum (lMAN). lMAN then projects back into the descending motor circuit at RA (Nottebohm et al. 1976; Bottjer et al. 1989). Lesions of the accessory loop in adults have little effect on song, suggesting that these nuclei play little role in motor control of song per se. However, lesions of lMAN or area X in young male zebra finches prevent song learning, and such lesioned males eventually sing highly abnormal songs (Bottjer et al. 1984; Sohrabji et al. 1990; Scharff and Nottebohm 1991. Thus, it appears that the accessory loop plays a critical role in song learning, but not in maintenance of song in adults. Except for area X and possibly DLM, all of the brain regions mentioned thus far contain cells that accumulate androgen (Arnold 1980b). Therefore, they are thought to express androgen receptors, implying that androgens act directly on them to modulate their control of song.

Auditory feedback is required for song learning (see below), and hence one would expect auditory pathways to communicate with the regions responsible for motor control of song. Field L, the primary telencephalic auditory area, projects to a shelf of neurons or neuropil around HVC, and to a similar shelf around RA (Kelley and Nottebohm 1979). In all of the brain regions in the main motor control path and the accessory path, neurons respond to auditory stimuli (Katz and Gurney 1981; Williams 1985; Williams and Nottebohm 1985; Margoliash 1986; Doupe and Konishi 1991). In HVC, RA, and lMAN, neurons respond better to the bird's own song than to other acoustic stimuli such as reversed song or pure tones. This selectivity implies important song-related auditory filtering. The sound-evoked responses recorded in RA and nXIIts appear to be relayed from HVC, since they are eliminated by lesions of HVC (Williams 1985; Williams and Nottebohm 1985; Doupe and Konishi 1991). The convergence of auditory and motor information in both loops of the song circuit is presumably the neural substrate for comparison of motor commands and their sensory consequences, which is thought to underlie portions of song learning (see below).

SEXUAL DIFFERENTIATION OF BRAIN AND BEHAVIOR

Song is usually the prerogative of the male. In zebra finches, females do not sing at all, even if given treatment with exogenous androgen (Arnold 1974). In a broad class of other species that includes canaries (*Serinus canarius*) and white-crowned sparrows (*Zonotrichia leucophrys*), females treated with androgen as adults will sing, but their song is simpler than the male's. Although sexual dimorphism in song is the rule, there are some species in which the female's song is as elaborate as the male's, such as the bay wren (*Thryothorus nigricapillus*) (Brenowitz et al. 1985). There are no known passerine species in which the female sings more complex songs than the male.

The large sex differences in song are the result of striking sex differences in the structure of the neural song system (fig. 9.2). Using virtually any measure of neuronal structure, large sex differences are detected in nearly all of the song regions. In zebra finches, song control regions are larger in males (up to fivefold, in HVC, RA, area X, and nXIIts: Nottebohm and Arnold 1976), and contain more neurons (in RA and HVC: Gurney 1981; Nordeen and Nordeen 1989), larger neurons (in HVC, RA, lMAN: Arnold 1980b; Gurney 1981; Konishi and Akutagawa 1985; Nordeen et al., 1986), longer neuronal dendrites (in RA: Gurney 1981), and more androgen target cells (in HVC and lMAN: Arnold and Saltiel 1979; Arnold 1980b; Nordeen et al. 1986). Clearly, the inability of female zebra finches to sing results from their atrophic song system.

Just as there is species diversity in the degree of sexual dimorphism in vocal ability, there is diversity in the degree of sex difference in the structure of the song system. Species can be grouped into those in which the female sings no song (zebra finch), sings some song but less than the male (canary; red-winged blackbird, *Agelaius phoeniceus*; rufous-and-white wren, *Thryothorus rufalbus*: Pesch and Güttinger 1985; Kirn et al. 1989; Brenowitz and Arnold 1986), or sings song as elaborate as that of the male (bay wren; buff-breasted wren, *Thryothorus leucotis*: Arnold et al. 1986). In general, the complexity of the song system correlates well with the complexity of the vocal repertoire. Sexual dimorphism in the volumes of vocal control regions is greatest in zebra finches, smaller in canaries, and smallest or absent in species in which females sing complex duets with males. Similarly, the number of neurons in RA is dimorphic in vocally dimorphic zebra finches and canaries, but monomorphic in species such as the bay wren that lack sexual dimorphism in song.

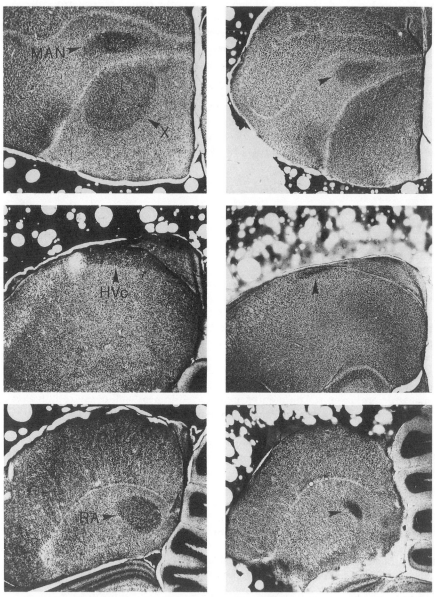

Fig. 9.2 Representative thionin-stained coronal sections of zebra finch telencephalons, 50 μm thick, illustrating the large sex differences in the neural song circuit. The sections on the left are from the male; on the right, from the female. Above, lMAN and area X; middle, HVC; below, RA. Area X is usually not visible in the female using this stain.

GENERAL PRINCIPLES OF SEXUAL DIFFERENTIATION

It is important to try to understand how sexual differences develop, not only for the understanding of neural and behavioral development themselves, but also for an appreciation of the species differences in programs of song development. If we can understand how the neural song system differentiates, we may be able to develop hypotheses regarding the kinds of genetic mutations that give rise to the different programs of behavioral song development in the various passerine species.

In general, all sex differences in behavioral and neural phenotype are thought to arise because of the differential secretion of sex steroid hormones (Goy and McEwen 1980). This conclusion has come predominantly from studies of sexual differentiation in mammals, which have been studied much more extensively than other vertebrates. The mammalian Y chromosome includes a gene known as testis determining factor (tdf), which directs the development of the testis (Blyth and Duckett 1991). In the absence of this factor, an ovary differentiates. Once the gonads have differentiated, all subsequent differential development of males and females is thought to be controlled by hormonal secretions of the gonads. Behavioral and neural differentiation results from the differential secretion of testosterone, which is higher in males than in females. This testosterone reaches the brain and quickly enters cells. Within the cell, testosterone may be converted to one of two major active metabolites, estradiol (catalyzed by the enzyme aromatase) or 5 α dihydrotestosterone (catalyzed by 5 α reductase). Testosterone or one of its metabolites then binds to a ligand-specific receptor within the cell. The steroid-receptor complex binds to DNA and alters the transcription of specific genes. The differential activation or repression of specific genes in hormone target cells in the CNS leads to differential growth or atrophy of selected neural circuits that mediate sex-specific behaviors. Testosterone (or its metabolites) *masculinizes* neural development—that is, it creates or prevents atrophy of circuits responsible for masculine functions—and it also *defeminizes* neural development—that is, it prevents the development of circuits responsible for feminine behaviors. Estrogenic metabolites are particularly important. In mammals, the aromatase enzyme is present in many of the brain regions that are masculinized by testosterone. Various lines of evidence indicate that testosterone must be converted to estradiol before it masculinizes the brain, and that brain masculinization is mediated by estradiol's action via estrogen receptors (MacLusky and Naftolin 1981; McEwen 1981; McEwen et al. 1979).

Two general concepts have influenced attempts to understand sexual differentiation of the song system in birds. The first concept is that the mechanisms of sexual differentiation, in broad outline, follow the mammalian pattern just described. Specifically, we expect sexual differentiation of the song system to be controlled by sex differences in gonadal secretion of steroid hormones, and the mechanisms of steroid metabolism, steroid-receptor interactions, and transcriptional regulation of development to be similar. The second concept is that despite the overall similarity of sexual differentiation in birds and mammals, avian sexual differentiation has some characteristics that are unique. Most importantly, studies of nonpasserine species such as quail suggest that sexual differentiation results not from testicular secretion of androgen, but from ovarian secretion of estrogen (Adkins-Regan 1987). For example, when quail embryos are treated with estrogen, the males as adults show fewer masculine copulatory behaviors than control males not treated with estrogen (i.e., they are demasculinized). In contrast, estrogen or testosterone treatment of female embryos has little effect on differentiation of their behavior. However, if females are treated with antiestrogens, they show more masculine copulatory behaviors than control females (Adkins 1976; Balthazart et al. 1992). Thus, it appears that estrogenic secretions of the ovary induce feminine programs of behavioral development (i.e., they cause demasculinization and feminization), whereas masculine patterns develop when estrogen levels are low (see Adkins-Regan 1987 for review). In this view, avian sexual differentiation is clearly similar to the mammalian pattern in that similar cellular events underlie the process of differentiation, but estrogens inhibit rather than promote masculine differentiation.

SEXUAL DIFFERENTIATION IN ZEBRA FINCHES

The earliest studies of sexual differentiation of the zebra finch song system implied that zebra finches follow the mammalian pattern. Gurney and Konishi (1980; Gurney 1981) found that estradiol treatment of female hatchlings causes masculinization of the telencephalic system for song. As adults, estradiol-treated females sing and possess larger song regions and larger neurons. This finding has been replicated in several laboratories (e.g., Simpson and Vicario 1991b). Further work showed that estradiol also increases the number of neurons (in RA and HVC) and the incidence of androgen-accumulating cells (in lMAN and HVC) (Nordeen et al. 1986, 1987; Nordeen and Nordeen 1989; Konishi and Akutagawa 1990). Treatment of hatchling females with dihydrotestosterone, a nonaromatizable metabolite of testosterone, causes relatively little masculinization (Gurney 1981, 1982; Schlinger and Arnold 1991a).

From the studies discussed thus far, we infer that estrogen is the most potent of the steroids that can masculinize the neural song system in females. This inference implies that the male's song system is normally exposed to higher levels of estrogen during development than is the female's. Higher exposure could occur because of higher testicular synthesis of estrogen itself, or because of higher synthesis of testosterone, which would then be aromatized to estradiol in the brain. However, current data do not strongly support either of these ideas. Three studies have measured the plasma levels of sex steroids in male and female zebra finches in the 2 weeks after hatching, a time when the song system of females can be potently masculinized by estrogen (Hutchison et al. 1984; Adkins-Regan et al. 1990; Schlinger and Arnold 1991c). All the studies agree that there are either no detectable sex differences in the plasma levels of androgens, or that females have slightly higher androgen levels. There is disagreement among the studies concerning the plasma levels of estrogen. One study reported that males have much higher levels of estrogen between days 3 and 10 after hatching (Hutchison et al. 1984), but the other two studies failed to replicate the sex difference, and found much lower plasma levels of estrogen at these ages. It is impossible at present to reconcile these discrepant results. At best, we can conclude that any sex differences in plasma levels of estrogen have so far not been found to be robust enough to allow replication.

It is highly unlikely that the testes are secreting estrogen. The substantial levels of estrogen found in plasma of adult male zebra finches are not reduced by castration (Adkins-Regan et al. 1990; Schlinger and Arnold 1991b), indicating that there is an extragonadal source of estrogen in males. To look for this source of estrogen, Schlinger and Arnold (1991b) measured the activity of aromatase in a variety of tissues in hatchling and adult zebra finches. They found high expression of aromatase in brain (see also Vockel et al. 1988, 1990), some in pituitary, but none in testes, adrenals, or in a variety of other tissues including skin and liver. In females, the ovary contains high levels of aromatase, as expected. The levels found in male and female telencephalon are extremely high, almost as high as those found in ovary. Similar measurements during the period of sexual differentiation suggest that testes and adrenals do not synthesize estrogens, and that the source of plasma estrogen in males is the brain (Schlinger and Arnold 1991c).

Further evidence from adult male zebra finches strongly supports the concept of the cerebral source of plasma estrogens (Schlinger and Arnold 1992). In adult male zebra finches, tritiated androgen was injected into the telencephalon near HVC (where aromatase is high) or into the cerebellum (which lacks aromatase). Five minutes later, tritiated estrogen was found in plasma in the telencephalon-injected males, but not in the cerebellum-injected males. Tritiated estrogen was found in plasma after similar injections into ovary, but not into testis. These results imply that estrogen is formed in telencephalon and ovary, and that this estrogen enters the blood. Finally, after tritiated androgen injections into the systemic circulation, tritiated estrogen was found enriched in blood leaving the brain compared with blood entering the brain. This result indicates that the brain can take up androgen from the blood, convert it to estrogen, and release the estrogen back into the circulation. Because no other organs in males appear to contain aromatase, one is left with the conclusion that the most likely source of estrogen for masculinization of the song system is the brain itself.

The cerebral synthesis of estrogen has several important implications. The first is that the levels of estrogen in specific locations in brain might be much higher than in plasma. Thus, measurements of plasma levels of estrogen may not reflect the levels to which the song system is exposed. Secondly, if aromatase activity is differentially regulated in various brain regions, then different circuits may be exposed to different levels of estrogen at specific periods of development. As explained below, this could account for some of the disparate effects of estrogen on behavioral development in zebra finches. However, if the brain is the controlling synthetic source of plasma estrogens, then it may be difficult to have independent regulation of estrogen levels in different brain regions, since estrogen synthesized in one region may circulate back to influence others. Finally, the local synthesis of estrogen suggests three mechanisms by which the male's song system may be exposed to higher levels of estrogen than the female's. First of all, there may be more local estrogen in males if there is more androgenic substrate; this is unlikely, since three studies indicate that males do not have more androgen in plasma than females. Second, the levels of aromatase might be higher in males than in females. Two studies have failed to find such a sex difference in aromatase, but one study was done on 20-day-old birds (Vockel et al. 1988), probably beyond the period of sexual differentiation, and the other study measured whole telencephalic aromatase (Schlinger and Arnold 1991c), and thus might have missed important local differences in levels of aromatase. Thus, the question of sex differences in aromatase has not been resolved. Third, masculinization of the song system could result even if the levels of estrogen are similar in males and females, if the levels of estrogen receptor are higher in males. This possibility has not been tested adequately.

It would be easier to answer the questions regarding brain exposure to estrogen if we knew where estrogen acts to masculinize the song system, since we could then look in that location for sex differences in levels of aromatase activity, estrogen receptors, and so forth. Thus far, attempts to find the site of estrogen action have centered around the search for estrogen receptors, which are required for the genomic actions of estrogen. Paradoxically, autoradiographic studies of estrogen accumulation (a result of functional binding of estrogen to receptor) have failed to find high densities of estrogen-labeled cells in the song system of female zebra finches at the time when estrogen can have potent masculinizing effects on the system (Nordeen et al. 1987). Of the telencephalic song regions, only HVC contains some estrogen-labeled cells, and these are weakly labeled and represent a small percentage of cells. In other regions, there are few or no such cells. Immunohistochemical techniques find the same pattern (Gahr et al. 1987; Gahr and Konishi 1988; but see Walters et al. 1988). However, immediately adjacent to HVC is a large field of well-labeled cells. Their proximity to HVC makes them the most likely site of estrogen action in the masculinization of HVC. According to this idea, estrogen acts near HVC to induce a masculine program of development for HVC; HVC in turn has transsynaptic masculinizing effects on other nuclei with which it connects. This idea is compatible with the finding that lesions of HVC block the masculinizing effects of estradiol on area X and RA, the two nuclei to which HVC projects (Herrmann and Arnold 1990).

The preceding paragraphs indicate that there is a great deal of evidence to support the idea that the hormonal control of sexual differentiation in zebra finches has some similarities to the mammalian pattern, in that a metabolite (estradiol) of a testicular secretion (testosterone) induces the masculine program of neural and behavioral development. However, the details of this hormonal regulation of development need a great deal of further clarification. It is not clear that testicular secretions are themselves important for sexual differentiation. Castration of young males is difficult because of their small size, so that castration has not been successful in males younger than 9 days after hatching, after which time the testes are no longer necessary for masculine song development (Arnold 1975b). Therefore, there is no convincing evidence for or against the notion that the entire process of behavioral and neural sexual differentiation stems from differential secretion of the gonads. Obviously, this question must be answered if we are to understand sexual differentiation in songbirds. The cerebral control of plasma estrogen levels represents a considerable departure from the mammalian pattern, since the brain itself might synthesize estrogen that could have effects on distant target cells.

Because estrogens are thought to demasculinize other species of birds, it is surprising that estradiol masculinizes zebra finch song. Adkins-Regan and Ascenzi (1987) have suggested that this difference is not a species difference, but rather a difference in the behaviors that were studied. In studies demonstrating a demasculinizing action of estradiol in quail and other species, only courtship and copulatory behaviors were studied, whereas only song was originally studied in zebra finches. Indeed, when Adkins-Regan and Ascenzi (1987) examined the effects of estradiol treatment on development of a range of reproductive behaviors in zebra finches, they found that estradiol treatment of young males prevented differentiation of some masculine courtship and copulatory behaviors, in a pattern similar to quail. This result is paradoxical, in that it suggests that estradiol secretion is required in males to masculinize song behavior, but is also required in females to prevent the development of masculine copulatory behaviors. If so, then one must explain how males are protected from the demasculinizing effects of estradiol, and how females are protected from the masculinizing effects. One scenario is that males and females both secrete estradiol, but at different periods of development. Neural circuits involved in copulation might then be sensitive to the demasculinizing effects of estradiol at a different period of development than the neural song circuit. An alternative idea is that sexual dimorphisms in aromatase levels exist in brain regions controlling both song and copulation, but the direction of the sex difference is reversed (higher in females in hypothalamus and other areas related to copulation, higher in males near the song circuit). Although further work is needed to resolve the mixed masculinizing and demasculinizing actions of estrogen, it is clear that estrogen plays a crucial role in the development of song and other reproductive behaviors.

SENSITIVE PERIODS FOR SONG LEARNING

In all oscine passerine birds studied to date, song is learned. Males must hear conspecific song in order to develop normal adult song. For example, a young male zebra finch copies the song of his father (Immelmann 1969; Böhner 1983). In other species, such as the white-crowned sparrow and the chaffinch (*Fringilla coelebs*), the young male copies songs heard from other males in the environment (Konishi and Nottebohm 1969). A male acoustically isolated from others of its own species will develop an abnormally simple song. If such isolated males are played songs early in development, they will eventually copy the songs that they hear.

Song is often learned during a *sensitive period*. For example, male swamp sparrows (*Melospiza georgiana*) most readily copy song elements heard between days 20 and 65 after hatching (Marler and Peters 1988). The end of this sensitive period is not absolute, since some song elements heard much later (up to 300 days of age) can also be copied, but the probability of learning is much higher before 65 days. This early period of learning is called the *memorization* or *sensory learning* phase, and represents the time at which the male learns the acoustic characteristics of the song elements to be copied. The operational definition of the sensory phase is the period during which exposure to song elements results in their eventual inclusion in the song repertoire. The memory of sounds acquired during this period is often called the *template* for later vocal development. However, learning also involves a second phase of *rehearsal* or *sensory-motor integration*, in which the male begins to sing progressively more accurate copies of the song elements. At first, the male sings quiet *subsong*, which is a rambling sequence of poorly formed sounds that bear little resemblance to the model song. Then the male enters the stage of *plastic song*, which includes diverse, recognizable song elements rendered in highly variable form. Finally, the fully adult, highly stereotyped song is *crystallized*, and little further song development occurs. During the sensory-motor integration phase, the male must be able to hear himself, presumably because he is comparing his own vocal output with the memory of the song elements that he is copying. This comparison requires auditory feedback, and birds deafened before the sensory-motor phase never produce a copy of any song elements (e.g., Konishi 1965). The sensory-motor phase may begin before the sensory learning phase is complete. For example, in chaffinches and zebra finches, song elements heard after the onset of rehearsal are often incorporated into the repertoire (Konishi and Nottebohm 1969; Immelmann 1969). In other species, such as the swamp sparrow, song elements are not practiced until after the main phase of sensory learning (Marler et al. 1987).

In some species, learning is not restricted to a sensitive period; rather, the song repertoire continues to change throughout the life of the bird. These species are often called *open-ended* learners, in contrast to sensitive-period learners. Canaries and red-winged blackbirds represent two species in which new song elements are added each year (Nottebohm et al. 1986; Kirn et al. 1989). Such learning in adulthood can occur particularly during specific seasons. This species diversity in learning programs (sensitive-period vs. open-ended) offers an interesting opportunity for comparison of species as a means to make deductions about the neural events responsible for learning. If learning is restricted to a single period in some species, it may be possible to find changes in the neural song system that occur exclusively in that period, which may therefore underlie the learning process. If the same neural events are then found not to be restricted to a sensitive period in species that are open-ended learners, then the species comparison offers support for the importance of those events in the learning process.

NEURAL CHANGES CORRELATED WITH SONG LEARNING

We turn now to studies that measure structural changes in the song system that occur during song learning. These changes are so fundamental that it is difficult to escape the conclusion that they have a role in the learning process itself. We focus first on the development of the song system in zebra finches, since this is the most studied sensitive-period species.

A young male zebra finch begins to memorize sounds that it hears at about day 25 after hatching (Immelmann 1969), and most sensory acquisition occurs between days 35 and 50, although some elements may be acquired as late as day 65 (Eales 1985, 1987). Singing itself begins at about 25–30 days of age, so that the memorization and rehearsal phases of learning overlap. By 60–70 days of age, most song elements have been added to the repertoire, and only minor modifications to song occur thereafter (Arnold 1974). This stabilization, or crystallization, of the adult song pattern can be taken as the end of song learning. Traditionally, it has been thought that after this time, the motor circuit for song undergoes no further change. This idea has been recently challenged, since adult male zebra finches can experience significant changes in the song pattern if they are deafened after this age (Nordeen and Nordeen 1992; see also Williams and McKibben 1992). The effect of deafening in adulthood is more modest than the effect of deafening early in life, and the changes in song occur slowly, over a period of weeks or months. Nevertheless, these findings demonstrate that the motor circuit for song is not immutable in fully adult zebra finches, and apparently requires auditory feedback for maintenance. Thus, some form of learning may continue throughout adult life, even in the so-called sensitive-period species. Despite this adult plasticity, there is still a valid dichotomy between the early sensitive period of song learning and the more stable adult phase of song production. In addition, sensitive-period learners still differ from open-ended learners because the degree of plasticity in the latter is much greater than in the former.

Morphological studies of the song system have concentrated on the period from 20 to 60 days after hatching, when the male is establishing his template and when he passes through most of the stages of motor production of song. During this period there are dramatic changes in the structure of the telencephalic song system. Neuroblasts continue to divide, and many newly formed neurons migrate into HVC and area X, contributing to a dramatic increase in the size of these regions (Bottjer et al. 1986; E. Nordeen and K. Nordeen 1988, 1989; K. Nordeen and E. Nordeen 1988). In HVC, the neurons born after day 20 project mainly to RA, whereas few project to area X. Thus, one important change during song learning is a major increase in the projection from HVC to RA. Axons from HVC are not found in RA at 25 days of age, but are found there after day 35, suggesting that there is abrupt ingrowth of HVC axons at about day 30 (Konishi and Akutagawa 1985). At the same time, RA grows in volume, both because of an increase in size of individual neurons and because of an increase in the spacing between neurons (Konishi and Akutagawa 1985; Bottjer et al. 1986; Herrmann and Bischoff 1986). This increase in the neuropil of RA correlates with a six-fold increase in the number of synapses in RA from HVC (Herrmann and Arnold 1991). In HVC, the number of androgen target neurons triples, in part because of the addition of newly generated HVC-to-RA neurons, many of which are androgen targets (Sohrabji et al. 1989).

Thus, major portions of the neural song circuit are formed and grow during the early phases of song learning. This growth may be required for song learning, or may at least allow learning to occur. For example, the formation of the HVC-to-RA pathway must be of fundamental importance for learning, because this is the major descending pathway for motor control of vocalization, and because HVC is thought to play a critical role in the feedback of auditory information into the song circuit. Because the HVC-to-RA pathway is laid down at a time when the bird is memorizing the template, one can imagine that the sounds the bird hears may influence the kinds of synaptic connections that are made or retained in RA, and hence influence either the subsequent perception or production of sounds. Indeed, much less growth of the HVC-to-RA circuit occurs in adult male and juvenile female zebra finches, which learn less than juvenile males (Konishi and Akutagawa 1985; Nordeen and Nordeen 1988; Williams 1985). Growth of area X may also be a critical part of the learning process, since area X receives auditory input, probably via HVC (Katz and Gurney 1981), and since lesions of area X prevent song learning (Sohrabji et al. 1990; Scharff and Nottebohm 1991).

In addition to the marked growth in the song system associated with learning, there is neuronal degeneration and death. Between days 25 and 53, lMAN declines in volume, loses about half of its neurons, and loses about half of its synaptic contacts in the neuropil of RA (Bottjer et al. 1985; Bottjer and Sengelaub 1989; Korsia and Bottjer 1990; Herrmann and Arnold 1991). Because lesions of lMAN prevent song learning, the developmental decline of the lMAN-to-RA circuit leads to the speculation that this input is required for song learning. According to this idea, the end of the sensitive period for song learning is controlled by the timing of the decline in lMAN. Although this hypothesis is attractive, current evidence suggests that it is too simple. Most of the death of lMAN neurons appears to have ended by about day 35 (Bottjer and Sengelaub 1989), at a time when song learning is still occurring, and at least 2 weeks before the decline in the ability of lMAN lesions to disrupt song learning (Bottjer et al. 1984). Moreover, the number of lMAN neurons does not change after day 35, even though there is a functional change in these neurons, revealed by the progressive decline in the effect of lMAN lesions. One change in the lMAN neuron population between day 40 and adulthood is that there is a decline in the number of androgen target cells, particularly in the subpopulation of neurons that do not project to RA (Korsia and Bottjer 1990). Thus, loss of sensitivity to androgen after day 40 might be correlated with the ability to learn.

The neural changes correlated with learning have also been studied in several species in which the program for learning differs from that of zebra finches (Nottebohm et al. 1990). In the canary, an open-ended learner, remarkable growth in HVC and RA accompanies the learning of song in the first year (Nottebohm et al. 1986). Moreover, the addition of newly generated neurons to HVC during this period involves primarily neurons that project to RA (Alvarez-Buylla et al. 1988; Kirn et al.

1991). Thus, song learning coincides with the formation of the HVC-to-RA pathway, as in zebra finches (Nottebohm et al. 1990). Because the canary is an open-ended learner, one would expect more plasticity in the neural song circuit in adulthood than has been observed in zebra finches. This expectation has been amply borne out. The addition of newly generated HVC-to-RA neurons does not end with the crystallization of song in a canary's first year, but rather continues beyond this point. The addition of new elements to canary song occurs more in late summer and fall than in other seasons, and the incorporation of new HVC-to-RA neurons is greater in fall than in spring (Nottebohm et al. 1986, 1987; Alvarez-Buylla et al. 1990). The plasticity of adult canaries has also been studied in females, which sing in response to testosterone treatment in adulthood. This testosterone treatment causes growth in the volumes of HVC, RA, and nXIIts, as well as increases in the lengths of dendrites and numbers of synapses in RA (Nottebohm 1980a, 1980b; DeVoogd and Nottebohm 1981; DeVoogd et al. 1985, 1991; Canady et al. 1988; Bottjer and Dignan 1988; Clower et al. 1989). This plasticity of the adult canary song system is much greater that that found in the adult zebra finch brain. In zebra finches, testosterone treatment of adult females does not increase the volumes of telencephalic nuclei (Arnold 1980a). Moreover, the rate of incorporation of new neurons into HVC in zebra finches declines at about the time that song is crystallized. Thus, the relative lack of behavioral change in adult zebra finches correlates well with the relative lack of neural plasticity. Although the correlation between vocal and neural plasticity is valid, there is not necessarily a precise relationship between the timing of neurogenesis and changes in vocal behavior in adult canaries. For example, addition of newly generated neurons to HVC occurs at day 240, when song is stable and there is no evidence for song learning. Perhaps the newly generated neurons are available for song learning that occurs later, or some perceptual learning may be occurring at this time.

The swamp sparrow represents another interesting contrast to the zebra finch. The swamp sparrow is a sensitive-period species like the zebra finch, but differs from it in that the two phases of song acquisition are temporally separated, so that it is possible to determine which neural changes correlate with each phase. As mentioned above, swamp sparrows memorize most of their song elements between 20 and 65 days of age, although some sensory learning can occur as late as 300 days of age. Some subsong begins between days 50 and 125, but production and rehearsal of elements of learned song occurs beginning at 250 days of age. Fully crystallized adult song is produced by 325–350 days (Marler and Peters 1988). As in zebra finches, both HVC and area X grow extensively during song learning because of a two- to three-fold increase in the number of neurons in these nuclei (Nordeen et al. 1989). This growth occurs between 23 and 61 days of age, at the time of the sensory phase of learning. In contrast, the numbers of neurons in lMAN and RA do not change during any phase of song learning. These data suggest that the addition of neurons to HVC and area X is important in determining when a male can memorize song elements that will be incorporated into the repertoire. Moreover, changes in the numbers of neurons in these nuclei are not correlated with song production, and hence are not needed to initiate the rehearsal phase of learning. Because lMAN was found not to lose neurons in swamp sparrows as it does in zebra finches, it is tempting to conclude that such loss is not involved in the timing of the sensitive period for learning. However, Nordeen et al. (1989) did find small decreases in the number of lMAN neurons during both phases of learning. Although these changes were not statistically significant, it is possible that they were real, so that it is difficult to use these data to rule out the importance of degeneration of lMAN for the learning process.

ROLE OF STEROID HORMONES IN SONG LEARNING

It is conceivable that song learning is controlled by sex steroid hormones. One idea is that because the ability to learn to sing is sexually dimorphic in some species, estradiol or other steroids that control sexual differentiation are necessary for the development of the mechanisms of song learning. A second idea is that the developmental changes in the levels of steroid hormones directly trigger the onset or offset of learning. Both hypotheses have been tested. The first has been tested in the zebra finch. Young females treated with estradiol eventually sing without the need for further hormonal treatment, and they learn song elements from tutors (Simpson and Vicario 1991a). Moreover, such treatment with estradiol triggers masculine patterns of neuronal development associated with learning, such as the addition of newly generated neurons to HVC (Nordeen and Nordeen 1989). Thus, it would appear that exposure of the brain to estrogen is necessary for this kind of learning. However, the learning that occurs in estradiol-treated female zebra finches occurs after the levels of exogenous estrogen have declined, suggesting that the timing of learning is not directly determined by the timing of the estrogen treatment. Rather, estrogen is thought to start a cascade of developmental events that results in song learning.

The second hypothesis, that changes in the levels of hormone directly cause changes in the ability to learn, has been tested in five different species. One approach has been to castrate young males to determine whether the removal of gonadal hormones alters the timing of learning. Nottebohm (1969) castrated a 6-month-old chaffinch, then treated him with testosterone at 2 years of age, long after the end of the normal period of song acquisition. This male, who ceased singing after castration, began singing after testosterone treatment, and copied songs played to him outside of the sensitive period. This experiment suggested that castration prevents the end of song learning, and that gonadal androgens might normally bring both memorization and rehearsal phases of learning to an end. This result has not been confirmed in other species. In zebra finches, castration at 9–17 days of age does not prevent males from copying sounds heard after castration (Arnold 1975b). Although castrated males develop song somewhat more slowly than normal, they nevertheless develop normal song without their gonads. Similarly, castration does not change the timing of learning or the amount learned by marsh wrens (*Cistothorus palustris*) (Kroodsma 1986; Kroodsma and Pickert 1980). However, castrated marsh wrens differ from normal males in that they fail to progress through the final stage of song crystallization. Because androgen treatment of castrates induces them to complete this final phase of song acquisition, it would appear that the only part of song acquisition that requires gonadal androgens is the final crystallization.

These studies of the effects of castration on song learning were conducted on the reasonable assumption that castration reduces the levels of sex steroids. However, none of these studies actually measured the levels of steroids. More recently it has become apparent that castration does not reduce plasma levels of estrogen in adult male zebra finches or juvenile male swamp sparrows, although the levels of androgens are reduced (Adkins-Regan et al. 1990; Schlinger and Arnold 1991b; Marler et al. 1989). Moreover, because estrogen is synthesized at high levels in brain, measures of plasma steroids may not fully reflect the levels of hormone at cerebral sites of action (Schlinger and Arnold 1991b). Nevertheless, it has become imperative to measure the plasma levels of steroids in any experiment of this kind, as a first approximation of the relevant levels of steroids. Marler et al. (1989) castrated young song sparrows (*Melospiza melodia*) and swamp sparrows and measured both song development and plasma levels of steroids. In these species, castrated males copied songs at about the same age as intact males. Like marsh wrens, they progressed to the plastic song phase, and they failed to crystallize song unless given exogenous androgen. Castration in these males effectively reduced plasma levels of androgen, indicating that the sensory phase of learning does not require gonadal androgen. However, the levels of estrogen were not reduced by castration.

Marler et al. (1987) studied the plasma levels of sex steroid hormones during development of intact male swamp sparrows as a further test of the importance of these hormones during specific phases of song learning. Testosterone levels in blood are normally elevated during both sensory learning (ages 20–85 days) and sensory-motor integration (after age 250 days), and are low at intermediate ages, when little learning is thought to take place. In light of the effects of castration on swamp sparrows discussed above, it would appear that the early phase of high testosterone secretion is not necessary for sensory learning, but that the later phase is important for final crystallization of song. Plasma levels of estradiol are elevated during the entire sensory learning phase, but are lower during the sensory-motor integration phase of learning. Marler et al. (1987) also found that plasma estradiol levels were higher during sensory acquisition in a subset of birds that copied songs from live tutors, compared with birds that did not copy the tutors. Thus, these data support the idea that estradiol may exert some control over the timing of sensory acquisition; when estradiol levels are high, sensory learning occurs, and when they fall, sensory learning declines. Assuming that the estradiol found in plasma of swamp sparrows, as in zebra finches, is synthesized in brain, the results of Marler et al. (1987) imply that we need to understand a great deal more about local levels of estradiol in brain at various periods of development, as well as the factors that regulate cerebral estradiol synthesis.

CONCLUSIONS

In this chapter I have emphasized the species diversity in song in passerine birds. There are different developmental programs of sexual differentiation that give rise to varying degrees of sexual dimorphism in the ability to sing. Similarly, different species have diverse developmental programs for song learning. Sensitive-period species differ in the time of life at which they learn song, and open-ended species appear to be able to learn throughout life. Although neural and behavioral development have been studied in only a few species, there is sufficient information to begin to shape ideas about the evolution of different programs of development.

The first realization is that even closely related species can have different degrees of sexual dimorphism in song. For example, the genus *Thryothorus* includes species such as the rufous-and-white wren (*T. rufalbus*), in which females sing less than males, and others in which there is little sexual dimorphism (the bay wren, *T. nigricapillus*, and the buff-breasted wren, *T. leucotis*) (Brenowitz et al. 1985; Brenowitz and Arnold 1985, 1986; Arnold et al. 1986). Because we assume that congeners share most of their genome, it would appear that the large differences in development of the song system in *Thryothorus* females can be explained on the basis of small genetic differences. One hypothesis is that small genetic changes could lead to different amounts of estrogen secretion in females. In monomorphic species, the song circuit would be exposed to high levels of estrogen in both sexes, so that the female's song system would be "masculinized"; whereas levels would be low in females of dimorphic species, so that the female's song circuit would remain feminine. However, generalized estrogen secretion into plasma should have undesirable masculinizing effects on other neural and behavioral systems, so that females of monomorphic species might become completely masculine in all traits. Thus, if this hypothesis is valid, one must postulate that the mechanisms of masculinization of song are independent of masculinization of other neural structures. This independence has already been discussed above with reference to the finding that estrogens appear to have both masculinizing and demasculinizing effects on different behavioral systems in zebra finches (Adkins-Regan and Ascenzi 1987). This independence could be achieved if estrogen acts at different times of development, or is secreted asynchronously at different loci in brain. A second hypothesis is that the absence of sex differences could be achieved by mutations that render song development insensitive to steroids. In other words, the genes that are developmentally regulated by estrogen in the song circuit could become constitutively expressed in both sexes, leading to sexual monomorphism in vocal ability. These ideas need to be tested by further study of ontogeny in different species.

We can speculate in a similar vein on the kinds of genetic changes that give rise to different developmental programs of song learning. However, this speculation is based on even less data than the discussion of sexual differentiation, since we know less about the factors that control the ability to learn. If the timing of learning is triggered by a hormone such as estradiol, then mutations in the genes controlling estradiol synthesis could again lead to different developmental programs of song learning. Alternatively, the timing of song learning may be controlled by the expression of specific trophic factors, for example factors that determine the timing of neuronal incorporation into HVC or the timing of growth of axons in RA, two events thought to determine the timing of song learning. Further speculation about genetic differences between species must await further information about the factors that regulate learning in songbirds.

REFERENCES

Adkins, E. K. 1976. The effects of the antiestrogen CI-628 on sexual behavior activated by androgen or estrogen in quail. *Hormones and Behavior* 7:417–29.

Adkins-Regan, E. 1987. Sexual differentiation in birds. *Trends in Neuroscience* 10:517–22.

Adkins-Regan, E., and M. Ascenzi. 1987. Social and sexual behaviour of male and female zebra finches treated with oestradiol during the nestling period. *Animal Behaviour* 35:1100–1112.

Adkins-Regan, E., M. Abdelnabi, M. Mobarak, and M. A. Ottinger. 1990. Sex steroid levels in developing and adult male and female zebra finches. *General and Comparative Endocrinology* 78:93–109.

Alvarez-Buylla, A., M. Theelen, and F. Nottebohm. 1988. Birth of projection neurons in the higher vocal center of the canary forebrain before, during, and after song learning. *Proceedings of the National Academy of Sciences U.S.A.* 85:8722–26.

Alvarez-Buylla, A., J. R. Kirn, and F. Nottebohm. 1990. Birth of projection neurons in adult avian brain may be related to perceptual or motor learning. *Science* 249:1444–46.

Arnold, A. P. 1974. Behavioral effects of androgen in zebra finches (*Poephila guttata*) and a search for its sites of action. Ph.D. thesis, Rockefeller University.

Arnold, A. P. 1975a. The effects of castration and androgen replacement on song, courtship, and aggression in zebra finches (*Poephila guttata*). *Journal of Experimental Zoology* 191: 309–26.

Arnold, A. P. 1975b. The effects of castration on song development in zebra finches (*Poephila guttata*). *Journal of Experimental Zoology* 191:261–78.

Arnold, A. P. 1980a. Effects of androgens on volumes of sexually dimorphic brain regions in the zebra finch. *Brain Research* 185:441–44.

Arnold, A. P. 1980b. Quantitative analysis of sex differences in hormone accumulation in the zebra finch brain: Methodological and theoretical issues. *Journal of Comparative Neurology* 189:421–36.

Arnold, A. P., and A. Saltiel. 1979. Sexual difference in pattern of hormone accumulation in the brain of a song bird. *Science* 205:702–5.

Arnold, A. P., S. W. Bottjer, E. A. Brenowitz, E. J. Nordeen, and K. W. Nordeen. 1986. Sexual dimorphisms in the neural vocal control system in song birds: Ontogeny and phylogeny. *Brain, Behavior and Evolution* 28:22–31.

Balthazart, J., A. De Clerck, and A. Foidart. 1992. Behavioral demasculinization of female quail is induced by estrogens: Studies with the new aromatase inhibitor, R76713. *Hormones and Behavior* 26:179–203.

Blyth, B., and J. W. Duckett, Jr. 1991. Gonadal differentiation: A review of the physiological process and influencing factors based on recent experimental evidence. *Journal of Urology* 145:689–94.

Böhner, J. 1983. Song learning in the zebra finch (*Taeniopygia guttata*): Selectivity in the choice of a tutor and accuracy of song copies. *Animal Behaviour* 31:231–37.

Bottjer, S. W., and T. P. Dignan. 1988. Joint hormonal and sensory stimulation modulate neuronal number in adult canary brains. *Journal of Neurobiology* 19:624–35.

Bottjer, S. W., and D. R. Sengelaub. 1989. Cell death during development of a forebrain nucleus involved with vocal learning in zebra finches. *Journal of Neurobiology* 20:609–18.

Bottjer, S. W., E. A. Miesner, and A. P. Arnold. 1984. Forebrain lesions disrupt development but not maintenance of song in passerine birds. *Science* 224:901–3.

Bottjer, S. W., S. L. Glaessner, and A. P. Arnold. 1985. Ontogeny of brain nuclei controlling song learning and behavior in zebra finches. *Journal of Neuroscience* 5:1556–62.

Bottjer, S. W., E. A. Miesner, and A. P. Arnold. 1986. Changes in neuronal number, density and size account for increases in volume of song-control nuclei during song development in zebra finches. *Neuroscience Letters* 67:263–68.

Bottjer, S. W., K. A. Halsema, S. A. Brown, and E. A. Miesner. 1989. Axonal connections of a forebrain nucleus involved with vocal learning in zebra finches. *Journal of Comparative Neurology* 279:312–26.

Brenowitz, E. A., and A. P. Arnold. 1985. Lack of sexual dimorphism in steroid accumulation in vocal control brain regions of duetting song birds. *Brain Research* 344:172–75.

Brenowitz, E. A., and A. P. Arnold. 1986. Interspecific comparisons of the size of neural song control regions and song complexity in duetting birds: Evolutionary implications. *Journal of Neuroscience* 6:2875–79.

Brenowitz, E. A., A. P. Arnold, and R. N. Levin. 1985. Neural correlates of female song in tropical duetting birds. *Brain Research* 343:104–12.

Canady, R. A., G. D. Burd, T. J. DeVoogd, and F. Nottebohm. 1988. Effects of testosterone on input received by an identified neuron type of the canary song system: A golgi/electron microscopy/degeneration study. *Journal of Neuroscience* 8:3770–84.

Clower, R. P., B. E. Nixdorf, and T. J. DeVoogd. 1989. Synaptic plasticity in the hypoglossal nucleus of female canaries: Structural correlates of season, hemisphere, and testosterone treatment. *Behavioral and Neural Biology* 52:63–77.

DeVoogd, T. J., and F. Nottebohm. 1981. Gonadal hormones induce dendritic growth in the adult avian brain. *Science* 214:202–4.

DeVoogd, T. J., B. Nixdorf, and F. Nottebohm. 1985. Synaptogenesis and changes in synaptic morphology related to acquisition of a new behavior. *Brain Research* 329:304–8.

DeVoogd, T. J., D. J. Pyskaty, and F. Nottebohm. 1991. Lateral asymmetries and testosterone-induced changes in the gross morphology of the hypoglossal nucleus in adult canaries. *Journal of Comparative Neurology* 307:65–76.

Doupe, A. J., and M. Konishi. 1991. Song-selective auditory circuits in the vocal control system of the zebra finch. *Proceedings of the National Academy of Sciences U.S.A.* 88:11339–43.

Eales, L. A. 1985. Song learning in zebra finches: Some effects of song model availability on what is learnt and when. *Animal Behaviour* 33:1293–1300.

Eales, L. A. 1987. Song learning in female-raised zebra finches: Another look at the sensitive phase. *Animal Behaviour* 35:1356–65.

Gahr, M., and M. Konishi. 1988. Developmental changes in estrogen-sensitive neurons in the forebrain of the zebra finch. *Proceedings of the National Academy of Sciences U.S.A.* 85:7380–83.

Gahr, M., G. Flugge, and H.-R. Güttinger. 1987. Immunocytochemical localization of estrogen binding neurons in the songbird brain. *Brain Research* 402:173–77.

Goy, R. W., and B. S. McEwen, eds. 1980. *Sexual differentiation of the brain.* Cambridge, Mass.: MIT Press.

Gurney, M. 1981. Hormonal control of cell form and number in the zebra finch song system. *Journal of Neuroscience* 1:658–73.

Gurney, M. 1982. Behavioral correlates of sexual differentiation in the zebra finch song system. *Brain Research* 231:153–72.

Gurney, M., and M. Konishi. 1980. Hormone induced sexual differentiation of brain and behavior in zebra finches. *Science* 208:1380–82.

Harding, C. F., K. Sheridan, and M. J. Walters. 1983. Hormonal specificity and activation of sexual behavior in male zebra finches. *Hormones and Behavior* 17:111–33.

Herrmann, K., and A. P. Arnold. 1990. Lesions of Hvc block the developmental masculinizing effects of estradiol in the female song system. *Journal of Neurobiology* 22:29–39.

Herrmann, K., and A. P. Arnold. 1991. The development of afferent projections to robust archistriatal nucleus in male zebra finches: A quantitative electron microscopic study. *Journal of Neuroscience* 11:2063–74.

Herrmann, K., and H. J. Bischoff. 1986. Delayed development of song control nuclei in the zebra finch is related to behavioral development. *Journal of Comparative Neurology* 245:167–75.

Hutchison, J. B., J. C. Wingfield, and R. E. Hutchison. 1984. Sex differences in plasma concentrations of steroids during the sensitive period for brain differentiation in the zebra finch. *Journal of Endocrinology* 103:363–69.

Immelmann, K. 1969. Song development in the zebra finch and other estrildid finches. In R. A. Hinde, ed., *Bird vocalizations,* 61–74. Cambridge: Cambridge University Press.

Katz, L. C., and M. E. Gurney. 1981. Auditory responses in the zebra finch's motor system for song. *Brain Research* 221:192–97.

Kelley, D. B., and F. Nottebohm. 1979. Projections of a telencephalic auditory nucleus—field L—in the canary. *Journal of Comparative Neurology* 183:455–69.

Kirn, J. R., R. P. Clower, D. E. Kroodsma, and T. J. DeVoogd. 1989. Song-related brain regions in the red-winged blackbird are affected by sex and season but not repertoire size. *Journal of Neurobiology* 20:139–63.

Kirn, J. R., Alvarez-Buylla, and F. Nottebohm. 1991. Production and survival of projection neurons in a forebrain vocal center of adult male canaries. *Journal of Neuroscience* 11:1756–62.

Konishi, M. 1965. The role of auditory feedback in the control of vocalization in the white-crowned sparrow. *Zeitschrift für Tierpsychologie* 22:770–83.

Konishi, M., and E. Akutagawa. 1985. Neuronal growth, atrophy, and death in a sexually dimorphic song nucleus in zebra finches. *Nature* 315:145–47.

Konishi, M., and E. Akutagawa. 1990. Growth and atrophy of neurons labeled at their birth in a song nucleus of the zebra finch. *Proceedings of the National Academy of Sciences U.S.A.* 87:3538–41.

Konishi, M., and F. Nottebohm. 1969. Experimental studies in the ontogeny of avian vocalizations. In R. A. Hinde, ed., *Bird vocalizations*, 29–48. Cambridge: Cambridge University Press.

Korsia, S., and S. W. Bottjer. 1990. Developmental changes in the cellular composition of a brain nucleus involved with song learning in zebra finches. *Neuron* 3:451–60.

Kroodsma, D. W. 1986. Song development by castrated marsh wrens. *Animal Behaviour* 34: 1573–75.

Kroodsma, D. E., and R. Pickert. 1980. Environmentally dependent sensitive periods for avian vocal learning. *Nature* 288:477–79.

McCasland, J. S. 1987. Neuronal control of bird song production. *Journal of Neuroscience* 7:23–39.

McCasland, J. S., and M. Konishi. 1981. Interaction between auditory and motor activities in an avian song control nucleus. *Proceedings of the National Academy of Sciences U.S.A.* 78:7815–19.

McEwen, B. S. 1981. Neural gonadal steroid actions. *Science* 211:1303–11.

McEwen, B. S., P. G. Davis, B. Parsons, and D. W. Pfaff. 1979. The brain as a target for steroid hormone action. *Annual Review of Neuroscience* 2:65–112.

MacLusky, N. J., and F. Naftolin. 1981. Sexual differentiation of the central nervous system. *Science* 211:1294–1303.

Margoliash, D. 1986. Preference for autogenous song by auditory neurons in a song system nucleus of the white-crowned sparrow. *Journal of Neuroscience* 6:1643–61.

Marler, P., and S. Peters. 1988. Sensitive periods for song acquisition from tape recordings and live tutors in the swamp sparrow, *Melospiza georgiana*. *Ethology* 77:76–84.

Marler, P., S. Peters, and J. Wingfield. 1987. Correlations between song acquisition, song production, and plasma levels of testosterone and estradiol in sparrows. *Journal of Neurobiology* 18:531–48.

Marler, P., S. Peters, G. F. Ball, A. M. Dufty, Jr., and J. C. Wingfield. 1989. The role of sex steroids in the acquisition and production of birdsong. *Nature* 336:770–71.

Nordeen, E. J., and K. W. Nordeen. 1988. Sex and regional differences in the incorporation of neurons born during song learning in zebra finches. *Journal of Neuroscience* 8:2869–74.

Nordeen, E. J., and K. W. Nordeen. 1989. Estrogen stimulates the incorporation of new neurons into avian song nuclei during adolescence. *Developmental Brain Research* 49:27–32.

Nordeen, E. J., and K. W. Nordeen. 1990. Neurogenesis and sensitive periods in avian song learning. *Trends in Neuroscience* 13:31–36.

Nordeen, E. J., K. W. Nordeen, and A. P. Arnold. 1987. Sexual differentiation of androgen accumulation within the zebra finch brain through selective cell loss and addition. *Journal of Comparative Neurology* 259:393–99.

Nordeen, K. W., and E. J. Nordeen. 1988. Projection neurons within a vocal motor pathway are born during song learning in zebra finches. *Nature* 334:149–51.

Nordeen, K. W., and E. J. Nordeen. 1992. Auditory feedback is necessary for the maintenance of stereotyped song in adult zebra finch. *Behavioral and Neural Biology* 57:58–66.

Nordeen, K. W., E. J. Nordeen, and A. P. Arnold. 1986. Estrogen establishes sex differences in androgen accumulation in zebra finch brain. *Journal of Neuroscience* 6:734–38.

Nordeen, K. W., E. J. Nordeen, and A. P. Arnold. 1987. Estrogen accumulation in zebra finch song control nuclei: Implications for sexual differentiation and adult activation of song behavior. *Journal of Neurobiology* 18:569–82.

Nordeen, K. W., P. Marler, and E. Nordeen, 1989. Addition of song-related neurons in swamp sparrows coincides with memorization, not production, of learned songs. *Journal of Neurobiology* 20:651–61.

Nottebohm, F. 1969. The "critical period" for song learning. *Ibis* 111:386–87.

Nottebohm, F. 1980a. Brain pathways for vocal learning in birds: A review of the first 10 years. *Progress in Psychobiology and Physiological Psychology* 9:85–125.

Nottebohm, F. 1980b. Testosterone triggers growth of brain vocal control nuclei in adult female canaries. *Brain Research* 189:429–36.

Nottebohm, F., and A. P. Arnold. 1976. Sexual dimorphism in vocal control areas of the song bird brain. *Science* 194:211–13.

Nottebohm, F., T. M. Stokes, and C. M. Leonard. 1976. Central control of song in the canary (*Serinus canarius*). *Journal of Comparative Neurology* 165:457–86.

Nottebohm, F., D. B. Kelley, and J. A. Paton. 1982. Connections of vocal control nuclei in the canary telencephalon. *Journal of Comparative Neurology* 207:344–57.

Nottebohm, F., M. E. Nottebohm, and L. Crane. 1986. Developmental and seasonal changes in canary song and their relation to changes in the anatomy of song-control nuclei. *Behavioral and Neural Biology* 46:445–71.

Nottebohm, F., M. E. Nottebohm, L. A. Crane, and J. C. Wingfield. 1987. Seasonal changes in gonadal hormone levels of adult male canaries and their relation to song. *Behavioral and Neural Biology* 47:197–211.

Nottebohm, F., A. Alvarez-Buylla, J. Cynx, J. Kirn, C.-Y. Ling, M. Nottebohm, R. Suter, A. Tolles, and H. Williams. 1990. Song learning in birds: The relation between perception and production. *Philosophical Transactions of the Royal Society of London* B: *Biological Sciences* 329:115–24.

Pesch, A., and H.-R. Güttinger. 1985. Der Gesang des weiblichen Kanarienvogels. *Journal Ornithologie* 126:108–10.

Scharff, C., and F. Nottebohm. 1991. A comparative study of the behavioral deficits following lesions of various parts of the zebra finch song system. *Journal of Neuroscience* 11:2896–2913.

Schlinger, B. A., and A. P. Arnold. 1991a. Androgen effects on the development of the zebra finch song system. *Brain Research* 561:99–105.

Schlinger, B. A., and A. P. Arnold. 1991b. The brain is the major site of estrogen synthesis in the male zebra finch. *Proceedings of the National Academy of Sciences U.S.A.* 88:4191–94.

Schlinger, B. A., and A. P. Arnold. 1991c. Plasma sex steroids and tissue aromatization in hatching zebra finches: Implications for the sexual differentiation of singing behavior. *Endocrinology* 130:289–99.

Schlinger, B. A., and A. P. Arnold. 1992. Circulating estrogens in a male songbird originate in the brain. *Proceedings of the National Academy of Sciences U.S.A.* 89:7650–53.

Simpson, H. B., and D. S. Vicario. 1991a. Early estrogen treatment alone causes female zebra finches to produce learned, male-like vocalizations. *Journal of Neurobiology* 22:755–76.

Simpson, H. B., and D. S. Vicario. 1991b. Early estrogen treatment of female zebra finches masculinizes the brain pathway for learned vocalizations. *Journal of Neurobiology* 22:777–93.

Sohrabji, F., K. W. Nordeen, and E. J. Nordeen. 1989. Projections of androgen-accumulating neurons in a nucleus controlling avian song. *Brain Research* 488:253–59.

Sohrabji, F., E. J. Nordeen, and K. W. Nordeen. 1990. Selective impairment of song learning following lesions of a forebrain nucleus in the juvenile zebra finch. *Behavioral and Neural Biology* 53:51–63.

Vockel, A., E. Pröve, and J. Balthazart. 1988. Changes in the activity of testosterone-metabolizing enzymes in the brain of male and female zebra finches during the post-hatching period. *Brain Research* 463:330–40.

Vockel, A., E. Pröve, and J. Balthazart. 1990. Sex- and age-related differences in the activity of testosterone-metabolizing enzymes in microdissected nuclei of the zebra finch brain. *Brain Research* 511:291–302.

Walters, M. J., B. S. McEwen, and C. F. Harding. 1988. Estrogen receptor levels in hypothalamic and vocal control nuclei in the male zebra finch. *Brain Research* 459:37–43.

Williams, H. 1985. Sexual dimorphism of auditory activity in the zebra finch song system. *Behavioral and Neural Biology* 44:470–84.

Williams, H. 1989. Multiple representations and auditory-motor interactions in the avian song system. *Annals of the New York Academy of Science* 563:148–64.

Williams, H., and J. R. McKibben. 1992. Changes in stereotyped central motor patterns controlling vocalizations are induced by peripheral nerve injury. *Behavioral and Neural Biology* 57:67–78.

Williams, H., and F. Nottebohm. 1985. Auditory responses in avian vocal motor neurons: A motor theory for song perception in birds. *Science* 229:279–82.

TEN

The Nature and Nurture of Neo-phenotypes: A Case History

MEREDITH J. WEST, ANDREW P. KING,
AND TODD M. FREEBERG

IN ONE OF THE MORE VIVID metaphors about the costs and benefits of plasticity, the biologist F. W. Warburton proclaimed its positive value, reasoning that a biological system lacking plasticity left an organism in the uncomfortable state of being "sewn into one's winter underwear" (1995, 136). C. H. Waddington captured the essence of the opposing argument, suggesting that "if one lives in the Arctic, it may indeed be preferable to be sewn into one's winter underwear than to risk having it blown away by an unusual gust of wind!" (1957, 162). The plan of this chapter is to delve into the topic of the evolutionary coupling of developmental processes and environmental influences. At issue is whether the possession of plasticity potentially strips an organism or equips it to deal with its surroundings. How do organisms balance the need for conserving time-tested behaviors against the need for change in the face of environmental variation? We argue that differences in opinion on this issue rest primarily on ignorance about environmental surroundings, rendering evaluations of the various positions difficult. We also suspect that, in the absence of functional measures of species-typical environments, it has been easier to favor Waddington's view of nature as an unpredictable and harsh force. But, as Warburton recognized, environments also have constructive properties. And thus, our final aim is to illustrate the need to study inherited alignments between environments and organisms.

The concept of plasticity has a long history in the field of psychology, with especially important contributions deriving from the students and colleagues of T. C. Schneirla. One of the best-known contributors was Daniel Lehrman, whose writings have both general and specific relevance to the case history to follow (1970). The concept of behavioral plasticity has more recently come into favor among evolutionary biologists and ecologists (West-Eberhard 1989). Plasticity is accepted as a genetic process, and is defined as "production by a single genome of a diversity of potentially adaptive responses, whose timing, structural relations, and environmental sensitivity are subject to natural selection" (West-Eberhard 1989, 251).

Although some of the causes of consternation about plasticity, alluded to above by Waddington, have been resolved, the concept is still "stigmatized by poorly understood environmental influences and the ghost of Lamarck" (West-Eberhard 1989, 249). The question we wish to raise concerns this stigma: Why do the environmental influences continue to be poorly understood? Is it because such influences have yet to be measured with the same degree of scientific acumen as genetic processes? Or, is it, as West-Eberhard muses, that such influences possess a "curse" because "such processes are in some sense 'non-genetic'; and to be evolutionary is to be genetic" (in press, 2)? By describing environmental processes, we thus hope to aid in exorcising the curse of their supposedly "non-genetic" origins. Environments are inherited along with genes. It is the correlated nature of environmental and genetic inheritance that must be explicated to provide the proper developmental perspective by which to evaluate the construct of plasticity.

We have chosen to illustrate these points by examining the vocal habits of a certain songbird. Few areas of developmental study have as rich or relevant a history for the purposes of discussing theories and methods of ontogenetic analyses of behavior as the study of bird song. The paradigm rests squarely on the assumption that behavioral plasticity is fundamental to proximate and ultimate explanations. At a proximate level, all songbird species are presumed to achieve species-typical songs by environmental shaping of initially variable acoustic material into stereotyped forms (Kroodsma and Baylis 1982). At an ultimate level, such plasticity is presumed to have played a role in avian speciation, with learned differences in vocal patterns constituting a means of maintaining or facilitating reproductive isolation (Nottebohm 1972). Even though such theories clearly implicate environmental influences, students of bird song are divided by the same Warburton-Waddington split with respect to the costs and benefits of plasticity. On the one hand, theories of bird song assume that imitation is the primary mechanism of social transmission and thus that environmental input is required. At the same time, concern is expressed that imitation can lead to species-atypical transmission, as birds may copy the "wrong" species. Thus, "fail-safe mechanisms" (Kroodsma 1982) such as the "sensory template" (Marler 1976) have been proposed to explain how young song learners avoid potential environmental hazards.

The successes of bird song development as a paradigm for the study of ontogeny and phylogeny have been admirably reviewed, and we shall not repeat the evidence here (Hinde 1969; Kroodsma and Miller 1982; Thorpe 1961). Rather, we focus on how the role of environmental influences has been studied and how the methods have affected views on vocal plasticity. We use our own research extensively because some of what we say is critical, and we feel most comfortable examining our own errors. The aim, as stated earlier, is to address the question of the inheritance of environments to understand the development of behavior.

SONG DEVELOPMENT IN THE LABORATORY: ASSUMPTIONS AND PRACTICES

Simply said, what has become scientifically appealing about studying bird song is that you can separate the bird from the song. Once songs are recorded, they can be analyzed at multiple levels. And while all of this is going on, the bird can go about its business, often with a minimum of disturbance and without the investigator having to deal with any other behavior. But the bird's business can also be incorporated, providing potential syntheses of structure and function. And if desired, the bird can be separated from its natural surroundings as well, affording possibilities for developmental study.

Much of the paradigmatic power of the bird-song system came about in the last 40 years as the technology evolved to aid humans in seeing what they could not hear in avian song. A founder of the field, W. H. Thorpe, articulated the beginnings of the new field in an article in *Nature* introducing the sound spectrograph. Until this instrument was available, Thorpe argued, "it was scarcely possible to make much progress with the problem of song learning . . . the primary difficulty [being] perceiving accurately by the naked ear elaborate sound patterns of high frequency, high speed, and rapid modulation" (1954, 465).

Thorpe (1958) not only introduced the technology for analyzing sound, he also defined the conditions he thought necessary for dissecting song learning. He proposed that bird song represents "an elaborate integration of inborn and learned components, with the former constituting the basis of the latter" and that the "inborn component . . . can be revealed by hand-rearing from the early nestling life, either in acoustic isolation or at least out of contact with all . . . song" (1958, 554). To separate the inborn and learned components, Thorpe and those who followed carried out laboratory studies with a special scientific strain: "Kaspar Hausers," hand-reared birds raised in the laboratory out of contact with adult conspecifics. (The name derives from that of a German boy supposedly subjected to analogous deprivation.) The degree and length of social and/or acoustic deprivation varied with different species and different investigators. Sometimes only adult song was withheld, and sometimes isolation from all conspecifics or from all birds was also imposed on the subjects, depending on the question and the study (Thorpe 1958).

The use of such scientific strains has never been a matter of convenience. At a practical level, creating "Kaspar Hausers" is extraordinarily difficult. If there ever was an inconvenient preparation, nestling songbirds are it. Moreover, once a study is completed using such special birds, the quite plausible scientific charge to replicate or to modify the design using a new sample carries considerable cost. Even with the advent of sound analysis tools, the undertaking of developmental studies of songbirds requires extraordinary effort. To document song learning, a time course of months (or even years) is required before vocal learning is complete.

We emphasize the practical difficulties to underscore the theory-driven nature of the research. One must have a compelling rationale to justify such time-consuming procedures. The rationale seems quite simple. The songs of "Kaspar Hausers," often referred to as the "isolate" or "innate" songs of a species (see Konishi 1985; Searcy and Marler 1987), were thought to expose the genetic/neural precepts on which experience builds the adult's repertoire. A fundamental assumption for many was, and is, that such isolate songs bear a specifiable, evolutionarily grounded relationship to the species-typical songs of wild conspecifics (Konishi 1985; Marler 1990; Thorpe 1961). One might compare this to the genotype-phenotype relationship: phenotypes differ within a norm of reaction defined by a genotype. By housing birds apart from species-typical sounds, the endpoints in the norm of reaction should become visible.

Whether isolate song bears the same kind of relationship to normal song as a genotype does to a phenotype is debatable. Because songbirds are never exposed to such conditions in nature, is the behavior of isolated singbirds within the norm of reaction for the species? An alternative approach is to consider the "isolate song" an instance of "phenotypic engineering" in the sense defined by Ketterson and Nolan (chap. 14); that is, the deliberate creation of a *novel* phenotype based on manipulation of a natural feature of a complex behavior (in Ketterson and Nolan's case, increasing the testosterone levels of breeding dark-eyed juncos, *Junco hyemalis*). Although it is generally accepted and well documented that isolate songs are different in the sense that they are structurally abnormal, that is, unlike the songs of wild conspecifics, the source of the "engineering" is less clear. Isolate song is often presumed to expose the action of genes. As the bird has no source of instruction but himself (or sometimes other untutored peers), the resulting phenotype is presumed to reflect primarily genetic influences.

Whether any animal can exist in an environment and not respond to its proximal properties is a major point of disagreement among developmentalists interested in the ontogeny of behaviors such as bird song (Johnston 1987). Our view, following Kuo's (1967) lead, is that, like birds *not* in isolation, birds in isolation shape and are shaped by their surroundings. When those surroundings are species-atypical, the end result is a "neo-phenotype," to use Kuo's term, a behavior within the absolute norm of reaction of the organism but one rarely seen in nature because the conditions to stimulate its development typically do not occur. Hence, rats that cuddle with cats, and vice versa, are demonstrations of the breadth of ontogenetic responsivity bred into these species. Thus, songbirds producing isolate songs are exhibiting a neo-phenotypic response, but where it exists relative to the norm of reaction for the species cannot be predicted a priori.

Thus, a major issue that arises in the creation of vocal neo-phenotypes is calibration of the degree of departure from normal phenotypes. In the junco example, one can measure levels of testosterone before and after the manipulation. But what does one measure when the dose is an "anti-dose," that is, the removal of species-typical stimulation? Moreover, birds in isolation differ from wild conspecifics in many ways: they may sing more or less, hear their own songs more or less, attend to a different set of environmental sounds, and have less to remember and more time in which to do so. Thus, what part of the behavioral complex is considered most critical for the experimental question? And most important, how does one verify the nature of the manipulation? Testosterone can be implanted in quantifiable dosages, experience cannot. The less tangible quality of experience does not, however, mean that any form of measurement is precluded. In the research to be described below, our major goal has been to describe the nature of early experience during vocal development as a way of evaluating the nature of neo-phenotypic outcomes.

COWBIRDS: AN ONTOGENETIC ENDPOINT?

If there ever was a species that would seem to warrant Waddington's obligatory, unremovable underwear, it would be the brown-headed cowbird (*Molothrus ater*). Cowbirds are brood parasites, laying eggs in the nests of over 200 other species and subspecies, and providing no parental care whatsoever (Friedmann et al. 1977). Given that cowbirds are thus at the end of the avian continuum in degree of parental care, it has seemed reasonable to some to suppose that their ontogeny also would reveal the "end" of any putative ontogenetic continuum in terms of the possible hazards of behavioral plasticity. Thus, Lehrman (1970) and Mayr (1974), among others, proposed the cowbird as an exemplar of a "closed program" for the development of species-typical behaviors—closed to keep out any experimental "gusts" that could leave cowbirds out in the cold when it comes to matters of species or mate recognition. Thus, the circumstances of the cowbird's ontogeny afford a crucible for testing developmental rules. Does an unpredictable early environment require special genetic constraints? Here we explore only a subset of possible rules, those most concerned with vocal development, because of the aforementioned reliance on social and vocal experience with conspecifics in all other songbirds studied to date.

The ontogenetic question of most interest to us at first was whether cowbirds would prove to be exceptions to previously established patterns of songbird vocal development. On the surface, cowbirds' use of song cannot be differentiated from that of nonparasitic songbirds. Male cowbirds use song to interact with conspecifics, both males and females (Friedmann 1929). Males also share songs in local areas, and use songs when courting females. Young males in the wild go through stages of vocal development in generally the same seasonal time frame as other songbirds (Dufty 1985; King and West 1988). But, given the natural deprivation of species-typical songs experienced by young cowbirds relative to other songbirds, many questions remained about possibly different ontogenetic pathways leading to similar use of species-typical vocalizations.

To answer these questions, we have studied the song development of many young cowbirds. Although we hoped to answer the same questions about cowbirds that had been asked about other songbirds, we chose to modify some of the traditional methods. The major departure in the studies to be described was the consistent use of social housing for all birds. Thus, to learn whether males needed experience with adult males to develop appropriate song, we housed hand-reared, acoustically naive nestlings with female cowbirds, which do not sing. We also housed males with members of other species, which could vocalize, but not in a manner species-typical for cowbirds (see Dittus and Lemon 1969; Marler et al. 1972 for other examples of such housing). To mix two ethological terms, our "Kaspars" always had "Kumpans" (Lorenz 1970), social companions to provide for social needs, but not vocal ones. In this way, we hoped to clarify the nature of the intended manipulation, species-specific song deprivation. The use of females and other species as companions also appeared ecologically appropriate, as juvenile cowbirds typically overwinter in large flocks of mixed species composition.

The results of the first study on male vocal development were initially quite straightforward. When male cowbirds were deprived of species-typical song, they reacted, like all songbirds studied to date, by developing structurally atypical songs, which suggested that they must hear other males sing in order to produce species-typical song (King and West 1977). The most obvious difference made apparent by spectral analaysis was in the final phrase, the loud whistle (fig. 10.1), although subsequently other differences were also discovered (West et al. 1979). Thus, at this level of analysis, the "isolate" songs of cowbirds appeared to be similar to those of other songbirds in being structurally atypical.

"Isolate song?", the perceptive reader might ask? Why call them "isolate" songs when the males were housed socially? Because, at the time, like other song researchers, we were not thinking about the bird, but about the song. In terms of acoustic stimulation, the males had been isolated from the variable of interest, species-typical song. And so, even though behind every male cowbird was a female or a fellow songbird, we failed to break the verbal habit of "isolation-ese." In retrospect, it is quite remarkable that we clung to the terminology as long as we did. It appeared in a series of papers (even the titles!).

FUNCTIONAL ASSESSMENT OF DEVELOPMENTAL OUTCOMES

Our verbal habits seem all the more misguided, given that we had included a design feature in our first experiment in which females had a starring role. Because cowbirds use song to court females, we elected to use the females' ears, instead of our spectrographically aided eyes, as the best measure of song abnormality. Would structural and functional measures provide internal validity as to the effects of the manipula-

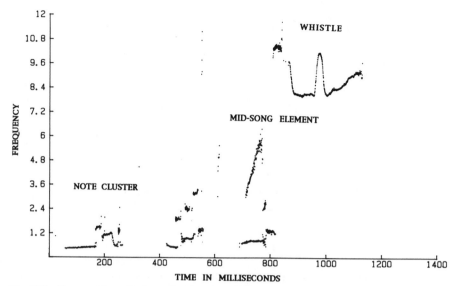

Fig. 10.1 Zero-crossings-displays of male song (see West et al. 1979 for description of analysis procedures). The terminal whistle is the song structure used most often to discriminate songs by song types (Dufty 1985). The midsong element does not occur naturally in the songs of *M. a. ater* from North Carolina or other eastern locations (King et al. 1980). Note clusters are acoustic features that appear quite distinctive to females: males sang more note clusters if housed with relatively unresponsive female cowbirds (King and West 1989).

tion? To answer this question, we housed female cowbirds in acoustic deprivation from males (but with other females) until they came into breeding condition. We then played songs to them to see whether they responded differently to the songs of males with different developmental histories. If a song elicited a copulatory posture while it was being played, we scored a positive response. If a song elicited no reaction, we scored a negative response. No intermediate conditions were observed (females either adopted postures or went about their business as if they heard nothing), so the procedure proved quite effective as an unambiguous bioassay of the stimulatory properties of the different songs (hereafter labeled song potency).

The first set of songs tested with this playback procedure included the isolate songs of acoustically naive males (King and West 1977). To our surprise, the songs of such males elicited significantly more postures than the songs of wild males. And both sets of songs reliably elicited more responses than songs of other species. Thus, female cowbirds appeared to be able to discriminate species identity, and appeared to be more provoked, for some reason, by the songs of males that had never heard another male sing. Did the data mean that acoustic isolation improved cowbird song? If so, why do males live in social groups in the wild? And why is song sharing a feature of natural song repertoires (Dufty 1985; West and King 1986)?

This finding led to a series of studies beyond the scope of the present chapter. But its major lesson concerned interpretation of structural measures of song. Our eyes had not revealed what the female cowbirds' ears discerned. The females not only made discriminations at finer levels of structural analysis than we did, but also appeared to categorize songs on a different basis. Consider the loud whistle of the male's song in figure 10.1. It was this feature that first indicated some abnormality in the songs of the acoustically naive males. It is also the feature by which cowbird songs are most easily compared, because males often share whistles in the same locale. But females did not appear to use the whistle to make their discriminations. Copulatory postures, when they occurred, reliably began before the onset of the whistle, about 400 milliseconds after the song's onset. Playback tests in which whistles were played alone or eliminated from the song confirmed the effect: the pre-whistle portion of the song elicited the female's response (West et al. 1979). Thus, the most conspicuous differences between songs as seen in spectrographic displays were erroneous guides to understanding the functional organization of song. After 17 years of recording and listening to cowbird song, we still cannot predict a song's potency based on inspection of its structural features. Other studies using females from different geographic populations have demonstrated that not all female cowbirds attend to the same acoustic parameters: distinct geographic preferences exist for different acoustic structures (King and West 1982b; King et al. 1986). Thus, not only is a bird's ear necessary, it has to be the "right" bird's ear. Such data cause us to question the continued reliance upon structural measures as straightforward assays of song function. While Thorpe correctly perceived the advantages that technology would bring, other measures appear necessary to assess the role of learning.

Although the playback studies simulated natural song deprivation, we were concerned about possible artifacts in perceptual discrimination induced by the absence of actual male companions. Were females housed in captivity and deprived of song attentive to the same features as females confronted with real males and real choices, in terms of mating? To study this possibility, we carrried out studies of mate choice in captivity, in large indoor-outdoor aviaries (West et al. 1981a, 1981b; West et al. 1983).

Under such circumstances, success at mating, as measured by number of copulations, is not random: certain males obtain far more copulations than others, even when there are unguarded and unpaired females available. We have now repeated this procedure with five cohorts. In each case, the same pattern of nonrandom mating emerged. We also obtained several other social measures on these males. First, we scored male dominance status in the aviaries throughout the preceding winter and spring. We found that dominance related to copulatory success. In four aviaries, the most dominant males obtained the most copulations (West et al. 1981a, 1981b; see West et al. 1983 for the exception). Second, we recorded the males' songs in the aviaries during the breeding season. We then played back these songs in sound attenuation chambers to females who were totally unfamiliar with the aviary males, such that all they heard were the males' recorded songs. The results indicated that the aviary and playback females attended to the same features: the males who engaged in the most copulations produced songs that elicited the most copulatory postures in the chambers (West et al. 1981a; West et al. 1983). Thus, we concluded that females can provide interpretable data for assessing the functional properties of songs, data more meaningful than that provided by even the most detailed acoustic analyses.

AUDITORY INEXPERIENCE: ENRICHMENT OR DEPRIVATION?

The experience of watching males court females in aviaries led to another important lesson regarding the assessment of neo-phenotypes, which dealt in particular with our earlier finding of the higher song potency of acoustically naive males (a finding

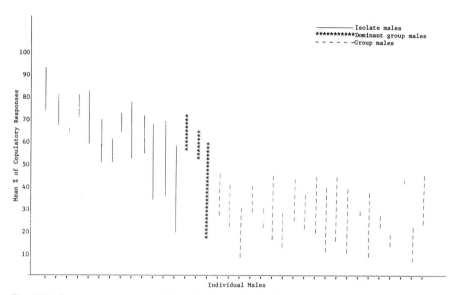

Fig. 10.2 Range in song potency for males housed with females or nonconspecifics (solid lines); males ranked as most dominant in a resident aviary (asterisks); and males housed in aviaries but of lower rank (dashed lines). Note that the maximum potency varies by group, with the males housed without males ("isolates") having more potent songs in their repertoires. (From West et al. 1981b).

we have replicated several times: King et al. 1980; West and King 1980). As noted earlier, not all males in the aviaries were equally successful at obtaining copulations; the range was extensive, from males who obtained 60 or more copulations to males who obtained none, even though they routinely sang to females. There also were always one or two males who sang, but not to females. Their songs were directed to other males or to no individual in particular. The latter males did not copulate and generally had quite ineffective song when it was played back to females. A few produced quite potent song, but those males' "problem" appears to have been that they did not sing it to females (West et al. 1981b). These data suggest that in this species, song potency is a necessary, but not sufficient, prerequisite for reproductive success. A male must also use his song appropriately, a finding in accordance with field observations (Yokel and Rothstein 1991).

When we compiled data on song potency and courtship success, we found that the songs of the most successful aviary males did not differ significantly in potency from the songs of the acoustically naive males tested earlier: the ranges overlapped (fig. 10.2). We also recognized that the potencies of the moderately successful aviary males were similar to the potencies of the songs of wild males used in the original experiment (King and West 1977) as normative song. The songs used in the original experiment had been recorded from several wild male cowbirds during the breeding season with no knowledge of their breeding status. Thus, we later wondered if we had unintentionally biased the 1977 experiment away from normative song by our use of a small sample of wild males. But more important, the findings suggested that we needed to reevaluate the status of the acoustically deprived males. Had we really put them at a disadvantage by depriving them of adult males? Or had we engineered a new phenotype, a male with instant dominance status, because he was the only male and had a captive audience?

We tested the possibility of a "dominant male" neo-phenotype in several ways. Our most important test was to introduce acoustically naive males to our resident colonies and watch their behavior and the reactions of the residents (West and King 1980). The males were introduced at first light and observed continuously for an entire morning, then removed. We also introduced socially experienced males from another aviary for comparison. We knew from examining our laboratory records of transfers of birds from one aviary to another that new males may be attacked and killed, especially if their songs are quite potent, and thus we could not risk leaving the naive birds to attempt to settle into the colony. The morning observations were sufficient to tell us what we wanted to know. First, the naive males behaved differently from the experienced males. The former showed no knowledge of "home-court" rules: they approached resident females and sang to them with no prior attention to the female's consort, who often reacted by physically attacking and chasing the naive male. Second, the experienced males maintained a low profile: they sang, but their songs were undirected. None was attacked. Thus, on the basis of these data, as well as data showing that potency can be manipulated from one year to the next by changing a male's companions (potency is increased with females, lowered with males: West and King 1980), we concluded that depriving young males of adult males had an enriching effect. The males' songs may have been structurally abnormal, but they were functionally effective, and the males themselves, by virtue of having been given a special audience, were socially naive but not socially incapacitated.

Most important, the data showed that talking about the males as "enriched" *or* "deprived" or "enriched" *and* "deprived" was developmental double-talk of little value. The males were different, having been "biased" (sensu Gibson 1969) or "alienated" (sensu Lewontin 1983) by their rearing in a novel social context. Since 1977, the isolate songs of several other songbirds have been tested using somewhat similar playback tests to females. In some cases, isolate songs had proved to be more potent than expected, eliciting more responses than the songs of deafened birds, but to our knowledge, none has proved to be more potent than the songs of nonisolated conspecifics. Although it is tempting to say that this is the case because the data from cowbirds are idiosyncratic, we cannot say it at present because correlative tests of the behavioral characteristics of the isolates of other species in functional contexts have only begun to be performed (Williams et al. 1993).

Of most importance to us was that the studies as a whole graphically illustrated the impossibility of separating organisms from their environments and reinforced the need to describe the actions of both in order to understand outcomes. Thus, one should *not* read the vocal slate of naive birds as having been written on only by the "inborn" part of the vocal system. In the same way that organisms can now be labeled by their DNA fingerprints, they can also be differentiated by environmental signatures. The naive male cowbirds bore the environmental marks of dominance (the ease with which we could change a male's status suggesting that genetic differences were not the reason), a signature we did not recognize until females provided the cues.

Lack of appreciation of the potential "positive" effects of isolation has been fostered by the tradition of not watching the animal that lives in such a context. As documented in the vast literature on the effects of "early experience" on learning, social behavior, or brain development, song researchers have not measured the nature of the behaviors of animals undergoing experimental enrichment or deprivation. We now turn to this issue and to the rudimentary steps we have made toward addressing this problem. We have much farther to go, but now possess a crude map as to what conceptual routes to follow.

WATCHING BIRDS WATCHING BIRDS: ENVISIONING SONG DEVELOPMENT

Years spent watching males and females interact in aviaries recently motivated us to reexamine the original "isolate song" experiment from 1977. This time, the goal was to provide a better perspective on the possibly interactive effects between the "subjects," the naive males, and their "audience," the females or nonconspecifics. "Audience effects" have been studied in animals in a variety of contexts since Zajonc (1965) coined the term, but typically not with a developmental perspective. In our original 1977 work, the two isolated males had potentially different kinds of audiences: one had been housed with a female cowbird, the other with a male cardinal (*Cardinalis cardinalis*). Both produced an equally effective range of songs, but with such a small sample, we did not know if the differences in their audience could have been detected. As a result, we chose to repeat the basic design with more birds. We raised thirteen juvenile males with companions: nine lived with individual female cowbirds, and four with pairs of canaries. With a larger sample, would we see an effect of living with females, which are perceptually experienced with cowbird song but have no acoustic vocabulary, as opposed to canaries, which have a vocabulary, but one of presumably little use to cowbirds?

The results indicated that the original findings were not wrong: the naive males sang highly potent songs. But the results showed a differentiation on the basis of companion: males housed with female cowbirds possessed a more potent set of songs, and although there was some overlap, none of the males housed with canaries sang songs as potent as those of the males housed with females (fig. 10.3).

Three of the female-housed males accounted for the most overlap in potency with the canary-housed males. But these males have been housed with atypical females. We possessed extensive knowledge of the perceptual responses to song for all nine of the females housed with the naive males (King and West 1989). We had studied them in playback experiments and in aviary mate choice tests, and for some, we had looked at the nature of their interactions with captive males during the winter and spring, when song learning changes most conspicuously (King and West 1988). Two of the three males whose potencies were most similar to those of the canary-housed males had female companions that had shown minimal responsiveness in playback studies and had obtained no copulations in an entire breeding season. The third female had been more successful in the aviary context, having copulated 15 times with the same male, but she had shown an aberrant playback profile: she responded to absolutely every song she heard (over 45 different songs). She was the only female out of over 200 females to show such a lack of discriminative responding within the category of cowbird songs. (On average, females from her natal area respond to 50% of song playbacks, depending on the nature of the songs.) Thus, although her deviance was in the opposite direction from the two extremely unresponsive females, it was more atypical. Thus, we assume that the three males housed with these females were exposed to audience characteristics closer to those of the canaries: the females' perceptual responsiveness to cowbird song contained little predictable or reliable information for males, thus all three were missing intraspecific selectivity. Without any female preferences, the males apparently received

DIFFERENCES IN RANGE OF SONG POTENCY

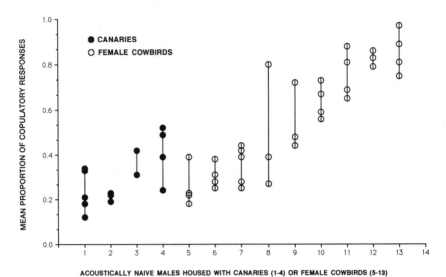

ACOUSTICALLY NAIVE MALES HOUSED WITH CANARIES (1-4) OR FEMALE COWBIRDS (5-13)

Fig. 10.3   Differences in range of song potency according to companion for juvenile male cowbirds. Each circle represents a song; solid circles represent the potencies of males housed with canaries (males 1–4); open circles, males housed with females (males 5–13). The companions of the three "unusual" females (see text) are represented by the first three columns of open circles (males 5–7).

no systematic reinforcement for specific vocal behaviors. Thus, the results support the original findings of the positive effects of female stimulation on males, and show that males are quite sensitive to the nature of the stimulation provided. When there is no interactive component—that is, when either the male or female does not respond systematically—communication does not take place, and learning from those communicative exchanges is precluded.

We have conducted several other studies to look at the degree of male sensitivity to females. To learn more about the developmental progression of different vocal phenotypes, we housed hand-reared, acoustically naive male cowbirds with one of three classes of companions: females from their natal area of North Carolina (NC), females from another part of the species range in Texas (TX), or with either canaries or starlings (*Sturnus vulgaris*). The two populations of female cowbirds were known to differ in their preferences for male song, with both populations preferring their native variants (King and West 1983c). Moreover, we knew the acoustic basis of their perceptual discrimination, as described in figure 10.1. Thus, we attempted to chart differences in the structural and functional properties of the songs of males presumably exposed to different kinds of feedback about particular song structures.

According to Thorpe's original reasoning, or that proposed later by others (e.g., Konishi 1985), neither the presence nor the geographic identity of the female companions should have made a difference in the structural ontogeny of song. Song learning has been consistently described as an instance of vocal imitation. If imitation were the only mechanism guiding song development, then the songs of males housed with females from different populations would not differ, because their opportunities to imitate themselves or rehearse from inherited templates were equivalent (the groups did not differ in how much they sang at any point in development:

King and West 1988). From as early as 150 days of age, however, the males with females from the two different populations produced structurally different repertoires. Moreover, their rates of vocal development differed, suggesting effects on other developmental parameters. Among the differences were consistent differences in the acoustic characteristics of the notes constituting the first half of a male's song, the phrase we knew to be most provocative to females. Males with TX females consistently sang fewer notes of higher frequency, but consistently produced them with more modulation. At the level of developmental rate, the males with TX females continued to sing more variable song later into the spring than either of the other two groups. Such variable song, often termed "plastic" song, has often been thought to be the means by which birds improvise on copied material. The NC males with TX females may have been attempting more variants in the face of different perceptual cues. The most striking structural difference concerned the presence of a set of notes termed the midsong element (MSE). We had established that the MSE occurs reliably among males in the TX population, but not in Eastern samples (see fig. 10.1). The difference had been verified in captive and field settings, involving comparisons of over twenty thousand songs and eight populations (Eastzer 1988). In the present study, the same pattern obtained: males with TX females produced MSEs, males with NC females did not.

The experiment included a third group of males that had been individually housed with members of other species (canaries or starlings). The males with other species produced "generic" repertoires containing features found in NC and TX, including MSEs, as well as other acoustic elements associated with other locales. These birds were the earliest to progress to the production of stereotyped song, suggesting that in the face of no relevant feedback, the males settled into acoustically satisfying patterns that they had no motivation to change. This group is vital to understanding the possible mechanism by which females influence song. If males normally possess a species-universal repertoire, the function of female stimulation may be to guide males toward song variants tuned more to local, geographic specifications.

Other studies using naive males whose social experiences were supplemented by hearing tutor tapes of normal male song provide additional evidence for a "tuning" scenario. When tutored with local song variants, NC males housed with other species sang significantly more of the tutor songs than did males individually housed with TX females and tutored with the same songs (West and King 1986). A similar effect was obtained with a new set of males, housed either with NC females or canaries. Although all the males had access to the same "inborn" program (sensu Thorpe 1958) and the same tutoring regime, their repertoires varied reliably by group from very early in development. In this case, males with NC females learned the tutor songs, as did the canary-housed males, but then went on to improvise and change the material enough so that the original songs were no longer easily identifiable. The canary-housed males did not deviate from the tutor songs: their repertoires were comparable in size to those of the males with NC females, but 97% of their vocalizations were clear copies of the tutor songs.

In sum, the three experiments involved 29 individual NC males and the analysis of 32,503 songs, none produced by females. We mention the latter figure not only to verify the nature of our sampling of the young males' repertoires, but because, as more and more evidence was gathered for the potential power of females to influence song development, it became important to verify that female cowbirds did not sing. Although females routinely produced food-related calls, no songs were ever recorded from the female companions. But the question remained: What were they doing? What could we say about the experience of the males who lived with female cowbirds during the period of song development?

FEMALES: MEASURING THE SOUNDS OF SILENCE

We examined several contexts to find answers to these questions. First, we studied a flock of overwintering cowbirds in NC to see whether juvenile males had access to females during the time of year when their songs undergo the most rapid and extensive changes. We found that the juvenile males did associate quite closely with females—much more closely, in fact, than with adult males. In a study of 197 color-banded birds, we found an overall sex ratio of 1.49 females per male across all observation sessions (King and West 1988). Juvenile males were captured with females in all sessions, yielding a higher ratio of juveniles to females (1.27) than of adults to females (.97). Thus, juveniles appear to have access to females, and vice versa. The categories of song produced by the wild juveniles were also compared with those of males housed with females in the laboratory at the same time of year: well over half of the juveniles' vocalizations were classified as plastic, variable song, a proportion indistinguishable from that of the laboratory-housed males. Thus, we were satisfied that males and females interacted with each other at a time when female influence might be important.

We also exploited a laboratory setting with males and females housed together in the same enclosure to videotape their interactions during the time period when flocks begin to disperse and males and females return to prospective breeding grounds. We videotaped 22 pairs for 328 hours, covering 82,080 songs. In looking at females' potential reactions to each of these songs, we found that most of the time, well over 90% (depending on the pair) of females appeared to do very little when a song occurred. But when a female did react, it was conspicuous, at least to us. We were often alerted to the change in the female's behavior by the behavior of males—every once in a while, when a male sang, he would change the pace of his singing, as well as sometimes abruptly moving toward the female.

With videotape, we retraced the male's steps and his songs (West and King 1988a). We found that such changes in the males' behavior were preceded by wing movements by females—movements we call wing strokes (if a bird could point, it would do so with a wing stroke). Wing strokes occurred infrequently. It is difficult to give a meaningful average for so many pairs and so many songs in sessions of different lengths and at different points in the spring. The average number of songs between wing strokes was 111, with a wide range (24–792). To be counted, wing strokes had to occur while a song was in progress.

In evaluating the data on wing strokes, it is essential to remember that a song lasts for only one second. Given that males sometimes sing a song every 20–30 seconds, the time between wing strokes can actually be very brief. Thus, in terms of absolute time, the ratio of female response to male singing is not as low as it might seem. The data also suggest that wing strokes may actually be clustered in time, with a series of wing strokes in quick succession, followed by a longer period of apparent nonresponsiveness. Thus, the data suggested to us that female cowbirds possess a visual, gestural signal that they use to communicate about vocal signals: a system within a system, as it were. Thus, males appeared to be influenced by watching females' reactions, which led to song modification not based on imitation.

To find out the functional significance of wing strokes, we employed the playback procedure during the breeding season. We wanted to know whether females would adopt copulatory postures to songs that had elicited wing strokes. We found that they did: a song that elicited a wing stroke from one set of females elicited significantly more copulatory postures from a different set (West and King 1988a).

The ability of the same songs to elicit wing strokes and copulatory postures leads us to speculate that wing strokes may be behavioral precursors to copulatory postures. We now also know that playback responsiveness and wing stroking tendencies, behaviors that take place at different times of year, are related in the same individual—females that are very responsive in the winter, as evidenced by wing stroking, tend to be most responsive to song in the breeding season. But more wing stroking and more responsiveness does not translate in a linear fashion into higher song potency. Although the data must be interpreted cautiously because they are based on only eight male-female pairs, the direction of effects appears to be that female responsiveness is correlated with song potency in a U-shaped function (King and West 1989). Thus, females who respond more than 50% but less then 90% of the time to song playback also appear to be in the mid-range for frequency of wing stroking. The males housed with such females produced the most potent songs, perhaps having been optimally motivated to improvise in order to stimulate their companions. And thus we conclude that wing stroking is positive reinforcement, dispensed on a partial reinforcement schedule serving to shape the male's singing toward more female-appropriate signals.

We are further tempted to draw linkages between the two kinds of female responses (copulatory postures and wing strokes) because we now know that plastic songs, the precursors to stereotyped song, and the most common vocalizations when wing stroking is most common, elicit copulatory postures from females in breeding condition (West and King 1988b). In the experiment just cited, we also played back the vocalizations of other species such as canaries and meadowlarks (*Sturnella magna*) to see whether females respond differentially to rudimentary vocal assemblies of cowbirdlike elements. We found that they did: there was a significantly greater response to plastic songs than to heterospecific songs. These data show that precursors to stereotyped song are potentially stimulating and thus potentially functional as communicative signals. These data do not mean, however, that "plastic" songs are sufficient for a male's reproductive success. Females responded significantly more to the "same" songs when they were sung in a more stereotyped (crystallized) manner, suggesting that as males continue to sing, females may become progressively more sensitized to the vocally stimulating qualities of different songs, such as the song's signal-to-noise ratio, which is much higher in the more stereotyped forms.

Thus, even the quite variable and inarticulate vocalizations that would be heard on wintering grounds or in the early spring when males and females congregate appear to be provocative to females. These data challenge another long-held assumption about song learning, the idea that until birds sing well, their songs have little, if any, communicative value. At least in cowbirds, this does not appear to be the case—males can react immediately to female signals by changing song content within a singing session, and females provide feedback to songs in a form that is hard for humans or machines to decipher. So again, the use of appropriate bioassays shows that human opinions of the quality or quantity of information must be calibrated by measuring the responses of appropriate recipients. The data also demonstrate the importance of viewing the timetable for social effects as a long one, perhaps equal in length to the period of song development. Indeed, what may define a "sensitive" period is whether the environment contains motivating or sensitizing influences during that period.

COWBIRDS: EXCEPTIONS OR RULES?

To summarize, song development in cowbirds is far from a closed system—males are modifiable, they copy and improvise on tutor songs, and they go through a clear developmental process to achieve a stable repertoire. Moreover, both imitation and trial-and-error learning are used to achieve vocal repertoires. These patterns occur in other well-studied, nonparasitic songbirds as well (Marler 1990; Marler and Peters 1982).

As we stated at the outset of this chapter, our intended goal was to focus attention on the environmental context surrounding song development. We have only discussed a small segment of that question, in particular, the laboratory paradigms most often used to define the parameters of song development. But we hope enough evidence has been presented to argue that (a) environments are analyzable and (2) their influences are essential to understanding plasticity.

These findings in cowbirds on the role of social interactions in the shaping of vocal repertoires should also illustrate why we are concerned about the inadequacy of studying vocal learning in narrow contexts. Too often, shutting the door of an isolation chamber cuts off the investigator, as much as the subject, from needed stimulation. In the case of the investigator, it results in deprivation of the experience of seeing developmental processes. A contributing factor in desensitizing investigators from describing experience is the aforementioned practice of separating birds from their songs. Unfortunately, the technology for housing and observing birds has not changed as dramatically as the technology for analyzing songs. But good developmental science demands that the lives of the singers of those songs receive equal time. Can the field of avian vocal learning rest so heavily on the singing of socially naive and probably socially disabled birds? That conditions of isolation produce such incompetencies is rarely questioned, but the effects have also rarely been described. Dufty and Wingfield (1986) documented higher corticosterone levels in captive male cowbirds raised alone as opposed to those housed with females. Such data indicate that certain housing conditions are stressful, but also that there are ways to assess social sensitivity.

Recent findings on the development of imprinting in ducklings reinforce the need for concern about these issues. The method of housing ducklings before and after exposure to the imprinting model appears to have had profound effects that render questionable many traditional beliefs about the nature of this supposedly well-studied learning system (Johnston and Gottlieb 1985; Lickliter and Gottlieb 1987; Gottlieb 1991). Briefly, the presence of siblings, a form of social inheritance for naturally reared ducklings, triggers a suite of physical effects, including more sleep and less distress calling. The presence of tactile stimulation appears to induce extreme malleability, such that ducklings come to prefer whatever call is presented during this time. This is not a case of the prompt recruitment of an alternative phenotype in the face of an environmental challenge. It seems doubtful that wood ducks would have in the range of normal phenotypes a "preference for mallard female" on the off-chance that ducklings would be thrust into such odd ecological circumstances. What permits the system to be malleable is the simultaneous action of many environmental and behavioral parameters, including the natural conditions of a nest that typically affords social, vocal, and tactile stimulation from broodmates.

To our knowledge, no one has measured whether isolated songbirds sleep more or less than socially housed peers. Nor has anyone measured differences in other possibly relevant physical parameters. The effects of sleep deprivation would be of considerable interest given their possible connection to memory consolidation, an important part of song learning (Thoman 1990). Therefore, normally occurring social conditions such as sibling-induced sleeping might produce memory interference or loss, a phenomenon of potential functional significance as it might facilitate new learning or refinement of perception of finer features of the social context.

We must begin to find ways to reconcile differences in the "condition" of the bird and the "condition" of his songs. In our view, the present tendency to separate the bird from his song has placed normally coupled ontogenetic processes seriously out of alignment in terms of methods and concepts. Consequently, the conditions of laboratory housing do not expose the bird to the conditions that presumably play highly instructive roles in nature. The influence of improbable agents such as nonsinging females is understated because it is understudied. Moreover, the absence of data from normally occurring social conditions may have led to an overemphasis on mechanisms such as long-term memory storage via templates, as well as an exclusive emphasis on imitation, that may not be representative of the range of processes normally occurring during species-typical development. Changes are in order on the songbird researchers' agenda, including our own, to remedy these problems.

First, we must abandon the idea that there are neutral or arbitrary environments with no biologically provocative features. Perhaps a better description for vocalization produced under typical laboratory rearing would be "alien" songs (sensu Lewontin 1983). Such a strong word might help to remind us that we are dealing with unknown degrees of phenotypic novelty. This is not to say that such novelty cannot be revealing of absolute capacities, but until the "alien" products are functionally assessed, their ontogenetic usefulness remains uninterpretable.

The second item on the agenda must be to develop a nomenclature of environmental conditions. Biologists since the time of Aristotle have found it useful to classify organisms and to debate the proper classification from many perspectives. Why should environments be treated differently? Moreover, given organism-environment interdependencies, it might make sense to begin with parallels to the classification of organisms. Thus, all organisms exist in a common biotic environment, but as the phyletic differentiation of organisms becomes specified, so does the specification of environment. Some progress has been made in this direction at the level of the individual organism, but not enough has been made at the level of populations, which constitute the main topics of interest for many evolutionary biologists (see Gottlieb 1992 for individual organisms and Odling-Smee 1988 for populations).

Such a taxonomy should also help to dispel the idea that environments are "non-genetic" and therefore "awkward" for the science of evolutionary biology (West-Eberhard, in press, 2). We suggest that it is useful to recall that the term "heredity" had a meaning before the discovery of genes, and that some of the principles regarding the inheritance of homes, land, resources, and birthright apply to social organisms other than humans. Thus, it is as likely that cowbirds inherit other cowbirds with which to communicate as it is that they inherit organs by which to reproduce. Conspecifics and reproductive organs are not inherited via the same processes, nor are they subject to the same rules of transmission, but they are correlated outcomes (West and King 1987). Thus, the challenge for developmental and evolutionary biology is not only to produce a taxonomy of environments, but also to study the degree of positive or negative correlation between environmental properties and organismal characteristics.

The science of ecology includes the idea of such correlations in its foundation, but that idea needs to be complemented by a knowledge of the behavioral processes affecting environmental legacies in order to be useful for developmentalists. The key to both tasks is to view species-typical environments as dependable, informative resources for normally behaving members of the species. Such an approach has radically affected the field of perception (Gibson 1966; Mace 1977), and has the possibility to do the same in the developmental domain (see West and King 1984 for a more complete explication of this view).

SUMMARY

We began the chapter with a discussion of contrasting views on the value of plasticity. We hope that the profile of cowbird vocal development attests to the idea that plasticity must be defined in relation to environmental properties. We also hope to have made a more important point regarding the nature of developmental analyses. In our view, the last 40 years of research on song learning under controlled conditions in the laboratory must be considered to be based on theoretically insupportable underpinnings. Environments and organisms cannot be separated in order to dissect inborn and acquired components. Rather, the opposite is true—organisms and environments must be united to uncover how plasticity is generated in response to the conditions encountered by the developing organism.

We also believe that the cowbird studies demonstrate that a bird's song is only one part of a communication system. Without functional assays of the structural properties of song, it is easy to get off the ontogenetic track with respect to a characterization of the nature of the developmental processes that contribute to the nature of plasticity. West-Eberhard framed the concept of plasticity in relation to potential "responses" of a genome to different conditions. Although song is a response in a behavioral sense, we are not convinced that it is a "response" in West-Eberhard's terminology. Part of the problem is that much of the research in phenotypic plasticity is based on morphological responses. Is a song comparable to changes in the length of a bird's wing? Or to changes in the vein in the wing? Or to its ability to enable flight? A bird's song represents the most conspicuous part of the bird's communicative competencies, but it is unclear whether it is sufficient to say that it constitutes a phenotypic answer. Based on our studies of cowbirds, we think not.

Given the present lack of connectedness between the behavioral ecology of songbirds and the developmental morphology of their songs, one must be very cautious in stating what is known about how birds learn to sing (see DeWolfe et al. 1991 for an example of a connnected approach). We can only be sure that further progress in understanding vocal plasticity requires new responses from investigators. We invite those less familiar with the concept of plasticity as defined by psychologists to review its history. It was Daniel Lehrman who best articulated such a view of development. That the species he nominated as a putatively closed developmental system, the cowbird, turned out to be an exemplar of, not an exception to, his rules should not be a deterrent to reading his words. His premature death denied him knowledge of his error and the chance to correct it. We cannot imagine a scientist who would have so enjoyed the chance to stand corrected. West-Eberhard voiced the fear of many biologists that a renewed emphasis on plasticity would revive the ghost of Larmarck. We remain unconvinced that such a resurrection would be counterproductive. If such a ghost (or that of Lehrman) heightens our sensitivity to the need for developmental analyses that dissect the environment in a fine-grained but functionally cohesive manner, then its influence would be a welcome one.

ACKNOWLEDGMENTS

The work reported here was supported by the NSF and the Center for Integrative Study of Animal Behavior at Indiana University.

REFERENCES

DeWolfe, B. B., L. F. Baptista, and L. Petrinovich. 1991. Song development and territory establishment in Nuttal's White-crowned Sparrows. *Condor* 91:397–407.

Dittus, W. P. J., and R. E. Lemon. 1969. Effects of song tutoring and acoustic isolation on the song repertoires of cardinals. *Animal Behaviour* 17:523–33.

Dufty, A. M., Jr. 1985. Song sharing in the brown-headed cowbird (*Molothrus ater*). *Ethology* 69:177–90.

Dufty, A. M., Jr., and J. C. Wingfield. 1986. The influence of social cues on the reproductive endocrinology of male brown-headed cowbirds: Field and laboratory studies. *Hormones and Behavior* 20:222–34.

Eastzer, D. H. 1988. Geographic variation in cowbird song. Ph.D. dissertation, University of North Carolina, Chapel Hill.

Friedmann, H. 1929. *The cowbirds.* Springfield, Ill.: C. C. Thomas.

Gibson, E. J. 1969. *Principles of perceptual learning and development.* New York: Appleton-Century-Crofts.

Gibson, J. J. 1966. *The senses considered as perceptual systems.* Boston: Houghton Mifflin.

Gottlieb, G. 1991. Social induction of malleability in ducklings. *Animal Behaviour* 41:953–62.

Gottlieb, G. 1992. *Individual development and evolution.* New York: Oxford University Press.

Hinde, R. A. 1969. *Bird vocalizations.* London: Cambridge University Press.

Johnston, T. D. 1987. The persistence of dichotomies in the study of behavioral development. *Developmental Review* 7:149–82.

Johnston, T. D., and G. Gottlieb. 1985. Effects of social experience on visually imprinted maternal preferences in Peking ducklings. *Developmental Psychobiology* 18:261–71.

King, A. P., and M. J. West. 1977. Species identification in the North American cowbird: Appropriate responses to abnormal song. *Science* 195:1002–4.

King, A. P., and M. J. West. 1983a. Dissecting cowbird song potency: Assessing a song's geographic identity and relative appeal. *Ethology* 63:37–50.

King, A. P., and M. J. West. 1983b. Epigenesis of cowbird song: A joint endeavor of males and females. *Nature* 305:704–6.

King, A. P., and M. J. West. 1983c. Female perception of cowbird song: A closed developmental program. *Developmental Psychobiology* 16:335–42.

King, A. P., and M. J. West. 1988. Searching for the functional origins of cowbird song in eastern brown-headed cowbirds (*Molothrus ater ater*). *Animal Behaviour* 36:1575–88.

King, A. P., and M. J. West. 1989. Presence of female cowbirds (*Molothrus ater ater*) affects vocal imitation and improvisation in males. *Journal of Comparative Psychology* 103:39–44.

King, A. P., M. J. West, and D. H. Eastzer. 1980. Song structure and song development as potential contributors to reproductive isolation in cowbirds (*Molothrus ater*). *Journal of Comparative and Physiological Psychology* 94:1028–39.

King, A. P., M. J. West, and D. H. Eastzer. 1986. Female cowbird song perception: Evidence for different developmental programs within the same subspecies. *Ethology* 72:89–98.

Konishi, M. 1985. Birdsong: From behavior to neuron. *Annual Review of Neuroscience* 8:125–70.

Kroodsma, D. E. 1982. Learning and the ontogeny of sound signals in birds. In D. E. Kroodsma and E. H. Miller, eds., *Acoustic communication in birds*, vol. 2, 1–24. New York: Academic Press.

Kroodsma, D. E., and J. R. Baylis. 1982. Appendix: A world survey of evidence for vocal learning in birds. In D. E. Kroodsma and E. H. Miller, eds., *Acoustic communication in birds*, vol. 2, 311–37. New York: Academic Press.

Kroodsma, D. E., and E. H. Miller. 1982. *Acoustic communication in birds*. Vols. 1 and 2. New York: Academic Press.

Kuo, Z. Y. 1967. *The dynamics of behavioral development: An epigenetic view.* New York: Random House.

Lehrman, D. S. 1970. Semenatic and conceptual issues in the nature-nurture problem. In L. R. Aronson, E. Tobach, D. S. Lehrman, and J. S. Rosenblatt, eds., *Development and evolution of behavior: Essays in memory of T. C. Schneirla*, 17–52. San Francisco: W. H. Freeman.

Lewontin, R. C. 1983. The organism as the subject and object of evolution. *Scientia* 118:65–82.

Lickliter, R., and G. Gottlieb. 1987. Social specificity: Interaction with own species is necessary to foster species-specific maternal preference in ducklings. *Developmental Psychobiology* 21:311–21.

Lorenz, K. 1970. *Studies in animal and human behaviour.* Vol. 1. Cambridge, Mass.: Harvard University Press.

Mace, W. M. 1977. James J. Gibson's strategy for perceiving: Ask not what's inside your head, but what your head's inside of. In R. Shaw and J. Bransford, eds., *Perceiving, acting, and knowing*, 43–66. Hillsdale, N.J.: Lawrence Erlbaum Associates.

Marler, P. 1976. Sensory templates in species-specific behavior. In. J. Fentress, ed., *Simpler networks and behavior*, 314–29. Sunderland, Mass.: Sinauer Associates.

Marler, P. 1990. Song learning: The interface between behaviour and neurothology. *Philosophical Transactions of the Royal Society of London*, Series B 329:109–14.

Marler, P., and S. Peters. 1982. Developmental overproduction and selective attrition: New processes in the epigenesis of birdsong. *Developmental Psychobiology* 15:369–78.

Marler, P., P. Mundinger, M. S. Waser, and A. Lutjen. 1972. Effects of acoustical deprivation on song development in red-winged blackbirds (*Agelaius phoeniceus*). *Animal Behaviour* 20:586–606.

Mayr, E. 1974. Behavior programs and evolutionary strategies. *American Scientist* 62:650–59.

Nottebohm, F. 1972. The origins of vocal learning. *American Naturalist* 106:116–40.

Odling-Smee, F. J. 1988. Niche-constructing phenotypes. In H. C. Plotkin, ed., *The role of behavior in evolution*, 73–132. Cambridge, Mass.: MIT Press.

Searcy, W. A., and P. Marler. 1987. Response of sparrows to songs of deaf and isolation-reared males: Further evidence for innate auditory templates. *Developmental Psychobiology* 20: 509–19.

Thoman, E. B. 1990. Sleeping and waking states in infants: A functional perspective. *Neuroscience and Biobehavioral Review* 14:93–107.

Thorpe, W. H. 1954. The process of song-learning in the chaffinch as studied by means of the sound spectrograph. *Nature* 173:465–69.

Thorpe, W. H. 1958. The learning of song patterns by birds, with especial reference to the song of the chaffinch, *Fringilla coelebs. Ibis* 100:535–70.

Thorpe, W. H. 1961. *Bird song.* London: Cambridge University Press.

Waddington, C. D. 1957. *The strategy of the genes: A discussion of some aspects of theoretical biology.* London: Allen and Unwin.

Warburton, F. 1955. Feedback in development and its evolutionary significance. *American Naturalist* 89:129–40.

West, M. J., and A. P. King. 1980. Enriching cowbird song by social deprivation. *Journal of Comparative and Physiological Psychology* 94:263–70.

West, M. J., and A. P. King. 1984. Learning by performing: An ecological theme for the study of vocal learning. In T. D. Johnston and A. T. Pietrewicz, eds., *Issues in the ecological study of learning,* 245–72. Hillsdale, N.J.: Lawrence Erlbaum Associates.

West, M. J., and A. P. King. 1985. Social guidance of vocal learning by female cowbirds: Validating its functional significance. *Ethology* 70:225–35.

West, M. J., and A. P. King. 1986. Repertoire development in male cowbirds (*Molothrus ater*): Its relation to female assessment of song. *Journal of Comparative Psychology* 100:296–303.

West, M. J., and A. P. King. 1987. Settling nature and nurture into an ontogenetic niche. *Developmental Psychobiology* 20:549–62.

West, M. J., and A. P. King. 1988a. Female visual displays affect the development of male song in the cowbird. *Nature* 334:244–46.

West, M. J., and A. P. King. 1988b. Vocalizations of juvenile cowbirds (*Molothrus ater ater*) evoke copulatory responses from females. *Developmental Psychobiology* 21:543–52.

West, M. J., A. P. King, D. H. Eastzer, and J. E. R. Staddon. 1979. A bioassay of isolate cowbird song. *Journal of Comparative and Physiological Psychology* 93:124–33.

West, M. J., A. P. King, and D. H. Eastzer. 1981a. The cowbird: Reflections on development from an unlikely source. *American Scientist* 69:57–66.

West, M. J., A. P. King, and D. H. Eastzer. 1981b. Validating the female bioassay of cowbird song: Relating differences in song potency to mating success. *Animal Behaviour* 29:490–501.

West, M. J., A. P. King, and T. H. Harrocks. 1983. Cultural transmission of cowbird song: Measuring its development and outcome. *Journal of Comparative Psychology* 97:327–37.

West-Eberhard, M. J. 1989. Phenotypic plasticity and the origins of diversity. *Annual Review of Ecology and Systematics* 20:249–78.

West-Eberhard, M. J. In press. Behavior and evolution. In P. R. Grant and H. Horn, eds., *Evolution* [Lectures honoring John T. Bonner]. Princeton: Princeton University Press.

Williams, H., K. Kilander, and M. L. Sotanski. 1993. Untutored song, reproductive success and song learning. *Animal Behaviour* 45:695–705.

Yokel, D. A., and S. I. Rothstein. 1991. The basis for female choice in an avian brood parasite. *Behavioral Ecology and Sociobiology* 29:39–45.

Zajonc, R. D. 1965. Social facilitation. *Science* 149:269–74.

ELEVEN

Constraints on Phenotypic Evolution

Stevan J. Arnold

CONSTRAINTS ON PHENOTYPIC EVOLUTION can take a variety of forms. Constraints can arise from inheritance, selection, development, and design limits. Contemporary visions of the evolutionary process often focus on one or two of these varieties of constraint and ignore the others. A unifying framework that considers all four major varieties is emerging within the discipline of quantitative genetics. In this chapter I attempt to sketch that emerging framework and summarize recent efforts toward unification. Although couched in the technical language of quantitative genetics, the ongoing search for a common framework promises a rapprochement between the approaches of optimality theorists, population geneticists, and developmental biologists.

This chapter has two aims. The first is to review the rapidly expanding literature dealing with constraints on phenotypic evolution. The second is to briefly discuss some aspects of constraints as they affect and are affected by behavioral evolution. No attempt is made to review the empirical literature on genetic variances and covariances for behavioral traits, a subject treated in two recent books (Hahn et al. 1990; Boake 1993).

Evolutionary constraints are restrictions or limitations on the course or outcome of evolution. Discussions of evolutionary constraints are often difficult to follow because of a failure to distinguish among underlying concepts. In particular, it is useful to distinguish between genetic, selective, functional, and developmental constraints. Definitions of these varieties of constraint and their interrelations are discussed in the sections that follow.

It is also useful to recognize four key properties of constraints: source, strength, consequence, and persistence (Maynard Smith et al. 1985). By *source* I mean the most proximate causes of constraint: for example, statistical distributions of allelic effects, in the case of genetic constraints; relationships among ontogenetic precursors, in the case of developmental constraints. *Strength* is an attribute that can sometimes be measured in statistical or mathematical terms (e.g., variance, covariance, regression slope, first and second derivatives). The *consequences* of a constraint are ultimately on evolutionary process and outcome. The consequences may be direct or may be mediated through other kinds of constraint. *Persistence* refers to the stability of a constraint over evolutionary time and can be assessed by longitudinal or comparative studies.

This chapter was previously published in a slightly different form in *American Naturalist* 140, Supplement (November 1992), S85–S107. © 1992 by The University of Chicago.

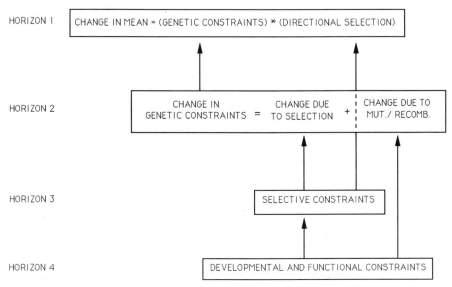

Fig. 11.1   A framework for the main theoretical connections between genetic, selective, developmental, and functional constraints. Immediate consequences are shown by arrows.

AN OVERVIEW OF THEORY

A current challenge is to find a unifying framework that accounts for all four basic properties of genetic and other kinds of constraint. Progress has been made, but we still do not have a conceptual framework that serves all purposes equally well. The most tangible progress has been made by building on the statistical framework provided by the field of quantitative genetics, the study of the inheritance of traits affected by many loci. A quantitative genetic framework does a good job of describing the strength and short-term consequences of genetic constraints. It has been less successful in dealing with the sources, long-term consequences, and persistence of genetic constraints. Much theoretical work in the last decade has been focused on correcting these deficiencies. To demonstrate both the triumphs and shortcomings of this recent work, a brief historical summary may be useful. For a fuller account of historical developments in quantitative genetics, see Hill (1984a, 1984b).

The theoretical structure provided by quantitative genetics consists of four horizons, shown diagrammatically in figure 11.1 with the oldest horizon on the top and the most recent on the bottom. The main message in the diagram is that selective, developmental, and functional constraints exert their evolutionary effects by affecting genetic constraints.

We may define *genetic constraints* as the pattern of genetic variation and covariation for a set of traits. The most useful definition of genetic variation and covariation has proved to be the additive genetic variances and covariances that describe resemblance between parents and offspring and that consequently enable us to predict responses to selection. In theoretical work these variances and covariances are often arranged in a so-called **G**-matrix in which the row and column labels refer to particular phenotypic attributes or traits (table 11.1). Thus, the elements on the main diago-

Table 11.1 Estimated Additive Genetic Variance-Covariance Matrix for a Bristle
Count and Four Measurements in an Experimental Population of
Drosophila melanogaster

Trait	BB	TX	WL	WW	TB
BB, Bristle number	0.9370	0.0042	−0.0119	−0.0022	0.0492
TX, Thorax length	0.0042	0.4844	0.2367	0.3713	0.1868
WL, Wing length	−0.0119	0.2367	0.3238	0.3696	0.1564
WW, Wing width	−0.0022	0.3713	0.3696	1.0241	0.2511
TB, Tibia length	0.0492	0.1868	0.1564	0.2511	0.3733

Source: Wilkinson et al. 1990.
Note: The estimates are based on an analysis of 181 sets of parents and 6 offspring from a line selected for large thorax length.

nal are genetic variances, and the other elements are genetic covariances. The heritability of a trait is simply its standardized additive genetic variance: $h_i^2 = G_{ii}/P_{ii}$, where
G_{ii} and P_{ii} refer, respectively, to the additive genetic variance and phenotypic variance of the ith trait. The genetic correlation between two traits is simply a standardized genetic covariance: $G_{ij}/\sqrt{G_{ii}\,G_{jj}}$, where G_{ij} is the additive genetic covariance
between the ith and jth traits. The magnitudes of the genetic variances and covariances (or their standardized analogues) measure the strength of genetic constraint.

The pattern of constraints embodied in the G-matrix can be visualized by trans-
forming it to a diagonal form. The elements on the main diagonal (eigenvalues)
of the transformed G-matrix are the genetic variances for trait combinations; the
off-diagonal elements are all zero. The rows of the transformation matrix (eigenvec-
tors) give the trait combinations or the new axes in multivariate space. Such diago-
nalized G-matrices can be used to identify directions in phenotypic space that are
most genetically constrained in the sense that they have the least genetic variance.
Examples of diagonalization of G-matrices using principal component analysis are
given by Gale and Eaves (1972), Arnold (1981), Atchley et al. (1981), Cheverud
(1982), and Leamy and Cheverud (1984).

The First Horizon: Response of the Mean to Selection

The first horizon consists of a family of equations for predicting the change in trait
means from one generation to the next. The univariate member of the family was
in place by the 1930s (see, e.g., Lush 1937). This earliest version says that the mean
of a single trait will be shifted across generations by an amount equal to the product
of heritability and the force of directional selection ($\Delta\bar{z} = h^2 S$). We may call this
version the breeder's equation, because its principal use is to predict how much
improvement to expect in some attribute when deliberate selection is practiced on
that attribute. The view that comes from the breeder's equation is that genetic constraint, expressed as heritability, acts as a simple brake on evolutionary change.
Under perfect inheritance the full force of selection is translated into evolutionary
change, but in a world of imperfect inheritance only a fraction of the selective force
is translated into change.

The inadequacies of the breeder's equation, even for practical applications, were
appreciated by the early 1940s (see, e.g., Hazel 1943). When selection is exerted on
a particular trait, that trait may respond to selection, but so may other traits. These
secondary or correlated responses to selection may be undesirable and must be

accounted for in the breeding program. The solution, which remains today, was to expand the concept of genetic constraint to include connections between traits. The solution was rooted in observations of the following kind. When a poultry flock is selected for increased egg production, production tends to increase, but body size decreases (Dickerson 1955). The Mendelian basis for such observations can be derived using the algebraic model independently introduced by Weinberg (1910), Fisher (1918), and Wright (1921). (For a translation of Weinberg's article by Karin Meyer, see Hill 1984a). Resemblance between parents and offspring in the same trait can be ascribed to variance in additive genetic values for the trait (additive genetic variance) (Fisher 1918). Correspondence between one trait in parents and another trait in offspring can be ascribed to covariance in additive genetic values (additive genetic covariance). For example, a negative genetic covariance between egg production and body size helps explain nonintuitive responses to selection in chickens (Gyles et al. 1955). Thus, by the early 1940s we had the ingredients for a multivariate view of genetic constraints. Over the next few decades the multivariate view of genetic constraints focused on the practical problem of improving domestic animals and plants. During this period it was largely ignored by evolutionary biologists.

The most relevant conceptual advance in the practical realm was the development of selection indices. The breeder has multiple objectives, so the problem is to devise a selection program that will give the best results across the board. The standard approach, rooted in the work of Smith (1936) and Hazel (1943), is to devise a weighted sum of traits, the index, upon which selection is practiced. To find the index that gives the best aggregate genetic response to selection, one needs to know both the genetic and phenotypic variances and covariances of the traits as well as the economic value of each trait (Hazel 1943). By the late 1970s a general solution had been achieved using the convenient notation of matrix algebra (Young and Weiler 1960; Magee 1965; Yamada 1977).

These later papers also provided a solution to the inverse problem of retrospective selection analysis, which arises when observed responses to a selection program do not match expectations. In these circumstances it is desirable to estimate the selection that was actually imposed and compare it with the selection that was supposedly imposed. The ingredients in such a retrospective analysis are the phenotypic covariances of the traits with the index, as well as the phenotypic and genetic variances and covariances of the traits. If we move beyond the purely practical concerns of the authors in question and equate the selection index with fitness, then their equations (see, e.g., Yamada 1977, equations 3b and 11) converge on those later used by evolutionary biologists to predict responses to multivariate natural selection (Lande 1979). This connection, however, is only obvious with the clarity of hindsight. In summary, solutions to the problems of imposing multivariate selection and analyzing that selection in retrospect were characterized by increasing sophistication in statistical techniques and multivariate characterizations of genetic constraints during the period 1936–1979. Meanwhile, application of quantitative genetics in evolutionary biology remained stalled at the level of the univariate breeder's equation, despite expository efforts by Robertson (1955) and Falconer (1960).

Multivariate concepts of selection and genetic constraint were first applied to evolutionary problems in a pathbreaking article by Lande (1979). The problem posed by Lande was how to predict the genetic response of a population when natural selection acts simultaneously on multiple traits. The ingredients of the solution consisted of multiple genetic constraints, encapsulated in a **G**-matrix, and a multivariate characterization of selection, a vector of selection gradients, $\boldsymbol{\beta}$. More precisely, the change in means of each of a set of traits from one generation to the next was shown to be the matrix product of **G** and $\boldsymbol{\beta}$, $\Delta\bar{z} = \mathbf{G}\boldsymbol{\beta}$. The selection gradients were derived as partial derivatives of mean fitness with respect to average trait values in the population. The derivation of multivariate response to selection followed the tradition in quantitative genetics of using the regression of offspring values on the phenotypes of their parents as a launching point. What was new in Lande's article, aside from the application to multivariate evolutionary problems, was the formal characterization of selection as a gradient on an adaptive landscape. This gradient view of selection provided a bridge to Wright's adaptive landscape for a field of gene frequencies, as well as the starting point for a host of evolutionary models, and it will be discussed here as the third horizon in our classification scheme. Thus, by the late 1970s the first horizon in the conceptual scheme consisted of a multivariate version of the breeder's equation that is useful for predicting response to natural selection.

The Second Horizon: Genetic Constraints and Their Evolutionary Persistence

The second horizon in the theoretical structure (see fig. 11.1) deals with the evolution of the genetic constraints themselves. The equation in the first horizon tells how much the mean of each trait will shift from one generation to the next. To apply the equation to successive generations, one must know whether the system of genetic constraints (i.e., the **G**-matrix) has changed. Many evolutionary biologists are familiar only with the early argument, which traces to Fisher (1930), that directional and stabilizing selection will progressively erode the genetic variance of a quantitative trait and eventually eliminate all heritable variation. Lande (1976a) argued that the genetic variance of a trait affected by many loci will enjoy appreciable input of mutation each generation. Consequently, under weak stabilizing selection, a balance will be struck between loss due to selection and input from mutation, so that appreciable genetic variance might be maintained at equilibrium. Later, analogous models and arguments were advanced for the maintenance of genetic covariances between traits (Lande 1980a, 1984). In response, Turelli (1984, 1985, 1986) argued that the amount of genetic variance or covariance maintained at equilibrium might be minuscule or appreciable, depending on what distributional assumptions are made about mutational input. Turelli (1988) argued that the issue could not be settled on theoretical grounds and has appealed for further empirical work. The important point is that by the mid-1980s a formal equation for evolutionary change in genetic constraints was advanced that took into account both selection and mutation. I shall take up the theoretical aspects of change in genetic constraints in the next section after I introduce a characterization of selective constraints. For the moment, I will concentrate on the empirical aspect of the second horizon.

The persistence of genetic constraints is an empirical issue. In principle, genetic variances and covariances could change considerably from generation to generation because they are functions of underlying gene frequencies (Falconer 1989). The issue is whether the **G**-matrix actually changes during evolutionary excursion of the multivariate mean phenotype. Stability has been assessed by selection experiments and comparative studies. The two kinds of studies offer slightly different perspectives on the issue of stability in nature. A selection experiment can tell us whether and how much the **G**-matrix changes under a known selection regime. To extrapolate the results to the natural world, we need to know how the imposed selection compares with selection in nature. A comparative study can tell us whether and how much the **G**-matrix has changed under natural selection regimes and over long time intervals. But, without a companion study of selection, a comparative study of **G**-matrices cannot diagnose the selective causes of change. Only a handful of studies have used one or the other technique to compare large **G**-matrices that are estimated from one hundred or more families in each sample (Atchley et al. 1981; Lofsvold 1986; Kohn and Atchley 1988; Wilkinson et al. 1990).

The comparison of covariance matrices, such as **G**-matrices, can be viewed as a series of nested hypotheses: matrix identity (equality of corresponding elements), matrix proportionality (the elements in one matrix are a scalar multiple of elements in the other matrix), and similarity in principal component structures (Flury 1988). The first two hypotheses can be visualized by making scatter plots such as those shown in figure 11.2. Under matrix identity the points should be arrayed on a 45° line. Under matrix proportionality the points should be arrayed on some other line that passes through the origin. The optimistic view of results from the best studies is that the **G**-matrices, as well as the **P**-matrices, from divergent lines or taxa are proportional and sometimes virtually identical. The pessimistic view is that some plots show an uncomfortably large amount of scatter, with some matrix elements showing considerable divergence. Note that the sample sizes for the phenotypic matrices plotted in figure 11.2 are 3–10 times larger than the sample sizes for the genetic matrices. Thus, the higher correspondence of phenotypic matrices ($r = 0.82$–0.96 versus $r = 0.53$–0.94 for genetic matrices) suggests that an appreciable amount of the scatter in the plots for genetic matrices represents errors of estimation.

Although there are encouraging signs that the constraints imposed by **G**-matrices may be evolutionarily persistent, much remains to be done. Clearly we need more data if we are to determine which traits show stable genetic variances and covariances and which do not, and on what time scale. An empirical challenge is to move from comparisons of pairs of taxa to multiple comparisons on a known phylogeny so that we can trace the evolution of **G**-matrices. Also, statistical techniques for comparison of **G**-matrices need to be perfected (e.g., see Turelli 1988; Shaw 1991, 1992; Cowley and Atchley 1992).

Constancy or proportionality of **G**-matrices opens the door to several novel modes of data analysis. We can reconstruct the net forces of directional selection that have produced differentiation in the phenotypic means of a pair of sister taxa (Lande 1979; Price et al. 1984; Schluter 1984; Price and Grant 1985; Arnold 1988; Lofsvold 1988; Turelli 1988). We can also test the pattern of among-taxa variation and covariation in phenotypic means against the pattern expected under multivariate drift (Lande 1979; Lofsvold 1988). In addition, we may be able to determine the roles of genetic constraints and among-taxa covariance in selection in producing interspecific covariation in means and allometry (Felsenstein 1988; Zeng 1988). Equations underlying the last two exercises are predicated on star phylogenies and need to be extended to the case of arbitrary branching sequences.

The Third Horizon: Selective Constraints and Their Consequences

Selective constraints arise from the ecological relationship between the population and its environment (Endler 1986) and also from interactions between the parts of an organism (Riedl 1979; Cheverud 1984). Selective constraints exert their effects on evolution directly under the guise of directional selection and indirectly under the guise of stabilizing (nonlinear) selection that affects genetic constraints (see fig. 11.1). The direct and indirect effects of selective constraints can be visualized using the

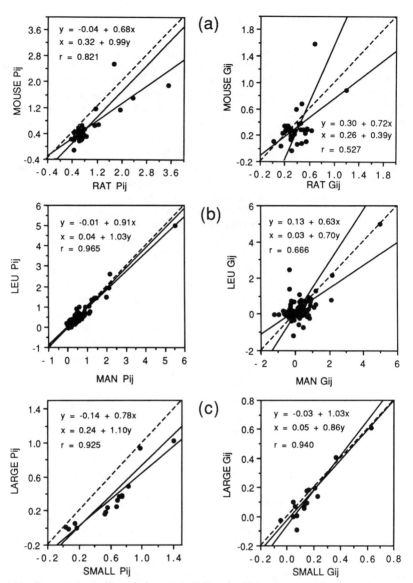

Fig. 11.2 Comparative studies of phenotypic (*left*) and additive genetic (*right*) variance-covariance matrices. The scatter plots show matrix elements from one population plotted against corresponding matrix elements from the other population. Sample sizes for phenotypic matrices are denoted n = number of individuals, and for genetic matrices are denoted N = number of families. The solid lines are least-squares regression lines. The dashed lines show the alignment of points expected under matrix identity. (*A*) Matrix comparisons for random-bred mice (*Mus musculus; n* = 707, N = 108) and selected lines of domestic rats (*Rattus norvegicus; n* = 522, N = 54) The matrices are based on eight pelvic measurements (yielding 36 data points in the plots.) (Data from Kohn and Atchley 1988.) (*B*) Matrix comparisons for white-footed mice (*Peromyscus leucopus; n* = ca. 315, N = 105) and deer mice (*Peromyscus maniculatus nebrascensis; n* = ca. 273; N = 91). The matrices are based on 15 skull and body measurements (yielding 120 data points in the plots.) (Data from Lofsvold 1986.) Estimates of genetic variance in the right-hand plot were constrained by the relationship $h^2 \leq 1.0$. (*C*) Matrix comparisons for two lines of *Drosophila melanogaster* after 23 generations of selection for large (*n* = 870, N = 145) and small (*n* = 762, N = 127) thorax size. The matrices are based on four body measurements and one bristle count (yielding 15 data points in the plots.) (Data from Wilkinson et al. 1990.)

idea of an adaptive landscape. More exactly, the strength and pattern of selective constraints can be measured as the slope and curvature of an adaptive landscape.

An adaptive landscape is simply the relationship between average fitness (or its logarithm) and average trait values. The most familiar adaptive landscape for a single trait is a curve that is bowed downward. Such a curve contracts the phenotypic variance of the trait and is said to exert stabilizing selection if the population mean is near the peak of the curve. The critical selection parameter for this effect is the curvature of the landscape (Lande and Arnold 1983). The stronger the curvature, the greater the contraction in variance, and the stronger the stabilizing effect. If we generalize this kind of landscape to two trait dimensions, we have a hill. In three trait dimensions, we have a spheroid. But, as we move to two or more dimensions, we need to consider the orientation of the hill, spheroid, and so forth. The selection parameters describing orientation, like the measures of curvatures, are second derivatives of fitness with respect to pairwise trait products. These orientation parameters are measures of so-called correlational selection. They describe effects on the phenotypic (and genetic) covariances of traits (Phillips and Arnold 1989). The set of curvature and orientation parameters can be conveniently arranged in a γ-matrix, with coefficients of stabilizing selection on the main diagonal and coefficients of correlational selection elsewhere. The γ-matrix can be estimated in a natural or experimental population by curvilinear regression of relative fitness on trait values and products of trait values (Lande and Arnold 1983). The important points are that we have a formal theory relating multivariate curvature of the adaptive landscape to changes in the G-matrix, and that we have statistical techniques for estimating those curvatures.

One useful equation from the second horizon relates the change in genetic constraints to the curvature of the adaptive landscape, and to mutation and recombination:

$$\Delta G = G(\gamma - \beta\beta^T)G + U,$$

where ΔG is the change in the G-matrix from one generation to the next; G is the additive genetic variance-covariance matrix before selection; the term in parentheses is the curvature of the adaptive landscape evaluated at the population's phenotypic mean; and U is a matrix describing mutational and recombinational contributions to genetic variances and covariances (Lande 1980a; Phillips and Arnold 1989). The first term on the right represents the change in G within a generation due to selection (before mutation and recombination). The matrix of genetic constraints, G, evolves toward an equilibrium pattern that is a compromise between the patterns imposed by selection and by mutation (Lande 1980a). The pattern of stabilizing selection, however, will evidently play the dominant role in shaping the G-matrix unless pleiotropic input from mutation is strong. At equilibrium, $\Delta G = 0$, and there is no directional selection ($\beta = 0$). Rearranging the equation, we find that $-\hat{G}\gamma\hat{G} = U$, where $\hat{G}$ is the equilibrium G-matrix. Solutions to this equation indicate that the pattern of the equilibrium G-matrix for two traits generally corresponds to the orientation of the adaptive landscape, which is described by the coefficients of stabilizing and correlational selection, γ. Sokal (1978) and Cheverud (1982, 1984) have argued for such a correspondence, but perhaps made the case stronger than it actually is. The correspondence between the equilibrium patterns of selection and of genetic constraint is strong in the special case considered by Cheverud (1984) in which there are no pleiotropic mutational inputs, or such inputs cancel, so that all off-diagonal elements in **U** are zero. In the face of strong pleiotropic inputs from mutation, however, the equilibrium pattern of genetic constraint can differ considerably from the pattern imposed by selection. Nevertheless, the theoretical results from the second and third horizons give hope that estimates of selection can illuminate the evolution of genetic constraints.

Although we can solve for the equilibrium **G**-matrix, solving for the evolutionary path of the **G**-matrix (its dynamics) is a difficult problem. As the phenotypic mean evolves on the adaptive landscape, the pattern of selection changes and, consequently, the rate and direction of evolution in the **G**-matrix change. Those changes in **G** can, in turn, alter the rate and direction of change in phenotypic mean. However, coupled equations from the second and third horizons can be used to numerically trace the evolution of the **G**-matrix. Via and Lande (1987) used this approach in a model for the evolution of phenotypic plasticity. In that model, even large displacements of the mean from an adaptive peak produced only small and transient changes in the **G**-matrix.

The long-term consequences of selective constraints on evolutionary outcome have been explored in a series of models for phenotypic evolution. These models assume constant patterns for genetic and selective constraints (i.e., an invariant **G**-matrix and a constant adaptive landscape). Even though the adaptive landscape is constant, as the population mean moves on the landscape, it experiences changing selection pressures. Typically, an adaptive landscape with a single peak is modeled (e.g., Lande 1980b; Via and Lande 1985), but some models with two peaks have also been constructed (Felsenstein 1979; Kirkpatrick 1982; Slatkin 1984; Lande 1986; Slatkin and Kirkpatrick 1987; Charlesworth and Rouhani 1988).

Evolutionary models with constant adaptive landscapes and **G**-matrices provide some generalizations about the effects of selective constraints. First, the phenotypic mean of the population tends to evolve in an uphill direction on the adaptive landscape (Lande 1979). When selection is frequency-independent and the landscape is Gaussian, the population mean equilibrates on an adaptive peak. Bürger (1986), however, modeled evolution on a landscape that was an ascending ridge with increasingly steep flanks. He found that the population would equilibrate on the ridge crest when stabilizing selection reached a critical level. Second, frequency-dependent selection generally causes the population to equilibrate some distance downslope from an adaptive peak (Lande 1976b, 1980b). Finally, when the number of selective constraints is less than the number of genetic constraints, so that some genetically variable traits are selectively neutral, there is a collection of possible equilibrium points (a line, plane, or hyperplane) rather than a single equilibrium point (Lande 1981; Lande and Arnold 1985). In the case of mating preferences and sexually selected traits, for example, the evolutionary outcome is changed dramatically depending on whether or not selection acts on mating preferences (Arnold 1987; Pomiankowski et al. 1991).

The most general effect of genetic covariances or unequal genetic variances is curved evolutionary trajectories (Lande 1980a; Via and Lande 1985). The population mean evolves on a straight path toward an adaptive peak only in the special case of zero genetic covariances and equal genetic variances. When genetic covariances are nonzero and/or genetic variances are unequal, evolution proceeds rapidly in some directions but only very slowly in other directions (fig. 11.3). The consequence is a curved evolutionary trajectory toward the adaptive peak with a slow final approach in a direction for which there is little genetic variation. When the landscape has two peaks and the population mean is situated in a critical boundary region,

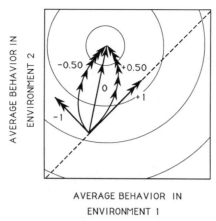

AVERAGE BEHAVIOR IN
ENVIRONMENT 1

Fig. 11.3   The Via-Lande model for the evolution of phenotypic plasticity. At the start of each generation, individuals disperse with equal probability into one of two environments and then spend their entire life in that environment. The contour lines show mean fitness as a function of average phenotype (behavior) in each environment. There is an adaptive peak near the top center of the plot. Populations with no behavioral plasticity would lie on the dashed line. Starting from the same behavioral average with no plasticity, evolutionary trajectories are shown for five populations with different values for genetic correlation in behavior. Arrows show evolutionary change in the mean at about 50-generation intervals. (Adapted from Via and Lande 1985, fig. 3C.)

genetic covariance can determine the peak toward which the population evolves (Bürger 1986; Slatkin and Kirkpatrick 1987).

The Fourth Horizon: Developmental and Functional Constraints and Their Consequences

A promising recent approach is to view genetic constraints as arising from underlying constraints (the lowest horizon in fig. 11.1). The search for order in genetic constraints has been pursued by students of morphological inheritance and by students of life history evolution. Both schools seek a manageable theory for the evolution of constraints that is cast in terms of underlying, proximate causes. The goal of both schools is to move beyond the earlier attitude of viewing genetic constraints as parameters that must be estimated on a case-by-case basis. The constraints of the morphological school are usually developmental. The functional constraints of the life history school deal with limitations on trait values, but the constraints are not described in developmental terms.

Developmental constraints. Developmental constraints are limitations on the set of possible developmental states and their morphological expressions (Atchley 1987). Developmental constraints arise from ontogenetic processes, especially from their ordering in time. Many authors have argued that developmental processes and constraints have important evolutionary consequences (e.g., Gould 1977; Alberch et al. 1979; Riedl 1979; Bonner 1982; Wake 1991). One difficulty in evaluating the evolutionary consequences of developmental constraints is that most arguments have been cast in nongenetic terms. We can ask, however, How does development impose genetic constraints? (Atchley 1987; Wagner 1989; Atchley and Hall 1991).

   One perspective on the genetic constraints that are imposed by development is gained by following the G-matrix through ontogeny. This approach has been explored by students of postnatal mammalian development for nearly three decades (reviewed in Atchley 1984). (For an introduction to an analogous tradition in the behavioral sciences, see Hahn et al. 1990). A breeding design is conducted to produce individuals of known relationship, then a set of measurements (mass, head length, tail length, etc.) is made on each individual at a series of ages. The many genetic variances and covariances that are estimated in such a study can be displayed in a variety of tabular and graphical formats. For the present discussion, it is useful to imagine them assembled in a single developmental G-matrix. The matrix includes genetic variances and covariances at each age, as well as genetic covariances between traits measured at different ages.

Developmental G-matrices commonly show regularities in their elements. These regularities reflect constraints that are imposed by developmental processes such as growth. For example, the genetic variance of a particular trait may increase with age (Cheverud et al. 1983a). Another common pattern is for the genetic correlation between the same trait measured at different ages (e.g., between head length at different ages) to be high when the two ages are close (e.g., ages 1 and 2) and to decline progressively as the two ages become more disparate (e.g., ages 1 and 10) (Cheverud et al. 1983b; Cheverud and Leamy 1985). Such regularities in the developmental G-matrix apparently reflect a greater commonality of underlying processes for traits expressed at adjoining ages than for traits expressed at widely separated ages. Atchley (1984) discusses how such regularities can be related to a model of compensatory (targeted) growth.

Inheritance, selection, and evolution of developmental trajectories can also be modeled with continuous functions (Kirkpatrick 1988; Kirkpatrick and Heckman 1989; Kirkpatrick et al. 1990). A character that changes with age can be viewed as a continuous function of age (an infinite-dimensional trait) rather than as a measurement made at a series of landmark ages. In the infinte-dimensional framework, the G-matrix becomes a G-function. By decomposing the G-function into its eigenvalues and eigenfunctions, one can determine possible directions of evolutionary change in growth trajectories for which there is little or no genetic variation. In other words, one can identify the directions of evolutionary change that are most genetically constrained.

Another way to find a bridge between development and genetic constraints is to construct a model for the relationship between two traits and their developmental precursors and examine the model's statistical consequences. The relationship between precursor and trait constitutes the developmental constraint. Perhaps the easiest consequence to model is the correlation that arises when two traits develop from a common precursor. For example, Riska (1986) used standard expressions for the variance and covariance of variables that are the sums or products of other variables to dissect the correlations expected under various simple developmental processes. To use such models directly we would need to assess the phenotypic and genetic variances and covariances of embryonic traits and developmental processes, as well as the statistics of endpoints. However, the quantitative genetics of embryos is a virtually unexplored realm.

Slatkin (1987) and Wagner (1989) have extended Riska's approach by constructing models for the evolution of developmentally coupled traits. Slatkin presented a modeling framework for the evolution of the developmental processes that underlie ontogenetic endpoints (adult traits). Slatkin solved for the phenotypic and genetic covariation of the endpoints in terms of the covariation of the rates and timing of the underlying developmental processes. Assuming that selection acts directly on the adult traits, and only indirectly on developmental processes, Slatkin was able to model the evolution of ontogeny as a correlated response to selection on adult traits. Wagner constructed his model on the premise that the pleiotropic effects of a gene are constrained by the developmental framework in which it is expressed. The system of developmental constraints was assumed to be evolutionarily constant, and mutational input was superimposed on that system. Wagner was able to show that the equilibrium G-matrix was a function of per-locus mutation rates, developmental constraints, and the curvature of the adaptive landscape.

The cause of genetic constraint is also a contemporary issue among students of life history evolution (Partridge and Sibly 1991). An important bridge between life history evolution and quantitative genetics was provided by the insight that life history trade-offs can be described by genetic covariances. If we wish to predict the short-term response of life history traits to selection, it is the genetic covariances between traits, rather than the more accessible phenotypic covariances, that most directly affect the prediction (Lande 1982; Reznick 1985). In recent years the focus has shifted to the issue of whether a trade-off inevitably implies a negative genetic covariance, and to the connection between functional and genetic constraints; in other words, the focus has shifted from the upper to the lower horizons in figure 11.1. Although this recent literature is directed at life history evolution, its lessons have broader implications.

Van Noordwijk and deJong (1986) made the important observation that variation in processes of resource acquisition can mask an underlying trade-off. The argument is easiest to understand by analogy with a problem in household economics. Within any given household there is likely to be a trade-off between the amount of money that can be spent on a car versus on the home. But, because households differ in income, when we look at the correlation between car and home expenditures across all households, the correlation may be positive rather than negative. More generally, variation in acquisition of resources can mask a trade-off between resources allocated to two competing alternatives (fig. 11.4). For example, the correlation between reproduction and subsequent survivorship may be positive rather than negative simply because individuals vary in the total amount of energy that can be devoted to these two functions. Houle (1991) cast the argument in genetic terms and explored its consequences. The bottom line is that a simple correlation is not an infallible indicator of underlying trade-offs. Interestingly, this point was appreciated by some empiricists, and circumvented using partial correlation (Stewart 1979), before the general theoretical argument was advanced. The van Noordwijk-deJong model, like Riska's (1986) models of developmental constraint, partitions the covariance between traits into the variation and covariation of underlying traits or processes. Limits on the values of trait combinations, which we will consider next, are a stronger form of constraint.

Functional constraints. Functional constraints are limitations on values of traits or of trait combinations. The limitations are imposed by time, energy, or the laws of physics. In models of life history equilibria, functional constraints take such forms as "total lifetime fecundity cannot exceed a fixed constant," or "there is a convex curve relating viability to fecundity" (Charnov 1989; Charlesworth 1990). Functional constraint is another label for the trade-off curves that are used in optimization models in behavioral ecology (e.g., Stephens and Krebs 1986; Mangel and Clark 1988). Under some rather strong assumptions, the genetic variances and covariances for a set of traits can be deduced from functional constraints (Charnov 1989;

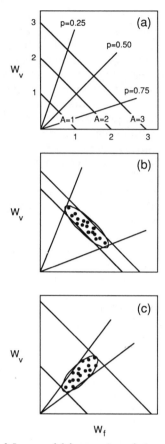

Fig. 11.4   The van Noordwijk-deJong model for covariance between two life history traits, w_f and w_v, in terms of variation in underlying processes of acquisition and allocation. (A) Acquisition, A, and allocation, p, constrain the values of viability, w_v, and fecundity, w_f, because $w_f = pA$ and $w_v = (1 - p)A$. Acquisition and allocation are independent random variables, but they determine the covariance between w_v and w_f. (B) Large variation in allocation and little variation in acquisition results in a negative covariance between w_v and w_f. (C) Large variation in acquisition and little variation in allocation results in a positive correlation between w_v and w_f.

Charlesworth 1990). The approach parallels Wagner's (1989) model of developmental constraints.

   Some general statements can be made about equilibrium genetic variances and covariances when a set of traits is bound together by functional constraints. A rendition of the Charnov-Charlesworth argument is given in figure 11.5 for a case involving only two traits that are bound together at equilibrium by a convex trade-off. When multiple traits are considered, genetic covariances are no longer a simple mapping of the underlying trade-off curves (Charlesworth 1990). Furthermore, negative genetic covariances are not an inevitable outcome; some genetic covariances can be positive (Pease and Bull 1988; Charlesworth 1990). In the case of multiple life history traits, the most general conclusion that has been reached is that at least one pair of traits will show a negative genetic covariance at equilibrium (Lande 1982; Charlesworth 1990). Note, however, that these arguments (e.g., fig. 11.5) may not

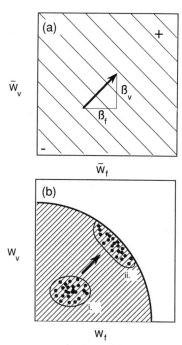

Fig. 11.5 The Charnov-Charlesworth model for equilibrium genetic covariance for a single pair of life history traits. (A) Two life history traits, fecundity and viability, with means $\bar{w}_f$ and $\bar{w}_v$, are under perpetual directional selection (*arrow*) for increasing values. In other words, the adaptive landscape slopes upward toward the upper right-hand corner. Contours for average total fitness are shown. (B) The heavy, curved line shows the trade-off curve that describes the functional constraint that binds the two traits, w_f and w_v. The constraint is such that w_f and w_v can only take values that lie inside the curve, in the cross-hatched region. Because w_f and w_v are always under directional selection for higher values, the population will tend to evolve outward from a position such as i, so that at equilibrium it lies on the trade-off curve, for example, at ii. Perpetual directional selection has the effect of pushing the bivariate distribution of genetic values up against the curve, so that a negative genetic covariance prevails at equilibrium.

apply if there is substantial curvature of the adaptive landscape. In that case, the population may equilibrate at an adaptive peak well inside the boundary imposed by a constraint.

Price and Schluter (1991) and Schluter et al. (1991) provide a different perspective on the genetic constraints affecting life history evolution. In the simplest rendition of the Price-Schluter argument (fig. 11.6), the two primary components of fitness, fecundity and viability, each exert stabilizing selection on a single morphological trait but toward different optima. The model assumes maintenance of genetic variance for the morphological trait. Genetic variance and covariance of fecundity and viability, and hence genetic variance for total fitness, are linear functions of directional selection on the morphological trait and its genetic variance. As the morphological trait evolves, the components of directional selection on it change, and consequently the genetic variances and covariances of the life history traits change in a simple pattern. In the Price-Schluter model, an evolving morphological trait constrains genetic variation in life history, while in the Charnov-Charlesworth model, functional limits on life history play that role. Despite their apparent differ-

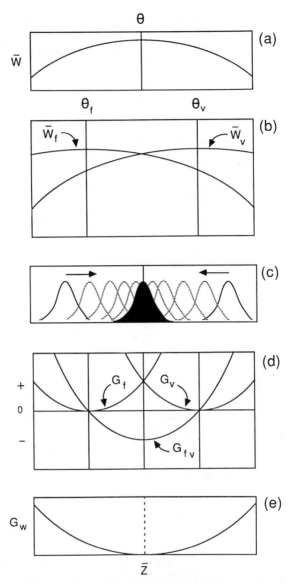

Fig. 11.6 The Price-Schluter model for the evolution of genetic variance and covariance of two fitness components in response to the evolution of the mean of a morphological trait, $\bar{z}$. (A) The morphological trait is under stabilizing selection toward an intermediate optimum, θ. (B) With respect to fecundity, the trait is selected toward an optimum, θ_f, that differs from the optimum under viability selection, θ_v. (C) The mean of the morphological trait and its phenotypic distribution (shown as normal curves) evolve toward an equilibrium position under the overall optimum, θ. (D) Because the two fitness components, fecundity and viability, are a function of directional selection on $\bar{z}$ and its genetic variance, the genetic variances of fecundity and viability, G_f and G_v, and their genetic covariance, G_{fv}, change as $\bar{z}$ evolves. (E) Because total fitness is a function of its two components, its genetic variance, G_w, also changes as $\bar{z}$ evolves.

ences, the two types of models yield similar predictions about the evolutionary genetics of life history.

BEHAVIOR AS A BUILDER AND BREAKER OF GENETIC CONSTRAINTS

In the preceding overview we have seen a trend toward seeking the causes of constraints in terms of lower-level constraints. Formal theory is developing more or less rapidly along these lines, and many aspects of behavioral evolution can be profitably viewed in this telescoping framework. In other words, when evolving behavior is viewed as a responder to constraints, a considerable amount of formal theory is available to us. Other aspects of behavioral evolution, however, do not fit so comfortably into the framework we have surveyed. In particular, the evolution of some behaviors will affect the evolution of constraints. Such effects, in which behavior acts as a builder and breaker of constraints, would constitute downward arrows in figure 11.1. Such effects do not challenge or threaten our framework, but behavior viewed as a builder and breaker of constraints has received less attention than behavior as a simple responder.

Behavior as a Builder of Constraints

Behaviors that affect the choice of mates and habitat have a special impact on population genetics. They can affect correlations between alleles at different loci and build genetic covariance through linkage disequilibrium. Such genetic covariance is fragile in the sense that it decays rapidly if selection is relaxed (Crow and Kimura 1970). But if selection is persistent, behaviors involved in mate and habitat choice may be capable of promoting genetic covariance that will in turn have important evolutionary consequences. Likewise, when optimal morphology or physiology varies with behavior, selection may promote genetic covariance between behavior and other phenotypic attributes.

Mating preferences within a population may help maintain a stable genetic covariance between the behavior of one sex and a sexually selected trait in the other sex (Lande 1981). Thus, mating preferences can create a genetic constraint that can in turn profoundly affect their evolution.

Correlational selection that favors certain combinations of behavior and morphology (or physiology) may be capable of creating genetic covariances by promoting linkage disequilibrium. Students of host plant polymorphism have argued that such selection may build genetic covariance between behaviors that function in host plant recognition and traits that affect efficiency in using host plants (e.g., Via 1991). Brodie (1989, 1992) has argued that correlational selection may favor certain combinations of coloration and antipredator behaviors, and has provided evidence for both correlational selection and genetic covariance in a polymorphic population of garter snakes.

Behavior as a Breaker of Constraints

Some kinds of behavior will have the effect of breaking prevailing constraints. Behavior that promotes acquisition of resources is one example; learning is another. Undoubtedly there are many others.

In the van Noordwijk-deJong (1986) model, two compartments for allocation show a negative covariance when there is little variation in acquisition of resources.

Variation in behaviors that promote acquisition of resources and time should be capable of breaking this kind of constraint. What limits variation in behavior that could rescue populations from constraints and limits on resource acquisition?

When behavior is not modified by maturation or experience, the same behavior is performed in all circumstances. Consequently, when the fitness consequences of a behavior change from moment to moment or from season to season, the best invariant behavior is a compromise between fluctuating optima in behavioral states. Temporally invariant behavior is constrained in the sense that there is perfect phenotypic (and presumably genetic) serial autocorrelation. Learning can break this type of constraint. In other words, learning represents an escape from the constraint of serial autocorrelation in behavior. We could probably model the evolution of learning, and its constraint-breaking effects, using the infinite-dimensional format recently introduced by Kirkpatrick and his colleagues (Kirkpatrick 1988; Kirkpatrick and Heckman 1989; Kirkpatrick et al. 1990).

CONCLUSIONS

Consideration of constraints is one of the most active areas in evolutionary biology, as evidenced by an ongoing explosion of literature. Constraints on evolution have been profitably viewed from many different perspectives. My approach has not been to challenge any of these diverse perspectives, but to seek their connections. I see genetics as the connecting principle because it provides the machinery for predicting the attributes of the next generation from those of its predecessor. Quantitative genetic theory emerges as the appropriate organizing principle for viewing evolutionary constraints because it deals with the inheritance and evolution of the polygenic, continuously distributed traits that figure in discussions of phenotypic evolution. The formal theory of quantitative genetics and its ongoing extensions can be conveniently grouped into a series of levels or horizons. The first two horizons are models for the evolution of the first two genetic moments of trait distributions (means and variance-covariances). Selective, developmental, and functional constraints exert their evolutionary effects via these two horizons. Consequently, quantitative genetics provides a formal mechanism for evaluating the short-term consequences of selective, developmental, and functional constraints.

Quantitative genetic theory does not make predictions about the long-term persistence of evolutionary constraints. The persistence of genetic and other kinds of constraints can be evaluated with comparative studies, but relatively little work has been pursued along these lines. The evolutionary persistence of adaptive landscapes, for example, is virtually unexplored. Thus, quantitative genetics alone is not a sufficient beacon to guide evolutionary studies. It needs to be combined with ecological, developmental, functional, and phylogenetic perspectives.

ACKNOWLEDGMENTS

I am grateful to L. Real and the American Society of Naturalists for inviting me to participate in the Vice-Presidential Symposium, "Behavioral Mechanisms in Evolutionary Ecology," at the Society's 1991 meeting. Among the many people who have patiently contributed to my education, I especially thank W. Atchley, E. Charnov, J. Cheverud, D. Houle, M. Kirkpatrick, R. Lande, R. Lewontin, T. Nagylaki, P.

Phillips, J. Maynard Smith, D. Schluter, M. Slatkin, G. Wagner, and D. Wake. Probably none of them, however, would subscribe to all of the views expressed here. I am also grateful to A. Abell and L. Real for comments on the manuscript and to J. Gladstone for graphics. The preparation of this chapter was supported by NSF Grants BSR 89-06703, BSR 89-18581, and BSR 91-19588.

REFERENCES

Alberch, P., S. J. Gould, G. Oster, and D. Wake. 1979. Size and shape in ontogeny and phylogeny. *Paleobiology* 5:296–317.
Arnold, S. J. 1981. Behavioral variation in natural populations. I. Phenotypic, genetic and environmental correlations among chemoreceptive responses to prey in the garter snake, *Thamnophis elegans*. *Evolution* 35:489–509.
Arnold, S. J. 1987. Quantitative genetic models of sexual selection: A review. In S. C. Stearns, ed., *The evolution of sex and its consequences*, 283–315. Basel: Birkhäuser.
Arnold, S. J. 1988. Quantitative genetics and selection in natural populations: Microevolution of vertebral numbers in the garter snake *Thamnophis elegans*. In B. S. Weir, E. J. Eisen, M. M. Goodman, and G. Namkoong, eds., *Proceedings of the second international conference on quantitative genetics*, 619–36. Sunderland, Mass.: Sinauer Associates.
Atchley, W. R. 1984. Ontogeny, timing of development, and genetic variance-covariance structure. *American Naturalist* 123:519–40.
Atchley, W. R. 1987. Developmental quantitative genetics and the evolution of ontogenies. *Evolution* 41:316–30.
Atchley, W. R., and B. K. Hall. 1991. A model for development and evolution of complex morphological structures. *Biological Reviews* 66:101–57.
Atchley, W. R., J. J. Rutledge, and D. E. Cowley. 1981. Genetic components of size and shape. II. Multivariate covariance patterns in the rat and mouse skull. *Evolution* 35:1037–55.
Boake, C. R. B., ed. 1994. *Quantitative genetic studies of behavioral evolution*. Chicago: University of Chicago Press. In press.
Bonner, J. T. 1982. *Evolution and development*. Berlin: Springer.
Brodie, E. D. III. 1989. Genetic correlations between morphology and antipredator behavior in natural populations of the garter snake *Thamnophis ordinoides*. *Nature* 342:542–43.
Brodie, E. D. III. 1992. Correlational selection for color pattern and antipredator behavior in the garter snake *Thamnophis ordinoides*. *Evolution* 46:1284–98.
Bürger, R. 1986. Constraints for the evolution of functionally coupled characters: A nonlinear analysis of a phenotypic model. *Evolution* 40:182–93.
Charlesworth, B. 1990. Optimization models, quantitative genetics, and mutation. *Evolution* 44:520–38.
Charlesworth, B., and S. Rouhani. 1988. The probability of peak shifts in a founder population. II. An additive polygenic trait. *Evolution* 42:1129–45.
Charnov, E. L. 1989. Phenotypic evolution under Fisher's fundamental theorem of natural selection. *Heredity* 62:113–16.
Cheverud, J. M. 1982. Phenotypic, genetic, and environmental morphological integration in the cranium. *Evolution* 36:499–516.
Cheverud, J. M. 1984. Quantitative genetics and developmental constraints on evolution by selection. *Journal of Theoretical Biology* 110:155–71.
Cheverud, J. M., and L. J. Leamy. 1985. Quantitative genetics and the evolution of ontogeny. III. Ontogenetic changes in correlation structure among live-body traits in randombred mice. *Genetical Research* 46:325–35.
Cheverud, J. M., L. J. Leamy, W. R. Atchley, and J. J. Rutledge. 1983a. Quantitative genetics and the evolution of ontogeny. I. Ontogenetic changes in quantitative genetic variance components in randombred mice. *Genetical Research* 42:65–75.

Cheverud, J. M., J. J. Rutledge, and W. R. Atchley. 1983b. Quantitative genetics of development: Genetic correlations among age-specific trait values and the evolution of ontogeny. *Evolution* 37:895–905.

Cowley, D. E., and W. R. Atchley. 1992. Comparison of quantitative genetic parameters. *Evolution* 46:1965–67.

Crow, J. F., and M. Kimura. 1970. *An introduction to population genetics theory.* New York: Harper and Row.

Dickerson, G. E. 1955. Genetic slippage in response to selection for multiple objectives. *Cold Spring Harbor Symposium in Quantitative Biology* 20:213–24.

Endler, J. A. 1986. *Natural selection in the wild.* Princeton: Princeton University Press.

Falconer, D. S. 1960. *Introduction to quantitative genetics.* New York: Ronald Press.

Falconer, D. S. 1989. *Introduction to quantitative genetics.* 3d ed. New York: Longman.

Felsenstein, J. 1979. Excursions along the interface between disruptive and stabilizing selection. *Genetics* 93:773–95.

Felsenstein, J. 1988. Phylogenies and quantitative characters. *Annual Review of Ecology and Systematics* 19:445–71.

Fisher, R. A. 1918. The correlation between relatives on the supposition of Mendelian inheritance. *Transactions of the Royal Society of Edinburgh* 52:399–433.

Fisher, R. A. 1930. *The genetic theory of natural selection.* Oxford: Clarendon Press.

Flury, B. K. 1988. *Common principal components and related multivariate models.* New York: Wiley.

Gale, J. S., and L. J. Eaves. 1972. Variation in wild populations of *Papaver dubium.* V. The application of factor analysis to the study of variation. *Heredity* 29:135–49.

Gould, S. J. 1977. *Ontogeny and phylogeny.* Cambridge, Mass.: Harvard University Press.

Gyles, N. R., G. E. Dickerson, G. B. Kinder, and H. L. Kempster. 1955. Initial and actual selection in poultry. *Poultry Science* 34:530–39.

Hahn, M. E., J. K. Hewitt, and N. D. Henderson. 1990. *Developmental behavior genetics: Neural, biometric, and evolutionary approaches.* Oxford: Oxford University Press.

Hazel, L. N. 1943. The genetic basis for constructing selection indexes. *Genetics* 28:476–90.

Hill, W. G. 1984a. *Quantitative genetics.* Part I. *Explanation and analysis of continuous variation.* New York: Van Nostrand Reinhold.

Hill, W. G. 1984b. *Quantitative genetics.* Part II. *Selection.* New York: Van Nostrand Reinhold.

Houle, D. 1991. Genetic covariance of fitness correlates: What genetic correlations are made of and why it matters. *Evolution* 45:630–48.

Kirkpatrick, M. 1982. Quantum evolution and punctuated equilibria in continuous genetic characters. *American Naturalist* 119:833–48.

Kirkpatrick, M. 1988. The evolution of size in size-structured populations. In B. Ebeman and L. Persson, eds., *The dynamics of size-structured populations,* 13–28. Berlin: Springer.

Kirkpatrick, M., and N. Heckman. 1989. A quantitative genetic model for growth, shape, reaction norms, and other infinite-dimensional characters. *Journal of Mathematical Biology* 27:429–50.

Kirkpatrick, M., and R. Lande. 1989. The evolution of maternal characters. *Evolution* 43: 485–503.

Kirkpatrick, M., D. Lofsvold, and M. Bulmer. 1990. Analysis of inheritance, selection and evolution of growth trajectories. *Genetics* 124:979–93.

Kohn, L. A. P., and W. R. Atchley. 1988. How similar are genetic correlation structures? Data from mice and rats. *Evolution* 42:467–81.

Lande, R. 1976a. The maintenance of genetic variability by mutation in a polygenic character with linked loci. *Genetical Research* 26:221–35.

Lande, R. 1976b. Natural selection and random genetic drift in phenotypic evolution. *Evolution* 30:314–34.

Lande, R. 1979. Quantitative genetic analysis of multivariate evolution, applied to brain:body size allometry. *Evolution* 33:402–16.

Lande, R. 1980a. The genetic covariance between characters maintained by pleiotropic mutations. *Genetics* 94:203–15.

Lande, R. 1980b. Sexual dimorphism, sexual selection, and adaptation in polygenic characters. *Evolution* 34:292–305.

Lande, R. 1981. Models of speciation by sexual selection on polygenic traits. *Proceedings of the National Academy of Sciences* U.S.A. 78:3721–25.

Lande, R. 1982. A quantitative genetic theory of life history evolution. *Ecology* 63:607–15.

Lande, R. 1984. The genetic correlation between characters maintained by selection, linkage and inbreeding. *Genetical Research* 44:309–20.

Lande, R. 1986. The dynamics of peak shifts and the pattern of morphological evolution. *Paleobiology* 12:343–54.

Lande, R., and S. J. Arnold. 1983. The measurement of selection on correlated characters. *Evolution* 37:1210–26.

Lande, R., and S. J. Arnold. 1985. Evolution of mating preference and sexual dimorphism. *Journal of Theoretical Biology* 117:651–64.

Leamy, L., and J. M. Cheverud. 1984. Quantitative genetics and the evolution of ontogeny. II. Genetic and environmental correlations among age-specific characters in randombred house mice. *Growth* 48:339–53.

Lofsvold, D. 1986. Quantitative genetics of morphological differentiation in *Peromyscus*. I. Test of the homogeneity of genetic covariance structure among species and subspecies. *Evolution* 40:559–73.

Lofsvold, D. 1988. Quantitative genetics of morphological differentiation in *Peromyscus*. II. Analysis of selection and drift. *Evolution* 42:54–67.

Lush, J. L. 1937. *Animal breeding plans*. Ames: Iowa State College Press.

Magee, W. T. 1965. Estimating response to selection. *Journal of Animal Science* 24:242–47.

Mangel, M., and C. W. Clark. 1988. *Dynamic modeling in behavioral ecology*. Princeton: Princeton University Press.

Maynard Smith, J., R. Burian, S. Kauffman, P. Alberch, J. Campbell, B. Goodwin, R. Lande, D. Raup, and L. Wolpert. 1985. Developmental constraints and evolution. *Quarterly Review of Biology* 60:265–87.

Partridge, L., and R. Sibly. 1991. Constraints in the evolution of life histories. *Philosophical Transactions of the Royal Society of London*, B 332:3–13.

Pease, C. M., and J. J. Bull. 1988. A critique of methods for measuring life history trade-offs. *Journal of Evolutionary Biology* 1:293–303.

Phillips, P. C., and S. J. Arnold. 1989. Visualizing multivariate selection. *Evolution* 43:1209–22.

Pomiankowski, A., Y. Iwasa, and S. Nee. 1991. The evolution of costly mate preferences. I. Fisher and biased mutation. *Evolution* 45:1422–30.

Price, T. D., and P. R. Grant. 1985. The evolution of ontogeny in Darwin's finches: A quantitative genetic approach. *American Naturalist* 125:169–88.

Price, T., and D. Schluter. 1991. On the low heritability of life-history traits. *Evolution* 45:853–61.

Price, T. D., P. R. Grant, and P. T. Boag. 1984. Genetic changes in the morphological differentiation of Darwin's ground finches. In K. Wohrmann and V. Loeschcke, eds., *Population biology and evolution*, 49–66. Berlin: Springer.

Reznick, D. 1985. Costs of reproduction: An evaluation of the empirical evidence. *Oikos* 44:257–67.

Riedl, R. 1979. *Order in living organisms*. New York: Wiley.

Riska, B. 1986. Some models for development, growth, and morphometric correlation. *Evolution* 40:1301–11.

Robertson, A. 1955. Selection in animals: Synthesis. *Cold Spring Harbor Symposium in Quantitative Biology* 20:225–29.

Schluter, D. 1984. Morphological and phylogenetic relations among the Darwin's finches. *Evolution* 38:921–30.

Schluter, D., T. D. Price, and L. Rowe. 1991. Conflicting selection pressures and life history trade-offs. *Proceedings of the Royal Society of London*, B 246:11–17.

Shaw, R. G. 1991. The comparison of quantitative genetic parameters between populations. *Evolution* 45:143–51.

Shaw, R. G. 1992. Comparison of quantitative genetic parameters: Reply to Cowley and Atchley. *Evolution* 46:1967–69.

Slatkin, M. 1984. Ecological causes of sexual dimorphism. *Evolution* 38:622–30.

Slatkin, M. 1987. Quantitative genetics of heterochrony. *Evolution* 41:799–811.

Slatkin, M., and M. Kirkpatrick. 1987. Extrapolating quantitative genetic theory to evolutionary problems. In M. D. Huetel, ed., *Evolutionary genetics of invertebrate behavior*, 283–93. New York: Plenum.

Smith, H. F. 1936. A discriminant function for plant selection. *Annals of Eugenics* 7:240–50.

Sokal, R. R. 1978. Population differentiation: Something new or more of the same? In P. F. Brussard, ed., *Ecological genetics: The interface*, 215–39. Berlin: Springer.

Stephens, D. W., and J. R. Krebs. 1986. *Foraging theory*. Princeton: Princeton University Press.

Stewart, J. R. 1979. The balance between number and size of young in the live-bearing lizard *Gerrhonotus coeruleus*. *Herpetologica* 35:342–50.

Turelli, M. 1984. Heritable genetic variation via mutation-selection balance: Lerch's zeta meets the abdominal bristle. *Theoretical Population Biology* 25:138–93.

Turelli, M. 1985. Effects of pleiotropy on predictions concerning mutation-selection balance for polygenic traits. *Genetics* 111:165–95.

Turelli, M. 1986. Gaussian versus non-Gaussian genetic analyses of polygenic mutation-selection balance. In S. Karlin and E. Nevo, eds., *Evolutionary processes and theory*, 607–26. New York: Academic Press.

Turelli, M. 1988. Phenotypic evolution, constant covariances, and the maintenance of additive genetic variance. *Evolution* 42:1342–47.

van Noordwijk, A. J., and G. deJong. 1986. Acquisition and allocation of resources: Their influence on variation in life history tactics. *American Naturalist* 128:137–42.

Via, S. 1991. The genetic structure of host plant adaptation in a spatial patchwork: Demographic variability among reciprocally transplanted pea aphid clones. *Evolution* 45:827–52.

Via, S., and R. Lande. 1985. Genotype-environment interaction and the evolution of phenotypic plasticity. *Evolution* 39:505–22.

Via, S., and R. Lande. 1987. Evolution of genetic variability in a spatially heterogeneous environment: Effects of genotype-environment interaction. *Genetical Research* 49:147–56.

Wagner, G. P. 1989. Multivariate mutation-selection balance with constrained pleiotropic effects. *Genetics* 122:223–34.

Wake, D. B. 1991. Homoplasy: The result of natural selection or evidence of design limitations? *American Naturalist* 138:543–67.

Weinberg, W. 1910. Weitere Beiträge zur Theorie der Verebung. *Archiv für Rassen-und Gesellschafts-Biologie* 7:35–49.

Wilkinson, G. S., K. Fowler, and L. Partridge. 1990. Resistance of genetic correlation structure to directional selection in *Drosophila melanogaster*. *Evolution* 44:1990–2003.

Wright, S. 1921. Systems of mating. I. The biometric relations between parent and offspring. *Genetics* 6:111–23.

Yamada, Y. 1977. Evaluation of the culling variate used by breeders in actual selection. *Genetics* 86:885–99.

Young, S. S. Y., and H. Weiler. 1960. Selection for two correlated traits by independent culling levels. *Journal of Genetics* 57:329–38.

Zeng, Z.-B. 1988. Long-term correlated response, interpopulation covariation, and interspecific allometry. *Evolution* 42:363–74.

TWELVE

Behavioral Constraints on the Evolutionary Expansion of Insect Diet: A Case History from Checkerspot Butterflies

MICHAEL C. SINGER

THE OUTSTANDING ECOLOGICAL TRAIT of herbivorous insects is the general narrowness of their diets. Each time we observe a specialist insect crawling through the greenery, starving because it cannot find an acceptable plant, we seem to see an animal that would survive if only it could feed on a wider range of plant species. Observations of starving specialists (e.g., Dethier 1959) imply selection for broadening the diet, and have triggered a hunt for the sources of counterbalancing forces that would tend to oppose or constrain evolutionary expansion of diet. The early work in this area centered on the search for physiological trade-offs; for example, genotypes that perform well on one host might perform poorly on another (Via 1984), or specialist genotypes on their natural hosts might outperform generalists on the same hosts. Physiological trade-off hypotheses were not supported by early experiments (reviewed in Jaenike 1990 and Via 1990). However, there were logical problems with some of the experimental designs (Singer 1984; Rausher 1988; Jaenike 1990), and recent, carefully designed work has lent the trade-off hypothesis cautious support (Jaenike 1990; Via 1991; but see Futuyma and Philippi 1987 for a counter-example).

Surprisingly, specialist insects may be able to grow well on host plants that they normally decline to eat, provided that we can experimentally persuade them to consume those plants (Bernays 1990). This suggests that at least some of the constraints on diet expansion are not connected with the physiological difficulty of utilizing alternative hosts. Bernays (1989) argues that diet specialization is principally related to predator avoidance, and that the advantages gained from restricted diet stem not from physiological efficiency but from such antipredator devices as efficient camouflage or sequestration of toxic host chemicals. Evidence to support this hypothesis emerges from analyses showing the greater vulnerability to predation of generalist insects, when compared with specialists on their own hosts (Bernays 1989).

I suspect that both the predation hypothesis and the physiological trade-off hypothesis will prove to be important evolutionary explanations for restricted insect diets. Recent evidence also suggests that phenological constraints on diet breadth may be important in seasonal environments, since insects may be restricted to hosts on which they can complete an integral number of generations in a growing season (Scriber and Lederhouse 1992). Bacterial parasites may also exert selection on insect diet breadth (Rossiter et al. 1990). Clearly, as research progresses, the diversity of known factors accounting for insect host specialization continues to increase. How can we tell which of them are important in any particular case? One way is to investigate empirically the obstacles that a specialist insect population does, in fact, overcome when it achieves evolutionary expansion of its diet to include a novel host species. If we can understand these obstacles, we may learn why such diet expansions are apparently rare or unstable. Although this type of study can explain why an already narrow diet tends to remain narrow, it does not explain how the diet came to be narrow in the first place. As Futuyma (1991) rightly comments, that is a separate question.

Winged herbivorous insects possess suites of behavioral traits that adapt them to the visual, chemical, and morphological properties of their hosts (Singer 1986; Harris and Miller 1991). The first individuals that use a novel host species may retain not only a preference for their long-standing host (Singer 1983; Prokopy et al. 1988; Thompson, in press), but the entire suite of behavioral adaptations to this species. This retention of adaptations to current hosts should constrain evolutionary diet change and lead in general to conservation of the existing diet (Futuyma 1991). In this chapter, I describe such constraints and show that they stem from behavioral traits of our study insect.

Although I take an entomocentric perspective here, I recognize that plant evolution is also a mechanistic cause of diet breadth evolution. For example, the diet of an insect population may become broader with no change in any insect trait if several of the plant populations in the habitat undergo convergent evolution to phenotypes that are accepted by the insects and can support their development.

STUDY ORGANISMS

Our study organisms are the nymphaline butterfly *Euphydryas editha* and its oviposition host plants in the families Scrophulariaceae and Plantaginaceae. This insect has a single generation each year, with adults flying in March–April at sea level, April–May at 1,000–2,000 m elevation, May–June at 2,400 m, and July–August at 3,500 m. Eggs are laid in clusters during the flight season and hatch after 1–2 weeks. Young larvae feed gregariously on the plants that their mothers chose, becoming more independent and capable of searching for hosts when they enter the third instar at 10–14 days of age. If second instar or very early third instar larvae cannot find food, they starve, but after a few days' feeding in the third instar, they reach a stage at which they respond to lack of food by diapause. If food remains plentiful, larvae enter an obligatory 9-month diapause at the end of third or fourth instar. After diapause is broken, either by winter rains (at sea level) or by snowmelt (at high elevation), larvae feed for a few weeks before wandering to find pupation sites under litter or inside pine cones. Mating is not associated with host plants. Oviposition usually begins on the second or third day of the female's adult life. In most of our study populations, each female lays one egg cluster per day, but in populations where cluster size is small, oviposition is more frequent.

Oviposition Behavior of Study Insect

Female *E. editha* that are searching for oviposition sites alight on all plant species present, both host and non-hosts. C. Parmesan (unpubl.) recorded these alightings and calculated an alighting bias for each plant species as the ratio of observed to expected alighting frequencies, with the expected frequency calculated from the proportional representation of the plant in the vegetation (random transects). Alightings were not random; alighting biases were often strong, and varied among plant species. More than 70% of this variation among plants in alighting bias was explained by the physical traits on the plant: in the population studied by Parmesan, insects were attracted to plants with complex leaves and rosette growth form, whether or not they were hosts. Although some butterflies respond to odor in their alighting behavior (Feeny et al. 1989), the present evidence indicates that searching *E. editha* are attracted to plants on the basis of visual stimuli.

After alighting, females taste the plant surface by drumming it with their atrophied foretarsi. A positive response to this tasting consists of curling of the abdomen and extrusion of the ovipositor, which is then pressed hard against the lower surface of the leaf or flower. We have shown that this abdominal curling is a response to plant chemicals, because differences among butterfly populations in responses to intact plants are maintained in responses to extracts of these same plants on filter paper (M. C. Singer and M. Perez, unpubl.). If a plant is chemically acceptable, the final stage in the oviposition sequence is response to physical stimuli detected by the ovipositor. We deduce that the ovipositor is not sensitive to chemical stimuli, since we observe that only the tarsi, not the ovipositor, must be in contact with the host for oviposition to occur. In consequence, eggs are sometimes laid adjacent to hosts rather than on them. We can sum up the sequence of responses thus: If it looks good, alight and taste; if it tastes good, feel it; if it feels good, lay eggs!

Estimating Oviposition Preference

As described above, we can estimate pre-alighting oviposition preferences by observing natural alightings and measuring the degree of alighting bias toward or away from particular plant species (Mackay 1985b). Post-alighting preferences have been obtained by taking advantage of the docility of *E. editha*, which duplicate their natural behavior when placed gently on a plant (Singer 1982; Singer et al. 1992a). When so manipulated, a female *E. editha* tastes the plant, drumming it with the atrophied foretarsi. Then, if the plant is chemically accepted, the butterfly curls its abdomen, extrudes its ovipositor, and presses it against the lower surface of a leaf. If the insect is then quickly removed from the plant, its acceptance can be recorded without oviposition actually occurring. This enables the experimenter to ask the insect which plants are chemically acceptable to it at different times. When we do this, we find that the range of acceptable plants expands with passing time. Each plant is initially rejected, then becomes acceptable, and remains acceptable until oviposition is allowed by the experimenter. Different plants become acceptable at different times. If the insect is placed repeatedly on two plants in alternation, a record such as that in figure 12.1 is usually obtained (Singer 1982; Singer et al. 1992a). This record produces both a rank order of preference (the order in which the plants are first accepted) and a quantitative estimate of strength of preference, or specificity, represented by the length of the discrimination phase during which one plant is consistently accepted and the other consistently rejected. If no such discrimination phase can be detected, then the insect is classified as showing no preference for the test plants.

THE ROLE OF BEHAVIOR IN VARIATION OF DIET

Before discussing behavioral constraints on the evolution of diet, I would like to address the ways in which behavior interacts with other factors to produce diet

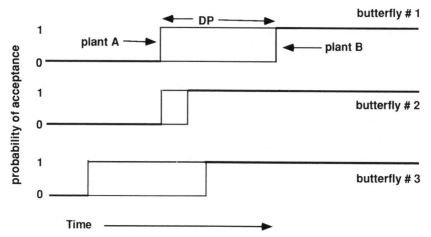

Fig. 12.1 Stylized depiction of changes over time in acceptance of two host species by three butter-flies. Each butterfly shows a refractory phase, when neither host is accepted, followed by a discrimina-tion phase (DP), when host A is consistently accepted and host B consistently rejected, followed by an acceptance phase that lasts until oviposition is allowed by the experimenter. The refractory phases extend back to ovipositions on previous days that are not shown. Butterflies 1 and 2 share the same preference rank (A > B) but differ in specificity, as estimated from the length of the discrimination phase. Butterflies 1 and 3 share the same preference rank and specificity, but differ in motivation to oviposit at any particular time. (From Singer et al. 1992a.)

variation. We have used this approach to investigate the mechanisms underlying the extensive geographical variation of diet in *E. editha* (fig. 12.2). The role of plant availability in diet variation can be demonstrated by taking a plant species to a site where it does not naturally occur and showing that at least some of the local insects accept that plant as readily as they do their principal host (Thomas et al. 1987). In this case, the absence of that plant species from the site is the sole factor excluding it from the diet, and we need not invoke traits of either plant or insect. However, this mechanism will not account for variation of diet among sites with similar plant availability. The map in figure 12.2 does not show plant availability, but I have indicated three pairs of sites between which the diet is different in spite of almost identical plant availabilities. These pairs are Tamarack and Colony in the western Sierra Nevada, Frenchman and Sonora on the eastern (Great Basin) slopes of the same mountain range, and Jasper Ridge and Pozo in the Coastal Ranges.

Observed diet variation among sites with similar vegetation could be directly caused by host-associated mortality. For example, eggs could be initially distributed in the same way at each site, then differentially removed by predators or killed by the host plant. However, our observations indicate that initial distributions of eggs are different between sites in each of the three population pairs. In order to illustrate the role of behavioral variation in producing these egg distributions, I will discuss one of the population pairs in detail. As the map (fig. 12.2) shows, the principal host at Sonora was *Collinsia*, while that at Frenchman was *Penstemon*. At both sites, the order of plant abundance was *Collinsia parviflora* > *Penstemon rydbergii* > *Castilleja pilosa*. The *Castilleja* was highly preferred by insects at both localities, but usually received only a small proportion of the eggs because of its rarity (Singer et al. 1989). So, the principal difference between the sites was the relative use of *Collinsia* and *Penstemon*. In four years of work at Sonora and two at Frenchman, we found only

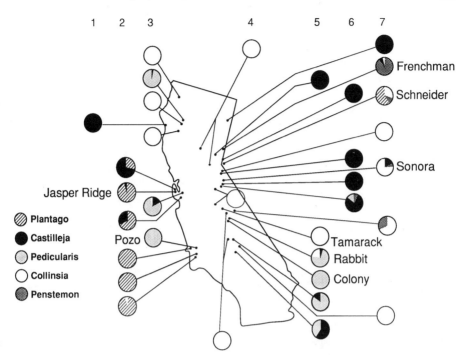

Fig. 12.2 Map of California showing geographical variation of diet in *Euphydryas editha*. Each pie diagram shows the estimated proportion of its eggs that the butterfly population laid on each host genus. These data are not preferences. For example, at Sonora, the most preferred host, *Castilleja*, received few eggs because of its rarity (Singer et al. 1989). The numbered vertical columns indicate habitat type and elevation. Column 1 is a sea cliff face. Populations in column 2 live in coastal grassland. Column 3 is chaparral vegetation in the inner coast ranges. Column 4 shows a set of populations at 1,000–5,000 feet elevation in the western Sierra Nevada. Column 5 represents 6,000–8,000 feet, still on the western slopes, while column 6 is at the crest of the mountain range (9,000–12,000 ft), and column 7 represents Great Basin sites on the eastern slopes of the Sierra Nevada at 4,000–8,000 ft.

a single egg cluster on *Penstemon* at Sonora and a single cluster on *Collinsia* at Frenchman.

When butterflies were placed on plants from their own habitats, insects from Frenchman ranked *Penstemon* over *Collinsia*, while those from Sonora (with one exception) showed the opposite preference. This result could stem from differences between the sites either in insect oviposition preference and/or in plant acceptability. For example, the difference could be caused by the *Penstemon* at Frenchman being less resistant to attack (i.e., more acceptable to ovipositing butterflies) than the same species at Sonora. It could also be caused by Frenchman butterflies being more *Penstemon*-preferring than Sonora butterflies. In fact, both of these effects existed, and they interacted in an additive manner to produce the observed variation of diet (Singer and Parmesan 1993). *Penstemon* at Sonora was avoided partly because it was less acceptable than *Penstemon* from Frenchman and partly because Sonora butterflies were averse to *Penstemon* in general. When insects from the two populations were raised from the egg on a common diet, the preference difference persisted undiminished. When *Penstemon* from the two sites were grown from seed in common soil (from either site), the difference in resistance also persisted (Singer and Parmesan 1993). Both plant and insect variation were therefore likely to be genetic.

This example illustrates the role of behavior in generating diet by showing how variation in insect preference interacts with variation in plant resistance to produce ecological patterns of plant/insect association. The diet itself is an emergent property resulting from patterns of encounter between insects with particular preferences and plants with particular acceptabilities.

Insect behavior plays a role in diet variation within as well as among populations of *E. editha*. At the Schneider and Rabbit sites (see fig. 12.2), where rapid diet evolution is occurring, differences among individuals in host choice were caused principally by variation in rank order of post-alighting oviposition preference (Singer 1983; Singer et al. 1989). Preference tests performed on insects caught in the act of oviposition showed that insects using particular hosts usually did so because they actively preferred these hosts after alighting. Within other populations that used more than one host species, preference rank was not variable, but preferences were so weak that many individuals failed to find their preferred hosts and accepted their second- or third-ranked choices (Singer et al., in press). When the most preferred host was rare, as *Castilleja* was at Sonora, the butterfly population deposited the majority of its eggs on lower-ranked species (Singer et al. 1989). At Sonora, variation in host use was not associated with variation in preference, so there was no real behavioral variation. In general, the observation that different individuals use different hosts does not tell us that preference is variable, and deduction of preference from naturally observed patterns of resource use is inadvisable (Singer and Thomas 1992).

The physiological ability of *E. editha* larvae to grow and survive on particular host species was also variable among populations (Rausher 1982). If adults frequently deposited eggs on plants that would not support larval growth, then larval performance would be an important mechanism generating the diet, when diet is estimated from distribution of larvae. However, this is not the case, since adult *E. editha* oviposit only on (or adjacent to) plant species that can support larval growth. We do not observe the "mistakes" in lepidopteran oviposition reported by other authors (Chew 1977; other references in Jaenike 1990). So, larval performance is not an important proximate cause of diet variation, though it must be one of the factors influencing the evolution of oviposition preference.

EVIDENCE FOR RAPID DIET EVOLUTION

Our evidence that the diet of *E. editha* can evolve rapidly stems from studies at two sites: Schneider's Meadow, Carson City, Nevada, and Rabbit Meadow, Tulare County, California. At Schneider the novel host is a European weed, *Plantago lanceolata*, which was incorporated into the diet of the insects sometime between 35 years ago and 105 years ago (Thomas et al. 1987). Experiments in which butterflies were manipulated to oviposit on different plants in the field showed that egg and prediapause larval survival were about ten times higher on this novel host than on the long-standing host, *Collinsia parviflora* (Singer 1984; Singer, unpubl.). Post-alighting oviposition preference for the two plant species was variable within the butterfly population, and insects captured ovipositing on different host species subsequently proved to have different preferences (Singer et al. 1989). Oviposition preference was both heritable within the insect population and associated with offspring performance. Insects that strongly preferred to oviposit on *Collinsia* produced offspring that grew faster on this plant and more slowly on *Plantago* than the offspring of females with other preferences (Singer et al. 1988; cf. Via 1986).

Comparisons with nearby populations where the weed *Plantago* had not been introduced indicated that the probable ancestral condition of the Schneider population did not include insects that preferred *Plantago*, but that about 10% of the insects may have had no preference, accepting *Plantago* as readily as *Collinsia* (Thomas et al. 1987). However, some of the insects at Schneider now actively prefer their novel host, *Plantago*. The estimated proportion of these insects rose sharply, from about 6% in 1982 ($n = 44$) to over 50% in 1990 ($n = 27$) (Singer et al. 1993). Since these data were gathered in the field with contemporary plants, the change could have been caused by temporal variation of either insect preference and/or plant acceptability. We were unable to test whether plant acceptability had changed. Investigation of potential changes in insect preference did prove tractable, because we had seed available from *Plantago* and *Collinsia* that had been used in preference trials to test the laboratory-raised offspring of butterflies caught in the field in 1983. From these seeds we grew plants that, we assumed, resembled those used in the earlier trials, and used them to test the preferences of laboratory-raised offspring of 1990 insects. The comparison between laboratory-raised offspring of 1983 insects and offspring of 1990 butterflies showed an increase over time in the proportion of family mean preferences favoring *Plantago* over *Collinsia*, in parallel with the changes estimated in the field. Thus, the natural increase in frequency of preference for *Plantago* was due at least in part to a genetic change within the insect population (Singer et al. 1993).

In the second rapidly evolving population, at Rabbit Meadow, the novel host, in this case *Collinsia*, became available sometime after 1967 and had been colonized before the start of our work in 1979. As at Schneider's Meadow, fitness (estimated in this case through the whole life cycle) was higher on the novel than on the long-standing host over the 3-year period from May 1984 to April 1987 (Moore 1989). Among butterflies that developed naturally on the novel host, the proportion preferring it rose from an estimated 6% in 1982–84 ($n = 31$) through 14% in 1985–86 ($n = 35$) to 23% in 1989 ($n = 43$; Singer et al. 1993). Below, I describe in detail the natural history of this very recent host colonization, preparatory to discussing what we can deduce about constraints on diet expansion from the interactions between the butterflies and their novel host.

NATURAL HISTORY OF A RECENT DIET EXPANSION

Along the Generals' Highway above Fresno, California, a series of populations of *E. editha* occurs in open woodland at about 2,400 m elevation. The butterflies have a single generation each year, flying in May–June. The diet of these butterflies differed among habitat patches with different histories of human disturbance. In most undisturbed areas, such as Colony Meadow in Sequoia National Park, all eggs were laid on the perennial hemiparasite *Pedicularis semibarbata*, though in some cases *Castilleja disticha* and its close relative *C. applegatei* also received eggs (Singer 1983). Another potential host, the ephemeral annual *Collinsia torreyi*, was abundant but was not utilized by the ovipositing butterflies. *Collinsia* underwent senescence 1–3 weeks after most eggs were laid, so that oviposition on this species would have resulted in larval starvation. However, when snow melted in spring, the germinating

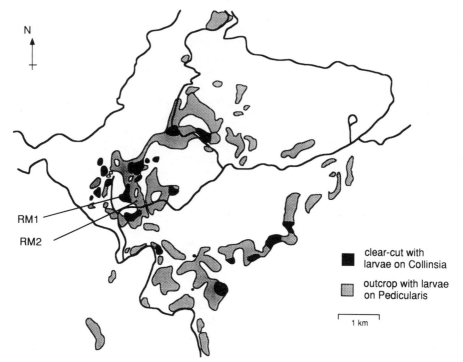

Fig. 12.3 Metapopulation structure and distribution of host use by *E. editha* at Generals' Highway. *Collinsia torreyi* was used exclusively in clear-cuts. *Pedicularis semibarbata* was used exclusively on outcrops. The two populations named (RM1 and RM2) are at Rabbit Meadow. (After C. D. Thomas, unpubl.)

seedlings of *Collinsia* were eaten by the postdiapause larvae, to the extent that in the presence of *E. editha* larvae, the survival of *Collinsia* seedlings varied with distance from the nearest *Pedicularis* individual (Thomas 1986).

In Sequoia National Forest, the practice of clear-cutting and burning the resulting debris was initiated at 2,400 m elevation in about 1967. In clear-cut areas *Pedicularis* was removed at a stroke, since it is a parasite of coniferous trees. At the same time the life span of *Collinsia* was extended for about 2 weeks, rendering many (but not all) individuals phenologically suitable for the insects. With a single act, the loggers had removed the butterflies' host and provided a new one. When we started work in the Generals' Highway region in 1979, the insects had already colonized *Collinsia* in at least one clear-cut patch (Singer 1983). By 1985, a complex patchwork distribution of host use had developed, with oviposition principally on *Collinsia* in the clear-cut patches and principally on *Pedicularis* on sparsely wooded granitic outcrops, which the loggers had left mostly undisturbed (fig. 12.3). Mark-recapture experiments have shown substantial insect movement among the two patch types (Thomas and Singer 1987; D. Boughton, unpubl.).

Study Populations

Within the metapopulation shown in figure 12.3, we have concentrated our work in two adjacent habitat patches at Rabbit Meadow, indicated on the figure as RM1 and RM2. RM1 was a typical clear-cut in which *Collinsia* was the principal host, with minor oviposition on *Castilleja* and *Mimulus*, while RM2 was a typical outcrop where the insects were monophagous on *Pedicularis*. Two other populations have been useful in our studies of diet evolution in the Generals' Highway region. One, Colony Meadow, is in undisturbed habitat in Sequoia National Park, about 10 km south of Rabbit Meadow. It represents the putative ancestral condition of Rabbit Meadow: monophagy on *Pedicularis*. The other population of interest, Tamarack Ridge, is about 60 km north of Rabbit Meadow at the same altitude as Rabbit, in habitat with similar vegetation and soil type (granitic sand). Tamarack insects are almost monophagous on *Collinsia torreyi*, the same species that has recently been colonized at Rabbit. *Pedicularis semibarbata* grows here, but is completely avoided by *E. editha*, and used by its sister species, *Euphydryas chalcedona*. (There is evidence of general diet displacement between these two insect species: Thomas et al. 1990).

Tamarack *E. editha* show evidence of much closer adaptation to *Collinsia* than the Rabbit insects (see below), and provide a possible target condition toward which the *Collinsia*-feeding insects in RM1 may be evolving. Thus we can examine ongoing evolution of diet by comparing insects from RM1 with their putative ancestral condition (Colony), a possible future condition (Tamarack), and an adjacent population (RM2) that retains the ancestral diet, but with which the RM1 population exchanges genes (Singer et al. 1992b). In order to make these comparisons among populations, we must first identify traits to compare. Below, I describe some of the behavioral traits that are involved in adaptation to *Collinsia* and *Pedicularis* at Rabbit Meadow.

Identification of Behavioral Traits Involved in Diet Evolution

Alighting Bias within the Habitat Patch. The efficiency of host search can be estimated from the alighting bias toward hosts. Searching female *E. editha* are readily identifiable from their habit of drumming leaves with their atrophied foretarsi, which are used only for tasting potential hosts. We compared the frequencies of alightings on hosts with those predicted from random transect data. At Rabbit Meadow, search for *Pedicularis* in RM2 was very efficient, with this species receiving about 50% of alightings although it constituted only 4% of the vegetation (Mackay 1985b). In contrast, alightings on *Collinsia* in RM1 occurred at random in one estimate (Mackay 1985a), and less often than random in a second estimate (C. Parmesan, unpubl.). The plant species toward which the alightings of RM1 insects were most biased were those that most closely resembled *Pedicularis* in shape, particularly *Chaenactis douglasii* (C. Parmesan, unpubl.). Since these pediculariform plants were not hosts, they were rejected after the insects tasted them, so they received many alightings but no eggs. The alighting bias toward plants resembling *Pedicularis* appeared in a sample of the very first alightings of the very first searches by naive insects (C. Parmesan, unpubl.). Since the insects used in this experiment had developed as larvae on *Collinsia* from eggs laid naturally on that host in RM1, it appears that the butterflies in RM1 had undergone little or no change in their search behavior from the ancestral condition (alightings biased toward *Pedicularis*). An anecdote may help to emphasize that their host-finding ability was poor. I watched two females searching for hosts without success for more than 5 minutes. They then collided head-on in midair, spun out of control, and both landed on *Collinsia*.

Post-Alighting Chemical Preference: Choosing a Host Species. Until recently, most insects in both RM1 and RM2 showed post-alighting preference for their long-standing host, *Pedicularis* (Singer 1983). Even though there has been a recent evolutionary increase in acceptance of *Collinsia* in RM1 (Singer et al., 1993), many of the butterflies that emerged there were still likely to reject *Collinsia* after alighting on it. With one exception (the year 1989), preference for *Pedicularis* was more frequent than that for *Collinsia* among insects that developed naturally on their novel host, *Collinsia*. Post-alighting preferences were associated with reciprocal movements among habitat patches, such that insects with no preference tended to move from outcrops to clear-cuts, and insects with *Pedicularis* preference moved in the opposite direction (Thomas and Singer 1987). The tendency for *Pedicularis*-preferring insects to leave clear-cuts, coupled with the high initial frequency of this preference among emerging butterflies, resulted in net emigration from this habitat type (C. D. Thomas et al., unpubl.).

A *Pedicularis*-preferring butterfly will oviposit on *Collinsia* if it searches within a clear-cut (where all the *Pedicularis* have been killed) until it reaches the high state of oviposition motivation at which *Collinsia* becomes acceptable (Singer 1983; Singer et al. 1992a). Since prolonged search in the same patch is more likely to happen in a large clear-cut than a small one, this aspect of search behavior should help to explain the observed responses to habitat patch size: the insects failed to colonize small *Collinsia* patches (C. D. Thomas, unpubl.), while small outcrops with *Pedicularis* did receive eggs (see map in Singer 1983).

Post-Alighting Chemical Preference: Choosing a Host Individual. Having alighted on a *Pedicularis*, those butterflies in RM2 that discriminated among individual hosts selected the plants on which their offspring's survival would be highest (Ng 1988). Those that did not discriminate among *Pedicularis* plants produced offspring that survived equally well on plants accepted and rejected by the discriminating insects (Ng 1988). Thus, the insect population was polymorphic, with an association between variable oviposition preference and offspring performance. The intraspecific discrimination that occurred was adaptive. In contrast, insects that used *Collinsia* in RM1 failed to select the most suitable individuals within the host species (Mackay 1985a). These insects tended to prefer old over young *Collinsia*, resulting in offspring mortality from host senescence (M. C. Singer, unpubl.).

Finding an Oviposition Site on the Chosen Plant. Having chosen a *Pedicularis* on which to oviposit, RM2 insects usually dropped to the base of the plant and searched there for an oviposition site. Observations of this process suggest that they require a site where they can press the ovipositor against the underside of a leaf or flower and meet physical resistance. On this rosette-shaped host, the butterflies had no difficulty satisfying these criteria. Butterflies in RM1 showed the same behavior, dropping in response to the chemical stimulus from the leaves and then probing with their ovipositors at the base of the plant. However, the thin stem of the erect *Collinsia* did not possess physically acceptable oviposition sites (such as stiff leaves) at its base, and the failure of the insects to find oviposition sites at the base of the plants sometimes led them to oviposit on stones or logs, resulting in high egg mortality from desiccation or overheating. When they did find sites that provided physical resistance to ovipositor probing, this resistance sometimes resulted from plant crowding, which in turn predicted early host senescence and high offspring mortality. Conversely, physical acceptability sometimes resulted from plant sturdiness, so that eggs were also laid on uncrowded, sturdy plants that were not likely to senesce. The result was that eggs tended to be bimodally distributed on the most suitable and the most unsuitable *Collinsia* plants (M. C. Singer, unpubl.).

Partitioning Eggs into Clusters. Having chosen an exact site on or near a host to lay eggs, RM1 insects laid larger clusters than those in RM2 (S. D. Moore 1987). Captive insects also laid larger clusters on *Collinsia* than on *Pedicularis* (Singer 1986). However, experiments in which group size was manipulated in the field suggested that the optimal cluster size for survival of individual offspring was lower on *Collinsia* than on *Pedicularis* (R. A. Moore 1987).

Behavior of Hatchling Larvae. Many (but not all) newly hatched larvae at Rabbit Meadow were positively geotropic. On *Pedicularis*, this behavior led them to spin webs between the ground and the leaves that rested on the ground. They then fed on the leaves that formed the upper surface of the nest. On *Collinsia*, positive geotropism tended to lead the larvae away from the edible parts of the plant. It is possible for larvae to spin a web around the base of the plant and commute up and down the stem to feed, but we have also observed larvae that starved before they were able to establish such a pattern.

CONSTRAINTS ON EVOLUTIONARY DIET EXPANSION

Retention of Adaptations to Long-Standing Hosts

The descriptions above indicate that there is negative covariance (Arnold, chap. 11) between adaptations to *Pedicularis* and adaptations to *Collinsia*, with many insects that used *Collinsia* apparently suffering difficulties that stemmed from their retention of traits adapting them to use *Pedicularis*. It seems that this retention of adaptations to *Pedicularis* caused Rabbit Meadow insects that used *Collinsia* to colonize only large host plant patches, within which they searched inefficiently, selected individual plants that would senesce, laid large clusters at the base of the plant instead of small ones at the top of the plant, and produced positively instead of negatively geotropic larvae. Studies of other populations adapted to *Collinsia* showed that none of these behaviors constitute long-term constraints to adaptation to this genus. *E. editha* at Schneider's Meadow searched efficiently for *Collinsia parviflora*, which closely resembles *C. torreyi* (Parmesan 1991). *E. editha* at Tamarack Ridge, which fed almost exclusively on *C. torreyi*, colonized small patches, tended to avoid oviposition on plants that would senesce, and balanced delicately on the tips of the plants, where they laid small clusters averaging 6 eggs rather than the 50 typical at Rabbit Meadow (M. C. Singer, unpubl.). The newly hatched larvae remained at the tops of the plants.

The differences between insects in RM1 and those at Tamarack and Schneider tell us that none of the apparent maladaptations to *Collinsia* by Rabbit Meadow insects represent long-term evolutionary constraints. The inefficient search for *Collinsia* does not stem from a physiological inability of the insects to perceive this small plant (Parmesan 1991). Colonization of small *Collinsia* patches can also be achieved by *E. editha*. Likewise, retention of post-alighting preference for *Pedicularis* among RM1 insects does not reflect a physiological constraint on preference rank: strong preference for *Collinsia* over *Pedicularis* is possible, and is characteristic of Tamarack insects. Egg cluster size is not constrained by fecundity: Tamarack insects have high fecundity, but can lay small clusters by ovipositing more frequently than Rabbit Meadow butterflies.

Although the behavioral adaptations to *Pedicularis* are not long-term constraints, they may nonetheless be very important in the initial colonization of novel hosts. Because search for *Collinsia* is inefficient, the insects require very high host density in order to find oviposition sites in a *Collinsia* patch. In order to maintain a population on *Collinsia* in the face of loss of individuals by migration to patches on the more preferred *Pedicularis* (Thomas and Singer 1987), the insects require very large patches of *Collinsia* within which *Pedicularis*-preferring insects will search unsuccessfully for such long periods that they eventually accept lower-ranked hosts such as *Collinsia*. For the butterflies to maintain positive population growth on *Collinsia* in spite of net emigration and maladaptive host choice, cluster size, geotropism, and so forth, the *Collinsia* must be extremely suitable. Because the populations using *Collinsia* are not isolated from those using *Pedicularis*, substantial colonization of *Collinsia* requires that fitness on *Collinsia* be higher than that on *Pedicularis*, even though the insects are better adapted behaviorally to *Pedicularis* than to *Collinsia*. These are stringent conditions. However, fitness on *Collinsia* has indeed been higher than that on *Pedicularis* at Rabbit Meadow (Moore 1989), so these conditions appear to have been met. In retrospect, they seem such stringent conditions that other forces may be at work: for example, the *Pedicularis* population at RM2 may have evolved resistance to *E. editha* attack as a result of long-term parasitism by the insects.

Evolutionary Dimensionality of Preference

Futuyma (1983) wrote that "the neurological sophistication of insects is probably insufficient to distinguish among all possible palatable and unpalatable plants." This statement implies constraints on the evolution of diet that stem from dimensionality of host choice behavior. It suggests that the number of independently evolving affinities for particular plants is constrained, and that incorporation of one host type into the diet may necessitate addition or deletion of other types (cf. Arnold 1981). An extreme version of this hypothesis has been articulated independently by Wasserman (1986) and Courtney et al. (1989). These authors provide experimental evidence that in their study organisms, *Callosobruchus* and *Drosophila*, the preference hierarchy is fixed, and any variation among individuals is in specificity (strength of preference), not rank order. So, preference is essentially unidimensional, at least in the short term. Both Courtney et al. and Wasserman suggested that if plant A is preferred over B, and B over C, in the fixed hierarchy, then evolution from monophagy on A to incorporation of C in the diet necessitates the incorporation of B also. In other words, the order of A with respect to B cannot evolve independently of the rank order of A versus C.

The unidimensional preference hypothesis does not apply to *E. editha*. We found genetic variation in rank order of preference within populations (Singer et al. 1988) as well as among them (Singer et al. 1991; Singer and Parmesan 1993). In one population with four potential host species, we tested 36 insects on the same plant individuals (one plant of each species) and found 15 of the 16 possible rank orders of pairs of plant species (Singer et al. 1989). However, constraints on dimensionality of preference do seem to exist in other butterfly species, notably papilionids (Wiklund 1981; Thompson, in press), and they are likely to apply at some level even to *E. editha*. It is, after all, unlikely that *E. editha* can possess independently evolving affinities for *n* potential resources when *n* is very large. The important question is, do associations among preferences constitute constraints on diet evolution in practice?

We have used our ability to compare evolving populations with their putative ancestral conditions in order to ask whether restricted dimensionality of preference has been influential in the recent diet expansion at Rabbit Meadow. We conducted experiments on a special case to ask whether associations between preferences within and among host species may constrain the evolution of diet. Our interest in this question stemmed from an early (1985) finding of an association among preferences within the RM1 butterfly population. At RM1, those butterflies that accepted their novel host, *Collinsia*, were likely to be nondiscriminators among individuals of the long-standing host, *Pedicularis*. This association led us to hypothesize that evolutionary expansion of diet in terms of numbers of host species may lead to loss of discrimination within species that are already used. When intraspecific discrimination is adaptive and important, such a relationship would cause selection against incorporation of novel species into the diet. Our experiments showed clearly that such a constraint was not operating in our study populations. RM1 butterflies were much more accepting of *Collinsia* than those from Colony (the ancestral condition), but no different in their level of discrimination among individual *Pedicularis* (Singer et al. 1992b). The association among preferences that we found in 1985 at RM1 was confirmed at that site by retesting in 1989, but did not exist at Colony in either 1988 or 1989. We concluded that the association among preferences was itself rapidly evolving and did not constrain diet evolution in this case.

Possible Long-Term Constraints: Limits to Polymorphism

All the constraints on diet breadth that I have discussed in *E. editha* are short-term, since they comprise maladaptations to novel hosts that are removed once the host has been in the diet for sufficient time. One source of long-term constraints can be deduced from the observation that all North American *Euphydryas* are confined to hosts that contain iridoid glycosides, whether or not these hosts are closely related (Bowers 1983). A second possible source of long-term constraints lies in limits to the number of suites of host adaptations that can be maintained in a population. In insects with flexible behavior, different individuals in a population can learn to search for different host species. Such learning is responsible for individual variation in foraging for leaf shape by the pipevine swallowtail butterfly, *Battus philenor* (Rausher 1978; Papaj 1986b). Exactly the same learning abilities exist in different, widely separated populations of this insect that use different hosts (Papaj 1986a). Post-alighting preferences may also be modified by learning, and exposure to a particular host may induce both increased preference for it by ovipositing adults and increased performance on it by larvae. In insects that learn, constraints on learning may limit the diet breadth of a population (Lewis 1986). For example, *B. philenor* found hosts more frequently when they were offered a single host species than when three species were offered together (Papaj 1990).

Studies of *E. editha* have provided no evidence for learning in any aspect of host choice behavior (Singer 1986; Thomas and Singer 1987; C. Parmesan, unpubl.). In the absence of learning, there are two mechanisms that could produce efficient use of several host species. The first is a set of switches by means of which an individual insect can recognize hosts of different types and respond appropriately to them. For example, both *B. philenor* butterflies (Pilson and Rausher 1988) and onion flies (Miller and Strickler 1984) laid different egg cluster sizes on hosts of different quality.

The second mechanism is the maintenance in a polymorphic population of suites of correlated adaptations to different hosts, such as those suites seen in polymorphic mimetic insects that mimic different models in color, flight speed, and diurnal activity rhythms. Although variation of host adaptations in *E. editha* is usually continuous rather than polymorphic, we are nonetheless beginning to accumulate evidence that these butterflies can generate suites of correlated adaptations, perhaps in response to correlational selection (Arnold, chap. 11). Twice, we have found correlations between adult oviposition preference and offspring performance (Ng 1988; Singer et al. 1988). At Schneider's Meadow, variation in maternal preference for novel versus long-standing hosts was correlated with variation in offspring growth rate on those hosts (Singer et al. 1988). Insects with strong oviposition preference for *Collinsia* tended to produce offspring that grew fast on this species. At the same site, insects searched efficiently for both novel and long-standing host species (Parmesan 1991). This finding contrasts with that at Rabbit, where the diet expansion is more recent and where search is efficient for the long-standing host but inefficient for the novel host (Mackay 1985b; Parmesan et al., in press).

We do not yet know how the Schneider insects achieve efficient search for two host species of very different appearance, but it is possible that search behavior has been incorporated into a suite of correlated traits along with oviposition preference and larval performance. If this phenomenon occurs, the diet breadth of a population may be limited by the number of such suites that can be maintained. In theory, these suites could be very complex, and could include several behavioral categories of adaptation to hosts, as well as physiological adaptations. However, up to now, our evidence suggests that broader diets in *E. editha* populations are not generally achieved by suites of polymorphisms. In the two populations in which evolutionary diet change is currently occurring, Rabbit and Schneider, insects that use different plant species do tend to have different host preferences (Singer 1983; Singer et al. 1988, 1989). However, in other populations, the use of several host species arises principally from weakness of preference (low specificity) rather than from diversity of preference, so that insects using different hosts are not following different behavioral rules (Singer et al. 1989; Singer et al., in press).

The colonization of a novel host species by an insect population has been viewed as incipient sympatric speciation (Bush 1975). If this were so, then some upper limit to population diet breadth would be set by the tendency to speciate. *Euphydryas* do not seem to behave in this way. Other species in the genus have a great diversity of both diet and diet breadth among populations (Mazel 1982, 1986), yet the number of sympatric species rarely exceeds two. Molecular evidence (Radtkey and Singer, in press) suggests that diet evolution in *E. editha* has been reticulate, with repeated colonization of each host genus. A mtDNA phylogeny that includes twenty-four populations of *E. editha* and one of its sister species, *E. chalcedona*, contains an estimated ten diet changes, yet the single speciation event is not associated with a diet change (Radtkey and Singer, in press).

CONCLUSION

Biologists interested in microevolution of insect diets have stressed three important areas: trade-offs in performance on different hosts, dimensionality of preference, and the role of the third trophic level, predators and/or parasites. Here, I add to this list a cluster of behavioral traits connected with the search for hosts, the handling and assessment of hosts prior to oviposition, and the actual act of oviposition. Their role has been deduced by comparing the interactions of insects with their novel and with their long-standing host species while the diet is actually undergoing evolutionary expansion. Use of this approach identifies suites of behavioral adaptations to the long-standing host that persist among insects using the novel host, and that severely restrict the ecological conditions under which evolutionary diet expansion can occur. These constraints are technically short-term, because they soon disappear after a diet shift. However, they may prevent the initiation of diet expansion, thereby helping to maintain narrow diets in specialist insects. Most of the behavioral adaptations described here have not been previously considered in the search for understanding of the evolution of insect diet breadth. Their importance in this particular case history suggests that detailed studies of insect behavior deserve more attention than they have traditionally received in studies of diet breadth evolution.

ACKNOWLEDGMENTS

Work described in this chapter was supported by grants from the University of Texas and by NSF grants to M. C. Singer, whose interest in this project began under the supervision of Paul Ehrlich. Permits for work at Colony Meadow and Tuolumne Meadow were granted by Sequoia National Park and Yosemite National Park, respectively. Schneider's Meadow was kindly lent to this project by Becky, Joe, and Troy Schneider. John Emmel, Sterling Mattoon, Dennis Murphy, Oakley Shields, and Bill Swisher provided guidance to some of the study sites. Generous permission to cite unpublished work was given by D. Boughton, R. A. Moore, S. D. Moore, D. Ng, C. Parmesan, M. Perez, H. Petursson, and C. D. Thomas. The manuscript was read by K. Agnew, D. Boughton, G. Perry, and V. Veit.

REFERENCES

Arnold, S. J. 1981. Behavioral variation in natural populations. I. Phenotypic, genetic and environmental correlations between chemoreceptive responses to prey in the garter snake. *Thamnophis elegans*. *Evolution* 35:489–509.

Bernays, E. A. 1989. Host range in phytophagous insects: The potential role of generalist predators. *Evolutionary Ecology* 3:299–311.

Bernays, E. A. 1990. Plant secondary compounds deterrent but not toxic to the grass specialist acridid *Locusta migratoria*: Implications for the evolution of gramnivory. *Entomologia experimentalis et applicata* 54:53–56.

Bowers, M. D. 1983. The role of iridoid glycosides in host plant specificity of checkerspot butterflies. *Journal of Chemical Ecology* 9:475–93.

Bush, G. L. 1975. Sympatric speciation in phytophagous parasitic insects. In P. W. Price, ed., *Evolutionary strategies of parasitic insects and mites*, 187–206. New York: Plenum.

Chew, F. S. 1977. Coevolution of Pierid butterflies and their cruciferous foodplants. II. The distribution of eggs on potential foodplants. *Evolution* 31:568–79.

Courtney, S. P., G. K. Chen, and A. Gardner. 1989. A general model for individual host selection. *Oikos* 55:55–65.

Dethier, V. G. 1959. Food-plant distribution and density and larval dispersal as factors affecting insect populations. *Canadian Entomologist* 88:581–96.

Endler, J. A. 1986. *Natural selection in the wild*. Princeton: Princeton University Press.

Feeny, P. P., E. Stadler, I. Ahman, and M. Carter. 1989. Effects of plant odor on oviposition by the black swallowtail butterfly, *Papilio polyxenes*. *Journal of Insect Behavior* 2:803–27.

Futuyma, D. J. 1983. Evolutionary interactions among herbivorous insects and plants. In D. J. Futuyma and M. Slatkin, eds., *Coevolution*, 207–31. Sunderland, Mass.: Sinauer Associates.

Futuyma, D. J. 1991. Evolution of host specificity in herbivorous insects: Genetic, ecological and phylogenetic aspects. In P. W. Price, T. M. Lewinsohn, G. W. Fernandes and W. W. Benson, eds. *Plant-animal interactions: Evolutionary ecology in tropical and temperate regions*, 431–54. John Wiley and Sons, New York.

Futuyma, D. J., and T. E. Philippi. 1987. Genetic variation and covariation in responses to host plants by *Alsophila pometaria*. *Evolution* 41:269–79.

Harris, M. O., and J. R. Miller. 1991. Quantitative analysis of ovipositional behavior: Effects of a host-plant chemical on the onion fly (Diptera: Anthomyiidae). *Journal of Insect Behavior* 4:773–92.

Jaenike, J. 1986. Genetic complexity of host-selection behavior in *Drosophila*. *Proceedings of the National Academy of Sciences U.S.A.* 83:2148–51.

Jaenike, J. 1988. Effects of early adult experience on host selection in insects: Some experimental and theoretical results. *Journal of Insect Behavior* 1:3–15.

Jaenike, J. 1989. Genetic population structure of *Drosophila tripunctata*: Patterns of variation and covariation of traits affecting resource use. *Evolution* 43:1467–82.

Jaenike, J. 1990. Host specialization in phytophagous insects. *Annual Review of Ecology and Systematics* 21:243–73.

Jaenike, J., and D. Grimaldi. 1983. Genetic variation for host preference within and among populations of *Drosophila tripunctata*. *Evolution* 37:1023–33.

Jaenike, J., and D. R. Papaj. In press. Learning and patterns of host use by insects. In M. Isman, and B. D. Roitberg, eds., *Chemical ecology: An evolutionary perspective*. London: Chapman and Hall.

Lewis, A. C. 1986. Memory constraints and flower choice in *Pieris rapae*. *Science* 232:863–65.

Mackay, D. A. 1985a. Conspecific host discrimination by ovipositing *Euphydryas editha* butterflies and its consequences for offspring survivorship. *Researches on Population Ecology* (Kyoto) 27:87–98.

Mackay, D. A. 1985b. Prealighting search behavior and host plant selection by ovipositing *Euphydryas editha* butterflies. *Ecology* 66:142–51.

Mazel, R. 1982. Exigences trophiques et évolution dans les genres *Euphydryas* et *Melitaea sensu lato* (Lep. Nympalidae). *Annales de la Societé Entomologique de France* 18:211–27.

Mazel, R. 1986. Structure et évolution du peuplement d'*Euphydryas aurinia* (Lepidoptera) dans le sud-ouest européen. *Vie Milieu* 36:205–25.

Miller, J. R., and K. L. Strickler. 1984. Finding and accepting host plants. In W. J. Bell and R. J. Cardé, eds., *Chemical Ecology of Insects*, 127–57. London: Chapman and Hall.

Moore, R. A. 1987. Patterns and consequences of within-population variation in reproductive strategies. Ph.D. dissertation, University of Texas, Austin.

Moore, S. D. 1987. Growth and survival of the butterfly *Euphydryas editha*: Intra-population variation associated with host plant species. Ph.D. dissertation, University of Texas, Austin.

Moore, S. D. 1989. Patterns of juvenile mortality within an oligophagous insect population. *Ecology* 70:1726–37.

Ng, D. 1988. A novel level of interactions in plant-insect systems. *Nature* 334:611–12.

Papaj, D. R. 1986a. Interpopulation differences in host preference and the evolution of learning in the butterfly *Battus philenor*. *Ecology* 40:518–30.

Papaj, D. R. 1986b. Shifts in foraging behavior by a *Battus philenor* population: Field evidence for switching by individual butterflies. *Behavioral Ecology and Sociobiology* 19:31–39.

Papaj, D. R. 1990. Interference with learning in Pipevine Swallowtail butterflies: Behavioral constraint or possible adaptation? *Symposia Biologia Hungarica* 39:89–101.

Parmesan, C. 1991. Evidence against plant "apparency" as a constraint on evolution of insect search efficiency. *Journal of Insect Behavior* 4:417–30.

Pilson, D., and M. D. Rausher. 1988. Clutch size adjustment by a Swallowtail butterfly. *Nature* 333:361–63.

Prokopy, R. J., S. R. Diehl, and S. S. Cooley. 1988. Behavioral evidence for host races in *Rhagoletis pomonella* flies. *Oecologia* 76:138–47.

Radtkey, R., and M. C. Singer. In press. Repeated reversals of host preference evolution in a specialist insect herbivore. *Evolution*.

Rausher, M. D. 1978. Search image for leaf shape in a butterfly. *Science* 200:1071–73.

Rausher, M. D. 1982. Population differentiation in *Euphydryas editha* butterflies: Larval adaptations to different hosts. *Evolution* 36:581–90.

Rausher, M. D. 1988. Is coevolution dead? *Ecology* 69:898–901.

Rossiter, M. C., W. G. Yendol, and N. R. Dubois. 1990. Resistance to *Bacillus thuringiensis* in gypsy moth: Genetic and environmental causes. *Journal of Economic Entomology* 86:2211–18.

Scriber, J. M., and R. C. Lederhouse. 1992. The thermal environment as a resource dictating geographic patterns of feeding specialization in insect herbivores. In M. D. Hunter, T. Ohgushi, and P. W. Price, eds., *Effects of resource distribution on animal-plant interactions*, 430–66. Orlando, Fla.: Academic Press.

Singer, M. C. 1982. Quantification of host preference by manipulation of oviposition behavior in the butterfly *Euphydryas editha*. *Oecologia* 52:224–29.

Singer, M. C. 1983. Determinants of multiple host use by a phytophagous insect population. *Evolution* 37:389–403.

Singer, M. C. 1984. Butterfly-hostplant relationships. In R. I. Vane-Wright and P. R. Ackery, eds., *The biology of butterflies*, 81–88. Orlando, Fla.: Academic Press.

Singer, M. C. 1986. The definition and measurement of oviposition preference. In J. Miller and T. A. Miller, eds., *Plant-insect interactions*, 65–94. Berlin: Springer-Verlag.

Singer, M. C., and C. Parmesan. 1993. Sources of variation in patterns of plant-insect association. *Nature* 361:251–53.

Singer, M. C., and C. D. Thomas. 1992. The difficulty of deducing behavior from resource use: An example from hilltopping in checkerspot butterflies. *American Naturalist* 140:654–64.

Singer, M. C., D. Ng, and C. D. Thomas. 1988. Heritability of oviposition preference and its relationship to offspring performance within a single insect population. *Evolution* 42: 977–85.

Singer, M. C., C. D. Thomas, H. L. Billington, and C. Parmesan. 1989. Variation among conspecific insect populations in the mechanistic basis of diet breadth. *Animal Behaviour* 37:751–59.

Singer, M. C., R. A. Moore, and D. Ng. 1991. Genetic variation in oviposition preference between butterfly populations. *Journal of Insect Behavior* 4:531–35.

Singer, M. C., D. A. Vasco, C. Parmesan, C. D. Thomas, and D. Ng. 1992a. Distinguishing between "preference" and "motivation" in food choice: An example of insect oviposition. *Animal Behaviour* 44:463–71.

Singer, M. C., D. Ng, D. Vasco, and C. D. Thomas. 1992b. Rapidly evolving associations among oviposition preferences fail to constrain evolution of insect diet. *American Naturalist* 139:9–20.

Singer, M. C., C. D. Thomas, and C. Parmesan. 1993. Rapid human-induced evolution of insect-host associations. *Nature* 366:681–83.

Singer, M. C., C. D. Thomas, H. L. Billington, and C. Parmesan. In press. Correlates of speed of evolution of host preference in a set of twelve populations of the butterfly. *Euphydryas editha*. *Ecoscience*.

Thomas, C. D. 1986. Butterfly larvae reduce host plant survival in the vicinity of an alternative host. *Oecologia* 70:113–17.

Thomas, C. D., and M. C. Singer. 1987. Variation in host preference affects movement patterns within a butterfly population. *Ecology* 68:1262–67.

Thomas, C. D., M. C. Singer, J. L. B. Mallet, C. Parmesan, and H. L. Billington. 1987. Incorporation of a European weed into the diet of a North American herbivore. *Evolution* 41: 892–901.

Thomas, C. D., D. A. Vasco, M. C. Singer, and D. Ng. 1990. Diet divergence in two sympatric congeneric butterflies: Community or species level phenomenon? *Evolutionary Ecology* 4: 62–74.

Thompson, J. N. In press. Preference hierarchies and the geographic structure of host use in swallowtail butterflies. *Evolution*.

Via, S. 1984. The quantitative genetics of polyphagy in an insect herbivore. II. Genetic correlations in larval performance within and across host plants. *Evolution* 38:896–905.

Via, S. 1986. Genetic covariance between oviposition preference and larval performance in an insect herbivore. *Evolution* 40:778–85.

Via, S. 1990. Ecological genetics and host adaptation in herbivorous insects: The experimental study of evolution in natural and agricultural systems. *Annual Review of Entomology* 35: 421–26.

Via, S. 1991. The genetic structure of host plant adaptation in a spatial patchwork: Demographic variability among reciprocally transplanted pea aphid clones. *Evolution* 45:827–52.

Wasserman, S. S. 1986. Genetic variation in adaptation to foodplants among populations of the southern cowpea weevil, *Callosobruchus maculatis:* Evolution of oviposition preference. *Entomologia experimentalis et applicata* 42:201–12.

Wiklund, C. 1981. Generalist vs. specialist oviposition behaviour in *Papilio machaon* (Lepidoptera) and the hierarchy of oviposition preferences. *Oikos* 36:163–70.

THIRTEEN

Individual Behavior and Higher-Order Species Interactions

EARL E. WERNER

BEHAVIOR IS ESSENTIALLY a set of devices for responding to environmental stimuli and tracking changes in those stimuli. This book contrasts different approaches to the study of behavioral mechanisms at different levels of integration in animal biology. In this chapter I address the manner in which ecological and developmental factors shape behavioral mechanisms and the consequences of these behaviors for species interactions. There would be little argument among ecologists that behavior is critical to the specification of interactions between species. Particular behaviors determine the nature and magnitude of interactions, be they among competitors, mutualists, parasites and their hosts, or predators and their victims. The vast majority of theory concerned with species interactions, however, essentially ignores behavior, at best employing simple density relations that reflect fixed behaviors. It is not clear that population and community theorists feel it will be necessary or practical to employ mechanistic statements of adaptive behavior to specify the dynamics of species interactions. Though the 1985 British Ecological Society symposium was devoted to exploring "the implications for population dynamics of the many recent advances in behavioral ecology," very few papers in that symposium actually addressed the linkage of adaptive behavior and population dynamics (Gilliam 1987), and this remains the case today.

In this chapter I argue that in order to understand species interactions in many systems, we will need to consider adaptive behavioral mechanisms. To illustrate this point I examine the consequences of two broad classes of individual behavioral responses, choice of activity level and choice of habitat, to species interactions. I show how these behaviors can generate higher-order effects where presence of one species qualitatively alters the interaction between two others. I further argue that theory at the individual level predicting behavioral responses affords an excellent opportunity to bridge mechanism and population phenomena. In the cases of both activity level and habitat choice, existing theory and experimental results at the behavioral ecology level could be employed to generate explicit mechanisms for species interaction theory. I further show how behavioral responses are constrained or influenced by size changes over the ontogeny of individuals and the consequences of this for species interactions. Finally, I speculate on the manner in which behaviors that are constrained by ontogeny may direct natural selection on species characteristics such as morphology and the evolution of complex life cycles.

This chapter was previously published in a slightly different form in *American Naturalist* 140, Supplement (November 1992), S5–S32. © 1992 by The University of Chicago.

BEHAVIOR, ALLOMETRY, AND HIGHER-ORDER INTERACTIONS

Species interact in the context of complex food webs, and their abundances are therefore affected by a variety of direct and indirect effects. Indirect effects arise in food webs of three or more species, and may be defined as the effects of one species on the interaction between two others. Keystone predators, trophic cascades, and exploitation competition are all examples of indirect effects. These types of indirect effects are relatively long-term responses and arise from changes in the densities of intervening species in a food web. Such indirect effects have been extensively discussed in the theoretical literature (e.g., Levine 1976; Holt 1977; Bender et al. 1984; Yodzis 1988) and documented empirically (e.g., Paine 1966; Dethier and Duggins 1984; Brown et al. 1986; Dungan 1986; Kerfoot and Sih 1987; Morin 1987). As Abrams (1991b) notes, these effects require no essential change in the form of population dynamics equations; if all direct effects are described properly, the long-term indirect effects are logical consequences. In principle, community dynamics could be constructed from pairwise experiments of species interactions quantifying per capita effects. Among the ecologist's challenges, then, are the bookkeeping problem (due to the complexity of ecological systems) and the problem of how to conduct appropriate experiments to estimate the strength of direct effects (e.g., Bender et al. 1984).

Increasing the complexity of food webs, however, also may precipitate other types of indirect effects (e.g., higher-order interactions) that introduce qualitatively different challenges. Higher-order interactions have been variously defined (Miller and Kerfoot 1987). Mathematically, if per capita growth rate of a species is not simply the sum of terms each involving the population size of a single species (e.g., interspecific nonlinearities), such relationships have been termed higher-order interactions (Case and Bender 1981). Statistically, higher-order interactions have been identified with significant interaction terms in experiments designed to assess the additivity of effects of several species on a target species (e.g., Wilbur 1972; Morin et al. 1988; Wilbur and Fauth 1990). Abrams (1983) defines a higher-order interaction as the circumstance in which presence of a species alters the nature of the interaction between two others (e.g., if the competition coefficient is a function of species other than the target pair). These definitions do not always identify the same phenomena. I use the term *higher-order interaction* to represent the subset of indirect effects where a third species qualitatively affects the nature of the interaction between two others by changing per capita effects. These are typically short-term responses that do not require changes in population densities of intermediate species (contra the indirect effects above). If higher-order interactions are important, adding additional species to a food web can change the essential form of population dynamic equations; that is, one cannot construct the dynamics of complex communities as a logical consequence of pairwise interactions. Consequently, ecologists have long been concerned with whether higher-order interactions are important in ecological communities (e.g., Vandermeer 1969; Neill 1974; Case and Bender 1981; Wilbur and Fauth 1990). Despite this concern, there is little empirical resolution of this question, as remarkably few experiments have addressed the issue (e.g., see reviews in Wilbur and Fauth 1990; Worthen and Moore 1991). Of course, indirect effects mediated by density changes and by higher-order effects occur in most communities, though generally on different time scales. Careful attention to methodology is required to separate these two types of indirect effects (see below).

In principle, higher-order interactions may occur for a number of reasons (Case and Bender 1981; Abrams 1983; Roughgarden and Diamond 1986). Abrams (1983) presents strong deductive arguments that nonlogistic growth of resources and non-linear functional responses of competitors would result in higher-order interactions among competitors. Interactions among herbivorous insects may be influenced by herbivory-induced changes in plant quality (i.e., an inducible defense, see Edelstein-Keshet and Rausher 1989). Presence of a third species may change the physical environment and thereby alter a predator's foraging efficiency on another species (e.g., Wootton 1993). In this chapter I focus on higher-order interactions that are behaviorally mediated through responses to the presence of predators (though this is only one of many behavioral mechanisms that could be examined to illustrate this potential). For example, responses to predators can levy nonlethal costs to prey that have effects on their competitive or resource acquisition abilities (e.g., Kerfoot and Pastorok 1978; Werner 1991). Despite the imposing literature on responses of prey to presence of predators (reviewed in Sih 1987; Lima and Dill 1990), we still have few examples of the consequences of those responses for interactions among prey species. Induced responses in morphology or life history due to presence of a preda-tor (e.g., Lively 1986; Stemberger and Gilbert 1987; Dodson and Havel 1988; Crowl and Covich 1990; Skelly and Werner 1990) may, of course, have effects similar to those of behavioral responses (e.g., Kerfoot 1977; Kerfoot and Pastorok 1978). In-duced responses (behavioral, life historical, or morphological) present valuable op-portunities to study higher-order interactions because we can experimentally present prey with the perception of predation risk and measure the effects of their responses on other interactions unconfounded by the direct mortality effects of the predators (see below).

The flexibility in behavioral responses that causes interactions to be context dependent may also depend on the state of the animal. Behavioral expression de-pends on morphological features, the integration/organization achieved over devel-opment, physiology, and so on. One of the most important such state dependencies involves the effects of ontogenetic change in size on behavior (i.e., behavioral al-lometries). Allometric studies indicate that much of the diversity in animal architec-ture, physiology, and ecology is related to differences in body size (Peters 1983; Schmidt-Nielsen 1984; Calder 1984). Behavior is no different. Examples of the allo-metric scaling of behavior show that, within taxa, larger species have greater abilities to learn, are socially dominant, show more group behavior, and are less cryptic and nocturnal than small species (Peters 1983; Jungers 1985). Allometric relations, however, are discussed almost exclusively in the context of interspecific or static allometry (comparisons made across species). Because size in many taxa increases by one to four orders of magnitude over ontogeny, intraspecific size changes often transcend those acknowledged to have extensive effects across species. Thus growth and development will have significant effects on behavioral expression over the life cycle, especially on responses to predators. I will show below that some of the most dramatic examples of higher-order interactions involve ontogenetic constraints on behavioral expression.

The importance of higher-order interactions is basically an empirical question. It is easy to imagine induced responses (or other mechanisms) that could substan-tially affect species interactions and give rise to higher-order interactions (e.g., Abrams 1983, 1991b). Deductive or plausibility arguments, however, say nothing of the magnitude of these interactions in nature. It may be the case that they are generally negligible and can be safely ignored. Given the widespread demonstration of facultative responses of species to their predators (reviewed in Kerfoot and Sih 1987; Lima and Dill 1990), this would appear to be a good place to evaluate the potential role of higher-order interactions. The critical question would seem to be not whether these responses cause higher-order interactions, but whether the higher-order effects are quantitatively important enough to warrant inclusion in basic population dynamic equations (e.g., whether inclusion of behavioral mechanisms qualitatively changes the predictions of species interaction theory).

BEHAVIOR, ALLOMETRY, AND HIGHER-ORDER INTERACTIONS: SOME THEORY AND EXAMPLES

In the following two sections I examine activity level and habitat choice responses of prey to the presence of predators. Both of these behaviors have important effects on species interactions and population dynamics. In each case I first introduce the important trade-off that underlies adaptive individual behavior. I then review theory available at the individual level that makes predictions concerning that behavior and the empirical evidence related to those predictions. I next review evidence that the behavioral responses indeed have an important higher-order effect on species interactions. Finally, because most of the latter evidence is based on short-term effects on components of fitness (e.g., individual growth rate), I evaluate to the extent possible whether the effects of the behavior will have significant population dynamic consequences.

Trade-Offs Mediated through Activity Level

Mobility is a fundamental characteristic of animals. The advantages of mobility in acquiring resources, selecting suitable habitats, and obtaining mates are obvious. At least over some range, it has been empirically demonstrated that activity level (speed, proportion of time active, foraging effort) is positively related to resource acquisition in a number of different taxa (see Werner and Anholt 1993 for a review). Activity level, however, is also positively related to mortality risk, and therefore mediates one of the more important and general trade-offs faced by animals. Encounter rates with predators (and other mortality factors) increase with activity, and many predators are able to detect only moving prey (Werner and Anholt 1993). The relations between speed and metabolic costs are also well documented (e.g., Schmidt-Nielsen 1984). Because activity level is related to resource acquisition and therefore competitive abilities, as well as risk of predation, it may be a prominent mechanism underlying trade-offs between competitive abilities and vulnerability to predators. Such trade-offs are fundamental to ideas concerning the interaction of competition and predation in structuring ecological communities (e.g., Brooks and Dodson 1965; Lubchenco 1978; Vance 1978) and how they affect species' distributions (e.g., McPeek 1990a, 1990b).

Individual-Level Theory. The adaptive control of activity or effort devoted to foraging (i.e., balancing costs and benefits related to activity levels) has been theoretically explored by Abrams (1982, 1990, 1991a) and Werner and Anholt (1993). The former emphasizes population dynamic consequences; the latter, individual behavioral responses. Abrams has generally assumed that time active (or foraging effort) has positive effects on reproductive rates but negative effects on survival rates; he determines optimal activity times by maximizing the difference between birth and death rates, or reproductive times survival rate. Werner and Anholt (1993) use a different fitness criterion developed for size-structured populations. They assume that growing individuals minimize the ratio of size-specific mortality rate (u) to growth rate (g)—the u/g criterion (Gilliam 1982; Werner and Gilliam 1984).

This emerging body of theory predicts that when gain and risk functions are linear with activity, animals should be either inactive or fully active (e.g., "sit and wait" or "searching") and that switches between these modes should occur as resource levels change (Abrams 1982; Werner and Anholt 1993). When mortality risk is linear and growth or birth rates have a negative second derivative (i.e., show diminishing returns), then activity can increase or decrease with resource levels. Decreases become more likely as the curvature of the growth or birth rate functions (with activity) increases (Abrams 1991a). The latter prediction is opposite to that of optimal speed arguments considering the trade-off between gains and metabolic costs (e.g., Ware 1975; Pyke 1981; Dunbrack and Giguere 1987). Activity also should decrease as mortality risk increases with speed or time active. Werner and Anholt (1993), however, show that as background mortality unrelated to activity increases, activity should increase. They also discuss how speed and time active may be mutually adjusted. For example, it is likely that the incremental cost of predation risk is higher with increasing speed than with time active; that is, detectability often increases with speed nonlinearly, whereas the increase with time active is more likely linear. All else being equal, this should favor adjustments in time active rather than in speed. Other constraints, however, may greatly alter this picture; for example, adjusting time active may be the most effective strategy until a day/night boundary is encountered that can markedly alter the risk of predation (see Mangel and Clark 1988). Werner and Anholt (1993) also examine predictions of activity rate when there are time constraints on the organism, such as those due to seasonality or the drying of temporary ponds. These sorts of models have broad application to the organization of activity patterns in many taxa and provide an initial framework for understanding behavioral responses to varying levels of predation risk and resources.

Evidence for Individual Responses. Qualitatively, numerous studies show that both speed and time active are reduced in the presence of predators as predicted (see review in Lima and Dill 1990). The magnitude of such responses is illustrated in figure 13.1A, which shows the large reduction in activity by anuran larvae in the presence of an odonate predator (see also Lawler 1989). A number of studies also report that searching activity or time active (or some conflation of these measures) declines with resource level (fig. 13.1B, reviewed in Werner and Anholt 1993). In a few cases we can actually make quantitative comparisons: when mortality functions are linear and resource acquisition functions are type II functional responses, Werner and Anholt (1993) predict that activity should decline as 1/square root of the factor of increase in resources. In several aquatic insect and tadpole species there is good correspondence between the predictions and the behavior of the animals (e.g., fig. 13.1, Werner and Anholt 1993). However, there are also species in these taxa that show little response to food levels (e.g., Skelly and Werner 1990) and some whose responses are in the opposite direction (e.g., Kohler and McPeek 1989). The latter may represent situations in which the animals are trading off speed and time active (see Werner and Anholt 1993). Nevertheless, the wide range of examples from many taxa that exhibit such responses suggests that adaptive behavior of this sort could be an important cause of higher-order interactions.

Effects on Species Interactions. What are the consequences of adaptive responses in activity level for species interactions? There is little empirical evidence on which

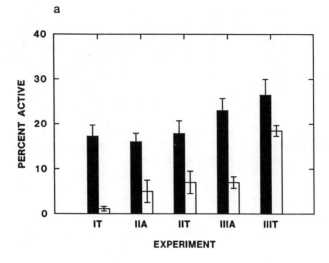

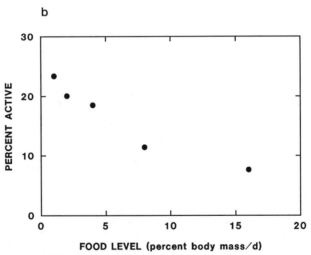

Fig. 13.1 (A) Mean percentage of larval green frog populations actively swimming when in the presence and absence of the odonate predator *Anax junius,* in a series of three different experiments (A, alone; T, together with equal numbers of bullfrog larvae). The solid bars represent activity in the absence of the predator; the open bars, activity in the presence of the predator. The predator effect is highly significant in each case (see Werner 1991). (B) Mean percentage of larval green frog populations actively swimming at different resource levels (percentage of body mass available per day). Each resource level is a doubling of the previous level. Resources were a 3:1 mixture of rabbit chow and fish flakes. Data are unpublished results of M. Breitsprecher. Activity declined an average of 0.77 ± 0.15 times that of the previous level with a doubling of resources. The decline predicted by the theory of Werner and Anholt (1993) is 0.71 times that of the previous level.

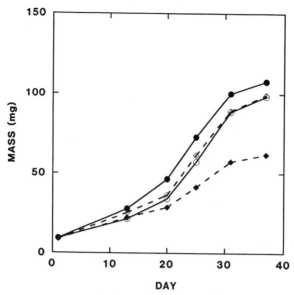

Fig. 13.2 Average mass (mg) of bullfrog and green frog larvae in the competition experiment, plotted against time. Solid lines (with circles) represent growth of the bullfrog; dashed lines (with diamonds), growth of the green frog. Open symbols represent species performance in the absence of *Anax*; solid symbols, species performance in the presence of *Anax*. Each data point is the mean of four replicates. The predator effect (difference in species mass on each date) was significant ($F_{1,6} = 13.4$, $P = .011$) by repeated measures analysis of variance. (Adapted from Werner 1991.)

to evaluate this question. I have conducted laboratory experiments with the larvae of two anuran species (the bullfrog, *Rana catesbeiana*, and the green frog, *R. clamitans*) competing in the presence and absence of a caged odonate predator (*Anax junius*) in order to examine such effects on species interactions (Werner 1991). Both competitive ability and predation risk appear to be related to activity level in these and related species (Werner 1991, 1992; E. E. Werner, unpubl.). In this experiment, any higher-order effects of the predator on competitive relations between prey species were isolated from the direct mortality effects of the predator. The two anurans exhibited similar competitive abilities in the absence of the predator, but very different abilities in the presence of the predator (fig. 13.2). Thus the mere presence of the predator greatly affected competitive relations between the two species. Both species reduced activity (e.g., fig. 13.1A) in the presence of the predator, which resulted in an increase in the relative activity of the bullfrog compared with that of the green frog. This change in relative activity level in the presence of the predator appears to be the mechanism causing the change in competitive relations between species. I am currently attempting to estimate the magnitude of the direct and higher-order effects under field conditions (see Werner 1991 for estimates of net effects of both based on laboratory data). Similarly, Feener (1981) experimentally demonstrated how the competitive interactions between two ant species were influenced by the presence of a parasitic phorid fly. The fly caused behavioral changes in the soldiers of one species, interfering with its defensive behavior. Huang and Sih (1990) experimentally demonstrated how interactions among a salamander, an

Table 13.1   Mean Activity Levels for Bullfrog and Green Frog Larvae
in Five Experiments Initiated with Different-Sized Larvae

	MASS (MG)				
	52	53	130	710	1015
Activity[a]					
Bullfrog	22.1	24.8	36.3	38.0	54.3
Green frog	16.9	16.1	24.8	32.7	30.0

Note: The data were analyzed as a mixed model analysis of variance with mass as a random factor. Both species and mass had significant effects (species $F_{1,4} = 16.4$, $P = .04$; mass $F_{1,68} = 23.9$, $P < .001$). Post hoc contrasts indicated that activity levels at each mass differed, with the exception of those at 52 and 53 mg.
[a] Percentage of population swimming.

isopod, and their mutual fish predator depended on behavioral responses in activity rate and time active outside a refuge. We can expect these sorts of interactions to be quite widespread.

As noted earlier, behavioral responses also may change with size and further exacerbate the higher-order effects. Foraging rate and predation risk both show strong allometries with size over ontogeny in most organisms (Werner and Gilliam 1984; Sebens 1987; Werner 1988), and therefore predictions of optimal activity will change over ontogeny (Werner and Anholt 1993). For example, vulnerability to predators decreases with size in many taxa (e.g., fishes: Werner and Hall 1988; amphibians: Wilbur 1980; Werner 1986), and therefore activity levels should increase over ontogeny. Similar effects can be expected because of the strong allometries of food-collecting abilities with size (Werner and Gilliam 1984; Sebens 1987). In several taxa increases in activity have been documented that appear to be related to the decrease in vulnerability with size (for examples with tadpoles see table 13.1; with crayfish, see Stein and Magnuson 1976). I have speculated that some of the reported advantages of size in larval anuran competition studies (e.g., Wilbur 1984) could be due to ontogenetic increases in activity that arise because of decreases in vulnerability to predators rather than size per se (Werner 1991). I know of no experiments addressing the effects of changing activity with size on species interactions. We will see clearly in the following section, however, how important ontogenetic responses to habitat choice can be to the creation of higher-order effects.

Unfortunately, there are no experiments that follow the population consequences of activity responses. For example, the anuran experiment measured short-term effects on individual growth rates, which can only be extrapolated to population consequences. Consequently, we must rely on the plausibility arguments of theory to make the case that such responses will have strong quantitative effects on populations. Several studies indicate that adaptive behavioral responses can markedly alter predictions relative to models without these effects. For example, Ives and Dobson (1987) developed a predator-prey model incorporating antipredator behavior (behavior that decreased the chance of capture). The cost of the behavior was exacted through decreased fecundity, and its efficiency was varied (but at each efficiency prey chose the optimal investment in the behavior). Increased efficiency of the antipredator behavior increased prey density and decreased the ratio of predator to prey densities. Behavioral responses tended to decrease the oscillatory dynamics of the model, thereby contributing to the stability of predator-prey interactions. Abrams (1982, 1990) has shown that if animals adaptively vary parameters of the functional response (time active or foraging, capture success, and handling time) with prey density, this can greatly alter details of the functional response and have a considerable effect on the dynamics of both predator and prey. The above studies begin to show the effects of including behavioral responses on predator-prey dynamics. Kotler and Holt (1989) also speculate that the presence of a predator may cause prey to move more slowly or spend less time active, thereby changing competitive relations between previously inferior and superior prey. Adaptive responses that change encounter rates with resources may change dietary overlap among competitors as well. Alternatively, the density of a prey species may affect the activity level of its predator, which in turn may affect the per capita predation rate on other prey species in the predator's diet, for example, through changes in hunger or satiation levels. Optimal foraging models are thus relevant to predicting higher-order effects in these cases.

Abrams (1984, and a more general extension in 1991b) has considered the question of the strength of higher-order interactions generated by optimally foraging consumers. His model was a three-species food chain in which the middle species adjusted foraging effort (activity) based on the individual-level gain/risk trade-off; the behaviors of top and bottom species were fixed. The relative magnitudes of effects were evaluated by comparing, for example, the proportionate effect of species 3 on species 1 with that of species 2 on species 1 (i.e., the indirect vs. the direct effect). The former was mediated through the behavioral responses of species 2. Abrams (1991b) argues that the wide range of circumstances leading to large-magnitude indirect effects indicates that behavioral responses must be incorporated into population dynamic models.

Thus we have evidence for strong adaptive activity level responses to the presence of predators and abundance of resources. Activity responses to predators in particular are well documented and of large magnitude (Lima and Dill 1990; Werner and Anholt 1993). Theory at the individual level qualitatively predicts such responses, though few quantitative tests of the theory are available. Strong behavioral responses in activity level should result in strong higher-order interactions, and the experiment with the amphibian larvae shows such an effect. These observations, along with Abrams' theoretical results, suggest that higher-order effects will be large and critical to understanding the nature of interactions in many ecological communities. In general, however, the empirical evidence is slim, and is limited to effects on variables such as individual growth rate rather than population consequences.

Trade-Offs Mediated through Habitat Choice

I now turn to the ways in which behavioral responses related to habitat choice can result in higher-order interactions. Habitat choice richly integrates behavior with population dynamics and community structure. Habitat is perhaps the prominent dimension along which competitive interactions are organized (e.g., Schoener 1974; Diamond and Case 1986). A number of models illustrate that competition can generate and maintain adaptations for habitat segregation, and empirical studies clearly document the effects of competitors on habitat use (Schoener 1983; Connell 1983; Werner and Hall 1976, 1979). More recently it has become apparent that predators also have a large impact on the habitat distribution of their prey, and that prey may partition "enemy-free space" (e.g., Jeffries and Lawton 1984). That is, a shared predator(s) presents problems of coexistence parallel to those presented by exploitation of shared limiting resources, and can lead to patterns of habitat partitioning similar to those generated by competition (Holt 1977, 1984; Schmitt 1987). Consequently, behavioral responses related to habitat use may have large higher-order effects on species interactions; for example, behavioral responses to predators can alter overlap in habitat use by competing prey species. Habitat use behavior, like activity, may be a prominent mechanism underlying the competitive ability/vulnerability trade-offs that are so important to the interaction of competition and predation in many systems.

I again examine systems in which species respond to the presence of predators, in this case by altering habitat use based on gain/risk trade-offs. In many systems, the physical characteristics of different habitats (e.g., structural complexity, refuges) render them more or less dangerous for prey. More dangerous habitats are often richer for the simple reason that they are less utilized by consumers because of the risks, and therefore resources achieve higher densities. Consequently, habitat choice decisions are often founded on foraging rate/risk trade-offs (see Werner and Gilliam 1984; Lima and Dill 1990). Such gain/risk trade-offs are nowhere more apparent than when viewed across the ontogeny of animals. Because habitat-specific foraging abilities and risks from predators strongly scale with size, the trade-off changes with size, prompting marked ontogenetic habitat shifts (see review in Werner and Gilliam 1984). Accordingly, when habitat choice behavior is flexible and gains and risks differ among species or change with size within species, strong and varied higher-order interactions can ensue.

Individual-Level Theory. At the individual behavioral level, the theory of habitat choice has a longer and richer history than does that of activity rate. Foraging theorists have presented a number of energy-based models to predict habitat choice, many built on the precepts of the marginal value theorem (Charnov 1976; reviewed in Stephens and Krebs 1986). Density dependence has been incorporated using notions of the ideal free distribution (Fretwell and Lucas 1970; reviewed in Rosenzweig 1991). Many of these ideas have been extended to address competitive interactions among species (e.g., MacArthur and Pianka 1966; Lawlor and Maynard Smith 1976; Brown 1990; Rosenzweig 1991), and some of the predictions have been empirically tested (e.g., Werner and Hall 1976; Rosenzweig 1991). Incorporation of predation risk into this theory and consideration of gain/risk trade-offs is more recent (Gilliam 1982; Brown 1988). There have been few attempts to incorporate these ideas into dynamical models of population consequences of the sort that Abrams has provided for the activity trade-off. Schwinning and Rosenzweig (1990) show that sustained oscillations in habitat distributions can be produced when predator and prey both choose habitats to try to achieve an ideal free distribution. Abrams and Shen (1989) incorporate adaptive choice into a population dynamic model of two consumers competing for two complementary resources that may be distributed in different habitats. The behavior of this model differs markedly from that of models with nonadapting consumers; that is, it provides a number of new mechanisms for population cycles.

The u/g criterion described in the activity section provides a general framework for decisions on habitat choice. Juveniles in a stationary population should choose the habitat from the array available that minimizes u/g (the criterion is somewhat more complex for adults and for changing populations: see Gilliam 1982). Because u (mortality rate) and g (growth rate) are functions of size, one can further predict the size at which ontogenetic shifts in habitat should occur (Gilliam 1982; Werner

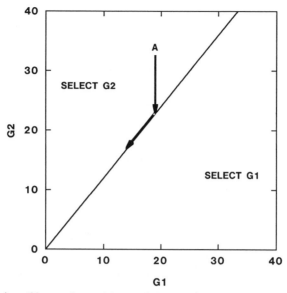

Fig. 13.3   Plane of possible growth rates (g) in two habitats (g_1) and (g_2) with a hypothetical switching curve ($g_2 = (u_2/u_1)g_1$ plotted indicating indifference to use of the two habitats. This curve separates regions where individuals should use habitat 1 or habitat 2 given habitat-specific growth and mortality (u) rates. Point A represents hypothetical initial resource levels (growth rates) in the two habitats. If a population using habitat 2 then depresses resources, growth rates will decline as portrayed by the heavy arrows. When the switching curve is intercepted, individuals should ideal free distribute between both habitats and maintain equal utility (u/g) in both. (Adapted from Gilliam and Fraser 1988.)

and Gilliam 1984). Gilliam and Fraser (1988) introduced density dependence into the u/g formulation and predicted the joint effects of resources and predators on habitat choice of competing species (see also Leibold 1988). That is, they sought the curve depicting combinations of growth and mortality rates at which a consumer would be indifferent to the use of the two habitats (i.e., where the two habitats are of equal value, $u_1/g_1 = u_2/g_2$, or $g_2 = (u_2/u_1)g_1$, equivalent to Rosenzweig's isolegs: Rosenzweig 1987, 1991). This curve (fig. 13.3) defines combinations of growth and mortality rates at which habitat shifts should occur. (Gilliam and Fraser actually present the density-dependent predictions in terms of u over feeding rate, but this should be broadly equivalent to u/g.) In general, predictions for two competitors indicate that either habitat segregation or complete overlap can result as a function of different resource and risk levels (Gilliam and Fraser 1988). If the consumer affects resources, then when one habitat is used, the resource trajectory is parallel to the resource axis for that habitat. If the switching curve is intercepted, however, both habitats should be used in proportions that maintain equivalent utility (u/g) in the two habitats, and the resource trajectory will then follow the switching curve (Gilliam and Fraser 1988). In this case the asymptotic distribution will be specialization by one species and relative generalization by the other (Parker and Sutherland 1986; Gilliam and Fraser 1988; Brown 1990). This theory illustrates how competitive effects can be very dynamic and depend on behaviors jointly influenced by resources and predators.

Evidence for Individual Responses. Abundant experimental evidence indicates that individuals respond to changes in both resources and predators in making habitat choices (see reviews in Werner and Gilliam 1984; Sih 1987; Lima and Dill 1990). I illustrate this point with examples from the aquatic environment, but there is a rich set of examples from terrestrial habitats as well (Stamps 1983; Kotler 1984; Lima 1985; Brown 1988; Kotler et al. 1991). For example, we have documented that the bluegill sunfish (*Lepomis macrochirus*) is adept at discerning differences in resource levels in different habitats and shifting between habitats accordingly (Mittelbach 1981; Werner et al. 1981; 1983b; Ehlinger 1990). The bluegill also alters habitat use in response to risk from piscivores such as the largemouth bass (*Micropterus salmoides*; Mittelbach 1981; Werner et al. 1983a; Werner and Hall 1988; Turner and Mittelbach 1990). There are many such examples in other species of fish (e.g., Power et al. 1985; Schmitt and Holbrook 1985; Gilliam and Fraser 1987). Further, it appears that fishes may be capable of adaptively balancing the trade-off between risks and resources (Werner et al. 1983a; Schmitt and Holbrook 1985; Gilliam and Fraser 1987). For example, Gilliam and Fraser (1987) tested a version of the u/g model and found that creek chubs (*Semotilus atromaculatus*) choose habitats in a way that balances gains and risks in good quantitative correspondence with the predictions of the model. What are the consequences of these well-documented behavioral responses to population and community dynamics? To begin to address this question, I examine some of the consequences of habitat shifts in the presence of predators at the intraspecific, interspecific, and trophic or ecosystem levels.

Effects on Species Interactions: Intraspecific Effects. Because risk from predators decreases as size increases, at some point in its life history the bluegill moves into more open (more dangerous) but richer habitats (Mittelbach 1981; Werner and Hall 1988). Large numbers of species undergo such habitat shifts during their ontogeny, apparently influenced by size-specific growth/risk trade-offs (Werner and Gilliam 1984). Werner and Hall (1988) showed that bluegill shift to the open-water regions of small lakes at sizes ranging from 51 to 83 mm (standard length; a variation of 4.4-fold in weight). The shift size was directly correlated with the density of piscivorous largemouth bass in these lakes, and growth rates often increased dramatically when the shift to the pelagic region was made (i.e., a trade-off existed: Werner and Hall 1988; Osenberg et al. 1988). Werner et al. (1983a) experimentally documented that bluegill respond behaviorally to largemouth bass in making such habitat shifts, and that this behavior alters the nature of intraspecific interactions among bluegill size classes. In the presence of largemouth bass, the smallest size class in the experiment spent more time in the less productive vegetated regions of the pond, and growth increment was 27% less than in the absence of the predator. This response mitigated intraspecific competition with the largest size class (which was immune to predation) in the open-water areas, and the latter size class grew faster in the presence of the predator (the two responses nearly compensated in terms of biomass produced). The experiment was unconfounded by direct mortality effects of the predator (individuals were replaced: Werner et al. 1983a). Thus, in the absence of the predator there was no habitat separation among size classes, but in its presence, size classes were separated. Patterns in natural lakes support the interpretation that responses to predators alter intraspecific interactions. Predators appear to control the sizes relegated to the littoral zone (Werner and Hall 1988), and bluegill compete when in the littoral zone (Mittelbach 1988). Growth rates of different size classes of bluegill across nine lakes showed that small bluegill growth rates were negatively density dependent, whereas those of larger classes were actually positively related (the latter appeared to be limited by the productivity of the open water habitat: Osenberg et al. 1988). Rankings of small and large class growth rates across lakes often reversed as the habitat-specific relations changed. Thus, behavioral responses to predators should introduce strong nonlinearities in intraspecific competitive interactions. Nonlinear intraspecific effects do not constitute higher-order interactions, but the interaction of these with interspecific effects are likely to lead to higher-order effects.

Effects on Species Interactions: Interspecific Effects. Predators confine smaller size classes, not only of bluegill but of many other species, to the littoral zone in small lakes. Mittelbach (1988) has experimentally demonstrated that small bluegill affect the growth rates of juveniles of a congener (pumpkinseed sunfish, *L. gibbosus*) in such lakes. Osenberg et al. (1988) showed that patterns across lakes were consistent with this result: small pumpkinseed and bluegill growth rates were negatively correlated with total density of small fish, and responses of small and large classes were more similar between than within species (pumpkinseed adults feed on different resources than bluegill). Osenberg et al. (1992) also showed a large effect of removal of bluegill (due to a winterkill) on the patterns of size-specific growth rates and food habits of pumpkinseed. Thus competition between early life stages of these species in the littoral zone (i.e., a competitive bottleneck) appears to be a function of behavioral responses to predators. Competitive effects have been measured in terms of individual growth responses; we have no empirical data on the population consequences of these effects. Theoretical analyses suggest that the population consequences of predators segregating size classes within species but causing overlap of some classes between species can be large and varied. Mittelbach and Chesson (1987) explored some of the ramifications of such interactions and showed how competitive effects can be transmitted between adults of the two species even when they use different resources. Because the size at which habitat shifts occur depends on the density (and no doubt size distribution and type) of predators (Werner and Hall 1988), the strength and balance of intraspecific and interspecific competition will depend on the immediate behavioral reactions of the prey to predators as well as on direct mortality effects, and we can expect very strong higher-order interactions.

Persson (1991) has constructed a laboratory analogue of the vegetated and open-water habitat case in large aquaria employing juveniles of two species of fish that differ in foraging abilities in the two habitats. For the particular configuration of habitats employed, the open-water specialist had a marked feeding rate (growth) advantage over the vegetation specialist when competing for a limited quantity of zooplankton. In the presence of a predator that forced both species to use the vegetation more, however, the feeding rate advantage was reversed. Thus the behavioral reactions to the predator reversed a feeding rate (and by inference a competitive) advantage between the species. (See Lima and Valone 1991 for results with birds that suggest effects of predators that are similar to the ones illustrated for fish.)

Effects on Species Interactions: Trophic Level Effects. Now consider the potential ecosystem level effects of these behavioral responses to predators. Removal or addition of top predators can have cascading effects across trophic levels—through planktivores, zooplankton, to at least the algae and perhaps to nutrient dynamics (e.g., Carpenter and Kitchell 1988; McQueen et al. 1989). Such effects traditionally have been viewed as direct effects mediated through changes in abundance of trophic level elements, and the ideas go back at least as far as Hairston et al. (1960). Several experimental studies, however, illustrate the potential for behavioral responses to affect such trophic cascades. For example, in an experimental pond study, Turner and Mittelbach (1990) demonstrated that the habitat response of small bluegill to the presence of largemouth bass had a significant effect on zooplankton dynamics (abundance and composition). There was little actual predation by bass on bluegill; thus the higher-order effects of bass on zooplankton through bluegill behavior were isolated from direct mortality effects. There are a number of other excellent studies showing such effects, though direct and indirect effects are confounded in most of these studies (e.g., Power et al. 1985; Carpenter et al. 1987).

Effects on Species Interactions: Reactions to Different Predators. I now present several examples to show how differential reactions by prey species to different predators may generate higher-order interactions, either by exacerbating or ameliorating overlap in habitat use. I have shown in both laboratory and field experiments that larvae of the bullfrog and the green frog are strong competitors (Werner 1991; E. E. Werner, unpubl.). The bullfrog is more vulnerable to invertebrate and salamander predators than the green frog because it is more active (an illustration of the activity-mediated trade-off; Werner 1991), but it is less palatable to fish than the green frog (table 13.2). The densities of the two suites of predators tend to be inversely related in ponds because of the effects of fish on the other predators (table 13.3; Crowder and Cooper 1982), though invertebrate predators often remain abundant in the extreme shallows of ponds with bluegill. In the absence of fish, invertebrate predators are often more abundant in deeper submerged vegetation than in the shallows (E. E. Werner and M. A. Peek, unpubl.).

Both anuran larvae respond spatially to predators, but do so differently in a way consistent with their relative vulnerabilities to predators. Laboratory experiments showed that bullfrogs when 1–2 g in size shift away from the major invertebrate predator in the ponds, *Anax junius*, but not from bluegill, whereas 1–2 g green frogs respond in the opposite way (fig. 13.4; E. E. Werner and M. A. McPeek, unpubl.). (Green frogs do show spatial responses to *Anax* when smaller: see Werner 1991). Both species are found in very shallow water after hatching and move into deeper water as they grow larger. The size at which these ontogenetic shifts occur, however,

Table 13.2   Results of Predator Preference Experiments

PREDATOR	n	MANLY'S INDEX	t-VALUE
Lepomis	7	0.93 ± 0.05	8.43***
Ambystoma	16	0.35 ± 0.07	2.18*
Anax	10	0.16 ± 0.03	10.76***

Source: E. E. Werner and M. A. McPeek, unpubl.
Note: Predators were offered equal numbers of size-matched bullfrog and green frog larvae. Data are Manly's index for the case in which prey are depleted (± SE; Chesson 1983), calculated for preference for green frog larvae. *n* is the number of replicate experiments; *t* values are for two-sided *t*-tests that Manly's index was different from random selection (i.e., Manly's index = 0.5). Predators were bluegill sunfish (*Lepomis macrochirus*), tiger salamanders (*Ambystoma tigrinum*), and a dragonfly larva (*Anax junius*).
* $P < .05$
*** $P < .001$

Table 13.3   Number (± SE) of Potential Invertebrate and
Salamander Predators on Anuran Larvae in Permanent Ponds
With and Without Bluegill

	WITH BLUEGILL	WITHOUT BLUEGILL
Invertebrate predators		
Odonata		
Aeshnidae	0.7 ± 0.7	235.0 ± 75.9
Libellulidae	2.4 ± 1.4	142.8 ± 39.3
Hemiptera		
Notonectidae	2.4 ± 1.4	320.2 ± 71.8
Other Hemiptera[a]	0.7 ± 0.7	7.9 ± 2.1
Coleoptera		
Dytiscidae	0.7 ± 0.7	45.9 ± 28.0
Annelidae		
Hirudinae	0.0	94.9 ± 53.1
Salamander predators		
Ambystoma tigrinum	0.0	6.8 ± 2.9
Notophthalmus viridescens	0.0	17.5 ± 9.5

Source: E. E. Werner and M. A. McPeek, unpubl.
Note: Numbers are means across ponds of the mean of two replicate hauls of a 16.2-m
seine in each pond. Three permanent ponds with bluegill and four without bluegill were
sampled for invertebrate predators; three ponds with bluegill and eleven without were sampled for salamander predators.
[a] Belostomatidae, Naucoridae, Nepidae

appears to vary with predator types and densities. In a preliminary experiment I
added bluegill to one side of a divided pond. On the side without fish, both anuran
species were found in very shallow water until they were quite large, presumably
in response to the higher levels of invertebrate predators in deeper water. On the
side with fish, however, the bullfrog tadpoles migrated to the center of the pond
after the bluegill had reduced invertebrate predators there, whereas the green frog
tadpoles remained in the shallows (E. E. Werner and M. A. McPeek, unpubl.). Thus
the differences in the spatial (depth) distribution of the other predators in the presence and absence of fish greatly affected tadpole habitat overlap. Due to their behavioral responses, the two species overlapped in habitat use for much of the life cycle
in the absence of bluegill, whereas the two species segregated early in the life cycle
in the presence of bluegill.

These are preliminary results, and we are presently repeating this experiment
and attempting to estimate the relative importance of the effects of behavioral responses compared with direct mortality effects on competitive interactions between
these two species. But such differential responses by prey to predators would appear
to hold much potential to alter competitive interactions among prey. In terms of the
theory presented in figure 13.3, if invertebrate predators and salamanders are more
abundant in deeper regions, the switching curves for the two species would be
similar, and both would be expected to use shallow water until quite large. Predators
in this case enforce overlap in habitat use. If fish remove other predators from
deeper regions, the switching curves for the two species rotate in opposite directions
because of their differences in vulnerabilities. The addition of fish therefore can

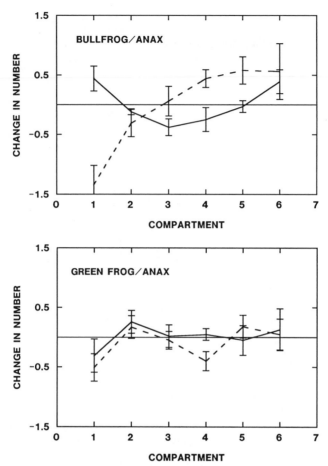

Fig. 13.4 Mean (± SE) change in number of bullfrog (*left*) and green frog (*right*) larvae in each of six linearly arranged compartments of control (*solid line*) and experimental (*dashed line*) aquaria after addition of caged bluegill (*Lepomis macrochirus*) or *Ajax junius* to one end (compartment 1, randomly determined) of the experimental aquaria. Treatments were replicated 10 (bluegill) or 11 (*Anax*) times. Data are numbers in each compartment after minus before predator addition. The bullfrog shift in

enforce marked habitat segregation of the species. These sorts of responses illustrate the potential complexity of the higher-order effects arising from species responding to predators and their distributions according to unique (i.e., species-specific) u/g relations.

Finally, there may be interactions mediated through both the activity schedule and habitat choice trade-offs that determine how activity in different habitats varies over some time horizon. For example, there is growing evidence that the diel vertical migration of some zooplankton can be behaviorally mediated in the presence of predators or their exudates (e.g., Bollens and Frost 1989; Leibold 1990; Neill 1990, 1992; Ringelberg 1991). Consider the system in small temperate lakes where two important cladoceran zooplankton are *Daphnia pulicaria* and *D. galeata*. *D. pulicaria* is generally larger, more visible, and therefore more vulnerable to fish. The two

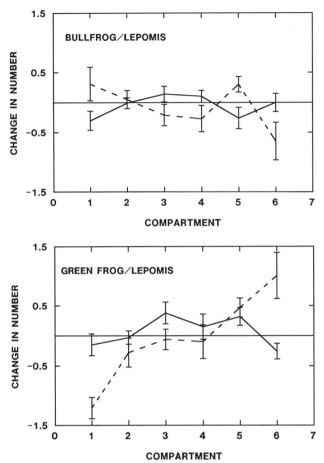

the presence of *Anax* and the green frog shift in the presence of bluegill were significant (multivariate analysis of variance, $F_{5,15} = 3.86$, $P = .02$ and $F_{5,14} = 5.5$, $P = .005$ respectively; response vector was defined as the number of respective species in the first five compartments, as the sixth was linearly dependent). The remaining treatments were not significant. (From E. E. Werner and M. A. McPeek, unpubl.)

species often compete (Leibold 1991). At low fish predation levels, *D. pulicaria* appears to be competitively superior and does not migrate extensively (Leibold and Tessier 1991). At high fish predation levels, *D. pulicaria* migrates to the hypolimnion during the day, whereas *D. galeata* migrates less extensively. The latter change appears to swing the competitive balance to *D. galeata* because it can sustain populations in the food-rich epilimnion at lower resource levels in the presence of high fish populations than can *D. pulicaria*. The phantom midge *Chaoborus* is a common invertebrate predator in such systems; it strongly migrates (to the surface at night, down during the day) in the presence of fish. The migration of both *Chaoborus* and *D. pulicaria* to the hypolimnion to avoid fish exposes the latter to greater *Chaoborus* predation. In contrast to fish, *Chaoborus* is negatively size-selective. Because of their habitat responses to predators, then, the two zooplankters may be under very different selective pressures on body size, which reinforces their differential vulnerability to the two predator types (Leibold and Tessier 1991). If migratory behavior is to some extent facultative, which seems to be the case with *D. pulicaria* (Leibold 1990), higher-order interactions of the sort postulated for the bullfrog/green frog system above are likely.

As in the previous examples, ontogeny may figure prominently in such interactions. Juvenile zooplankton often do not migrate as extensively as the adults, presumably because they achieve the same measure of protection from fish (i.e., lower visibility) as do smaller species. Thus inducible ontogenetic changes in diel habitat use in the presence of predators could greatly affect species interactions. Neill (1992) has experimentally demonstrated that induced diel vertical migration of a copepod (*Diaptomus kenai*) changes dramatically over ontogeny. Nauplii showed no migratory behavior and remained at the surface. Copepodids exhibited strong reverse diel vertical migration (migration to the surface during the day) in the presence of *Chaoborus*. This induced migration (by chemical exudates of *Chaoborus*) occurred through adulthood in copepods from a lake with no fish, but ceased at adulthood in copepods from a lake with fish. The latter population underwent the normal diel vertical migration (to the surface at night) when adult (i.e., when vulnerable to fish). Thus, ontogenetic patterns of diel migration are inducible and are locally adapted to the suite of size/stage-selective predators affecting each population. As in the case of the fish and amphibian systems above, marked behavioral responses to predators that alter the time or ontogenetic schedules of activity in different habitats may have large higher-order effects on competitive interactions in this system. This will especially be the case when species differ in their relative vulnerabilities to different predators, as seen with the anuran and *Daphnia* species. To my knowledge no experiments have yet been conducted to address the higher-order effects in this system.

Synthesis. The above examples from three different taxa (fish, frogs, and zooplankton) illustrate the potential of behavioral responses to alter the interaction of competition and predation in these taxa. Each example also illustrates how ontogenetic constraints on behavioral expression critically affect these interactions. There is little theoretical or empirical work exploring the consequences of these individual behavioral responses for population-level phenomena. The available theory (e.g., Abrams 1991b) suggests that such responses could be quite important in changing equilibrial population densities and even qualitatively changing the direction of responses expected from direct effects. The empirical studies are few, and typically have measured responses in terms of components of fitness (e.g., growth, size at metamorphosis) rather than population responses. These effects often are quite large, but at this point it is difficult to directly assess the importance of these interactions to population dynamics. We can only say that the adaptive behavioral responses are often striking, and that they can have large effects on individual performance.

The responses that I have reviewed have important implications for the experimental protocol of community ecologists. Bender et al. (1984) discuss the manner in which the direct effects of species in a complex community can be experimentally determined. They argue that these effects can be measured by responses occurring shortly after a perturbation, that is, before indirect effects due to changes in density are propagated through the food web and confound measurement of direct effects. The behavioral effects that I have examined, however, occur on the same scale on which direct effects are operating. Abrams (1991b) argues that if higher-order interactions due to adaptive behavioral responses are important, then there will be no separation of time scale, and any perturbation will reflect both direct and indirect effects. For example, because the activity response of frog larvae to predators is dose dependent (E. E. Werner, unpubl.), any perturbation of predator densities affects competitive interactions simultaneously through direct effects altering relative and absolute densities of species and higher-order effects altering relative competitive abilities (cf. fig. 13.2). These effects need to be experimentally separated to evaluate their relative importance.

Past attempts to evaluate higher-order effects have often confused higher-order interactions with longer-term, numerically propagated indirect effects. For example, significant interaction terms in analysis of variance were often perceived as equivalent to higher-order interactions in experiments examining the effects of a species on the interaction of two others (Worthen and Moore 1991). However, indirect effects mediated through density changes can result in significant interaction terms, with no change in per capita effects. This is particularly likely because most such analyses are of fairly long-term effects in which density changes of intermediate species occur. In order for an interaction term to indicate a higher-order effect, we must first remove any effects of intraspecific density changes, or the analyses must be of short-term per capita effects. Worthen and Moore (1991) present a schema in the context of analysis of variance tests to distinguish indirect from higher-order effects. Because these effects have so often been confused in the literature, it is difficult to evaluate their relative importance. Morin et al. (1988) partitioned the variance explained in their experiments due to direct and higher-order effects and found that the higher-order effects are significant, but account for far less of the variance than the direct effects. Wilbur and Fauth (1990) found that predictability of complex community responses from pairwise experiments is often poor, and present several methods for evaluating indirect effects in food webs.

The potential importance of higher-order interactions is perhaps most easily assessed in the context of known mechanisms, as I have attempted to do here. It is then possible to speculate on the exact nature of the higher-order effects and design appropriate experiments to test the mechanisms. In the case of the predator effects, for example, prey densities can be held constant and the perception of risk manipulated so that we can quantify the change in per capita effects of prey on each other. We can then systematically conduct experiments adding the direct effects and evaluate the relative importance of different components of the interaction. This approach largely circumvents problems that arise when higher-order interactions are deduced through lack of fit to a general additive model (mathematical or statistical; e.g., controversies over whether lack of fit to additive models is accounted for by nonlinear intraspecific density dependence: see Case and Bender 1981). Further, deductions arising from the lack of fit to additive models are likely to leave us with a catalogue of presence and absence of nonadditive effects, without a conceptual framework for attacking these questions. Approached as suggested above, mechanisms can be directly incorporated into theory so that specific predictions can be generated to guide experimentation. The interface between behavioral ecology and population biology can be exploited to a great extent in this enterprise.

THE ACTIVE ROLE OF BEHAVIOR IN EVOLUTIONARY CHANGE

I have discussed how trade-offs mediated through activity level and habitat characteristics affect adaptive behaviors, and the consequences of these behaviors for species interactions. I now briefly discuss how these behaviors additionally have the potential to influence the evolution of species' characteristics and life cycle forms, particularly when growth over ontogeny interacts with these trade-offs to constrain behavioral expression.

Consider the facultative changes in size at metamorphosis in amphibians. Wilbur and Fauth (1990), in a series of experiments, noted that size at metamorphosis in the American toad (*Bufo americanus*) was smaller in the presence of a predator. Skelly and Werner (1990) demonstrated that this reduction in size was a facultative response to predators, as I theoretically proposed (Werner 1986). In the nonlethal (caged) presence of an odonate predator (*Anax junius*), the American toad metamorphosed at a significantly smaller size. Toad larvae reduced activity, and therefore growth rates, in the presence of the predator, which in turn appeared to be the predominant mechanism affecting size at metamorphosis (Skelly and Werner 1990). Activity responses due to the growth rate/risk trade-off therefore can affect the organization of the life cycle. Changes in size at metamorphosis can have extensive effects on fitness in frogs (Berven and Gill 1983; Smith 1987), and clearly the time (sizes) spent in each stage will affect species interactions with the different organisms encountered in the two stages (and therefore will have higher-order effects). In many ways this response parallels the ontogenetic habitat shifts of the bluegill. I expect that nonlethal effects of predators on life histories will be widespread; they have been documented thus far, for example, in amphibians, snails (Crowl and Covich 1990), and zooplankton (Dodson and Havel 1988).

How might such behavioral responses affect the evolution of life cycles? That is, what are the relations between behavior, development, and evolution? Recently, considerable attention has been given to the relation between ontogeny and evolution, with most studies stressing the constraining role of ontogeny or development on evolution (Maynard Smith et al. 1985). Ontogeny, however, can also focus evolution or facilitate change. As Bateson (1988) notes, when developmental issues are coupled to questions about evolution, it becomes much easier to perceive how an organism's behavior can initiate and direct lines of evolution. For example, behaviors expressed during development can extensively influence morphology (West-Eberhard 1989). There are examples of food choice behavior interacting with developmental systems to alter the morphological phenotype (Meyer 1987; Wimberger 1991). These are not genetic changes, but alter the selective context the animal experiences so that selection is focused on genetic changes contributing to fitness in relation to these alterations.

There are numerous examples in which the adoption of a novel behavior has led to the subsequent evolution of elaborate morphological features in support of that behavior (e.g., Wcislo 1989). That is, the behavioral choices of an animal can diminish or exacerbate environmental heterogeneity, or determine the selective environment that the animal experiences, and therefore generate evolutionary novelty (see recent reviews in Bateson 1988; Wcislo 1989; West-Eberhard 1989). Behavior can also change physical and social conditions or expose animals to novel conditions that may reveal heritable variability (Bateson 1988). Though the role of behavior (habitat choice) has been widely discussed in the context of sympatric speciation (e.g., Diehl and Bush 1989) and multiple-niche polymorphism questions in population genetics (e.g., Endler 1986), it has rarely been examined in regard to ontogeny and life cycle phenomena.

Animals exercising habitat choice over ontogeny expose themselves to different patterns of natural selection for ecologically important traits at different stages during development. There are numerous examples of developmental programs cued by the environment that alter the nature of the organism (i.e., polyphenisms). Polyphenisms (or polymorphisms) illustrate the malleability of the developmental program as distinct from its constraining properties. If such polymorphisms are consistently evoked in certain environments, this can focus selection and limit the polyphenisms, for example, by exacerbating differentiation among populations. Phenotypic expressions thus may enhance the independent evolution of alternative tactics. In some cases these alternative tactics can exhibit phenotypes with a macroevolutionary degree of divergence (West-Eberhard 1989), which is certainly the case in many complex life cycles. The size-specific use of particular habitats would seem to hold considerable potential for this sort of feedback on evolution of life cycle structure. If the sorts of trade-offs explored above cause an organism to switch habitats with size (e.g., the bluegill), this behavior may reliably focus selection on alterations during development that increase fitness in the respective habitats. That is, an organism that makes a habitat shift at a consistent size during ontogeny focuses selection on the developmental program to alter the phenotype at the transition size such that the phenotype better matches the subsequent environment. Therefore, ontogenetic constraints on behavioral decisions may induce differentiation in the developmental program and affect evolution of the life cycle. I have proposed a general hypothesis for the evolution of complex life cycles based on the constraints imposed by the scaling of ecological factors with ontogenetic increases in size (Werner 1988). This scaling can affect the expression of certain behaviors over the course of development and lead to strong ontogenetic allometries in behavior. Behavior, then, may be an important mechanism facilitating or directing the evolution of aspects of life cycle form.

CONCLUSIONS

I have examined two major classes of adaptive behavioral choices, activity level and habitat use, in relation to trade-offs between resource acquisition and mortality risk. A large number of behavioral ecology studies demonstrate adaptive alterations in the behavior of animals in response to various factors (e.g., Krebs and Davies 1984), and especially to the presence of predators (e.g., Lima and Dill 1990). The large and reliable effects of predators on prey behavior indicate that behavioral responses will alter interactions of the prey with their resources, other prey, and other predators. The examples presented suggest that such behavioral responses will lead to qualitative modification of species interactions over very short time scales, but we have few experiments that explicitly address the higher-order effects of such behaviors on interactions among species. Because such behavioral responses are occurring on the same time scale as the direct effects of interest, they must be considered simultaneously with the direct effects; that is, they cannot be abstracted (sensu Schaffer 1981). It appears unlikely that we will have a predictive theory of species interactions in many systems without explicitly incorporating adaptive behavioral mechanisms in the trophic models that dominate population and community biology. This speculation remains tentative, however, as there are few studies that evaluate the relative importance of higher-order effects, especially with vertebrates, where behavioral responses tend to be more sophisticated.

The trade-offs influencing the behaviors that I have examined further indicate how developmental processes are inextricably entwined with species interactions at all levels. The ecological circumstances of an animal change dramatically over ontogeny due to the enormous changes in size that occur (Werner and Gilliam 1984). Thus behavioral responses (or constraints on those responses) change over ontogeny, and these changes in responses can be expected to be as great as the differences occurring between adults of different species. Ontogenetic shifts in habitat are a case in point; the ontogenetic constraints on behavioral responses dramatically alter the interactions that occur between many species. Increasing numbers of these ontogenetic shifts (as well as life history traits) can be shown to be facultative responses to factors such as presence of predators. The potential for these responses to lead to higher-order interactions therefore appears to be widespread, and suggests that developmental responses must be considered in species interaction theory.

Attention to developmental phenomena also will help clarify the active role of behavior in the evolution of species characteristics and life cycle phenomena. I have argued that fundamental life cycle properties such as metamorphosis and complex life cycles may be the result of the size scaling of ecological interactions over ontogeny (Werner 1986, 1988). Alterations in species interactions over ontogeny and the related behavioral responses may have large effects on selection on life cycle phenomena. Study of these scaling relations with respect to behavior and development should provide insight into the adaptive significance and origins of many life cycle features.

I have stressed that in the case of both activity level and habitat choice we have theory at the behavioral ecology level that attempts to predict individual responses. This theory affords the opportunity for a rich interaction between theory at the individual and at the population interaction level. There have been recent calls for a more mechanistic approach to population and community biology (Schoener 1986; Tilman 1987), and in many systems the mechanisms of species interactions have large behavioral components. At some level, then, behavioral ecology theory should provide a mechanistic basis for theories of community interactions. We should attempt to identify common population consequences that follow from specific, but generalizable, behaviors. It is rare, however, for ecologists to map the consequences of a behavior onto population and/or community phenomena. The behavioral models provide direction for the appropriate experiments to discover higher-order interactions, and the experiments then provide a forum for testing the behavioral models as well.

The importance of higher-order interactions in ecological communities remains largely an empirical question. Thought experiments can generate a host of plausible mechanisms for the existence of higher-order interactions. The examples that I have presented in this chapter suggest that higher-order interactions could be very significant in ecological communities, but in general the data to assess this proposition are meager. Careful experiments that show quantitatively how important such interactions are to the resolution of species interactions will have a large influence on the direction of both theory and experiment in ecology.

ACKNOWLEDGMENTS

I thank Les Real for the opportunity to participate in the American Society of Naturalists' symposium prompting this chapter. I am indebted to P. Abrams, G. Mittelbach, S. Naeem, L. Real, D. Skelly, and G. Wellborn, whose comments greatly improved the chapter. Preparation of the chapter was supported by National Science Foundation Grant DEB-9119948.

REFERENCES

Abrams, P. A. 1982. Functional responses of optimal foragers. *American Naturalist* 120:382–90.

Abrams, P. A. 1983. Arguments in favor of higher order interactions. *American Naturalist* 121: 887–91.

Abrams, P. A. 1984. Foraging time optimization and interactions in food webs. *American Naturalist* 124:80–96.

Abrams, P. A. 1990. The effects of adaptive behavior on the type-2 functional response. *Ecology* 71:877–85.

Abrams, P. A. 1991a. Life history and the relationship between food availbility and foraging effort. *Ecology* 72:1242–52.

Abrams, P. A. 1991b. Strengths of indirect effects generated by optimal foraging. *Oikos* 62: 167–76.

Abrams, P. A., and L. Shen. 1989. Population dynamics of systems with consumers that maintain a constant ratio of intake rates of two resources. *Theoretical Population Biology* 35:51–89.

Bateson, P. 1988. The active role of behavior in evolution. In W. M. Ho and S. W. Fox, eds., *Evolutionary processes and metaphors*, 190–207. New York: Wiley.

Bender, E. A., T. J. Case, and M. E. Gilpin. 1984. Perturbation experiments in community ecology: Theory and practice. *Ecology* 65:1–13.

Berven, K. A., and D. E. Gill. 1983. Interpreting geographic variation in life history traits. *American Zoologist* 23:85–97.

Bollens, S. M., and B. W. Frost. 1989. Predator-induced diel vertical migration in a planktonic copepod. *Journal of Plankton Research* 11:1047–65.

Brooks, J. L., and S. I. Dodson. 1965. Predation, body size, and composition of plankton. *Science* 150:28–35.

Brown, J. H., D. W. Davidson, J. C. Munger, and R. E. Inouye. 1986. Experimental community ecology: The desert granivore system. In J. Diamond and T. J. Case, eds., *Community ecology*, 41–61. New York: Harper and Row.

Brown, J. S. 1988. Patch use as an indicator of habitat preference, predation risk, and competition. *Behavioral Ecology and Sociobiology* 22:37–47.

Brown, J. S. 1990. Habitat selection as an evolutionary game. *Evolution* 44:732–46.

Calder, W. A. 1984. *Size, function, and life history*. Cambridge, Mass.: Harvard University Press.

Carpenter, S. R., and J. F. Kitchell. 1988. Consumer control of lake productivity. *BioScience* 38:764–69.

Carpenter, S. R., J. F. Kitchell, J. R. Hodgson, P. A. Cochran, J. J. Elser, M. M. Elser, D. M. Lodge, D. Kretchmer, and X He. 1987. Regulation of lake primary productivity by food web structure. *Ecology* 68:1863–76.

Case, T. J., and E. A. Bender. 1981. Testing for higher order interactions. *American Naturalist* 118:920–29.

Charnov, E. L. 1976. Optimal foraging: The marginal value theorem. *Theoretical Population Biology* 9:129–36.

Chesson, J. 1983. The estimation and analysis of preference and its relation to foraging models. *Ecology* 64:1297–1304.

Connell, J. H. 1983. On the prevalence and relative importance of interspecific competition: Evidence from field experiments. *American Naturalist* 122:661–96.

Crowder, L. B., and W. E. Cooper. 1982. Habitat structural complexity and the interaction between bluegills and their prey. *Ecology* 63:1802–13.

Crowl, T. A., and A. P. Covich. 1990. Predator-induced life-history shifts in a freshwater snail. *Science* 247:949–51.

Dethier, M. N., and D. O. Duggins. 1984. An "indirect commensalism" between marine herbivores and the importance of competitive hierarchies. *American Naturalist* 124:205–19.

Diamond, J., and T. J. Case. 1986. *Community ecology*. New York: Harper and Row.

Diehl, S. R., and G. L. Bush. 1989. The role of habitat preference in adaptation and speciation. In D. Otte and J. A. Endler, eds., *Speciation and its consequences*, 345–65. Sunderland, Mass.: Sinauer Associates.

Dodson, S. I., and J. E. Havel. 1988. Indirect prey effects: Some morphological and life history responses of *Daphnia pulex* exposed to *Notonecta undulata*. *Limnology and Oceanography* 33: 1274–85.

Dunbrack, R. L., and L. A. Giguere. 1987. Adaptive responses to accelerating costs of movement: A bioenergetic basis for the type-III functional response. *American Naturalist* 130: 147–60.

Dungan, M. L. 1986. Three-way interactions: Barnacles, limpets, and algae in a Sonoran Desert rocky intertidal zone. *American Naturalist* 127:292–316.

Edelstein-Keshet, L., and M. Rausher. 1989. The effects of individual plant defenses on herbivore populations. I. Mobile herbivores in continuous time. *American Naturalist* 133: 787–810.

Ehlinger, T. J. 1990. Habitat choice and phenotype-limited feeding efficiency in bluegill: Individual differences and trophic polymorphism. *Ecology* 71:886–96.

Endler, J. A. 1986. *Natural selection in the wild*. Princeton: Princeton University Press.

Feener, D. H. 1981. Competition between ant species: Outcome controlled by parasitic flies. *Science* 214:815–17.

Fretwell, S. D., and H. L. Lucas. 1970. On territorial behavior and other factors influencing habitat distribution in birds. *Acta Biotheoretica* 19:16–36.

Gilliam, J. F. 1982. Habitat use and competitive bottlenecks in size-structured fish populations. Ph.D. dissertation, Michigan State University, East Lansing.

Gilliam, J. F. 1987. Individual behavior and population dynamics. *Ecology* 68:456–57.

Gilliam, J. F., and D. F. Fraser. 1987. Habitat selection under predation hazard: Test of a model with foraging minnows. *Ecology* 68:1856–62.

Gilliam, J. F., and D. F. Fraser. 1988. Resource depletion and habitat segregation by competitors under predation hazard. In Ebenmann, B., and L. Persson, eds., *Size-structured populations*, 173–84. Berlin: Springer-Verlag.

Hairston, N. G., F. E. Smith, and L. B. Slobodkin. 1960. Community structure, population control, and competition. *American Naturalist* 94:421–25.

Holt, R. D. 1977. Predation, apparent competition, and the structure of prey communities. *Theoretical Population Biology* 12:197–229.

Holt, R. D. 1984. Spatial heterogeneity, indirect interactions, and the coexistence of prey species. *American Naturalist* 124:377–406.

Huang, C., and A. Sih. 1990. Experimental studies on behaviorally mediated, indirect interactions through a shared predator. *Ecology* 71:1515–22.

Ives, A. R., and A. P. Dodson. 1987. Antipredator behavior and the population dynamics of simple predator-prey systems. *American Naturalist* 130:431–47.

Jeffries, M. J., and J. H. Lawton. 1984. Enemy-free space and the structure of ecological communities. *Biological Journal of the Linnean Society* 23:269–86.

Jungers, W. L., ed. 1985. *Size and scaling in primate biology*. New York: Plenum Press.

Kerfoot, W. C. 1977. Competition in cladoceran communities: The cost of evolving defenses against copepod predation. *Ecology* 58:303–13.

Kerfoot, W. C., and R. A. Pastorok. 1978. Survival versus competition: Ecolutionary compromises and diversity in the zooplankton. *Verhandlungen Internationale Vereinigung Limnologie* 20:362–74.

Kerfoot, W. C., and A. Sih. 1987. *Predation: Direct and indirect impacts on aquatic communities*. Hanover, N.H.: University Press of New England.

Kohler, S. L., and M. A. McPeek. 1989. Predation risk and the foraging behavior of competing stream insects. *Ecology* 70:1811–25.

Kotler, B. P. 1984. Risk of predation and the structure of desert rodent communities. *Ecology* 65:689–701.

Kotler, B. P., and R. D. Holt. 1989. Predation and competition: The interaction of two types of species interactions. *Oikos* 54:256–60.

Kotler, B. P., J. S. Brown, and O. Hasson. 1991. Factors affecting gerbil foraging behavior and rates of owl predation. *Ecology* 72:2249–60.

Krebs, J. R., and N. B. Davies. 1984. *Behavioral ecology.* Oxford: Blackwell Scientific Publications.

Lawler, S. P. 1989. Behavioral responses to predators and predation risk in four species of larval anurans. *Animal Behaviour* 38:1039–47.

Lawlor, L. R., and J. Maynard Smith. 1976. The coevolution and stability of competing species. *American Naturalist* 110:79–99.

Leibold, M. A. 1988. Habitat structure and species interactions in plankton communities of stratified lakes. Ph.D. dissertation, Michigan State University, East Lansing.

Leibold, M. A. 1990. Resources and predators can affect the vertical distributions of zooplankton. *Limnology and Oceanography* 35:938–44.

Leibold, M. A. 1991. Trophic interactions and habitat segregation between competing *Daphnia*. *Oecologia* 86:510–20.

Leibold, M. A., and A. J. Tessier. 1991. Contrasting patterns of body size for *Daphnia* species that segregate by habitat. *Oecologia* 86:342–48.

Levine, S. H. 1976. Competitive interactions in ecosystems. *American Naturalist* 110:903–10.

Lima, S. L. 1985. Maximizing feeding efficiency and minimizing time exposed to predators: A trade-off in the black-capped chickadee. *Oecologia* 66:60–67.

Lima, S. L., and L. M. Dill. 1990. Behavioral decisions made under the risk of predation: A review and prospectus. *Canadian Journal of Zoology* 68:619–40.

Lima, S. L., and T. J. Valone. 1991. Predators and avian community organization: An experiment in a semi-desert grassland. *Oecologia* 86:105–12.

Lively, C. M. 1986. Canalization versus developmental conversion in a spatially heterogeneous environment. *American Naturalist* 128:561–72.

Lubchenco, J. 1978. Plant species diversity in a marine intertidal community: Importance of herbivore food preference and algal competitive abilities. *American Naturalist* 112:23–39.

MacArthur, R. H., and E. R. Pianka. 1966. On optimal use of a patchy environment. *American Naturalist* 100:603–9.

McPeek, M. A. 1990a. Behavioral differences between *Enallagma* species (Odonata) influencing differential vulnerability to predators. *Ecology* 71:714–26.

McPeek, M. A. 1990b. Determination of species composition in the *Enallagma* damselfly assemblages of permanent lakes. *Ecology* 71:83–98.

McQueen, D. J., M. R. S. Johannes, J. R. Post, T. J. Stewart, and D. R. S. Lean. 1989. Bottom-up and top-down impacts on freshwater pelagic community structure. *Ecological Monographs* 59:289–309.

Mangel, M., and C. W. Clark. 1988. *Dynamic modeling in behavioral ecology.* Princeton: Princeton University Press.

Maynard Smith, J., R. Burian, S. Kauffman, P. Alberch, J. Campbell, B. Goodwin, R. Lande, D. Raup, and L. Wolpert. 1985. Developmental constraints and evolution. *Quarterly Review of Biology* 60:265–87.

Meyer, A. 1987. Phenotypic plasticity and heterochrony in *Cichlasoma managuense* (Pisces, Cichlidae) and their implications for speciation in cichlid fishes. *Evolution* 41:1357–69.

Miller, T. E., and W. C. Kerfoot. 1987. Redefining indirect effects. In W. C. Kerfoot and A. Sih, eds., *Predation: Direct and indirect impacts on aquatic communities,* 33–37. Hanover, N.H.: University Press of New England.

Mittelbach, G. G. 1981. Foraging efficiency and body size: A study of optimal diet and habitat use by bluegills. *Ecology* 62:1370–86.

Mittelbach, G. G. 1988. Competition among refuging sunfishes and effects of fish density on littoral zone invertebrates. *Ecology* 69:614–23.

Mittelbach, G. G., and P. L. Chesson. 1987. Indirect effects on fish populations. In W. C.

Kerfoot and A. Sih, eds., *Predations: Direct and indirect impacts on aquatic communities*, 315–32. Hanover, N.H.: University Press of New England.

Morin, P. J. 1987. Salamander predation, prey facilitation, and seasonal succession in microcrustacean communities. In W. C. Kerfoot and A. Sih, eds., *Predation: Direct and indirect impacts on aquatic communities*, 174–87. Hanover, N.H.: University Press of New England.

Morin, P. J., S. P. Lawler, and E. A. Johnson. 1988. Competition between aquatic insects and vertebrates: Interaction strength and higher order interactions. *Ecology* 69:1401–9.

Neill, W. E. 1974. The community matrix and interdependence of the competition coefficients. *American Naturalist* 108:399–408.

Neill, W. E. 1990. Induced vertical migration in copepods as a defence against invertebrate predation. *Nature* 345:524–26.

Neill, W. E. 1992. Developmental and geographical variation in predator induced vertical migration of copepods. *Nature* 356:54–57.

Osenberg, C. W., E. E. Werner, G. G. Mittelbach, and D. J. Hall. 1988. Growth patterns in bluegill (*Lepomis macrochiris*) and pumpkinseed (*L. gibbosus*) sunfish: Environmental variation and the importance of ontogenetic niche shifts. *Canadian Journal of Fisheries and Aquatic Sciences* 45:17–26.

Osenberg, C. W., G. G. Mittelbach, and P. C. Wainwright. 1992. Two-stage life histories in fish: The interaction between juvenile competition and adult performance. *Ecology* 73:255–67.

Paine, R. T. 1966. Food web complexity and species diversity. *American Naturalist* 100:65–75.

Parker, G., and W. Sutherland. 1986. Ideal free distributions when individuals differ in competitive ability: Phenotype-limited ideal free models. *Animal Behaviour* 34:1222–42.

Persson, L. 1991. Behavioral response to predators reverses the outcome of competition between species. *Behavioral Ecology and Sociobiology* 28:101–5.

Peters, R. H. 1983. *The ecological implications of body size.* Cambridge: Cambridge University Press.

Power, M. E., W. J. Mathews, and A. J. Stewart. 1985. Grazing minnows, piscivorous bass, and stream algae: Dynamics of a strong interaction. *Ecology* 66:1448–56.

Pyke, G. H. 1981. Optimal travel speeds of animals. *American Naturalist* 118:475–87.

Ringelberg, J. 1991. Enhancement of the phototactic reaction of *Daphnia hyalina* by a chemical mediated by juvenile perch (*Perca fluviatailis*). *Journal of Plankton Research* 13:17–25.

Rosenzweig, M. L. 1987. Habitat selection as a source of biological diversity. *Evolutionary Ecology* 1:315–30.

Rosenzweig, M. L. 1991. Habitat selection and population interactions: The search for mechanisms. *American Naturalist* 137:S5–S38.

Roughgarden, J., and J. Diamond. 1986. Overview: The role of species interactions in community ecology. In J. Diamond and T. J. Case, eds., *Community ecology*, 333–43. New York: Harper and Row.

Schaffer, W. M. 1981. Ecological abstraction: The consequences of reduced dimensionality in ecological models. *Ecological Monographs* 51:383–401.

Schmidt-Nielsen, K. 1984. *Scaling: Why is animal size so important?* Cambridge: Cambridge University Press.

Schmitt, R. J. 1987. Indirect interactions between prey: Apparent competition, predator aggregation, and habitat segregation. *Ecology* 68:1887–97.

Schmitt, R. J., and S. J. Holbrook. 1985. Patch selection by juvenile black surfperch (Embiotochidae) under variable risk: Interactive influence of food quality and structural complexity. *Journal of Experimental Marine Biology and Ecology* 85:269–85.

Schoener, T. W. 1974. Resource partitioning in ecological communities. *Science* 185:27–39.

Schoener, T. W. 1983. Field experiments on interspecific competition. *American Naturalist* 122:240–85.

Schoener, T. W. 1986. Mechanistic approaches to community ecology: A new reductionism? *American Zoologist* 26:81–106.

Schwinning, S., and M. L. Rosenzweig. 1990. Periodic oscillations in an ideal-free predator-prey system. *Oikos* 59:85–91.

Sebens, K. P. 1987. The ecology of indeterminate growth in animals. *Annual Review of Ecology and Systematics* 18:371–407.

Sih, A. 1987. Predators and prey lifestyles: An evolutionary and ecological overview. In W. C. Kerfoot and A. Sih, eds., *Predation: Direct and indirect impacts on aquatic communities*, 203–24. Hanover, N.H.: University Press of New England.

Skelly, D. K., and E. E. Werner. 1990. Behavioral and life historical responses of larval American toads to an odonate predator. *Ecology* 71:2313–22.

Smith, D. C. 1987. Adult recruitment in chorus frogs: Effects of size and date of metamorphosis. *Ecology* 68:344–50.

Stamps, J. A. 1983. The relationship between ontogenetic habitat shifts, competition and predator avoidance in a juvenile lizard (*Anolis aeneus*). *Behavioral Ecology and Sociobiology* 12: 19–33.

Stein, R. A., and J. J. Magnuson. 1976. Behavioral response of crayfish to a fish predator. *Ecology* 57:751–61.

Stemberger, R. S., and J. J. Gilbert. 1987. Defenses of planktonic rotifers against predators. In W. C. Kerfoot and A. Sih, eds. *Predation: Direct and indirect impacts on aquatic communities*, 227–39. Hanover, N.H.: University Press of New England.

Stephens, D. W., and J. R. Krebs. 1986. *Foraging theory*. Princeton: Princeton University Press.

Tilman, D. 1987. The importance of the mechanisms of interspecific competition. *American Naturalist* 129:769–74.

Turner, A. M., and G. G. Mittelbach. 1990. Predator avoidance and community structure: Interactions among piscivores, planktivores, and plankton. *Ecology* 71:2241–54.

Vance, R. R. 1978. Predation and resource partitioning in one predator-two prey model communities. *American Naturalist* 112:797–813.

Vandermeer, J. H. 1969. The competitive structure of communities: An experimental approach with Protozoa. *Ecology* 50:362–72.

Ware, D. M. 1975. Bioenergetics of pelagic fish: Theoretical changes in swimming speed and ration with body size. *Journal of the Fisheries Research Board of Canada* 35:220–28.

Wcislo, W. T. 1989. Behavioral environments and evolutionary change. *Annual Review of Ecology and Systematics* 20:137–69.

Werner, E. E. 1986. Amphibian metamorphosis: Growth rate, predation risk, and the optimal size at transformation. *American Naturalist* 128:319–41.

Werner, E. E. 1988. Size, scaling, and the evolution of complex life cycles. In B. Ebenman and L. Persson, eds., *Size-structured populations: Ecology and evolution*, 60–81. Berlin: Springer-Verlag.

Werner, E. E. 1991. Nonlethal effects of a predator on competitive interactions between two anuran larvae. *Ecology* 72:1709–20.

Werner, E. E. 1992. Competitive interactions between wood frog and northern leopard frog larvae: The role of size and activity. *Copeia* 1992:26–35.

Werner, E. E., and B. Anholt. 1993. Ecological consequences of the tradeoff between growth and mortality rates mediated by foraging activity. *American Naturalist* 142:242–72.

Werner, E. E., and J. F. Gilliam. 1984. The ontogenetic niche and species interactions in size-structured populations. *Annual Review of Ecology and Systematics* 15:393–425.

Werner, E. E., and D. J. Hall. 1976. Niche shifts in sunfishes: Experimental evidence and significance. *Science* 191:404–6.

Werner, E. E., and D. J. Hall. 1979. Foraging efficiency and habitat switching in competing sunfishes. *Ecology* 60:256–64.

Werner, E. E., and D. J. Hall. 1988. Ontogenetic habitat shifts in the bluegill sunfish (*Lepomis macrochirus*): The foraging rate-predation risk tradeoff. *Ecology* 69:1352–66.

Werner, E. E., G. G. Mittelbach, and D. J. Hall. 1981. The role of foraging profitability and experience in habitat use by bluegill sunfish. *Ecology* 62:116–25.

Werner, E. E., J. F. Gilliam, D. J. Hall, and G. G. Mittelbach. 1983a. An experimental test of the effects of predation risk on habitat use in fish. *Ecology* 64:1540–48.

Werner, E. E., G. G. Mittelbach, D. J. Hall, and J. F. Gilliam. 1983b. Experimental tests of optimal habitat use in fish: The role of relative habitat profitability. *Ecology* 64:1525–39.

West-Eberhard, M. J. 1989. Phenotypic plasticity and the origins of diversity. *Annual Review of Ecology and Systematics* 20:249–78.

Wilbur, H. M. 1972. Competition, predation, and the structure of the *Ambystoma-Rana sylvatica* community. *Ecology* 53:3–20.

Wilbur, H. M. 1980. Complex life cycles. *Annual Review of Ecology and Systematics* 11:67–93.

Wilbur, H. M. 1984. Complex life cycles and community organization in amphibians. In P. W. Price, C. N. Slobodchikoff, and W. S. Gaud, eds., *A new ecology: Novel approaches to interactive systems*, 195–224. New York: Wiley.

Wilbur, H. M., and J. E. Fauth. 1990. Experimental aquatic food webs: Interactions between two predators and two prey. *American Naturalist* 135:176–204.

Wimberger, P. 1991. Causes of morphological plasticity in cichlid fishes. Ph.D. thesis. Cornell University, Ithaca, N.Y.

Wootton, J. T. 1993. Indirect effects and habitat use in an intertidal community: Interaction chains and interaction modifications. *American Naturalist* 141:71–89.

Worthen, W. B., and J. L. Moore. 1991. Higher-order interactions and indirect effects: A resolution using laboratory *Drosophila* communities. *American Naturalist* 138:1092–1104.

Yodzis, P. 1988. The indeterminacy of ecological interactions as perceived through perturbation experiments. *Ecology* 69:508–15.

Part IV

HORMONAL PROCESSES

FOURTEEN

Hormones and Life Histories: An Integrative Approach

Ellen D. Ketterson and Val Nolan Jr.

THE OBJECTIVE OF THIS CHAPTER is to describe the potential that studies of hormones have for increasing understanding of the consequences of variation in behavior and physiology and thus understanding of the evolution of life histories, social organization, and mating systems. We begin by reviewing some basic facts about the action of one hormone, testosterone, in birds. Our aim is to illustrate a few principles, including (1) that hormones are secreted in response to stimuli that are generated both internally and externally, (2) that hormones have multiple targets and diverse effects, and (3) that the complexity of hormone action leaves ample room for variation on which natural selection can act.

We then pose four interconnected questions that relate hormones to the evolutionary process and that provide fruitful avenues for research: (1) How might hormones form the mechanistic bases for trade-offs in life histories? (2) What is the role of hormones in the evolution of inter- and intraspecific variation in social organization and mating systems? (3) Does natural selection act on organisms or on traits? (4) What is the evolutionary significance of phenotypic variation in hormonally mediated traits? We give examples of studies from the recent literature that address, or could be used to address, each question.

Finally, we describe an experimental method that we call "phenotypic engineering" (Ketterson et al., 1992) and show how we and others have used this method to address the fourth question, the evolutionary significance of phenotypic variation. Our model organism is a small passerine bird, the dark-eyed junco (*Junco hyemalis*). We have manipulated plasma concentrations of testosterone in free-living male juncos in order to alter their behavioral and physiological phenotypes. Some of the changes we have induced are potentially beneficial, while others are potentially detrimental. Our objective has been to determine the overall effect these phenotypic changes have on fitness.

ABOUT TESTOSTERONE

Testosterone, like other hormones, is remarkable because it is effective in minute quantities and has a wide array of behavioral, physiological, and morphological consequences. To an evolutionary biologist, testosterone may be one of the most

This chapter was previously published in a slightly different form in *American Naturalist* 140, Supplement (November 1992), S33–S62. © 1992 by The University of Chicago.

TARGETS FOR TESTOSTERONE

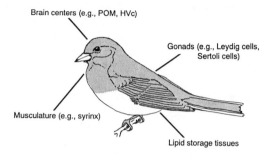

Fig. 14.1   Targets for the hormone testosterone. Although the drawing is of a dark-eyed junco, the targets are taken from studies of a variety of bird species and have not necessarily been identified in the junco.

interesting hormones because it is so intimately associated with reproduction and affects such a wide variety of traits. What is presented here is necessarily oversimplified. (For general reviews of the action and effects of testosterone, see Adler 1981; Knobil and Neill 1988; for treatment of testosterone in birds, see Nalbandanov 1976; Epple and Stetson 1980; Balthazart 1983; and Silver and Ball 1989.)

Testosterone is a steroid hormone secreted by the testes in response to gonadotropins, which are produced by the anterior pituitary gland in response to stimulation by releasing hormones from the hypothalamus. The stimuli that trigger secretion of testosterone are diverse and include day length, aggressive encounters with conspecifics, and female sexual behavior (Wingfield and Moore 1988). Testosterone contributes to its own regulation by suppressing further secretion of releasing hormones and gonadotropins.

Testosterone enters the circulation and influences cellular activity at target tissues in the central nervous system (e.g., many nuclei in the brain: see Balthazart and Ball 1993) and in the periphery (e.g., gonads, syrinx, integument, musculature) (fig. 14.1). Individuals may vary in responsiveness to environmental stimuli, releasing hormones, and gonadotropins; rate of secretion of testosterone; half-life in the circulation; density of receptors at the targets; affinity of receptors for the hormone; and activity of the enzymes responsible for metabolic conversion of the hormone to its active form. Thus, the potential for complex interactions is great and an understanding of these interactions is critical to a full understanding of integrated organismal function. Values for many of the regulatory variables, such as density of receptors, cannot be measured in animals that remain alive in the field, although it is possible to flood receptors with compounds that block hormone action (antiandrogens: e.g., Hegner and Wingfield 1987; Beletsky et al. 1990; Schwabl and Kriner 1991). Fortunately, plasma concentrations of testosterone in free-living animals can be determined relatively easily, and they have been shown to correlate with many functional aspects of behaviors and physiology.

As for behavior of birds, testosterone affects male incubation and feeding of young (Silverin 1980; Hegner and Wingfield 1987; Oring et al. 1989; Ketterson et al.

1992), vocalizations (Arnold 1982; Wada 1981, 1982, 1986; Harding et al. 1988; Gyger et al. 1988; Nowicki and Ball 1989; Ketterson et al. 1992), and aggressiveness (Balthazart 1983; Wingfield et al. 1987, 1990; Archawaranon and Wiley 1988; Beletsky et al. 1990). However, it is important to recall that the effect of steroids is not the direct production of behavior; rather, steroids affect the likelihood that behavior will be expressed (Feder 1984; Moore 1991). In terms of physiology, testosterone increases locomotor activity (Wada 1982, 1986; Massa and Bottoni 1987) and metabolic rate (Hännsler and Prinzinger 1979; Feuerbacher and Prinzinger 1981), and it suppresses lipid storage (Wingfield 1984; Ketterson et al. 1991a) and the onset of prebasic molt (Runfeldt and Wingfield 1985; Schleussner et al. 1985; Nolan et al. 1992). At the intraspecific level, very little is known about the quantitative relationship between natural variation in levels of testosterone and behavior. However, a recent study of male red-winged blackbirds (*Agelaius phoeniceus*) has shown that individual variation in testosterone correlates both with song rate and with frequency of aggressive interactions (Johnsen 1991; fig. 14.2).

The question whether intraspecific variation in plasma testosterone has a genetic basis has received little study, but there is indirect evidence for an additive genetic component to variation in aggressiveness, which is influenced by testosterone (e.g., Moss et al. 1982; Boag 1982; Maxson et al. 1983). In quail (*Coturnix japonica*), artificial selection for high mating frequency, which is also associated with testosterone, leads to significant correlated responses in aggressiveness and in a morphological trait, the size of the cloacal gland (Sefton and Siegal 1975; Cunningham and Siegal 1978). Therefore, genetic variation for testosterone-mediated traits seems likely. In the rest of this chapter we will assume that such variation exists and that a response to selection is possible. Clearly, this point deserves further study.

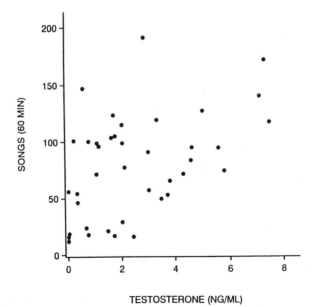

Fig. 14.2 Song rate as a function of plasma concentration of testosterone in free-living red-winged blackbirds (*Agelaius phoeniceus*) observed during breeding. (Redrawn with permission from Johnsen 1991.)

HORMONES AND THE EVOLUTIONARY PROCESS

Question 1: How Might Hormones Form the
Mechanistic Basis for Trade-Offs in Life Histories?

According to Stearns (1989), trade-offs are fitness costs that occur when a beneficial change in one trait is linked to a detrimental change in another. Changes in fitness that occur when two traits depend upon the same hormone and are co-expressed, but only one is beneficial, may also be regarded as trade-offs. In this view, hormones are one of the important mechanistic links between organismal biology and evolutionary ecology.

Despite the considerable attention that has been given to trade-offs (see reviews in Dingle and Hegmann 1982; Reznick 1985; Partridge and Harvey 1988; Stearns 1989), more effort has been devoted to measuring the cost of reproduction (e.g., De Steven 1980; Nur 1984) and documenting the genetic basis for variation in life history traits (e.g., Dingle and Hegmann 1982) than to uncovering the physiological mechanisms that underlie trade-offs, particularly those associated with the endocrine system. Several investigators have recently called for more work in this area (Stearns 1989; West-Eberhard 1989, 1992; Zuk et al. 1990; Ligon et al. 1990).

The dearth of hormonal studies is surprising because hormones can produce correlated effects with potentially antagonistic fitness consequences, precisely the situation we seek when attempting to demonstrate the existence of trade-offs (Williams 1957). Thus, while it is obviously important to study genetic correlations among characters and to determine how these correlations vary across environments, it is equally important to study the proximate causes of genetic correlations. Only by knowing their physiological bases can we hope to understand how such correlations originate or how they can be disrupted by selection. Finally, the shortage of hormonal studies is surprising because, in terms of levels of organization, hormones lie between genes and life history traits but are much easier to manipulate than genes.

Two recent examples of an experimental approach to hormones and life histories addressed the potential trade-offs between clutch size and egg size (Sinervo and Licht 1991a, 1991b) and energy devoted to growth or to maintenance (Derting 1989). Sinervo and Licht (1991a, 1991b) studied egg size and egg number (clutch size) in female side-blotched lizards (*Uta stansburiana*). Females treated with follicle-stimulating hormone (FSH) produced clutches that were larger than normal (6.0 vs. 4.2 eggs), but their eggs were of smaller than normal size (0.33 g vs. 0.43 g). In contrast, females whose follicles were partially ablated with a needle produced smaller than normal clutches, but their eggs were larger than normal. Variation in clutch size and variation in egg size in this species may be linked physiologically, and a major factor regulating both may be FSH. Comparisons of the survival of hatchlings produced from small and large eggs and then released into the field have indicated that selection on offspring size is optimizing in these lizards (Sinervo et al. 1992).

Derting (1989) investigated the potential for a trade-off between growth and maintenance in young cotton rats (*Sigmodon hispidus*). A basic assumption about animal life histories has been that energy is limiting and that energy devoted to maintenance is not available for production (growth and reproduction) (Cody 1966; Hirshfield and Tinkle 1975; van Noordwijk and de Jong 1986). However, Derting has shown that thyroid hormone can promote both production *and* maintenance, and that the advantages of high or low thyroid activity may depend upon environmental factors. Derting administered thyroid hormone (T_4) to young cotton rats in the laboratory. To one set of animals (A) she provided food ad libitum; in another set (R), food was restricted. Within each level of food, some subjects (E) were treated with T_4 to elevate their metabolic rate, while controls (N) were left untreated and had normal metabolism. The results (fig. 14.3) revealed a striking and unexpected interaction among maintenance (metabolic rate), production (growth), and food availability. Not surprisingly, the NA rats grew faster than the NR rats. However, the EA rats grew even faster, while the ER rats barely grew at all. It is tempting to speculate, as Derting did, that food-rich environments might select for rats with greater thyroid activity, while other environments might favor animals with lower activity. The interesting outcome was that, given sufficient food, cotton rats treated with thyroid hormone apparently can increase production despite higher maintenance costs, thus avoiding the expected trade-off.

Moore (1991) considered the role of hormones in the determination of alternative male strategies, another form of life history trade-off. He drew a parallel between the organizational and activational effects of hormones during development and the decision rules employed by males for the adoption of alternative strategies. When strategies are fixed (males assume one form or another, but do not shift, as in salmonid fishes: Gross 1984, 1985), coordination of behavior and morphology is likely to be determined early in development through the organizing effects of hormones. When strategies are plastic and males may shift from one to another as the environment or their condition changes, switches of strategy are likely to be under activational control (e.g., sex change in the stoplight parrotfish, *Sparisoma viride:*

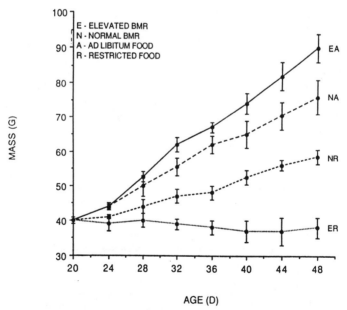

Fig. 14.3 Growth of young cotton rats (*Sigmodon hispidus*). The four groups are EA, experimental animals treated with thyroid hormones (T_4) and given free access to food; ER, animals treated with T_4 whose access to food was restricted; NA, control animals not treated with T_4 and given free access to food; NR, untreated control animals whose access to food was restricted. Adapted from Derting 1989, with permission.

Cardwell and Liley 1991a, 1991b). Moore provides numerous examples in support of his interesting thesis and strongly urges an experimental hormonal approach to development and alternative male strategies (see also West-Eberhard 1992).

Collectively, the studies reviewed in this section suggest that hormones can provide mechanistic explanations for the existence of life history trade-offs.

Question 2: What Is the Role of Hormones in the Evolution of Inter- and Intraspecific Variation in Social Organization and Mating Systems?

Several researchers have recently asserted that variation in patterns of secretion of a few key hormones, or variation in responsiveness of target tissues to these same hormones, might explain much inter- and intraspecific variation in vertebrate social behavior, social organization, and mating systems. Because these efforts to relate hormones to social evolution are relatively new, of necessity they are somewhat speculative. As examples we refer to the work of Wingfield et al. (1987, 1990), Boinski (in press), and Carter (1992).

It is widely known that males of polygynous bird species tend to provide less parental care than males of monogamous species (Verner and Willson 1969). (We use the terms *monogamous* and *polygynous* in the traditional sense to refer to species in which males form social pair bonds with one or more than one female.) It is also true that polygynous males, unlike males of most monogamous species, often continue to display after they have acquired their first mate (Nolan 1978; Vehrencamp and Bradbury 1984; Emlen and Oring 1977; Wingfield et al. 1990). At the physiological level, there is evidence to suggest that testosterone is at least partially responsible for this conflict between mating effort and parental effort (Wingfield et al. 1987, 1990).

Interspecific variation in patterns of testosterone secretion among birds correlates with interspecific variation in the mating system (Wingfield et al. 1987, 1990). Comparison of seasonal profiles of plasma testosterone in polygynous and monogamous species reveals that polygynous males tend to sustain secretion at peak levels for much of the breeding season, whereas monogamous males exhibit a single brief peak early in the breeding season (Wingfield et al. 1987). The ratio of summer to winter levels of testosterone is also significantly associated with mating system and parental behavior: the higher the ratio, the more likely a species is to be polygynous (Wingfield et al. 1990). However, an aggressive challenge (e.g., by an intruder) leads to a greater increase in testosterone in monogamous than in polygynous species (Wingfield 1991). Apparently testosterone levels of polygynous males are closer to the physiological maximum during most of the breeding season, while monogamous species maintain lower levels of testosterone but secrete higher levels "on demand."

Boinski (in press) has detected correspondence between geographical variation in social organization of free-living squirrel monkeys (*Saimiari* spp.) and geographical variation in adrenocortical function among laboratory-held *Saimiari*. Squirrel monkeys in Costa Rica and Peru differ in group size, levels of aggression, dominance relations between the sexes, and patterns of natal dispersal. Boinski speculates that the populations may also differ in disposition or emotional reactivity. The populations that are more aggressive (Peru) may have a higher setting to their adrenocortical axis, which would be reflected in higher levels of cortisol and slower recovery from episodes of stress. This could be the result of local adaptation or a few key (genetic?) changes in receptor density, availability of neurotransmitters, or levels of adrenocorticotropic hormone (ACTH). Boinski speculates further than this variation may be driven by geographical variation in the spatial distribution of the foods eaten by the monkeys. When the preferred food is abundant but concentrated (Peru), group size and social interaction increase; when food is dispersed (Costa Rica), social stress may be less a factor in driving social organization. This system is currently under study.

Finally, Carter (1992) has detected a relationship among small mammals in the frequency with which females nurse young and males initiate copulation, and she suspects that the pattern may be explained by a shared mechanism, the hormone oxytocin. Female rabbits nurse young once a day, and a single brief mating bout is sufficient to induce pregnancy. Female rats nurse several times a day, and mating activity lasts for several hours. Female prairie voles lactate continuously while in the nest, and matings may go on for more than a day. Only the prairie voles are monogamous. Carter speculates that oxytocin may regulate this pattern of interspecific covariation between the sexes in lactation frequency and duration of copulation. Oxytocin facilitates the muscular contractions that lead to milk release and to ejaculation. It may also act in the brain to produce the sensation of satiety that terminates bouts of both lactation and sexual behavior, and to promote the social bonds that form between mother and young and, in some species, between male and female. If lactation, sexual behavior, and social bonding have a common physiological basis, natural selection acting on regulation of hormone action could account for diverse aspects of interspecific differences in social organization. It could also account for correlated interspecific changes in the functioning of both males and females.

While primarily correlative, these studies of hormonal action and social behavior are intriguing and point to the potential importance of hormones and behavior in the evolution of social systems. The overwhelming diversity in behavior, physiology, and morphology that characterizes vertebrates may ultimately be explicable in terms of a relatively small number of physiological systems.

Question 3: Does Selection Act on Organisms or on Traits?

According to Travis (1989), the focus of selection is one of the central questions in evolutionary biology. When several traits are dependent upon the same hormone, it seems likely that even if selection acts on traits, the effect will be to shape organisms. That is, if selection favors individuals with high values of a trait whose expression depends on hormone concentration, then it will necessarily act on the other traits whose expression also depends on the same hormone. If changes in all traits are beneficial, we would expect individuals exhibiting favorable values to increase in frequency. But if some changes are advantageous while others are disadvantageous, then prediction becomes more difficult. Such negative correlations among traits might be viewed as design constraints that limit adaptation, or, as already seen, they may represent trade-offs that persist because they allow the organism to adjust its tactics to varying and unpredictable environmental variation.

Organisms also have the capacity to dissociate control of different hormone-dependent traits, that is, to compartmentalize their attributes and thus to insulate themselves against detrimental effects that might occur when selection favors only one among a complex of hormone-dependent traits. For example, modification of the number or affinity of receptors at the different targets of the hormone could disassociate responses, thereby reducing detrimental effects of a change in hormone concentration while enhancing beneficial effects. Similarly, a target may require that a hormone be converted from one form to another before it is active. If conversion were to depend upon an inducible enzyme, this dependence on induction could account for loss of a trait or for condition-dependent expression. Finally, behaviors that were once hormone-dependent can become uncoupled from hormones altogether (e.g., copulatory behavior: see Moore and Marler 1988). In quail, testosterone activates both copulatory behavior and crowing (see Balthazart and Ball, 1993), but the neuroanatomical sites for these behaviors are separated physically. Copulation (neck grab, mount, and cloacal contact movement) requires that testosterone be aromatized enzymatically into estradiol (E_2) at the preoptic medial nucleus (POM), while crowing depends upon enzymatic reduction of testosterone to 5-α-dihydrotestosterone (DHT) in the nucleus intercollicularis (ICo). According to Balthazart and Ball (1993), "Circulating steroids in the plasma provide a general signal for an increase in reproductive activity that is then modified and amplified at the various targets."

Many variables will determine the outcome of a situation in which selection favors one of a complex of hormone-dependent traits, and the answer to the question of whether selection ultimately acts on organisms or on traits is almost certainly that it acts on both. Although this kind of analysis seems critical to an understanding of the evolution of integrated organismal function, few endocrine studies address this point directly in an evolutionary context. The studies of correlated responses to selection for aggressiveness or sexual behavior in quail (see above) seem promising (Sefton and Siegal 1975; Cunningham and Siegal 1978). Such studies should provide ample opportunities for physiologists to determine the multiple changes that might accompany a response to selection (e.g., changes in plasma hormone concentrations, receptor levels and affinities, activity levels of enzymes that convert hormones from one form to another, and alterations in cellular activities at the target).

Question 4: What Is the Evolutionary Significance of Intrapopulational Phenotypic Variation?

Phenotypic Engineering as an Experimental Approach. Phenotypic engineering consists of manipulating the phenotype of an organism, quantifying the effects of the manipulation, and relating those effects to performance or fitness (Ketterson et al. 1992). This method permits us to explore the evolutionary significance of phenotypic variation by asking whether a rare or a novel phenotype would increase in frequency, assuming the requisite genetic variation (cf. Sinervo and Huey 1990).

The classic approach to understanding the consequences of variation in a trait is to modify the trait and determine the effect of fitness. An elegant example is the work of Andersson (1982). By varying the tail length of male widowbirds (*Euplectes progne*), Andersson found that females prefer males with longer tails, even when the tails are longer than those that occur naturally. This raises the obvious question, what prevents the evolution of long tails in male widowbirds? The answer is elusive because the classic approach has been to alter just a single trait and to measure just a few fitness consequences of the alteration. We cannot know whether, if males were to produce longer tails, they would also differ in other ways (traits as well as fitness consequences) that would be detrimental (see Selander's thoughts [1965] on the effect of tail length on survival of male grackles, *Quiscala* spp.).

Phenotypic Engineering with Hormones. Modifying phenotypes by varying circulating levels of hormones is a method of particular interest precisely because hormones influence so many interlocking traits. Thus hormonal manipulations enable the investigator to avoid the trait-by-trait study of organisms and to measure the effect of simultaneous changes in suites of characters. At the risk of exaggerating the point by ignoring for the moment the capacities of organisms to compartmentalize their responses, the method provides the potential to predict the outcome of generations of direct and indirect selection, all in a single experiment.

Thus, by comparing the fitness of hormonally produced rare or novel phenotypes with the fitness of unaltered controls, it is possible to probe the question why existing phenotypes persist despite the fact that alternative phenotypes are possible. In fact, without experimental intervention, relating variation in phenotype to variation in fitness is a difficult undertaking, in part because most traits are normally distributed and individuals that express extreme values of a trait are rare. Phenotypic engineering allows the investigator to create large numbers of extreme individuals and to see how they fare. At least one caveat seems important, however: the success of some phenotypes, particularly those involving social behavior, is likely to be frequency dependent.

In the future, investigators will almost certainly use hormones to manipulate both single traits and suites of traits in order to comprehend integrated organismal responses to selection. In the laboratory, neurobiologists are able to implant microquantities of hormones in targeted portions of the brain in order to study localized effects of hormones on behavior. In theory it should also be possible to employ these techniques in the field in order to determine the fitness consequences of uncoupling traits as well as coupling them.

Phenotypic Manipulations and Correlates of Fitness: Possible Outcomes. How might hormonal manipulations affect fitness? There are three possible outcomes, depicted in figure 14.4, which can be described as paradoxical, adaptive, and neutral. Curve 1, the paradoxical outcome, shows fitness as a positive linear function of the magnitude of a trait and indicates that individuals with high values perform better than those at or below the population mean; the paradox is that selection has not led to such higher values. The adaptive outcome is depicted in curve 2. Individuals that deviate from the mean in either direction have lower fitness than those at the mean. This is the situation we would expect under normalizing selection. In curve

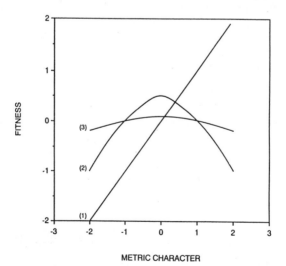

METRIC CHARACTER

Fig. 14.4   Fitness profiles showing possible relationships between variation in a quantitative trait and fitness. Numbers of the *x*-axis indicate standard deviations from the mean. 1, paradoxical outcome; 2, adaptive outcome; 3, neutral outcome. Adapted with permission from Falconer 1989.

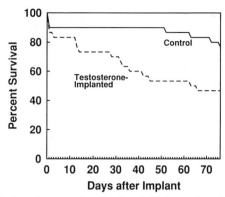

Fig. 14.5   Summer survival rate in mountain spiny lizards (*Sceloporus jarrovi*), according to treatment. (Redrawn with permission from Marler and Moore 1988b.)

3, fitness is relatively independent of the value of the trait; individuals at the norm and those that deviate from it in either direction are equally effective at leaving descendants. This outcome would imply that the character is not important to fitness; the model is compatible with the interpretation that fitness is largely determined by chance.

An example of this manipulative approach to the study of fitness is the work of Marler and Moore (1988a, 1988b, 1989, 1991) on the mountain spiny lizard, *Sceloporus jarrovi*. These investigators treated free-living male lizards with long-lasting implants of testosterone and observed their behavior, physiology, and survivorship. Testosterone-implanted males displayed more, spent more time above ground, and had higher field metabolic rates and lower energy reserves. They also suffered higher mortality than controls (fig. 14.5). It is not known whether the treatment affected their reproductive success.

Experimental studies of the effect of testosterone on fitness in free-living male birds have shown an influence on major components of fitness, namely, acquisition of mates (e.g., Watson and Parr 1981; Wingfield 1984), maintenance of territory (Beletsky et al. 1990), production of offspring (Hegner and Wingfield 1987; Oring et al. 1989), and survival (Dufty 1989, Nolan et al. 1992). Some of the effects on fitness are negatively correlated: testosterone increases the number of mates acquired by song sparrows (*Melospiza melodia*), but decreases survival of brown-headed cowbirds (*Molothrus ater*). In the house sparrow (*Passer domesticus*), males treated with testosterone rear fewer offspring to fledging than do controls because their mates are apparently unable to compensate for the reduction in male parental effort (Hegner and Wingfield 1987). Whether elevated testosterone would also have beneficial effects on male house sparrows is not known, although Hegner and Wingfield (1987) speculated that males with high testosterone might be more successful in extra-pair fertilizations.

PHENOTYPIC ENGINEERING IN JUNCOS: A CASE HISTORY

We made it our objective to apply phenotypic engineering in the field. We too manipulated testosterone and set out to quantify its phenotypic effects and to relate these changes to as many components of fitness as we were able to measure.

The Study Species

Dark-eyed juncos are members of the avian subfamily Emberizinae and are abundant and widely distributed in North America. The subspecies that we study, the Carolina junco (*J. h. carolinensis*), is an altitudinal migrant that breeds at high elevations in the Appalachian Mountains of southeastern North America. Our study area is the University of Virginia's Mountain Lake Biological Station in southwestern Virginia and its environs (see Wolf 1987; Chandler et al., in press, for a description). Our general field methods and demographic findings are summarized in Wolf et al. (1988, 1990, 1991), Ketterson et al. (1991a, 1991b, 1992), and Nolan et al. (1992). Since 1983, we have color-marked over 3,000 individuals.

During the breeding season male juncos are territorial and form a pair bond with a single female, although polygyny occasionally occurs (<10%, personal observations). Female juncos build the nest, which is typically on the ground, and perform all the incubation. Both sexes protect eggs, and both feed and protect nestlings and fledglings. Pairs tend to remain together for the entire breeding season, and fidelity to the previous year's breeding site is high among adults of both sexes. When both pair members survive winter, they usually re-form the pair. Yearlings tend to breed close to their natal sites, and some young males fail to obtain territories and become floaters (Ketterson et al. 1991b). The breeding sex ratio is somewhat male-biased.

Breeding begins in late April and continues until mid-August; clutch size ranges from three to five. Occasional pairs raise as many as three broods in a season, but nest predation in some years is very high (average across years = 66% loss between laying and fledging), so many pairs leave no offspring in a season. In winter, juncos live in flocks. Adult males tend to remain near their breeding territories, while females and yearlings normally move downslope (Ketterson et al. 1991b).

Hormone Treatments, Assays, and Rationale

We randomly assign males to treatment within two categories: age (first-year vs. older) and habitat (wet woods, dry woods, open areas). Thus we study approximately equal numbers of experimental males (T-males) and control (C-males) of each age and in each habitat.

Birds are implanted with two 10-mm lengths of silastic tubing (Dow Corning, i.d. = 1.47 mm, o.d. = 1.96 mm: Ketterson et al. 1991a), either packed with crystalline testosterone or left empty. To check the effectiveness of the implants as well as to measure natural variation in circulating levels of testosterone, we recapture the birds after treatment, take a blood sample, and use radioimmunoassay to determine plasma concentrations of steroid hormones. (Our assay methods are the same as those of Wingfield, as described in Wingfield and Farner 1975, Wingfield et al. 1982, and Ketterson et al. 1991a.

Time of implanting has varied from year to year (fig. 14.6). In one year (1989) we began implanting in early March, but in most years (1987, 1988, 1990), birds were treated in April and May. Implants were removed in August, except during 1990 when we delayed removal until October and left some implants in place all winter.

The rationale and effectiveness of our treatment can best be understood by comparing figures 14.6 and 14.7. Testosterone ordinarily peaks in male juncos in early spring, after which it falls and is maintained at lower levels for the remainder of the breeding season (fig. 14.7; unpublished observations in 1989), although males retain the capacity to increase testosterone throughout the breeding season (data not shown). In males with testosterone implants, testosterone is elevated to peak

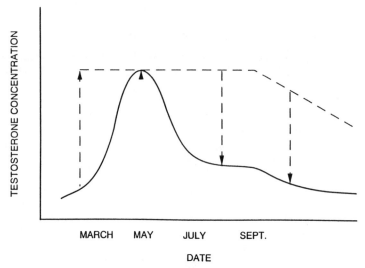

Fig. 14.6 Schematic diagram of the seasonal profile of plasma testosterone concentration in untreated male dark-eyed juncos (solid line). The dashed lines and arrows are also schematic and indicate the timing of insertion (upward arrows) and removal (downward arrows) of testosterone implants and their intended effects on plasma concentration of testosterone. The dashed line extending beyond October indicates the presumed testosterone level of males whose implants were not removed.

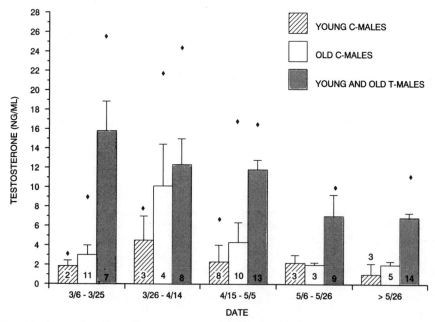

Fig. 14.7 Seasonal profile of plasma concentration of testosterone in treated and untreated dark-eyed juncos (mean ± 1 SE; n in histogram bar indicates size of sample). The untreated birds are classified according to their age: young birds were in their first breeding season; old birds, in their second or subsequent breeding season. Diamonds indicate the maximum value observed in each treatment group during each time period. Where no diamonds are plotted, the maximum lay within the standard error bar. (E. Ketterson and V. Nolan Jr., unpublished data from 1989, when implanting began in March.)

physiological levels for at least 40 days, and in some individuals remains higher than normal for an additional period of perhaps 100 days (Nolan et al. 1992). Thus, we have used our treatments in three ways: to maintain testosterone at its peak for the entire breeding season (1987, 1988); to raise testosterone in early spring, before it peaks in the natural population (1989); and to prolong a high level beyond the end of the breeding season (1990). The rationale for the first treatment was to mimic the seasonal profile of a typical polygynous species (Wingfield et al. 1987). The rationale for the second and third treatments was to probe the factors that might set the limits on the timing seasonal reproduction. Although we altered the seasonal pattern in these ways, we did not raise testosterone above levels that juncos are capable of achieving naturally. The maxima observed in untreated males in early spring were quite similar to those produced in T-males (fig. 14.7).

Effect of Testosterone on the Behavioral Phenotype: Vocal Behavior, Parental Behavior, Use of Space

Most of our observations of male behavior have been made at the nest, where we have noted two changes that could affect fitness. T-males sang more frequently than C-males (fig. 14.8), and they fed their nestlings less frequently (fig. 14.9). This could have decreased survival of offspring, but females mated to T-males compensated for their mates' reduced effort by increasing their own rates of feeding (fig. 14.9). Thus females are able to assess the level of parental effort of their mates. One might also expect that their added work would have consequences for females, such as poor physical condition or lower survival rates, or, perhaps, a greater tendency to desert an uncooperative male and find a new mate before attempting to reproduce again (see below).

With the aid of radiotransmitters attached to males' backs, we quantified their behavior away from the nest (Chandler et al., in press). We noted a male's location every 30 minutes throughout the day for 3 or 4 consecutive days and mapped the

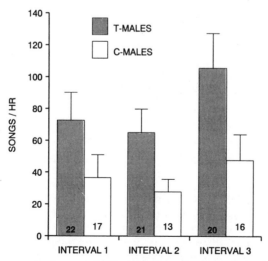

Fig. 14.8 Audible song rates (mean ± 1 SE) of T-males and C-males during intervals when their nestlings were 0–3 (interval 1), 4–7 (interval 2), and 8–10 (interval 3) days old (0 equals hatching day). (From Ketterson et al. 1992.)

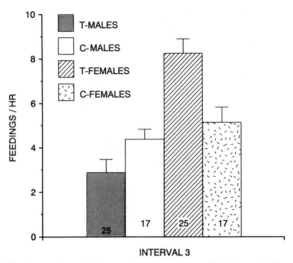

Fig. 14.9 Food deliveries per hour (mean ± 1 SE) to nestlings 8–10 days old (interval 3 of nestling life) by male dark-eyed juncos and their mates (mean ± 1 SE). Only the males were treated; females are categorized according to the treatment of their mates (see text). (From Ketterson et al. 1992.)

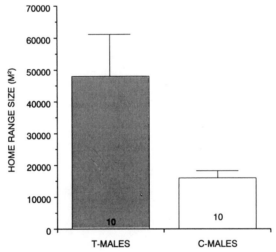

Fig. 14.10 Home range size (mean ± 1 SE) of male dark-eyed juncos according to treatment. (Data from Chandler et al., in press.)

distances separating the locations. T-males ranged more widely than C-males (Chandler et al., in press) (fig. 14.10). Laboratory studies of other bird species have shown that testosterone increases locomotor activity (Wada 1986), suggesting that T-males may be more "restless" than controls. Restlessness might increase the likelihood that such males would encounter fertile females mated to other males (Westneat et al. 1990).

Effect of Testosterone on the Physiological Phenotype:
Body Mass, Corticosterone, Molt Schedule

We have noted three physiological changes that might influence fitness of T-males. First, T-males treated in the early spring (1989) lost mass and visible subcutaneous body fat before the normal time when males lose winter fat (fig. 14.11). Later in the breeding season T-males and C-males did not differ in mass (Ketterson et al. 1991a). Acceleration of loss of winter fat might increase a T-male's vulnerability to late spring snowstorms, which occur in the mountains and pose a risk to early breeders.

Second, plasma levels of corticosterone were higher in T-males than in controls (fig. 14.12). Whether this indicates that elevated testosterone affects the ways in which males respond to potential stressors is not known (Siegel 1980), nor do we know whether prolonged exposure to elevated testosterone or corticosterone might lead to immunosuppression and increased susceptibility to disease (Chrousos et al. 1988; Zuk 1990). These questions are currently under investigation.

Finally, testosterone delayed molting (fig. 14.13). Of ten T-males still carrying implants and examined in October 1990, only six had begun to molt their primary (principal wing) feathers. All of nine C-males had renewed their primaries or were far advanced in molt at that time. A complete molt after breeding is an annual event in juncos, but T-males whose implants we did not remove and that returned to breed in 1991 had failed to molt; their plumage was very degraded. Worn plumage almost certainly provides less effective insulation and probably impairs maneuverability in flight.

Thus both premature and unusually prolonged exposure to high plasma levels of testosterone have clear effects on behavior and physiology, but do these translate into differences in fitness?

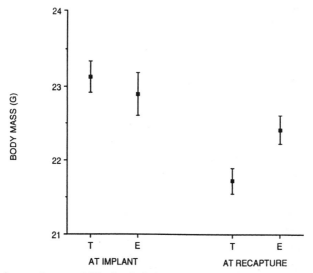

Fig. 14.11   Body mass (mean ± 1 SE) of male dark-eyed juncos at the time of implant in early spring and when recaptured, according to treatment (for both treatments, $n = 32$). (From Ketterson et al. 1991a.)

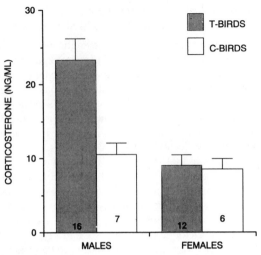

Fig. 14.12 Plasma concentration of corticosterone (mean ± 1 SE) in male and female dark-eyed juncos. Only males were treated; females are categorized according to treatment of their mates. (From Ketterson et al. 1991a.)

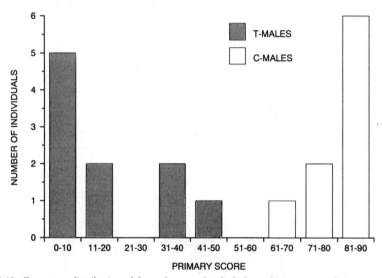

Fig. 14.13 Frequency distribution of the molt score of male dark-eyed juncos, according to treatment. Juncos have nine primaries on each wing. Individuals were given a score of 10 for each right-wing primary they had replaced; thus, individuals with a score of 90 had completed primary molt, those with a score of 0 had not begun, and those with intermediate scores were in the process of molting. (From Nolan et al. 1992.)

Table 14.1 Correlates of Fitness Measured: Summary of Trends

	T-MALES	C-MALES
Territory acquisition	+	
Mate acquisition		=
Apparent reproductive success		
Eggs hatched		=
Nest success[a]		=
Nestlings fledged		+
Mass at fledging		+
Fledgling survival	?	
Renesting interval		+
Size of replacement clutch		+
Eggs hatched, replacement clutch		+
Extra-pair fertilizations/fertility	?	
Mate condition, mate retention		
Female mass at fledging	+	
Within-season fidelity		+
Annual survival		+
Between-season fidelity	+	
Minimum overwinter survival		
Summer removal	+*	
Autumn removal		=
No removal		+*
Total	4	8

Note: + indicates treatment group with higher performance (which is the same as the treatment with the greater mean, except for renesting interval, where lower values indicate higher performance). Note that + does not indicate significant statistical differences, only trends, and that the trends were often weak (see tables 14.2 and 14.3). The symbol = indicates that two means were the same. * indicates variables for which treatment groups differed significantly ($P < .05$; see fig. 14.15).
[a]Nest success means that at least one fledgling left the nest.

Correlates of Fitness

We compared T-males with controls for various measures of fitness, including territory acquisition, mate acquisition, apparent reproductive success, extra-pair fertilizations, mate retention, overwinter survival, and the mate's physical condition and overwinter survival (table 14.1). These correlates of fitness are arranged approximately chronologically according to the sequence of events in the breeding season and thereafter and are the major components of fecundity selection and survival selection. However, not all the measures are independent. To take an obvious example, the number of young that leave the nest cannot be greater than the number of eggs hatched.

Territory Acquisition. For males present in early spring prior to territory formation, we compared the proportions of T-males and C-males that subsequently acquired territories on the study area (fig. 14.14; unpublished observations from 1989). For old

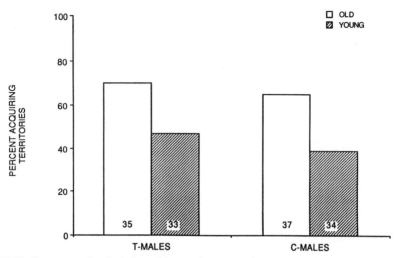

Fig. 14.14   Percentage of male dark-eyed juncos that acquired territories on the study area in 1989, according to age (see fig. 14.7) and treatment. (Data from V. Nolan Jr. and E. Ketterson, unpublished observations.)

males (second breeding season or beyond), in which fidelity to the former territory is nearly complete, the question was whether the hormone would affect the likelihood of reacquiring last year's territory; for young (first-year) males, the question was whether testosterone would influence the likelihood of succeeding in the attempt to gain a territory on the area (recall that there are "floater" nonterritorial males, almost always young: Ketterson et al. 1991b).

Among 35 adult T-males present on the study area before territories were established, 25 (71%) acquired territories there, as compared with 24 of 37 adult controls (65%). Among first-year males, 15 of 33 (46%) T-males acquired territories, as compared with 13 of 34 (38%) controls. Neither difference was significant; the trend slightly favored T-males in each age class.

Mate Acquisition. Based on Wingfield's (1984) observation of polygyny in male song sparrows that he implanted with testosterone, we predicted a similar result in juncos. However, it did not occur (Ketterson et al. 1992, unpublished observations). Two T-males whose implants we did not remove returned next spring in faded and much abraded plumage, having failed to undergo prebasic molt. Despite their abnormal appearance, both obtained mates.

Apparent Reproductive Success. We found no significant differences in the seven measures of apparent reproductive success listed in table 14.2. Success is "apparent" because we assumed that each male sired all young in his mate's nest (but see below). Because our sample sizes are relatively small, the power of these tests is not high (power reported in Ketterson et al. 1992). Still, the P values indicate that the treatment groups were quite similar in all these measures.

To relate reproductive success to the fact that T-males fed young less frequently than C-males, the two treatment groups differed only slightly in production of fledglings per brood, presumably because the T-females compensated for their mates'

Table 14.2   Correlates of Fitness: Apparent Reproductive Success

	T-PAIRS			CONTROLS			
	$\overline{X}$	(n)	SE	$\overline{X}$	(n)	SE	P
Eggs hatched, clutch A	3.5	(27)	0.12	3.5	(18)	0.22	.696
Nest success (%)	33.3	(81)	—	33.3	(93)	—	.999
Nestlings fledged, clutch A	3.1	(26)	0.18	3.3	(15)	0.25	.460
Mass at fledging (g)	16.9	(21)	0.26	17.2	(15)	0.28	.360
Renesting interval (days)	8.2	(16)	0.95	7.1	(15)	0.66	.350
Clutch size, clutch B	3.6	(14)	0.13	3.8	(10)	0.20	.560
Eggs hatched, clutch B	3.2	(9)	0.41	3.4	(11)	0.25	.870

Source: Data from Ketterson et al. 1992, except nest success (E. Ketterson and V. Nolan Jr., unpublished observations).

Note: Comparisons were by Mann-Whitney U-test (clutch size, number hatching, renesting interval), *t*-test (nestling mass, renesting interval), or χ^2 (% nesting success, i.e., proportion of nests producing at least one fledgling). All probability values are two-tailed.

reduced effort. We note that controls outperformed experimentals in five of the seven measures.

Genetic Analysis of Paternity. Reliable measures of reproductive success require determination of the paternity of young. Our collaborator, P. G. Parker, has completed a preliminary (unpublished) genetic analysis. She analyzed hypervariable DNA from the male, the female, and the nestlings in 13 families. Some pairs produced more than one brood, and the total number of broods analyzed was 18. There were 13 putative fathers (6 T-males and 7 C-males), 11 putative mothers (2 of the females had 2 mates in succession), and 52 offspring. She also analyzed DNA fingerprints of 6 males (1 control and 5 experimental) that held territories adjoining those of one or more of these families.

DNA was analyzed as in Rabenold et al. (1991). In offspring that were assignable to both putative parents, the average number of unattributable bands was 0.230, and the average number of bands on which exclusion was based was 10 (extremes, 5–16). Twelve of the 52 offspring (23%) were sired by males other than the putative father, and all of these young were assignable to neighboring males. In all cases the putative female parents were the mothers of the offspring; that is, none of the eggs had been introduced by parasitizing females (and we have no evidence that intraspecific parasitism occurs in our population).

T-males sired 16 of their 20 putative offspring (80%); C-males 24 of their 32 (75%). Interestingly, all 12 offspring that were produced by extra-pair fertilizations were sired by C-males ($n = 3$ males). None of the 11 T-males sired any young through extra-pair fertilizations, whereas 3 of 8 C-males did so (Fisher's exact $P = .107$). We cannot reconcile the tendency revealed by this preliminary genetic analysis with the behavioral data indicating that T-males move over a larger area than C-males.

Female Condition, Female Survival, and Mate Retention. We found no evidence that reduced male parental contribution and compensatory effort by the female led to loss of physical condition by the female or to significantly more frequent desertion of the male by the female (table 14.3). Females paired with T-males weighed slightly more than mates of C-males at the time their young left the nest, and they were slightly more likely to change mates within breeding seasons. They were somewhat less likely to return in the year following the experiment, but when both females and their mates returned, T-females were somewhat more likely to remate with their previous mates than were control females.

Survival of Males. Survival was affected by treatment, although in rather unexpected ways (fig. 14.15). For individuals whose implants we removed in summer (July–August 1987, 1988, 1989), T-males were significantly more likely than C-males to return in the year following treatment (Ketterson et al. 1992, supplemented with unpublished observations collected during 1990). Stated another way, males known to be alive in late summer were more likely to survive to the following spring if they

Table 14.3   Correlates of Fitness: Mate Condition, Survival and Retention

	T-PAIRS			CONTROLS			
	$\overline{X}$	(n)	SE	$\overline{X}$	(n)	SE	P
Female mass, day 10 (g)	20.4	(20)	0.26	20.3	(14)	0.22	.680
Mate fidelity, within season	95%	(20)	—	100%	(13)	—	.664
Female return rate, annual	42%	(60)	—	46%	(48)	—	.424
Mate fidelity, between seasons	86%	(14)	—	72%	(15)	—	.411

Source: Data from Ketterson et al. 1992, but female return rate and mate fidelity between seasons were supplemented with unpublished observations made during 1990.
Note: Comparisons were by t-test (female mass) and χ^2 or Fisher's exact (other variables); all probabilities are two-tailed.

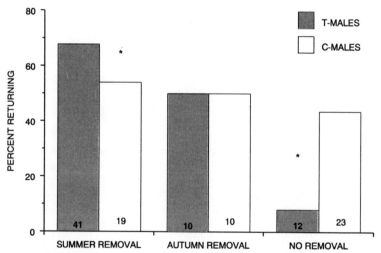

Fig. 14.15   Minimum survival (percentage returning to the study area in the year following treatment) of male dark-eyed juncos, according to treatment and time when the implants were removed. * indicates $P < .05$, χ^2 test. (Data from Ketterson et al. 1992, supplemented with unpublished observations [summer removal]; Nolan et al. 1992 [autumn removal, no removal].)

had been treated with testosterone in the preceding summer. We have no explanation for this result, which suggests a delayed effect of treatment.

Males whose implants we removed in autumn (October 1990) were equally likely to return the following spring, regardless of treatment (Nolan et al. 1992). Thus, even though testosterone delayed molting by at least a month in the experimental males, they molted completely after implant removal, and the delay had no detectable effect on their survival.

Males whose implants we did not remove at all were significantly less likely to return the following spring (Nolan et al. 1992), which we interpret as the effect of greater overwinter mortality. Such mortality may have been attributable to the loss of insulation or maneuverability owing to their failure to molt, but it also may have resulted from some effect of testosterone on the tendency to join flocks in winter, to migrate downslope for the winter, or to accumulate winter fat reserves.

Do Juncos Vary in Any Measure of Reproductive Fitness?

Annual Reproductive Success. The similarity in apparent reproductive success of T-males and C-males led us to ask how much individual male juncos vary in reproductive success. Figure 14.16 classifies males according to the number of fledglings that left the nest during single breeding seasons and according to treatment (Nolan et al., unpublished data from 1989 and 1990). Fifty-two percent of a total of 52 males left no offspring, 37% left one to three, and 12% left more than three. Interestingly, the individual that appeared to leave eleven fledglings was one of those analyzed by DNA fingerprinting, which revealed that five of the eleven offspring were sired by another male. Obviously, data like those in fig. 14.16 will ultimately need to be corrected after measures of true paternity have been made.

Nevertheless, it is clear that males vary considerably in offspring production, even though the two treatment groups did not differ significantly. This suggests to

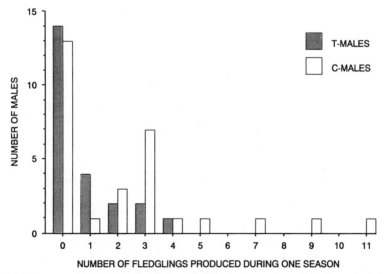

Fig. 14.16 Frequency distribution of numbers of fledglings apparently produced by male dark-eyed juncos during single breeding seasons, according to treatment. (Data from Nolan et al., unpublished observations from 1989 and 1990.)

us that the major determinants of reproductive success may be events beyond the control of the individual's choice of behavioral options; that is, that behavior and fitness are not tightly linked. Among Carolina juncos, nest predators (rodents, especially the chipmunk *Tamias striatus*) play an extremely important role in determining whether a pair produces offspring, and the density of predators can vary between years by as much as an order of magnitude (E. D. Ketterson and V. Nolan Jr., personal observations; J. Wolff, pers. comm.). This variation appears to be attributable to fluctuations in the annual production of acorns, a major food of rodents in our study area.

Conclusions

Treatment with testosterone has various phenotypic consequences in male juncos: it reduces the frequency of feeding nestlings, increases song rate, accelerates reduction of winter fat deposits to the low level characteristic of the breeding season, leads to elevated corticosterone, and postpones or prevents prebasic molt (Ketterson et al. 1991a, 1992; Nolan et al. 1992). It is almost certainly true that even for testosterone-dependent traits, there is more to the regulation of trait expression than simply the concentration of testosterone in the plasma, but our study shows that alteration of the normal seasonal pattern of plasma testosterone leads to alterations in the phenotype.

However, most of the correlates of fitness that we have investigated thus far do not appear to be much affected by the phenotypic alterations induced by elevated testosterone. Returning to table 14.1, controls weakly outperformed T-males on eight of the sixteen correlates measured, T-males somewhat outperformed controls on four, and on four the treatment groups were equal. However, only two of the comparisons were statistically significant, and in all but one of the nonsignificant comparisons the means of the treatment groups differed by less than 10%. One of the significant differences was paradoxical: males with elevated testosterone during the summer were more likely to survive the following winter if the implants were removed. The other significant difference was predictable from an adaptive standpoint: males that did not molt did not survive as well over the winter. Two important correlates of fitness that remain to be measured are true paternity as determined by genetic analysis and survival of offspring between nest-leaving and independence.

As caveats, we note, first, that the correlates we used are not true measures of fitness because they do not follow individuals from zygote to zygote, but instead cross generations—the parental and the filial. Second, not all of them are independent. Therefore we regard them simply as indicators of relative performance. Finally, other measures of fitness—not yet measured by us—might have been more sensitive to the behavioral and physiological changes we induced with testosterone than the ones we monitored. Nevertheless, we return to the question whether significant deviations from the normal seasonal pattern of plasma concentrations of testosterone in male juncos lead to results that are paradoxical, neutral, or, as would be expected from an adaptationist perspective, disadvantageous.

In some respects our results support an adaptive interpretation. With regard to the temporal pattern of high testosterone, wide deviations in the onset or the termination of peak levels would probably increase mortality. T-males were leaner in spring, when severe weather is still likely, and maintenance of high testosterone into autumn evidently increases mortality. The effect of mimicking the pattern seen in polygynous species, with testosterone elevated only throughout the normal breeding season, is less clear. To detect such effects we look to the measures of reproductive success in years when we elevated testosterone only during breeding: T-males and C-males did not differ significantly in any measure examined so far. We might argue that the nonsignificant differences (generally favoring C-males) that we observed in reproductive success would predict larger, biologically meaningful differences under more severe environmental circumstances. If this is correct, we would conclude that, in other years or at other times, prolonged high testosterone would produce measurable decreases in fitness. For example, if food were in short supply, reproductive success of T-males might be lower because their mates would be unable to compensate for reduced male parental care. In turn, under such conditions females might more frequently desert males that were poor parents. In this adaptive view, spatial and temporal variation in selective regimes would be responsible for maintaining the status quo.

Alternatively, if we take our observations to mean that the treatment groups would not differ, then we may conclude that a wide range of behavioral and physiological phenotypes is effectively neutral in the junco. If this is true, then many of the important determinants of survival and reproduction in the field may be random events that override individual variation in behavior and physiology, and fitness profiles in nature may be flatter than we have come to expect.

At this stage of investigation, we do not attempt to discriminate among these possibilities, and evolutionary biologists probably would disagree as to which is most likely to be correct. We continue to measure other potential fitness correlates of testosterone, especially the frequency of extra-pair fertilizations, to monitor reproductive success in order to determine how much annual variability there is, and to seek further phenotypic consequences, both behavioral and physiological, of elevating testosterone. We also remain open to the possibility that chance is a major determinant of reproductive success and survival in the juncos that we study.

ACKNOWLEDGMENTS

We thank our collaborators, postdoctoral and graduate students, and the many people who helped us with the fieldwork and in the laboratory. Collaborators and postdoctoral students include G. Ball, M. Cawthorn, C. R. Chandler, A. Dufty, P. G. Parker, W. Weathers, L. Wolf, J. Wingfield, and C. Ziegenfus. Graduate students and research assistants include P. Arguin, B. Beaty, K. Bruner, L. Callahan, D. Cristol, M. Crowe, D. Cullen, T. Johnsen, K. Jones, J. Kidd, K. Kimber, G. McPeek, M. Ragland, S. Radjy, S. Raouf, A. Richards, S. Robbins, L. Rowe, J. C. Martinez-Sanchez, M. Tavel, and R. Titus. Several people with whom we have not worked have influenced our thinking in important ways, including M. Moore, S. Skinner, S. Stearns, J. Travis, and M. J. West-Eberhard. We also thank our colleagues and students at Indiana University and the Mountain Lake Biological Station for their stimulating company, and the Mountain Lake Hotel for allowing us to work on their property. We are grateful to the following people who read and commented on an earlier draft of this manuscript: M. Cawthorn, D. Cristol, B. Cushing, C. Lively, S. Skinner, J. Walters, and H. Wiley. Our junco research has been supported by the NSF (BSR 87-18358 and BSR 91-11498), Indiana University, Biomedical Research Support Grants, and the American Philosophical Society. Finally, we thank Les Real for organizing the symposium that gave rise to this chapter.

REFERENCES

Adler, N. T., ed. 1981. *Neuroendocrinology of reproduction: Physiology and behavior.* New York: Plenum Press.

Andersson, M. 1982. Female choice selects for extreme tail length in a widowbird. *Nature* 299:818–20.

Archawaranon, M., and R. H. Wiley. 1988. Control of aggression and dominance in white-throated sparrows by testosterone and its metabolites. *Hormones and Behavior* 22:497–517.

Arnold. A. 1982. Neural control of passerine song. In D. E. Kroodsma, E. H. Miller, and H. Ouellet, eds., *Acoustic communication in birds,* vol. 1, 75–94. New York: Academic Press.

Balthazart, J. 1983. Hormonal correlates of behavior. In D. S. Farner, J. R. King, and K. C. Parkes, eds., *Avian biology,* vol. 7, 221–365. New York: Academic Press.

Balthazart, J., and G. F. Ball. 1993. Neurochemical differences in two steroid-sensitive areas mediating reproductive behaviors. In R. Gilles, ed., *Advances in comparative and environmental physiology,* vol. 5, 133–61. Berlin: Springer.

Beletsky, L. D., G. H. Orians, and J. C. Wingfield. 1990. Effects of exogenous androgen and antiandrogen on territorial and nonterritorial red-winged blackbirds (Aves:Icterinae). *Ethology* 85:58–72.

Boag, D. A. 1982. How dominance status of adult Japanese quail influences the viability and dominance status of their offspring. *Canadian Journal of Zoology* 60:1885–91.

Boinski, S. In press. Geographical variation in behavior of a primate taxon: Stress responses as a proximate mechanism in the evolution of social behavior. In S. A. Foster and J. A. Endler, eds., *Geographic diversification of behavior: An evolutionary approach.* Oxford: Oxford University Press.

Cardwell, J. R., and N. R. Liley. 1991a. Androgen control of social status in males of a wild population of the stoplight parrotfish, *Sparisoma viride. Hormones and Behavior* 25:1–18.

Cardwell, J. R., and N. R. Liley. 1991b. Hormonal control of sex and color change in the spotlight parrotfish, *Sparisoma viride. General and Comparative Endocrinology* 81:7–20.

Carter, S. 1992. Oxytocin and sexual behavior. *Neuroscience and BioBehavioral Reviews* 16:131–44.

Chandler, C. R., E. D. Ketterson, V. Nolan Jr., and C. Ziegenfus. In press. Effects of testosterone on spatial activity in free-ranging male dark-eyed juncos, *Junco hyemalis. Animal Behaviour.*

Chrousos, G. P., D. L. Loriaux, and P. W. Gold, eds. 1988. Mechanisms of physical and emotional stress. *Advances in Experimental Medicine and Biology,* vol. 245, 1–530. New York: Plenum Press.

Cody, M. L. 1966. A general theory of clutch size. *Evolution* 20:174–84.

Cunningham, D. L., and P. B. Siegal. 1978. Response to bidirectional and reverse selection for mating behavior in Japanese quail *Coturnix japonica. Behavioral Genetics* 8:387–97.

Derting, T. L. 1989. Metabolism and food availability as regulators of production in juvenile cotton rats. *Ecology* 70:587–95.

De Steven, D. 1980. Clutch size, breeding success and parental survival in the tree swallow (*Iridoprocne bicolor*). *Evolution* 34:278–91.

Dingle, H., and Hegmann, J. P. 1982. *Evolution and genetics of life histories.* Berlin: Springer.

Dufty, A. M., Jr. 1989. Testosterone and survival: A cost of aggressiveness? *Hormones and Behavior* 23:185–93.

Emlen, S. T., and L. W. Oring. 1977. Ecology, sexual selection, and the evolution of mating systems. *Science* 197:215–23.

Epple, A., and M. H. Stetson, eds. 1980. *Avian endocrinology.* New York: Academic Press.

Falconer, D. S. 1989. *Quantitative Genetics.* 3d ed. New York: Wiley.

Feder, H. 1984. Hormones and sexual behavior. *Annual Review of Psychology* 35:165–200.

Feuerbacher, I., and R. Prinzinger. 1981. The effects of the male sex-hormone testosterone on body temperature and energy metabolism in male Japanese quail (*Coturnix japonica*). *Comparative Biochemistry and Physiology* 70A:247–50.

Gross, M. R. 1984. Sunfish, salmon, and the evolution of alternative reproductive strategies and tactics in fishes. In G. W. Potts and R. J. Wooten, eds., *Fish reproduction: Strategies and tactics*, 55–75. New York: Academic Press.

Gross, M. R. 1985. Disruptive selection for alternative life histories in salmon. *Nature* 313:47–48.

Gyger, M., S. J. Karakashian, A. M. Dufty, Jr., and P. Marler. 1988. Alarm signals in birds: The role of testosterone. *Hormones and Behavior* 22:305–14.

Hännsler, I., and R. Prinzinger. 1979. The influence of the sex hormone testosterone on body temperature and metabolism of Japanese quail. *Experientia* 35:509–10.

Harding, C. F., M. J. Walters, D. Collado, and K. Sheridan. 1988. Hormonal specificity and activation of social behavior in male red-winged blackbirds. *Hormones and Behavior* 22: 402–18.

Hegner, R. E., and J. C. Wingfield. 1987. Effects of experimental manipulation of testosterone levels on parental investment and breeding success in male house sparrows. *Auk* 104: 470–80.

Hirshfield, M. F., and T. W. Tinkle. 1975. Natural selection and the evolution of the reproductive effort. *Proceedings of the National Academy of Sciences U.S.A.* 72:2227–31.

Johnsen, T. S. 1991. Steroid hormones and male reproductive behavior in red-winged blackbirds (*Agelaius phoeniceus*): Seasonal variations and behavioral correlates of testosterone. Ph.D. dissertation, Indiana University, Bloomington.

Ketterson, E. D., V. Nolan Jr., L. Wolf, C. Ziegenfus, A. M. Dufty, Jr., G. F. Ball, and T. S. Johnsen. 1991a. Testosterone and avian life histories: The effect of experimentally elevated testosterone on corticosterone and body mass in dark-eyed juncos. *Hormones and Behavior* 25:489–503.

Ketterson, E. D., V. Nolan Jr., C. Ziegenfus, D. P. Cullen, and M. Cawthorn. 1991b. Nonbreeding season attributes of male dark-eyed juncos that acquired breeding territories in their first year. *Acta XX Congressus Internationalis Ornithologici* 2:1229–39.

Ketterson, E. D., V. Nolan Jr., L. Wolf, and C. Ziegenfus. 1992. Testosterone and avian life histories: Effects of experimentally elevated testosterone on behavior and correlates of fitness in the dark-eyed junco (*Junco hyemalis*). *American Naturalist* 140:980–99.

Knobil, E., and J. D. Neill. 1988. *The physiology of reproduction*. Vols. 1 and 2. New York: Raven Press.

Ligon, J. D., R. Thornhill, M. Zuk, and K. Johnson. 1990. Male-male competition, ornamentation and the role of testosterone in sexual selection in red jungle fowl. *Animal Behaviour* 40:367–73.

Marler, C. A., and M. C. Moore. 1988a. Energetic costs of increased aggression in testosterone-implanted males. *American Zoologist* 28:186A.

Marler, C. A., and M. C. Moore. 1988b. Evolutionary costs of aggression revealed by testosterone manipulations in free-living male lizards. *Behavioral Ecology and Sociobiology* 23:21–26.

Marler, C. A., and M. C. Moore. 1989. Time and energy costs of aggression in testosterone-implanted free-living male mountain spiny lizards (*Sceloporus jarrovi*). *Physiological Zoology* 62:1334–50.

Marler, C. A., and M. C. Moore. 1991. Supplementary feeding compensates for testosterone-induced costs of aggression in male mountain spiny lizards (*Scleroporus jarrovi*). *Animal Behaviour* 42:209–19.

Massa, R., and L. Bottoni. 1987. Effect of steroidal hormones on locomotor activity of the male chaffinch (*Fringilla coelebs* L.). *Monitore zoologico italiano*, n.s. 21:69–76.

Maxson, S. C., P. Shrenker, and L. C. Vigue. 1983. Genetics, hormones, and aggression. In B. S. Svare, ed., *Hormones and aggressive behavior*, 179–96. New York: Plenum Press.

Moore, M. C. 1991. Application of organization-activation theory to alternative male reproductive strategies: A review. *Hormones and Behavior* 25:154–79.

Moore, M. C., and C. A. Marler. 1988. Hormones, behavior, and the environment: An evolutionary perspective. In M. H. Stetson, ed., *Processing of environmental information in vertebrates*, 71–83. Berlin: Springer.

Moss, R., A. Watson, P. Rothery, and W. Glennie. 1982. Inheritance of dominance and aggressiveness in captive red grouse *Lagopus scoticus*. *Aggressive Behaviour* 8:1–18.

Nalbandanov, A. V. 1976. *Reproductive physiology of mammals and birds.* San Francisco: W. H. Freeman.

Nolan, V., Jr. 1978. The ecology and behavior of the prairie warbler, *Dendroica discolor.* Ornithological Monographs no. 26. Washington, D.C.: American Ornithologists' Union.

Nolan, V., Jr., E. D. Ketterson, C. Ziegenfus, D. P. Cullen, and C. R. Chandler. 1992. Testosterone and avian life histories: Experimentally elevated testosterone delays postnuptial molt in male dark-eyed juncos. *Condor* 94:364–70.

Nowicki, S., and G. F. Ball. 1989. Testosterone induction of song in photosensitive and photorefractory male sparrows. *Hormones and Behavior* 23:514–25.

Nur, N. 1984. The consequences of brood size for breeding blue tits. I. Adult survival, weight change and the cost of reproduction. *Journal of Animal Ecology* 53:479–96.

Oring, L. W., A. J. Fiviizzani, and M. E. El Halawani. 1989. Testosterone-induced inhibition of incubation in the spotted sandpiper (*Actitis mecularia*) [sic]. *Hormones and Behavior* 23: 412–23.

Partridge, L., and P. H. Harvey. 1988. The ecological context of life history evolution. *Science* 241:1449–55.

Rabenold, P. P., K. N. Rabenold, W. H. Piper, J. Haydock, and S. W. Zack. 1991. Shared paternity revealed by genetic analysis in cooperatively breeding tropical wrens. *Nature* 348:538–40.

Reznick, D. 1985. Costs of reproduction: An evaluation of the empirical evidence. *Oikos* 44: 257–67.

Runfeldt, S., and J. C. Wingfield. 1985. Experimentally prolonged sexual activity in female sparrows delays termination of reproductive activity in their untreated mates. *Animal Behaviour* 33:403–10.

Schleussner, G., J. P. Dittami, and E. Gwinner. 1985. Testosterone implants affect molt in male European starlings, *Sturnus vulgaris. Physiological Zoology* 58:597–604.

Schwabl, H., and E. Kriner. 1991. Territorial aggression and song of male European robins (*Erithacus rubecula*) in autumn and spring: Effects of antiandrogen treatment. *Hormones and Behavior* 25:180–94.

Sefton, A. E., and P. B. Siegal. 1975. Selection for mating ability in Japanese quail. *Poultry Science* 54:788–94.

Selander, R. K. 1965. On mating systems and sexual selection. *American Naturalist* 99:129–41.

Siegel, H. S. 1980. Physiological stress in birds. *BioScience* 30:529–34.

Silver, R., and G. F. Ball. 1989. Brain, hormone and behavior interactions: Status and prospectus. *Condor* 91:966–78.

Silverin, B. 1980. Effects of long-acting testosterone treatment on free-living pied flycatchers, *Ficedula hypoleuca,* during the breeding period. *Animal Behaviour* 28:906–12.

Sinervo, B., and R. Huey. 1990. Allometric engineering: Testing the causes of interpopulational difference in performance. *Science* 278:1106–9.

Sinervo, B., and P. Licht. 1991a. Hormonal and physiological control of clutch size, egg size, and egg shape in side-blotched lizards (*Uta stansburiana*): Constraints on the evolution of lizard life histories. *Journal of Experimental Zoology* 257:252–64.

Sinervo, B., and P. Licht. 1991b. Proximate constraints on the evolution of egg size, number, and total clutch mass in lizards. *Science* 252:1300–1302.

Sinervo, B., P. Doughty, R. B. Huey, and K. Zamudio. 1992. Allometric engineering: A causal analysis of natural selection on offspring size. *Science* 258:1927–30.

Stearns, S. C. 1989. Trade-offs in life-history evolution. *Functional Ecology* 3:259–68.

Travis, J. 1989. The role of optimizing selection in natural populations. *Annual Review of Ecology and Systematics* 20:279–96.

van Noordwijk, A. J., and G. de Jong. 1986. Acquisition and allocation of resources: Their influence on variation in life-history tactics. *American Naturalist* 1238:137–42.

Vehrencamp, S. L., and J. W. Bradbury. 1984. Mating systems and ecology. In J. R. Krebs and N. B. Davis, eds., *Behavioural ecology: An evolutionary approach,* 2d ed., 215–23. Oxford: Blackwell Scientific Publications.

Verner, J., and M. F. Willson. 1969. Mating systems, sexual dimorphism, and the role of male

North American passerine birds in the nesting cycle. Ornithological Monographs no. 9. Washington, D.C.: American Ornithologists' Union.

Wada. M. 1981. Effects of photostimulation, castration, and testosterone replacement on daily patterns of calling and locomotor activity in Japanese quail. *Hormones and Bahavior* 15: 270–81.

Wada, M. 1982. Effects of sex steroids on calling, locomotor activity, and sexual behavior in castrated male Japanese quail. *Hormones and Behavior* 16:147–57.

Wada, M. 1986. Circadian rhythms of testosterone-dependent behaviors, crowing and locomotor activity, in male Japanese quail. *Journal of Comparative Physiology* A 158:17–25.

Watson, A., and R. Parr. 1981. Hormone implants affecting territory size and aggressive and sexual behaviour in red grouse. *Ornis Scandinavica* 12:55–61.

West-Eberhard, M. J. 1989. Phenotypic plasticity and the origins of diversity. *Annual Review of Ecology and Systematics* 20:249–78.

West-Eberhard, M. J. 1992. Behavior and evolution. In P. R. Grant and H. Horn, eds., Molds, molecules, and metazoa: Growing points in evolutionary biology, 57–75, Princeton: Princeton University Press.

Westneat, D. F., P. W. Sherman, and M. L. Morton. 1990. The ecology and evolution of extra-pair copulations in birds. *Current Ornithology* 7:331–69.

Williams, G. C. 1957. Pleiotropy, natural selection, and the evolution of senescence. *Evolution* 11:398–411.

Wingfield, J. C. 1984. Androgens and mating systems: Testosterone-induced polygyny in normally monogamous birds. *Auk* 101:665–71.

Wingfield, J. C. 1991. Mating systems and hormone-behavior interactions. *Acta XX Congressus Internationalis Ornithologici* 4:2055–62.

Wingfield, J. C., and D. S. Farner. 1975. The determination of five steroids in avian plasma by radioimmunoassay and competitive protein binding. *Steroids* 26:311–27,

Wingfield, J. C., and M. C. Moore. 1988. Hormonal, social, and environmental factors in the reproductive biology of free-living male birds. In D. Crews, ed., *Psychobiology of reproductive behavior*, 149–75. Englewood Cliffs, N.J.: Prentice Hall.

Wingfield, J. C., A. Newmann, G. L. Hunt, and D. S. Farner. 1982. Endocrine aspects of female-female pairing in the western gull (*Larus occidentalis wymani*). *Animal Behaviour* 30:9–22.

Wingfield, J. C., G. F. Ball, A. M. Dufty, Jr., R. E. Hegner, and M. Ramenofsky. 1987. Testosterone and aggression in birds. *American Scientist* 75:602–8.

Wingfield, J. C., R. E. Hegner, A. M. Dufty, Jr., and G. F. Ball. 1990. The "Challenge Hypothesis": Theoretical implications for patterns of testosterone secretion, mating systems, and breeding strategies. *American Naturalist* 136:829–46.

Wolf, L. 1987. Host-parasite interactions of brown-headed cowbirds and dark-eyed juncos in Virigina. *Wilson Bulletin* 99:338–50.

Wolf, L., E. D. Ketterson, and V. Nolan Jr. 1988. Paternal influence on growth and survival of dark-eyed junco young: Do parental males benefit? *Animal Behaviour* 36:1601–18.

Wolf, L., E. D. Ketterson, and V. Nolan Jr. 1990. Behavioural response of female dark-eyed juncos to the experimental removal of their mates: Implications for the evolution of male parental care. *Animal Behaviour* 39:125–34.

Wolf, L., E. D. Ketterson, and V. Nolan Jr. 1991. Female condition and delayed benefits to males that provide parental care: A removal study. *Auk* 108:371–80.

Zuk, M. 1990. Reproductive strategies and disease susceptibility: An evolutionary viewpoint. *Parasitology Today* 6:231–33.

Zuk, M., K. Johnson, R. Thornhill, and J. D. Ligon. 1990. Parasites and male ornaments in free-ranging and captive red jungle fowl. *Behaviour* 114:232–48.

FIFTEEN

Immunology and the Evolution of Behavior

Marlene Zuk

BEHAVIOR AND ITS EVOLUTION rely on many proximate physiological mechanisms. Neurological pathways in the brain and sense organs, for example, determine how information in an animal's environment is perceived, and to a certain extent, how it is processed. The presence or amount of particular hormones circulating in the bloodstream may regulate the onset of mating, egg-laying, or fighting behavior. Metabolic rate, in conjunction with body size parameters, often constrains the rate at which many activities are performed or the kind of climate in which animals are able to survive. As a result, neurobiology, endocrinology, and energetics are all commonly linked with behavioral studies, and fields such as behavioral neurobiology are the frequent outcome.

Immunology has rarely been considered in the context of behavior, and certainly not in the context of sociobiology or behavioral ecology, with their emphasis on the evolution of adaptive strategies and social interactions. At the same time, however, mainstream evolutionary biology and population ecology have begun to recognize the importance of parasites and diseases in host community and population studies (Dobson and Hudson 1986; Toft and Karter 1990). The two tracks have remained largely separate. Yet for a single organism, the immune response that occurs when its body is invaded by parasites or disease is equivalent to an individual-level strategy for dealing with the population-level phenomenon of infection.

Usually the immune system is invoked by ecologists only as a source of genetic resistance for the population under consideration, rather than as a variable affecting individual behavior. Hamilton and Zuk (1982) suggested that parasites might play an important role in sexual selection, but did not consider the mechanisms by which this might occur, or the feedback between the immune system and morphology that must mediate the interaction. One exception to this neglect is the research on self/nonself discrimination in colonial invertebrates; the marine ascidian *Botryllus schlosseri*, for example, appears to be able to detect kin based on the immunologically important histocompatibility locus (Grosberg and Quinn 1986). Indeed, much of the terminology of immunology is evocative of behavioral research, including as it does the concepts of "recognition," "surveillance," "selection," and "maturation."

It turns out that immunology has a great deal to offer behavioral ecology, and in this chapter I discuss some of the proximate and ultimate relationships between behavior and the immune system. The rapidly developing field of psychoneuroimmunology explores some of these connections, but mostly with regard to the effects of human behavioral states such as stress or depression on the immune system's ability to fight cancer and other diseases (Ader 1981; Ader et al., 1991). I would like to expand this viewpoint, not only to include nonhumans, but also by investigating both directions of the feedback: the psychological state of an individual influences its immune response, but equally important, aspects of immunity may have profoundly affected the evolution of such behaviorally important traits as social group formation, the development of secondary sex characters, parental care patterns, and the mating system. In addition, knowledge about the immune competence of individuals or populations may lead to a new and potentially more realistic assay of "condition" or "robustness" than measures such as length:weight ratios or subjective gauges of fur or feather wear. Below I treat several promising areas in "behavioral immunology," including sex differences in disease susceptibility, the role of sex hormones in mediating the relationship between parasites and sexual selection, and the use of immunological assays in population biology.

SEX DIFFERENCES IN DISEASE SUSCEPTIBILITY

Epidemiologists have long noted sex differences in patterns of infection with a wide variety of parasites and diseases (Bundy 1989; Alexander and Stimson 1989). Chagas' disease, for example, is much more prevalent in males than in females, both in mice and in humans (Hauschka 1947; Alexander and Stimson 1989). Among parasitologists it is common practice to use male rat or mouse hosts rather than females to rear "good crops" of various kinds of helminths (E. A. Platzer, pers. comm.). Although exceptions exist, this trend toward greater susceptibility to disease in males is widespread among vertebrates. Sociological differences between the sexes certainly account for some of the disparity in disease susceptibility, with men and women tending, as a hypothetical example, to be differentially exposed to waterborne vectors of some diseases because of sex-limited tasks such as fishing. More interesting for our purpose here, the greater vulnerability of males may be viewed as an evolved phenomenon, reinforced by both short- and long-term forces, and variable in ways that reflect natural selection acting differently on animals with different life histories (Zuk 1990).

The more short-term explanation is based on the increasingly well-documented relationship between stress and the immune system (Sapolsky 1987; Monjan 1981; Keller et al. 1991). In humans as well as other vertebrates, stress in the form of environmental unpredictability or physical challenge (food sources suddenly disappear or change, war breaks out, a late spring storm interrupts breeding), social pressures (a previously subordinate animal repeatedly challenges a dominant one, a business partner is promoted in lieu of oneself), or personal sources (death of a loved one) can lead to greater vulnerability to infection by parasites and diseases (Sapolsky 1987; Kiecolt-Glaser and Glaser 1991). Both the physiological products of the stress response itself, such as changes in corticosteroid levels, and the concomitant physical hardships of being cold, hungry, and so forth, can be debilitating.

No one could be more stressed than the males of many vertebrate species during the mating season, when courtship displays are exhausting, the environment must be constantly scrutinized for competitors and those competitors fought off, and secondary sex characters such as the elaborate antlers of many deer must be developed. As Sapolsky (1987) discovered in studies of olive baboons, neither dominant nor subordinate status means freedom from the physiological costs of stress. The vocal courtship of some frogs and toads has been shown to be up to three times more energetically costly than resting metabolism (Taigen and Wells 1985). This energy expenditure and stress probably renders males less able to resist invasion by pathogens and more likely to suffer greater damage when infected, leading to the observation of greater incidence of disease in males.

Females, of course, are expending energy in reproduction as well, especially mammals, in which pregnancy and lactation occur. But these stresses are predictable, and females are less stressed socially by competition during a short, intense time period. Because of the classic differences in male and female reproductive strategies, males are more likely to be risk takers, while variance in female reproductive success is relatively low (Trivers 1972). Most females are likely to produce some offspring even if they are not the most vigorous or healthy member of their group, but a sickly male will probably lose out entirely. The male strategy of "live hard, die young" is thus typical of many different species (Zuk 1990).

In a more long-term sense, testosterone, the male sex hormone responsible for many of the displays and structures males use in attracting mates, has its own insidious role. As will be detailed in the next section, testosterone has a suppressive effect on the immune system (Grossman 1985; McCruden and Stimson 1991). Males are thus necessarily more vulnerable to disease as they acquire the accoutrements of maleness, in what may be viewed as a cruel bind of nature. The deleterious effects of testosterone may be an unavoidable price paid by males for achieving reproductive success in a competitive environment. One can envision a mechanism here similar to the negative pleiotropy that evolutionary biologists suggest may account for the senescence and eventual death of most living organisms (Williams 1966). If a gene has a beneficial effect early in life, when reproduction is occurring, but a deleterious effect late in life, selection will still favor individuals with the gene because by the time it exerts its damage, its bearers will already have outreproduced individuals with longer life spans but smaller reproductive output (Daly and Wilson 1983; Alexander 1979). In addition, such genes exert their beneficial effects while more competitors are still alive. Testosterone may make males more vulnerable to disease, but because it enhances competitive and aggressive abilities in many animals during a crucial phase of life, selection for males with high androgen levels continues. Counterselection for males with high degrees of resistance to parasites, though expected to occur, will not be as effective as selection for greater mating success, since male reproductive strategy depends on such a high-risk, high-stakes game.

Most of the research on sex differences in disease susceptibility has focused on rodents and humans as subjects, as is typical of fields with biomedical goals. Here is the area where behavioral biologists can bring new insights, because if students of animal behavior have learned nothing else over the last several decades, we have learned that different species have evolved under different circumstances. This almost embarrassingly obvious recognition is still ignored by many medical researchers, but it may have profound implications for the use of rodents evolved under one set of conditions as model organisms for animals evolved under entirely different sets. Mating systems and the intensity of sexual competition differ greatly among species: some primates, such as gibbons, are monogamous, with males providing virtually all parental care, while elephant seals are exaggeratedly polygynous, and many other species are somewhere in between (Leighton 1987; Le Boeuf and Reiter 1988; Bronson 1979). Male reproductive success in monogamous species is unlikely to rely as heavily on testosterone-dependent traits as it does in polygynous species. One would expect, therefore, that differences between the sexes in disease susceptibility would reflect the strength of differential selection on males and females in the

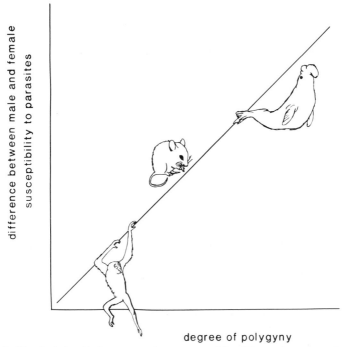

Fig. 15.1   Predicted relationship between mating system and the difference between male and female susceptibility to parasites. Gibbons are monogamous, elephant seals highly polygynous, and house mice probably intermediate. (From Zuk 1990.)

evolutionary sense (fig. 15.1). Species with monogamous mating systems, like the gibbons, should have relatively more similar selection on males and females, and the sexes should be relatively less different in disease susceptibility, than species with very dissimilar reproductive strategies for males and females. A good test species for this idea would be a polyandrous one, such as one of the jacanas, South American birds in which females are more aggressive, more brightly colored, and more competitive than the males, and also have higher testosterone titers (Johns 1964). Species with polyandry as the mating system are expected to show a reversal of the usual trend for males to be more susceptible to pathogens.

The rodents commonly used in laboratory studies of parasite infection did, of course, evolve in a particular natural environment; most Old World rats and mice from which experimental stocks derive were probably moderately polygynous, with both males and females mating with more than one individual but with variance in reproductive success being higher in males than in females (Bronson 1979). Still, generalizing to all other mammals from results obtained using species adapted to one set of ecological circumstances is probably risky.

TESTOSTERONE, IMMUNITY, AND THE EVOLUTION OF ORNAMENTS

The immunosuppressive effects of testosterone have recently been examined in the context of sexual selection and the evolution of male secondary sex characters by

Folstad and Karter (1992). They propose an "immunocompetence handicap" hypothesis in which testosterone-dependent signals are facultative, and males able to pay the price of lowered immune competence while maintaining elaborate ornaments are favored by females (Folstad and Karter 1992). Here I review their idea in a more general context, and suggest alternative mechanisms for the feedback between sex hormones and immune system function.

Parasites and Sexual Selection

As mentioned above, parasites and diseases have been implicated in sexual selection and reproductive behavior by several researchers. Freeland (1976) suggested that many aspects of primate social behavior might serve to avoid infection, and Hart (1990) reviewed the behavioral aspects of reduction of disease transmission. Hamilton and Zuk (1982) proposed that because of the unique genetic interactions possible between parasites and their hosts, females are expected to choose mates based on male resistance to pathogens. Male secondary sex characters or ornaments are seen to have evolved at least in part as indicators of this resistance, and females choosing males with well-developed ornaments are obtaining the genes for resistance for their offspring (Hamilton and Zuk 1982). Hamilton and Zuk (1982) predicted that within species, females should prefer males with fewer parasites, as indicated by the degree of development of secondary sex characters, whereas among species, sexual selection should appear to have been most intense, and males showiest, in those species with the heaviest parasite levels.

Intraspecific tests have been generally, though not entirely, supportive of the Hamilton and Zuk hypothesis (for recent reviews of both inter- and intraspecific studies see Møller 1990b; Sullivan 1991; Zuk 1992). A comparison of development of male morphological characters in unparasitized jungle fowl with that in roosters experimentally infected with *Ascaridia galli*, an intestinal nematode, showed that although chicks in the parasitized group grew more slowly, characters such as tarsus length and bill size were not significantly different in the two groups at sexual maturity, suggesting that the parasitized birds had compensated for the effects of infection (Zuk et al. 1990). In contrast, ornamental characters such as comb length and eye color remained different into adulthood, with the control roosters exhibiting brighter, showier characters. In mate choice tests allowing a hen to choose between a parasitized and a control rooster, controls were preferred by a margin of 2:1, and analysis of preference patterns revealed that the hens used the same characters to make their mating decisions as were important to the researchers in distinguishing control from parasitized males (Zuk et al. 1990).

Other relevant research includes Møller's work (1990a, 1990c) on tail length and ectoparasites in swallows, Spurrier et al.'s (1991) study of mate choice in sage grouse, and work by Pruett-Jones et al. on hematozoa in New Guinea birds of paradise (Pruett-Jones et al., 1990, 1991; Pruett-Jones and Pruett-Jones, 1991). These and other studies suggest that the connection between parasites and both sexually selective characters and female mate choice is a real one, although the mechanisms responsible for that link remain poorly understood (Zuk 1992).

Sex Hormones and Reproductive Characters

The close relationship between sex hormone levels and many reproductive characters, both morphological and behavioral, is well documented (see Ketterson and Nolan, chap. 14). Testosterone is responsible for the growth and development of many familiar secondary sex characters, including the combs of jungle fowl, chickens, and other gallinaceous birds, increasing male muscle mass in numerous birds and mammals, development of antlers in cervids, and facial and body hair in human males (Allee et al., 1939; Lofts and Murton 1973; Johnson 1986; Crews 1987; Wittenberger 1981; Folstad and Karter 1992). Social dominance has also been demonstrated to be testosterone dependent, with experimental castration generally reducing aggression and subsequent testosterone injections usually causing its return (Lofts and Murton 1973; Wittenberger 1981; Wingfield and Ramenofsky 1985). Artificial elevations of testosterone in usually monogamous white-crowned sparrows resulted in polygamy, and the territory size of testosterone-implanted song sparrows was significantly increased over that of controls (Wingfield 1984). Although hormones clearly interact with environmental factors such as age, experience, and habitat, which mediate the endocrine effects, hormones such as testosterone nonetheless are critical in the proximate mechanisms of sexual selection.

Sex Hormones and Immune Response

It has been known for some time that sex hormones affect the immune system, and more particularly that testosterone has a detrimental effect on the ability of the immune system to respond to infection (Alexander and Stimson 1989; Grossman 1985; McCruden and Stimson 1991). This suppression appears to be widespread among the vertebrates studied, and probably accounts for the previously mentioned greater susceptibility of sexually mature males to infections from a wide variety of parasites and diseases (Hauschka 1947; Goble and Konopka 1973; Purtilo and Sullivan 1979; Grossman 1985; Folstad et al. 1989; Zuk 1990). Evidence from golden hamsters even suggests that sexual behavior itself suppresses the immune response, an effect presumably mediated by the sex hormones (Ostrowski et al. 1989). Hormonal effects on the immune system range from suppression or enhancement of circulating immunoglobulin levels to tissue-level responses such as growth of the thymus gland, the site of T-lymphocyte differentiation (Grossman 1985). Empirical evidence from mammals, fishes, and birds ranges from demonstrations that immune competence decreases during the breeding season, when testosterone titers are peaking, to experimental findings of reduced natural killer cell activity after administration of testosterone (Folstad and Karter 1992). Immune system responses themselves, of course, are essential in the body's reaction to parasites and diseases of all types (Glick 1986; Roitt et al. 1990).

Links therefore exist between parasites and sexual selection, between sex hormones and sexual selection, and between the immune system and sex hormones. How do these three facets interact, and how is the interaction manifested in behavior? Folstad and Karter (1992) suggest that male ornaments such as combs and showy plumage are "honest" signals of a male's ability to withstand the obligatory immunosuppressive effects of high testosterone titers. Cheating is unlikely because the cost of doing so, of presenting showy characters when viability or fitness is low, is a loss of immune function that makes the male too vulnerable to the pathogenic effects of parasites for the character to be maintained. Males with highly developed secondary sex characters are therefore indicating their capacity to resist the effects of prevalent parasites even with a compromised immune system.

This idea, however, does not distinguish between the effects of innate defense mechanisms and those of acquired immunity. Although they interact, in general the host response to invasion by a foreign substance has two arms: genetic immunity, which as the name implies is inherited and is either present or capable of being activated at birth, or else absent or incapable of activation; and acquired immunity, the result of facultative cell-mediated and humoral processes, including antibody formation (Roitt et al. 1990). Testosterone, because it is transient in its presence and amount in the body, presumably affects the acquired aspect of immunity. Folstad and Karter's model suggests no explanations of how some males might be better able to afford expenditure of energy and deal with the effects of immune suppression.

An alternative, albeit related, explanation of how immunity and sexual selection are linked is that females benefit most by mating with males that are *genetically* resistant to parasites. Males with this "bonus" resistance are spared the effort of mounting as complete an acquired immune response as are males that lack such genetic capability. Females choosing such males are therefore gaining heritable benefits of parasite resistance for their offspring, as well as the more immediate benefits of mating with a vigorous male. This heritable fitness gain is of course the advantage originally postulated by Hamilton and Zuk (1982). Males with both genetic and acquired immunity might thus be most likely to produce well-developed secondary sex characters, which requires high testosterone production. The implication of at least one aspect of genetic immunity, the Major Histocompatibility Complex, in mate choice and sexual selection will be discussed in a later section.

Yet another possible mechanism for the interaction of the endocrine and immune systems in producing ornaments might be derived from the molecular processes initiated by host cellular immune responses to parasitic infections. Cellular immune responses are controlled by cytokines, such as the interleukins and tumor necrosis factors (TNF). These cytokines are transiently released by immune cells, and act like hormones to initiate inflammatory host responses to clear the body of infectious agents (Beutler and Cerami 1989). In chronic infections or malignancies, however, cytokine production is elevated, resulting in systemic deleterious changes in host physiology. Such chronic release of cytokines may lead to a generalized progressive wasting process known as cachexia, characterized by weight loss, anorexia, and anemia; the phenomenon is common in diseases such as AIDS. Accompanying these pronounced catabolic effects, and mediated by TNF, may be the downregulation of testosterone in sexually mature males (Beutler and Cerami 1989). Thus, chronic cytokine release due to recurrent parasite infection may result in reduced development of secondary sex characters. This hypothesis suggests that infection causes the immune system to depress hormone levels, rather than testosterone itself reducing immune competence. As detailed in the last section, it also suggests a way to monitor immune competence in animals. Whatever the mechanisms of endocrine-immune interaction turn out to be, it is clear that information about both the short-term and the long-term effects of disease on immune function, hormone levels, and the development of sexually selected characters is sorely needed if we are to integrate behavior and immunology.

MATING PREFERENCES AND THE MHC

The Major Histocompatibility Complex (MHC) is a gene linkage group found in all vertebrates and in some invertebrates that codes for the cell surface antigens important in antigen presentation and recognition, the cytokines involved in inflammatory processes, and the complement proteins that act in concert with antibody to destroy pathogens (Perkins 1980; Roitt et al. 1990). The gene complex is inherited in groups of highly polymorphic alleles called haplotypes, some of which may be associated with resistance to various pathogens (Perkins 1980; Roitt et al. 1990; Howard 1991).

The interest of evolutionary biologists in the MHC has centered on two areas: the factors responsible for maintaining the extraordinarily high diversity of alleles (Potts and Wakeland 1990), and the possible use of MHC haplotype as a criterion in mate choice (Yamazaki et al. 1976, 1988; Egid and Brown 1989; Boyse et al. 1991; Potts et al. 1991a, 1991b). The former has important implications for population genetics theory, while the latter is an illustration of the kind of relationship between immunology and behavior that may prove fruitful in understanding sexual selection.

In a variety of inbred strains of laboratory mice, and in an outbred population of *Mus musculus* living under seminatural conditions, males and females appear to prefer to mate with individuals of an MHC type different from their own (Yamazaki et al. 1976, 1988; Egid and Brown 1989; Potts et al. 1991b). By increasing the heterozygosity of their offspring, the mice are presumed to be making an adaptive mate choice, because highly heterozygous individuals may be able to fend off a wider variety of pathogens (Potts and Wakeland 1990; Howard 1991). Early work on strains inbred to differ only at the MHC and nowhere else on the genome yielded somewhat equivocal results, with some strains showing the tendency to prefer nonself MHC types and others not; in addition, the experiments were always conducted using highly inbred strains under laboratory conditions. More recently, Potts et al. (1991b) demonstrated that free-ranging *Mus* in a large barnlike enclosure were mating disassortatively with regard to MHC, and females appeared to be seeking extraterritorial matings with males of MHC types different from their own, resulting in a level of homozygosity in their offspring that was on average 27% lower than expected under random mating (fig. 15.2).

The phenomenon of MHC-based mate choice appears well established in mice and perhaps in rats; whether it occurs in other mammals or other vertebrates awaits discovery. An unexplored area of this research is the possible interaction of morphology and the MHC; mice clearly use odor cues for detecting the different haplotypes (Yamazaki et al. 1988; Boyse et al. 1991), but other vertebrates that rely more on visual or auditory signals in sexual selection or individual recognition may use MHC

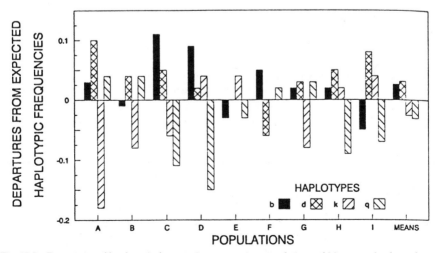

Fig. 15.2   Departures of haplotypic frequencies across nine populations of *Mus musculus* from those expected under random mating. Departure values were calculated as observed proportions minus expected proportions for each of the four MHC haplotypes. (From Potts et al. 1991b.)

information differently, or not at all. The volatile chemicals used by the mice to distinguish among haplotypes may be carried in intestinal microorganisms (Howard 1977; Boyse et al. 1991), which suggests that animals such as birds, which are not as reliant on odor, may be at a disadvantage in this regard.

IMMUNOLOGICAL GAUGES OF CONDITION

The final area I would like to highlight as a potentially valuable link between immunology and behavioral ecology is the use of immunological competence as a gauge of condition, both on an individual and a population level. Field techniques for estimating health and vigor of animals are generally fairly crude, consisting of such measures as body length: or wing chord:weight ratios, amount of visible fat, condition of fur or feathers, or some subjective assessment of "vigor." More precise analyses of body composition or fat consumption are valuable, but have the drawback of assuming a tight correlation between the presence of fat stores and fitness; certainly in the breeding season of many vertebrates, some of which do not even eat during this time, such a relationship may be absent.

Because all animals are exposed to parasites and disease, maintenance of a robust immune system is critical to survival and reproduction. A measure of immune system status might therefore be extremely useful for assessing vigor or fitness. In theory at least, one could monitor immunocompetence as a function of age, sex, time of year, or other environmental variables such as climatic events.

In collaboration with Doris Watt, I am examining the characteristics of blood cells from male and female American goldfinches captured throughout the breeding season in Indiana. We hope to determine whether the sexes differ in immune system parameters such as white blood cell composition (see below for details), and whether males show a drop in immunocompetence at the time when testosterone production is peaking, as the Folstad and Karter model would suggest. In addition, behavioral dominance data on some of the birds could help us detect relationships between immune function and the ability to maintain a high dominance rank. Following changes in blood cell characteristics throughout the nesting cycle could also suggest ways that the goldfinches are responding to different levels of stress; for example, females might be expected to be more energetically drained and immunologically vulnerable than males during egg laying and incubation, while males may be more stressed when competition for mates is at its height.

Whether or not any of these suggestions are borne out by the available data, the technique of examining immune function, both by blood cell characteristics and via the other techniques outlined below, is a potentially valuable one for behavioral ecologists. It allows evaluation of individual performance by a measure that is ecologically relevant and has undoubted importance to reproductive success. Virtually no baseline data are available on immune function in wild animals; almost all previous studies have been carried out on domestic animals, especially rodents (Glick 1986; Roitt et al. 1990; cf. Puerta et al. 1990). Adding some simple immunological assessments to other measures typically taken on individuals captured in the field could improve this situation, and might add much to our ability to understand and predict behavior.

The same measures could also yield insights at a population level. Population geneticists, conservation biologists, and many other evolutionary biologists are often concerned with assessing the health, vigor, or likelihood of persistence of a population (Gilpin and Soulé 1986). Such variables may depend on the degree of inbreeding depression, itself difficult to measure (Ralls et al. 1986; Allendorf and Leary 1986). Even when inbreeding depression can be assessed, it is difficult to evaluate the significance of, for example, changes in the levels of heterozygosity in certain allozymes (Templeton 1986). Because the in vivo function of the proteins used in electrophoretic work is often unknown, the conclusion that decreased genetic variability at particular loci is always deleterious may be ill-founded (Templeton 1986). Nonetheless, concern about increasingly fragmented native habitat and natural populations has led to a need for some means of comparing vigor among groups of animals.

As in the case of measurement of individual condition, assaying immune system status may provide at least a partial solution to this problem. By examining immunocompetence, it may be possible to determine whether small, isolated populations have retained the capacity to withstand disease outbreaks, both on an acquired and on a genetic level. Examination of the MHC holds particular promise for answering these questions, but other measures (detailed below) could also be useful, especially for looking at short-term ability to respond to challenge by infection. Work on MHC variability in cheetahs and other endangered mammals has already outlined levels of genetic variability within populations (O'Brien et al. 1983, 1985); it would be interesting and potentially valuable to focus even more attention on loci known to be of immunological importance.

WHAT TO MEASURE?

Suggesting an assessment of something as vague-sounding as "immune competence" is well and good, but what variables should be measured, and how? Below is a list of immune system components that might prove useful to examine in behavioral research. They range from the general to the specific, with concomitant costs in terms of ease and accuracy; blood smears are easy to obtain and read, but give far less specific information than assays of antibodies to targeted antigens, for example. A hoped-for side effect of increasing use of these techniques among behavioral biologists is a better understanding of the role of immune system components in wild animals living under natural conditions, circumstances unfortunately neglected by classic immunology.

Blood Cell Characteristics

A complete blood count and smear can be used to determine overt changes in cellular elements and thereby to assess an animal's general immune/hematological status. Information about white:red blood cell ratios and the presence or abundance of white blood cell types may be used as an indication of several different conditions. For example, increasing levels of eosinophils indicate marshaling of an immune response to challenge by helminth parasites (Wintrobe et al. 1974; Heyneman and Welch 1980; Roitt et al. 1990), and lowered hemoglobin levels suggest a depression in hematopoietic/bone marrow function (Wintrobe et al. 1974). The few studies carried out thus far suggest interesting variation in these components both among and within species (Chilgren and deGraw 1977; deGraw et al. 1979; Puerta et al. 1990).

Serum Immunoglobulin Levels and Triglycerides

Because immunoglobulins are the portion of the serum protein that contains antibodies, levels of immunoglobins reflect the production of antibodies to any foreign tissue invasion. Depending on the species studied, certain immunoglobulins may be of particular interest. Immunoglobulin M (IgM), for example, is often seen at elevated levels early in the human (and presumably mammalian) response to antigenically complex infectious organisms, whereas IgE is associated with response to helminth parasites (Roitt et al. 1990). Immunoglobulin assays can provide a gross measure of the status of B cell function, the B cells being the major component of one of the two arms of the acquired immune response (Roitt et al. 1990). Unfortunately, little is known about nonmammalian immunoglobulins, but some form of the molecules is known to exist in all vertebrate classes (Roitt et al. 1990). Tests for Ig levels are often commercially available or relatively simple to devise for a new subject.

Triglycerides are of interest because of the cachexia or wasting response mentioned above. One of the actions of TNF is to mobilize adipocytes and simultaneously suppress the enzyme lipoprotein lipase; the resulting decreased uptake of lipids leads to increased circulating levels of fat, particularly triglycerides, which therefore provide a relatively simple assay for systemic immune responsiveness (Glick 1986). In effect, fat unable to be utilized in the body can be measured in the serum. If indicated by elevated triglycerides, lipoprotein lipase activity in the liver can itself be measured. Because of their biomedical importance, triglyceride assays are commercially available and easy to use. They may also be integrated with commonly collected data on weight gain or loss.

Specific Antibody Tests

If a parasite or disease is of particular interest in the host species under study, tests for the presence and amount of antibody specific to that pathogen may be devised. Using samples from an unexposed host (or host egg or embryo tissue) and the pathogen, the level of serum (or egg-associated) antibodies specific to that pathogen may be measured in vitro. Specific antibodies can then be screened in samples of serum taken from host animals, using a highly sensitive technique called enzyme-linked immunosorbent assay (ELISA), which is a standard assay commonly used in immunology (Roitt et al. 1990).

MHC Haplotype Analysis

Heightened interest in DNA "fingerprinting" and other related analyses has drawn much recent attention to the use of molecular genetic techniques in ecology and evolutionary biology (Burke 1989). Because the MHC is so variable, it is a useful portion of the genome for these purposes. Analysis of MHC haplotypes via restriction fragment length polymorphism (RFLP) is now realizable, and information about the variability and frequencies of the various haplotypes could be used in conjunction with information about behavior, morphology, and susceptibility to disease to suggest ways in which the genetic aspects of immunity interact with reproduction and behavior. For instance, are certain haplotypes associated with a particular degree of development of a secondary sex character? Again, the technique is best developed in mammals, but probes for birds are available, and MHC analysis will probably soon be widely accessible to behavioral ecologists.

CONCLUSIONS

This chapter was designed to convince behavioral biologists that "behavioral immunology" is not as unlikely a combination of fields as it might appear. Some of the actual techniques suggested may prove impractical, but the basic premise should still hold: all animals in all environments must have an immune system that allows them to fend off the omnipresent pathogens, and therefore immune response is a physiological universal, more so than the need for food or rest, that affects all other aspects of the organism's biology.

Numerous questions, of course, remain. Much of this chapter has been cast toward vertebrates, for the simple reason that almost nothing is known about invertebrate immunology (cf. Gupta 1991). Invertebrates certainly suffer from a wide variety of diseases (Lackie 1986; Gupta 1991), but little information is available about the mechanisms involved in their response. At one time it was thought that insects lacked the humoral immune responses of vertebrates, but more recently this belief has been challenged by the discovery of B cell-like components in insect hemolymph (Gupta 1991). It may turn out that invertebrate immune systems are quite comparable to those of vertebrates; certainly no sharp line of demarcation between the two groups exists for behavioral ecologists, who have discovered mating and social systems, for example, to be remarkably similar (Thornhill and Alcock 1983).

Current biomedical research on humans is blurring the dichotomy between mind and body; the immune system may be the last to go in terms of ideas on the autonomous existence of organ systems (Ader et al. 1991). Perhaps by taking advantage of some of the biotechnology used in this work, behavioral biologists can achieve a similar integration of physiological mechanisms and evolution.

ACKNOWLEDGMENTS

I thank John T. Rotenberry, Carl Ware, and Ivar Folstad for helpful discussion.

REFERENCES

Ader, R., ed. 1981. *Psychoneuroimmunology*. New York: Academic Press.

Ader, R., D. L. Felten, and N. Cohen, eds. 1991. *Psychoneuroimmunology*. 2d ed. New York: Academic Press.

Allee, W. C., N. E. Collias, and C. Z. Lutherman. 1939. Modification of the social order in flocks of hens by the injection of testosterone proprionate. *Physiological Zoology* 12:412–39.

Allendorf, F. W., and R. F. Leary. 1986. Heterozygosity and fitness in natural populations of animals. In M. E. Soulé, ed., *Conservation biology: The science of scarcity and diversity*, 57–76. Sunderland, Mass.: Sinauer Associates.

Alexander, J., and W. H. Stimson. 1989. Sex hormones and the course of parasitic infection. *Parasitology Today* 4:189–93.

Alexander, R. D. 1979. *Darwinism and human affairs*. Seattle: University of Washington Press.

Beutler, B., and A. Cerami. 1989. The biology of cachectin/TNF—a primary mediator of the host response. *Annual Review of Immunology* 7:625–55.

Boyse, E. A., G. K. Beauchanp, J. Bard, and K. Yamazaki. 1991. Behavior and the major histocompatibility complex of the mouse. In R. Ader, D. L. Felten, and N. Cohen, eds., *Psychoneuroimmunology*, 2d ed., 831–46. New York: Academic Press.

Bronson, F. H. 1979. The reproductive ecology of the house mouse. *Quarterly Review of Biology* 54:265–99.

Bundy, D. A. P. 1989. Gender-dependent patterns of infection and disease. *Parasitology Today* 4:186–89.

Burke, T. 1989. DNA fingerprinting and other methods for the study of mating success. *Trends in Ecology and Evolution* 4:139–44.

Chilgren, J., and W. A. deGraw. 1977. Some blood characteristics of white-crowned sparrows during molt. *Auk* 94:169–71.

Crews, D., ed. 1987. *Psychobiology of reproductive behavior: An evolutionary perspective.* Englewood Cliffs, N.J.: Prentice-Hall.

Daly, M., and M. Wilson. 1983. *Sex, evolution, and behavior.* 2d ed. Boston: Willard Grant Press.

deGraw, W. A., M. D. Kern, and J. R. King. 1979. Seasonal changes in the blood composition of captive and free-living white-crowned sparrows. *Journal of Comparative Physiology* 129: 151–62.

Dobson, A. P., and P. J. Hudson. 1986. Parasites, disease and the structure of ecological communities. *Trends in Ecology and Evolution* 1:11–15.

Egid, K., and J. L. Brown. 1989. The major histocompatibility complex and female mating preferences in mice. *Animal Behaviour* 38:548–49.

Folstad, I., and A. J. Karter. 1992. Parasites, bright males, and the immunocompetence handicap. *American Naturalist* 139:603–22.

Folstad, I., A. C. Nilssen, O. Halvorsen, and J. Anderson. 1989. Why do male reindeer (*Rangifer t. tarandus*) have higher abundance of second and third instar larvae of *Hypoderma tarandi* than females? *Oikos* 55:87–92.

Freeland, W. J. 1976. Pathogens and the evolution of primate sociality. *Biotropica* 8:11–24.

Gilpin, M. E., and M. E. Soulé. 1986. Minimum viable populations: Processes of species extinction. In M. E. Soulé, ed., *Conservation biology: The science of scarcity and diversity*, 19–34. Sunderland, Mass.: Sinauer Associates.

Glick, B. 1986. Immunophysiology. In P. D. Sturkie, ed., *Avian physiology*, 87–101. New York: Springer-Verlag.

Goble, F. C., and E. A. Konopka. 1973. Sex as a factor in infectious disease. *Transactions of the New York Academy of Sciences* 35:325–46.

Grosberg, R. K., and J. F. Quinn. 1986. The genetic control and consequences of kin recognition by the larvae of a colonial marine invertebrate. *Nature* 322:457–59.

Grossman, C. J. 1985. Interaction between the gonadal steroids and the immune system. *Science* 227:257–61.

Gupta, A. P., ed. 1991. *Immunology of insects and other arthropods.* Boca Raton, Fla.: CRC Press.

Hamilton, W. D., and M. Zuk. 1982. Heritable true fitness and bright birds: A role for parasites? *Science* 218:384–87.

Hart, B. L. 1990. Behavioral adaptations to pathogens and parasites: Five strategies. *Neuroscience and Biobehavior Reviews* 14:273–94.

Hauschka, T. S. 1947. Sex of host as a factor in Chagas' disease. *Journal of Parasitology* 33: 399–404.

Heyneman, D., and T. M. Welch. 1980. Parasitic diseases. In H. H. Fundenberg, D. P. Stites, J. L. Caldwell, and J. V. Wells, eds., *Basic and clinical immunology*, 677–90. Los Altos, Calif.: Lange.

Howard, J. C. 1977. H-2 and mating preference. *Nature* 266:406–8.

Howard, J. C. 1991. Disease and evolution. *Nature* 352:565–67.

Johns, J. E. 1964. Testosterone-induced nuptial feathers in phalaropes. *Condor* 66:449–55.

Johnson, A. L. 1986. Reproduction in the male. In P. D. Sturkie, ed., *Avian physiology*, 432–51. New York: Springer-Verlag.

Keller, S. E., S. J. Schleifer, and M. K. Demetrikopoulos. 1991. Stress-induced changes in immune function in animals: Hypothalamo-pituitary-adrenal influences. In R. Ader, D. L. Felten, and N. Cohen, eds., *Psychoneuroimmunology*, 2d ed., 771–88. New York: Academic Press.

Kiecolt-Glaser, J. K., and R. Glaser. 1991. Stress and immune function in humans. In R. Ader, D. L. Felten, and N. Cohen, eds., *Psychoneuroimmunology*, 2d ed., 849–68. New York: Academic Press.

Lackie, A. M., ed. 1986. *Immune mechanisms in invertebrate vectors.* Symposia of the Zoological Society of London no. 56. Oxford: Clarendon Press.

Le Boeuf, B. J., and J. Reiter. 1988. Lifetime reproductive success in northern elephant seals. In T. H. Clutton-Brock, ed., *Reproductive success*, 344–62. Chicago: University of Chicago Press.

Leighton, D. R. 1987. Gibbons: Territoriality and monogamy. In B. B. Smuts, D. L. Cheney, R. M. Seyfarth, R. W. Wrangham, and T. T. Struhsaker, eds., *Primate societies*, 135–45. Chicago: University of Chicago Press.

Lofts, B., and R. K. Murton. 1973. Reproduction in birds. In D. S. Farner, J. R. King, and K. C. Parkes, eds., *Avian Biology*, vol. 3, 1–107. New York: Academic Press.

McCruden, A. B., and W. H. Stimson. 1991. Sex hormones and immune function. In R. Ader, D. L. Felten, and N. Cohen, eds., *Psychoneuroimmunology*, 2d ed., 475–94. New York: Academic Press.

Møller, A. P. 1990a. Effects of a haematophagous mite on the barn swallow: A test of the Hamilton and Zuk hypothesis. *Evolution* 44:771–84.

Møller, A. P. 1990b. Parasites and sexual selection: Current status of the Hamilton and Zuk hypothesis. *Journal of Evolutionary Biology* 3:319–28.

Møller, A. P. 1990c. Parasites, sexual ornaments, and mate choice in the barn swallow. In J. E. Loye, and M. Zuk, eds., *Bird-parasite interactions: Ecology, evolution, and behaviour*, 328–48. Oxford: Oxford University Press.

Monjan, A. A. 1981. Stress and immunologic competence: Studies in animals. In R. Ader, ed., *Psychoneuroimmunology*, 185–228. New York: Academic Press.

O'Brien, S. J., D. E. Wildt, D. Goldman, D. R. Merril, and M. Bush. 1983. The cheetah is depauperate in genetic variation. *Science* 221:459–62.

O'Brien, S. J., M. E. Roelke, L. Marker, A. Newman, C. A. Winkler, D. Meltzer, L. Colly, J. F. Evermann, M. Bush, and D. E. Wildt. 1985. Genetic basis for species vulnerability in the cheetah. *Science* 227:1428–34.

Ostrowski, N. L., D. W. Kress, P. K. Arora, and A. A. Hagan. 1989. Sexual behavior suppresses the primary antibody response in the golden hamster. *Brain, Behavior, and Immunity* 3: 61–71.

Perkins, H. A. 1980. The human major histocompatibility complex (MHC). In H. H. Fundenberg, D. P. Stites, J. L. Caldwell, and J. V. Wells, eds., *Basic and clinical immunology*, 181–90. Los Altos, Calif.: Lange.

Potts, W. K., and E. K. Wakeland. 1990. Evolution of diversity at the major histocompatibility complex. *Trends in Ecology and Evolution* 5:181–87.

Potts, W. K., C. J. Manning, and E. K. Wakeland. 1991a. The evolution of MHC-based mating preferences in *Mus*. In J. Klein and D. Klein, eds., *Molecular evolution of the major histocompatibility complex*, 421–34. NATO ASI Series, volume H 59. Berlin: Springer-Verlag.

Potts, W. K., C. J. Manning, and E. K. Wakeland. 1991b. Mating patterns in seminatural populations of mice influenced by MHC genotype. *Nature* 352:619–21.

Pruett-Jones, M. A., and S. G. Pruett-Jones. 1991. Analysis and ecological correlates of tick burdens in a New Guinea avifauna. In J. E. Loye and M. Zuk, eds., *Bird-parasite interactions: Ecology, evolution, and behaviour*, 154–76. Oxford: Oxford University Press.

Pruett-Jones, S. G., M. A. Pruett-Jones, and H. I. Jones. 1990. Parasites and sexual selection in birds of paradise. *American Zoologist* 30:287–98.

Pruett-Jones, S. G., M. A. Pruett-Jones, and H. I. Jones. 1991. Parasites and sexual selection in a New Guinea avifauna. *Current Ornithology* 8:213–45.

Puerta, M. L., J. C. Alonso, V. Huecas, J. A. Alonso, M. Abelenda, and R. Munoz-Pulido. 1990. Hematology and blood chemistry of wintering common cranes. *Condor* 92:210–14.

Purtilo, D. T., and J. L. Sullivan. 1979. Immunological bases for superior survival of females. *American Journal of Diseases of Children* 133:1251–53.

Ralls, K., P. H. Harvey, and A. M. Lyles. 1986. Inbreeding in natural populations of birds and mammals. In M. E. Soulé, ed., *Conservation biology: The science of scarcity and diversity,* 35–56. Sunderland, Mass.: Sinauer Associates.

Roitt, I., J. Brostoff, and D. K. Male. 1990. *Immunology.* 2d ed. New York: Harper and Row.

Sapolsky, R. M. 1987. Stress, social status, and reproductive physiology in free-living baboons. In D. Crews, ed., *Psychobiology of reproductive behavior: An evolutionary perspective,* 292–322. Englewood Cliffs, N.J.: Prentice-Hall.

Spurrier, M. F., M. S. Boyce, and B. F. J. Manly. 1991. Effects of parasites on mate choice by captive sage grouse. In J. Loye and M. Zuk, eds., *Bird-parasite interactions: Ecology, evolution and behaviour,* 389–91. Oxford: Oxford University Press.

Sullivan, B. K. 1991. Parasites and sexual selection: Separating causes and effects. *Herpetologica* 47:250–64.

Taigen, T. L., and K. D. Wells. 1985. Energetics of vocalization by an anuran amphibian (*Hyla versicolor*). *Journal of Comparative Physiology* 155:163–70.

Templeton, A. R. 1986. Coadaptation and outbreeding depression. In M. E. Soulé, ed., *Conservation biology: The science of scarcity and diversity,* 105–16. Sunderland, Mass.: Sinauer Associates.

Thornhill, R., and J. Alcock. 1983. *The evolution of insect mating systems.* Cambridge, Mass.: Harvard University Press.

Toft, C. A., and A. J. Karter. 1990. Parasite-host coevolution. *Trends in Ecology and Evolution* 5:326–29.

Trivers, R. L. 1972. Parental investment and sexual selection. In B. Campbell, ed., *Sexual selection and the descent of man, 1871–1971,* 136–79. Chicago: Aldine-Atherton.

Williams, G. C. 1966. *Adaptation and natural selection.* Princeton: Princeton University Press.

Wingfield, J. C. 1984. Androgens and mating systems: Testosterone-induced polygyny in normally monogamous birds. *Auk* 101:665–71.

Wingfield, J. C., and M. Ramenofsky. 1985. Testosterone and aggressive behavior during the reproductive cycle of male birds. In R. Gilles and J. Balthazart, eds., *Neurobiology,* 92–104. Berlin: Springer-Verlag.

Wintrobe, M. M., G. R. Lee, D. R. Boggs, T. C. Bithell, J. W. Athens, and J. Foerster. 1974. *Clinical hematology.* 7th ed. Philadelphia: Lea and Febiger.

Wittenberger, J. F. 1981. *Animal social behavior.* Boston: Duxbury Press.

Yamazaki, K., E. A. Boyse, V. Mike, H. T. Thaler, B. J. Mathieson, J. Abbot, J. Boyse, Z. A. Zayas, and L. Thomas. 1976. Control of mating preferences in mice by genes in the major histocompatibility complex. *Journal of Experimental Medicine* 144:1324–35.

Yamazaki, K., G. K. Beauchamp, D. Kupniewski, J. Bard, L. Thomas, and E. A. Boyse. 1988. Familial imprinting determines H-2 selective mating preferences. *Science* 240:1331–32.

Zuk, M. 1990. Reproductive strategies and sex differences in disease susceptibility: An evolutionary viewpoint. *Parasitology Today* 6:231–33.

Zuk, M. 1992. The role of parasites in sexual selection: Current evidence and future directions. *Advances in the Study of Behavior* 21:39–68. New York: Academic Press. In press.

Zuk, M., R. Thornhill, K. Johnson, and J. D. Ligon. 1990. Parasites and mate choice in red jungle fowl. *American Zoologist* 30:235–44.

Part V

THE SOCIAL CONTEXT OF BEHAVIOR

SIXTEEN

The Evolution of Social Cognition in Primates

ROBERT M. SEYFARTH

AND

DOROTHY L. CHENEY

IT IS OFTEN EASIER to explain behavior in terms of its adaptive significance than in terms of the mechanisms that underlie it. This is particularly true when the mechanisms concern learning, memory, and the processing of information. Many animals, for example, encounter their food in patches; under these conditions an animal must decide when to abandon its present patch and search for an alternative. The optimum strategy—that is, the strategy that yields food at the highest rate—can be established theoretically (Charnov 1976): leave a patch when the current rate of return reaches the average rate of return for the environment as a whole. Many animals, including great tits, paper wasps, wheatears, and starlings (Cowie 1977; Krebs and McCleery 1984) behave in a way that is surprisingly close to this optimum.

While the evolutionary advantages of such behavior are clear, the underlying mechanisms are not. To behave as they do the animals must have some representation of time and the number of items gathered for both their current patch and the environment as a whole. They must then use these data to calculate two rates, compare the rates, and make a decision (Gallistel 1989). Should we conclude, then, that birds and wasps calculate and compare in the way that humans do? Or is there some simpler rule of thumb that guides their behavior?

The need to clarify underlying mechanisms is nowhere more apparent than in the social behavior of primates, for which provocative anecdotes abound, systematic tests are few, and the animals' close evolutionary links with humans make "cognitive" explanations of behavior expecially tempting. In a study of captive chimpanzees (*Pan troglodytes*) at the Arnhem Zoo in The Netherlands, for example, de Waal (1982) once buried some grapefruit in the sand while the chimpanzees waited, out of sight, in the indoor part of their compound. As soon as they were released, the animals searched frantically for the fruit but were unable to find it. The group included a low-ranking male, Dandy, who would certainly have lost the fruit to higher-ranking aninmals had he found it. Dandy walked past the hiding place without slowing down or showing any special interest in it. Later that day, however, when the other chimps were out of sight, Dandy walked straight to the grapefruit, dug them up, and ate them all. One explanation for Dandy's behavior is that he recognized that he, alone among his companions, knew were the food was hidden and that he chose to keep them in ignorance in order to deceive them. A more parsimonious explanation argues that Dandy had simply learned a contingency rule: Never eat food while others are nearby or it will be taken away. Given the information at hand, we cannot distinguish between these two explanations.

In this chapter we examine the cognitive mechanisms underlying social behavior and communication in nonhuman primates. We begin by considering the mental abilities—skills in learning, memory, reasoning, or representing the world—that we must ascribe to monkeys and apes in order to account for their behavior. For any pattern of behavior there are a number of possible underlying mechanisms, each involving more or less of what we might call reasoning, intelligence, or cognition. The ethologist's job is to devise observations or experiments that test among these alternatives.

Though it is concerned initially with underlying mechanisms, research of this sort inevitably leads to questions about the evolution of cognition and intelligence. What are the problems animals face in their daily lives? What do they *need* to know, and how might one method of obtaining and storing information give some individuals a reproductive advantage over others? A second goal of this chapter, then, is to shed some light on the evolution of animal cognition.

THE REPRESENTATION OF SOCIAL RELATIONSHIPS

The Problem

Like many other species of Old World monkeys, East African vervet monkeys (*Cercopithecus aethiops*) live in groups containing a number of adult males, adult females, and juveniles, typically 10–30 individuals in all. Females remain throughout their lives in the group where they were born, and can be ranked in a linear dominance hierarchy. Offspring acquire ranks immediately below those of their mothers. Members of the same matriline interact with each other at high rates. They also form aggressive alliances against members of other matrilines (reviewed in Cheney and Seyfarth 1990b).

Early one morning in 1980, while the adult vervets fed, some juveniles were playing in a nearby bush. Macaulay, the son of a low-ranking female, wrestled Carlyle, the daughter of the highest-ranking female, to the ground. Carlyle screamed, chased Macaulay away, and then went to forage near her mother. Apparently the fight had been noticed by others, however, because 20 minutes later Shelley, Carlyle's sister, approached Austen, Macaulay's sister, and without provocation bit her on the tail.

Obviously, it is difficult even to describe this incident without implying that vervets understand the kin relations that exist among others in their group. Such descriptions, however, do not constitute an explanation. After all, baseball outfielders seem to know instantly, just after a ball has been hit, where it is going to land. Watching them, we might conclude that they are calculating the instantaneous rate of change of a parabolic function, but this would be wrong: a simpler heuristic is clearly at work. So how much do monkeys really know about what they are doing? Do they actually make use of concepts like "kinship" or "closely bonded," or are they simply responding on the basis of much simpler associations that they have formed between other group members?

Research on animal intelligence frequently attempts to distinguish between "knowing how" and "knowing that," a distinction first drawn by the philosopher Ryle (1949; see also Dickinson 1980; Whiten and Byrne 1988b). "Knowing how" refers to the ability to perform a specific, procedural task based on the recognition of a particular stimulus. Ants, for example, remove the carcasses of dead conspecifics from their nest. The function of this behavior is to rid the nest of bacteria, but ants certainly are not aware of the relation between corpses and disease; they are simply responding to the presence of oleic acid on the decaying corpse. Ants will remove anything that smells of oleic acid, regardless of whether it is dead or infected (Wilson 1971). Even a live ant dabbed with oleic acid will be dragged, struggling, out of the nest. Similarly, a vervet mother's response to her offspring's scream might be relatively unmodified; she might simply run to her offspring's aid whenever she heard a vocalization with a particular set of acoustic properties.

By contrast, "knowing that" refers to "declarative representations or knowledge" (Dickinson 1980), and implies an ability to make causal inferences about the world. Rather than simply running whenever her offspring screams, for example, the monkey mother might understand enough about the relation between dominance rank and kinship to recognize that she should only intervene on her offspring's behalf when the offspring is fighting with a member of a lower-ranking matriline. In other words, because it refers to more general knowledge about things and can be divorced from a particular response, "knowing that" allows greater flexibility in behavior depending upon changes in the social and physical environment.

In analyzing social knowledge, therefore, we must distinguish between knowledge that can be used in only a limited set of circumstances and knowledge that can be applied more broadly. A monkey may have formed an association between two members of the same matriline because the two animals are often encountered together. As a result, the monkey knows that whenever she approaches one individual she is also likely to be near the other. Such knowledge, however, might be limited to those two individuals, or to a small set of animals within the monkey's own group. It would prepare the monkey for some (indeed, many) sorts of interactions, but not for those outside her immediate group or for those that depend on the recognition of more differentiated relationships—for example, the difference between a relative and a "friend."

Alternatively, the monkey might have interacted with many different kin pairs, and she might have inferred, on the basis of her experiences and observations, that such relationships share similar properties regardless of the particular individuals involved. The monkey might even have labels, like "closely bonded" or "enemies," that help order relationships into types. The evolutionary advantages to be gained from such mental representations seem clear. The monkey's knowledge would be less constrained by particular stimuli, more general, and more abstract. It could also be applied in a much wider variety of circumstances.

Knowledge of Social Relationships

In order to understand a dominance hierarchy or to predict which individuals are likely to form alliances with each other, a monkey must step outside his own sphere of interactions and recognize the relationships that exist among others. Such knowledge can only be obtained by observing interactions in which one is not involved and making the appropriate inferences. There is growing evidence that monkeys do possess knowledge of other animals' social relationships, and that such knowledge affects their behavior.

Studies of hamadryas baboons (*Papio hamadryas*) were the first to show that nonhuman primates assess the relationships that exist among others. Under natural conditions, hamadryas baboons are organized into one-male units, each of which contains one fully adult male and two to nine adult females (Kummer 1968; Sigg et al. 1982; reviewed in Stammbach 1987). One-male units frequently come into contact with single, unattached males, and a male unit leader must constantly defend himself against attempts by other males to take over his females. Experiments on captive hamadryas have shown that rival males assess the strength of a leader's relationship with his females before competing to acquire them. Rival males do not attempt to take over a female if they have previously seen her interact with her leader. Such "respect of possession" holds even when the rival is dominant to the leader in other contexts (Kummer et al. 1974).

To test the hypothesis that rivals make judgments about the strength of bonds between a male and his females, Bachmann and Kummer (1980) studied six adult males and six adult females, using choice tests to determine how strongly each male preferred each female and how strongly each female preferred each male. A male-female pair was then placed in a large outdoor enclosure where they interacted freely. Different rival males were allowed to watch the pair and then given an opportunity to challenge the owner. Bachmann and Kummer found that the probability of a challenge was not correlated with either the rival's or the owner's preference for a particular female. The female's preference, however, did make a difference: if a female was with a male she strongly preferred, this inhibited challenges from middle- and low-ranking rivals. The two highest-ranking males challenged all owners regardless of how strongly females preferred them. Although Bachmann and Kummer could not rule out the possibility that rival males were simply responding to the females' actions rather than their relationships, the results suggested that males assessed the strength of the attraction between an owner and his female and avoided challenging owners when the pair's relationship was close. This seems an adaptive strategy, because aggressive challenges may be too costly if the contested female prefers to remain with her current mate.

Further evidence that monkeys recognize relationships among others comes from playback experiments with vervet monkeys. In many primate species, mothers will run to their offspring's aid when the offspring scream during a fight, which suggests that females can distinguish among the calls of different individuals. To test this hypothesis, we played the scream of a two-year-old juvenile from a concealed loudspeaker to its mother and two control females who also had offspring in the group. We found that mothers consistently either looked toward or approached the speaker for longer durations than did control females, indicating that they recognized the voice of their offspring (Cheney and Seyfarth 1980; Green and Marler 1979 review similar data from other species).

More interesting, however, was the behavior of control females. When the responses of control females were compared with their behavior before the scream was played, we found that playbacks significantly increased the likelihood that control females would look at the mother. By contrast, there was no change in the likelihood that control females would look at each other (Cheney and Seyfarth 1980, 1982). The females appeared to be able to associate particular screams with particular juveniles, and these juveniles with particular adult females. They behaved as if they recognized the kin relationships that existed among other group members.

At this point we should emphasize that, when speaking of kin recognition in primates, we define the term operationally as the recognition of a close social bond. The ability to recognize other animals' kin does not imply that monkeys have a concept of "kinship" or "genetic relatedness," but simply that they recognize the close associates of other group members. In most cases close associates are also kin, and this rule of thumb appears to be the primary mechanism underlying kin recognition in nonhuman primates (Waldman et al. 1987; see also Grafen 1990). There is at present no evidence that monkeys differentiate among kin relationships that are characterized by similar rates of interaction—say, sisters as opposed to mothers and daughters—simply because the relevant tests have not yet been conducted.

For additional evidence that monkeys recognize the kin relationships (or close associates) of other group members, consider the phenomenon of redirected aggression. In many primate species, an animal that has been involved in a fight will redirect aggression and threaten a third, previously uninvolved individual. In rhesus macaques (*Macaca mulatta*) (Judge 1982) and vervet monkeys (Cheney and Seyfarth 1986, 1989) such redirected aggression is not distributed randomly, but is directed toward a close relative of the prior opponent. In our study population, vervets were significantly more likely to threaten unrelated individuals following a fight' with those animals' close kin than during matched control periods. This was not because fights caused a general increase in aggression toward unrelated animals. Instead, aggression seemed to be directed specifically toward the kin of prior opponents (for similar data on kin-biased reconciliation, see de Waal and Roosmalen 1979; York and Rowell 1988; de Waal and Yoshihara 1983; de Waal and Ren 1988; Aureli et al. 1989).

Among monkeys, knowledge about other animals' relationships is not limited to the recognition of matrilineal kin. Male and female savanna baboons (*Papio cynocephalus*), for example, sometimes form long-term pair bonds, or "friendships," in which proximity and cooperative behavior are maintained throughout the female's reproductive cycle (Smuts 1985; see also Seyfarth 1978; Altmann 1980; Strum 1984). In some groups, friendships persist for years at a time. In the best-documented study of friendships, Smuts (1985) found that females and males often redirected aggression against their opponents' friends. Following a fight with another male, for example, a male frequently appeared to seek out his rival's female friend and chase her. The baboons, in other words, seemed to recognize friendships.

Comparing Social Relationships

In the laboratory, monkeys can readily be taught to solve problems that require the recognition of a *relation* between objects. In oddity tests, for instance, a subject is presented with three objects, two of which are the same and one of which is different. It receives a reward only if it chooses the different object. Many monkey species achieve scores of 80%–90% correct even when new stimuli are used for each problem and each set of stimuli is presented for only one trial (e.g., Strong and Hedges 1966; Davis et al. 1967). Such performance suggests that the animals are using an abstract hypothesis, "pick the odd object." The hypothesis is called abstract because "odd" does not refer to any specific stimulus dimension, as does "red" or "square." Instead, oddity is a concept that specifies a relation between objects independent of their specific stimulus attributes (Essock-Vitale and Seyfarth 1987).

Premack (1983, 1986) contends that tasks like oddity tests require only judgments about relations between elements, not relations between relations. By contrast, judgments about relations between relations are involved in tasks like analogical reasoning. They are less fundamental and universal than judgments about relations between elements, and they have thus far been demonstrated only in language-trained chimpanzees.

In his study of analogical reasoning in chimpanzees, Premack (1976, 1983) trained a subject, Sarah, to make same/different judgments between pairs of stimuli. Once Sarah could use these words correctly even when confronted with entirely new stimuli, she was shown two pairs of items arranged in the form A/A' and

B/B'. Her task was to judge whether the two relations were the same or different from each other. Alternatively, Sarah was given an incomplete analogy in the form A/A' *same as* B/? Her task then was to complete the analogy in a way that satisfied this relation.

In the most complex test, the objects shared no obvious physical similarity. For example, Sarah was asked, "Lock is to key as closed paint can is to _____," where the options for completing the analogy were a can opener and a paint brush. Here the identity between the two relations is not based on physical similarity (in fact they look quite different), but on the underlying relation *opening*, which both cases instantiate. It is not the stimuli themselves but this relation that must be represented in the subject's mind. To solve an analogy the chimpanzee must infer the appropriate relation for each stimulus pair and then compare these two relations to see if they are the same (Gillan et al. 1981; Premack 1983). Sarah solved the problems correctly.

Sarah's behavior prompts us to ask whether group-living primates might use abstract criteria to make same/different judgments about social relationships. The data that most directly address this question come from a study conducted by Dasser (1988b) on a group of forty captive longtailed macaques (*Macaca fascicularis*). Dasser trained two adult females so that they could be temporarily removed from the group and placed in a small test room to view slides of other group members. In one test that used a simultaneous discrimination procedure, the subject saw two slides. One showed a mother and her offspring, the other showed an unrelated pair of group members. The subject was rewarded for pressing a response button below the mother-offspring slide. Having been trained to respond to one mother-offspring pair (five different slides of the same mother and her juvenile daughter), the subject was tested using 14 novel slides of different mothers and offspring paired with 14 novel unrelated alternatives. The mother-offspring pairs varied widely in their physical characteristics. Some slides showed mothers and infant daughters, others showed mothers and juvenile sons or mothers and adult daughters. Nonetheless, in all 14 cases the subject correctly selected the mother-offspring pair.

In a second test that used a match-to-sample procedure, the mother was represented as the sample on a center screen, while one of her offspring and another stimulus animal of the same age and sex as the offspring were given as positive and negative alternatives, respectively. Having learned to select the offspring during training, the subject was presented with 22 novel combinations of mother, offspring, and unrelated individual. She chose correctly on 20 of 22 tests.

Finally, to test whether monkeys could recognize other categories of social affiliation, Dasser (1988a) trained a subject to identify a pair of siblings and then tested the subject's ability to distinguish novel sibling pairs from (*a*) mother-offspring pairs, (*b*) pairs of otherwise related group members, like aunts and nieces, and (*c*) pairs of unrelated group members. The subject correctly identified the sibling pair in 70% of tests. Seven of the eight errors occurred when she was asked to compare siblings with a mother-offspring pair; one occurred when she compared siblings with two less closely related members of the same matriline.

Data on redirected aggression and reconciliation in vervet monkeys provide additional evidence that animals classify social relationships into types, independent of the particular individuals involved. Recall that in some monkey species redirected aggression and reconciliation are kin-biased, such that animals often interact with the kin of their prior opponents. In vervet monkeys, moreover, redirected aggression and reconciliation can extend even to the previously uninvolved kin of prior opponents. Data gathered in two social groups over two different time periods showed that an animal was more likely to threaten another individual if one of her own close relatives and one of her opponent's close relatives had recently been involved in a fight (Cheney and Seyfarth 1986, 1989). The same was true of reconciliation. Two unrelated individuals were more likely to engage in an affinitive interaction following a fight between their close kin than during matched control periods.

Although these data on vervet monkeys are preliminary, they support Dasser's results in suggesting that monkeys recognize that certain types of social relationships share similar characteristics. When a vervet (say, A2) threatens an unrelated animal (B2) following a fight between one of her own relatives (A1) and one of her opponent's relatives (B1), A2 acts as if she recognizes that the relationship between B2 and B1 is similar to her own relationship with A1. In other words, we may think of A2 as having been presented with a natural problem in analogical reasoning: A1/B1 *same as* A2/? A2 correctly completes the analogy by directing aggression to another member of the B family.

What Are the Underlying Mechanisms?

Although the case for some kind of social concept is strong, we know virtually nothing about how such concepts are acquired or how they are coded in a monkey's mind. One possibility is that monkeys use physical resemblance as a cue, since members of the same matriline often look alike. Vervets and longtailed macaques treat bonds between kin as similar, however, even when they involve pairs of animals whose within-family resemblances, at least to a human observer, are markedly different. In Dasser's study, for example, subjects generalized to a diverse array of mother-offspring pairs (mothers and young infants, mothers and juvenile sons, mothers and adult daughters) even though they had been trained with only one example from this category (Dasser 1988b). Similarly, male and female baboon "friends" do not resemble each other, yet other baboons nevertheless recognize that certain males and females associate at high rates.

A second possible mechanism is simply "brute force" associative memory. According to this hypothesis, a monkey observes and remembers all possible dyadic interactions among group members until she is able to conclude that individuals A1, A2, and A3 go together, B1 and B2 go together, and so on. From this knowledge she is then able to predict, for example, who will form alliances with whom.

The brute force method involves processes of classical conditioning and memory that are widespread in animals. It also requires no special skills in reasoning or computation. The hypothesis is appealing, therefore, because it seems to explain much of what we see in a simple, straightforward manner. Recent evidence, however, suggests that hypotheses based on classical conditioning and memory are unable to explain many features of animal behavior (Gallistel 1989), including the social knowledge of primates (Cheney and Seyfarth 1990b).

Applied to primate social knowledge, theories of classical conditioning assume that a monkey forms associations between others based solely on rates of interaction: how often A is near B, grooms with B, or forms alliances with B compared with how often A and B are seen near, grooming, or in alliance with other individuals (Gallistel 1989; Cheney and Seyfarth 1990b). However, the classification of social relationships by monkeys cannot be explained simply in terms of the number and types of interactions, because monkeys also seem to take into account the quality of interactions—for example, the frequency with which two individuals reconcile after fighting (Hinde 1976, 1983). Monkeys' minds seem to have been selected to organize information about social behavior according to many different variables, including the identity of the individuals involved, their age, sex, and dominance rank, the type and quality of their interactions, their past history of behavior, and so on. Armed with this information, the individual constructs a mental representation of the social relationship that exists between two other animals. As a result, in the mind of a vervet or macaque, for example, a mother and her son are still classified as mother and offspring even when they interact rarely.

What Are the Evolutionary Advantages?

Social relations among monkeys and apes, like those in many other species, are simultaneously competitive and cooperative. The ability to form a mental representation of a social relationship may have evolved because it offers the most accurate means of predicting the behavior of others (see also Humphrey 1976; Whiten and Byrne 1988b). Because relationships conceived in this way are abstractions, they are simpler and more parsimonious than absolute judgments, which require learning the characteristics of every interaction (Allen 1989; Dasser 1985; Kummer 1982). Selection for the ability to form mental representations of social relationships may be particularly strong under three conditions.

Large Group Size. Vervets live in groups of 10–30 individuals. In such relatively small groups an animal could easily memorize all the interactions he had seen and form associations of different strength between different individuals. The brute force method would probably suffice. Baboons and many species of macaques, however, often live in groups of over 100 animals. As group size increases, the number of dyadic relationships increases algebraically, placing severe constraints on an individual's ability to remember the specific characteristics of every relationship.

Faced with the problem of remembering an increasing number of separate entities, both humans (e.g., Mandler 1967) and animals (Terrace 1987; Swartz et al. 1991) typically recode items into larger units called chunks. The result is a measurable improvement in recall and problem solving. Chunking may be based on any number of criteria. Visual stimuli, for example, may be grouped according to physical resemblance or color, while chunking of words may be based on phrase groupings or semantic relations. In the case of group-living primates like baboons or macaques, one obvious criterion to use when organizing conspecifics into chunks derives from the shared association of subsets of individuals with a common female—in other words, from matrilineal relatedness. Increasing group size may therefore place increasingly strong selection pressure on monkeys to organize relationships according to rank-ordered, matrilineal kin groups.

Where Individuals Transfer between Groups or Interact with Other Groups. Even in species in which group size is small, behavioral or demographic factors may increase the number of animals with whom an individual interacts. Among vervets, for example, competition between groups occurs at high rates and has a measurable effect on female reproductive success (Cheney and Seyfarth 1987; Cheney et al. 1988). Females compete regularly with up to six neighboring groups, and appear to recognize individuals and their dominance ranks in each of these groups (Cheney and Seyfarth 1982, 1990b). Similarly, male vervets leave their natal group at sexual maturity and join another group. A male may live in as many as four different groups during his lifetime. With each transfer a male meets aggression from both males and females in his new group (Cheney and Seyfarth 1983). In order to succeed and reproduce in his new group, a male must rapidly learn not only his new companions' identities but also how they behave and with whom they associate.

The brute force method of observe-and-memorize could, of course, be made to work under these conditions. Nevertheless, both males and females would gain an advantage in accuracy of memory and speed of retrieval if they were predisposed to classify relationships into types, organize their opponents into ranked matrilineal kin groups, and predict their behavior accordingly. If a monkey's knowledge of his rivals is based solely on the memory of specific experiences with particular individuals, he can predict future behavior only when interacting with the same individuals under roughly the same circumstances. This is hardly of much use when entering a new group. On the other hand, if a monkey organizes data on past experiences according to a set of general rules, knowledge gained in one group with one set of individuals can rapidly be applied to another, even when the individuals are unfamiliar.

Where Alliances Are Common. All animals must anticipate the imminent behavior of their competitors. The ability to judge whether a rival is about to attack or flee is clearly relevant to survival. It requires that animals be superb ethologists, noting minute changes in behavior and evaluating these in light of past interactions.

In some species, the assessment of opponents is complicated by the entry of a previously uninvolved animal who forms an alliance with one of the original antagonists. Alliances occur at particularly high rates in groups of monkeys and apes, and they produce varying outcomes. Among male chimpanzees and baboons, two lower-ranking individuals can consistently drive away one who outranks them both (e.g., Bercovitch 1988; Nishida and Hiraiwa-Hasegawa 1987). By contrast, among female vervets, baboons, and macaques, two allies can only defeat a third female if that female already ranks lower than at least one of the allies (Cheney 1983; Datta 1983; Chapais et al. 1991; see also chapters in Harcourt and de Waal 1992).

Many birds and mammals form alliances. The alliances formed by nonhuman primates, however, differ in a number of respects from those formed by other species. For example, alliances among adults of the same sex occur more often in primates than in other species, though this distinction is by no means absolute (see, e.g., alliances among related and unrelated male lions, Packer and Pusey 1982). Second, to date only primates have been observed to reciprocate support in alliances (Packer 1977; Harcourt 1989; Cheney and Seyfarth 1990b) and to retaliate against those who fail to support them (de Waal and Luttrell 1988).

Perhaps most important, only nonhuman primates are known to choose alliance partners on the basis of their relative quality. Dominance hierarchies occur in many species of animals. Only primates, however, seem to recognize that high-ranking individuals are potentially more powerful allies than low-ranking animals. Monkeys and apes regularly compete for and preferentially attempt to cultivate relationships with high-ranking individuals. They also selectively intervene and attempt to disrupt other individuals' alliances with these more powerful animals (Harcourt 1989, 1992).

As a result of the high frequency of alliances in groups of monkeys and apes, an individual who attempts to gain a social and reproductive advantage over others must be able not only to predict other animals' behavior but also to assess their relationships. It is not enough to know who is dominant or subordinate to oneself; one must also know who is allied to whom and who is likely to come to an opponent's aid.

We should expect knowledge of other individuals' social relationships to be particularly well developed in any animal species in which alliances are common. The ecological conditions that favor the evolution of alliances are not unique to nonhuman primates. However, with the possible exception of elephants (Moss 1988) and toothed whales (Connor et al., 1992), primates' use of alliances as a competitive strategy appears to be unique (Harcourt 1989, 1992; see also Kummer 1982).

THEORIES OF MIND

The Problem

In their analysis of the evolution of signaling in animal contests, Maynard Smith and Parker (1976) argue that under most conditions an individual's signals will not provide accurate information about its intentions, such as how long or with what intensity it is prepared to fight. Selection will therefore favor individuals who bluff, provided their bluff cannot be detected. "However," they continue, "it is equally clear that selection will favor individuals capable of distinguishing bluff from actual resource holding potential . . ." (1976, 174). The result will be competition between signalers who attempt to deceive and recipients who attempt to discriminate between signals that do and do not provide accurate information (Krebs and Dawkins 1984).

This argument, of course, does not mean that animals actually make conscious judgments about the motives of their opponents. The authors are concerned not with the proximate mechanisms involved, but with the ways in which natural selection might favor the evolution of certain sorts of behavior. In most species, the detection of deceit will be based on a simple contingency rule; for example, "If the signal is rarely followed by attack, devalue the signal or pick another cue" (Harper 1991). If an animal falsely signals imminent attack by lowering its head, for instance, selection will favor those opponents who ignore head lowering and use some other cue when estimating their rival's next move (Andersson 1980; for recent discussion see Grafen 1991; Dawkins and Guilford 1991).

It seems possible, however, that in some species the detection of deception does involve active assessment of an opponent's motives or intentions. As with judgments about social relationships, there may be many strategies for distinguishing bluff, ranging from ones based purely on behavioral contingencies to others based on deliberate judgments about mental states. Humans clearly protect themselves against deceit by attributing motives, knowledge, or beliefs to others and searching for any mismatch between an individual's behavior and his presumed mental state. For us, judging the motives of others provides a powerful means of detecting deception. Have similar abilities evolved in any other species?

To attribute beliefs, knowledge, and emotions to both oneself and others is to have what Premack and Woodruff (1978) term a "theory of mind." A theory of mind is a theory because, unlike behavior, mental states are not directly observable, although they can be used to make predictions about behavior. Monkeys are clearly adept at recognizing the similarities and differences between their own and other individuals' social relationships. What is not known is whether they are equally adept at recognizing the similarities and differences between their own and other individuals' mental states.

How should we characterize the mental states, if any, that underlie the behavior of animals? The philosopher Dennett (1983, 1987) argues that we can best understand the behavior of other species if we adopt what he calls an "intentional stance" and assume, at least for the purposes of analysis, that animals are "intentional" creatures, capable of mental states like believing, thinking, or wanting. The question then becomes: what kind of beliefs, thoughts, or desires? Here Dennett offers a number of different "levels of intentionality," each of which constitutes a testable hypothesis.

Consider, for example, the alarm calls that vervets give in response to leopards. When vervets spot a leopard, they give loud alarm calls that cause other individuals to run into trees (Seyfarth et al. 1980). In explaining this behavior, we must first entertain the possibility that vervets are *zero-order intentional systems*, with no beliefs or desires at all. A zero-order explanation holds that vervet monkeys give alarm calls simply because they are frightened.

Alternatively, vervets might be *first-order intentional systems*, with beliefs and desires but no beliefs *about* beliefs. At this level, vervet monkeys give leopard alarm calls, for example, because they believe that there is a leopard nearby or because they want others to run into trees. The caller need not have any conception of his audience's state of mind, nor need he recognize the distinction between his own and another animal's knowledge about the leopard's presence.

It is also possible that vervets are *second-, third-, or even higher-order intentional systems*, with some conception about both their own and other individuals' states of mind. A vervet monkey capable of second-order intentionality gives a leopard alarm call because he wants others to believe that there is a leopard nearby. At higher and increasingly baroque levels, both the signaler's and the audience's states of mind come into play. At the third level of intentionality, for example, a vervet gives an alarm call because he wants others to believe that he wants them to run into trees.

Higher-order intentionality demands the ability to recognize that two different individuals can have quite different beliefs about the world. Given the ease with which adult humans attribute mental states different from their own to others, it is rather surprising that young children often have considerable difficulty recognizing false beliefs in others. In one study, for example, Wimmer and Perner (1983) presented children aged 3 to 9 years with scenarios in which they had to describe the knowledge of others. In one case, the children watched a puppet show in which a boy, Maxi, puts a piece of chocolate into a blue cupboard. Maxi then leaves the room, and in his absence his mother removes the chocolate from the blue cupboard and places it in a green one. The children were then asked where Maxi would look for the chocolate. Children under 4 years of age consistently indicated the green cupboard, the cupboard in which they themselves knew the chocolate to be located. In contrast, about half of the 4- to 6-year-old children, and over 80% of the 6- to 9-year-old children, correctly pointed out that Maxi would still think that the chocolate was in the blue cupboard. The younger children's errors were not due to a failure of memory, because most of the children who gave an incorrect answer to the question nevertheless gave a correct answer when asked if they remembered where Maxi had put the chocolate. Rather, it seems that children's ability to represent two incompatible beliefs does not become established until around the ages of 4 to 6 years (for further discussion see the chapters in Astington et al. 1988).

Studies of Animals

Most of the controversy surrounding animal communication (including the detection of deceit) centers on second- and third-order intentionality: whether or not animals are capable of acting as if they want others to believe that they know or believe something (Cheney and Seyfarth 1990b; Whiten 1990). It is at this level that the most intriguing anecdotes surface, and it is at this level that future experiments in behavioral ecology must focus.

When a signaler deceives an opponent, he creates or supports a false belief in his opponent. Hence any evidence of deceptive signaling in animals potentially provides evidence for a theory of mind. A signaler who creates a false belief, however, need not necessarily know that he has done so. In fact, most examples of deceptive signaling in animals can probably be explained in terms of behavioral contingencies, without recourse to higher-order intentionality. For present purposes, then, it becomes crucial to distinguish between the function of deception and the mechanisms that underlie it.

A male scorpionfly can only copulate with a female if he first provides her with a nuptial gift of a dead insect. Some males catch their own insects, while other males adopt the behavior and posture of females to steal insects from males that already have them (Thornhill 1979). While the males' behavior certainly functions to deceive others, we cannot conclude that scorpionflies attribute beliefs to their rivals. Scorpionflies apparently never attempt to deceive rivals in any other context or with any other pattern of behavior. Because their deception is relatively inflexible and occurs in only a narrow range of contexts, it probably does not imply a theory of mind. What revisions to our thinking would be required, however, if we were to find that scorpionflies also occasionally give false alarm signals to drive their rivals from females? And what would we conclude if we learned that a particular individual scorpionfly, having once deceived a rival by mimicking the behavior of a female, could no longer deceive him by giving a false alarm call?

Monkeys occasionally act as if they recognize that other individuals have beliefs, but even the most compelling examples can just as easily be explained in terms of learned behavioral contingencies, without recourse to a theory of mind. For example, Kummer (1982) reports observing a female hamadryas baboon who spent 20 minutes gradually shifting her way in a seated position toward a rock, where she began to groom a subadult male—an act that would not normally be tolerated by the dominant adult male. From his resting position, the dominant male could see the back and head of the female, but not her arms. The subadult male sat in a bent position and was also invisible to the dominant male. What made Kummer doubt that this arrangement was accidental was the exceptionally slow, inch-by-inch shifting of the female toward the rock (for many more examples, see Byrne and Whiten 1988, Whiten and Byrne 1988a).

Though clearly not accidental, the baboon's behavior could nonetheless be guided by one of two quite different mechanisms. It is possible that the female was able to recognize that she could manipulate the dominant male's beliefs and acted in order to keep him in ignorance. It is also possible, however, that she had simply learned that she would be attacked if she groomed a subordinate male out in the open. This problem of interpretation is common to all examples of apparent deception in nonhuman primates: although the observation is consistent with higher-order intentionality, it cannot be explained solely in terms of this factor (see Byrne and Whiten 1988 and comments thereon, Whiten and Byrne 1988a; see also Cheney and Seyfarth 1990b).

In fact, there is very little evidence that monkeys attribute to others mental states different from their own (reviewed in Cheney and Seyfarth 1990b). For example, although their vocalizations certainly function to alert others to the presence of food, danger, or one another, monkeys do not modify their behavior according to whether or not their audience is ignorant or already informed. Similarly, while monkeys are clearly able to acquire novel skills from others through observation, social enhancement, and trial and error learning, there is little evidence that they imitate each other, again perhaps because they are unable to impute motive (Galef 1988; Cheney and Seyfarth 1990b).

There is also no conclusive evidence that monkeys teach each other. For example, when infant vervets first begin to give alarm calls or respond to the alarms of others, they make many mistakes. Some, like an infant's eagle alarm to a pigeon, are relatively harmless. Other mistakes—an infant who looks up in the air when he hears a leopard alarm, for example—actually increase the infant's risk of being taken (Seyfarth and Cheney 1986). Under these conditions one might expect adults to intervene actively and help their infants learn about predators. Somewhat surprisingly, however, they do not. Despite extensive observations and experiments, there is no evidence that adults selectively encourage infants who give alarm calls to the appropriate predators, nor do adults correct infants who have responded to an alarm inappropriately (Seyfarth and Cheney 1986). As a result, infant vervets are left to learn by observation alone, without explicit tutelage. This reliance on observational learning is widespread among animals (Nishida 1987; Galef 1988) and severely constrains both the rate and the efficiency of cultural transmission (Boyd and Richerson 1985). It seems possible that this lack of explicit pedagogy can ultimately be traced to adults' failure to recognize that their offspring's knowledge is different from their own.

Furthermore, although monkeys experience emotions like fear and grief, they show no evidence of compassion or empathy and do not seem to recognize emotions in others. Moreover, while monkeys are adept at recognizing their own position in a social network or dominance hierarchy, they show little self-awareness (see e.g., Gallup 1982; Cheney and Seyfarth 1990b). This, too, is consistent with the view that monkeys do not know what they know and cannot reflect upon their knowledge, their emotions, or their beliefs.

Many of these generalizations may apply more to monkeys than to apes. Although most of the data are anecdotal, numerous observations suggest that chimpanzees, if not other apes, have at least a rudimentary theory of mind. They deceive each other in more ways and in more contexts than monkeys, and they seem better than monkeys at recognizing both their own and other individuals' knowledge and limitations (see discussions by de Waal 1982; Whiten and Byrne 1988a; Byrne and Whiten 1990; Cheney and Seyfarth 1990b; Povinelli et al. 1990). There is also intriguing evidence from the Ivory Coast that chimpanzee mothers may teach their infants how to crack open palm nuts. Not only do mothers in this population facilitate their infants' learning by providing them with the appropriate tools and demonstrating the correct movements for opening nuts, but on several occasions they have also been observed to correct their infants' technique (Boesch 1991). It remains for future research to determine the extent to which chimpanzees and other apes attribute mental states different from their own to each other.

What Are the Evolutionary Advantages?

When we label behavior in monkeys, chimpanzees, or children as examples of second- or third-order intentionality, this is more than just an exercise in behavioral classification. As with the representation of social relationships, there are evolutionary advantages to be gained by an individual who can attribute mental states to others, recognize that such states can be different from his own, and recognize that what another individual thinks can have a causal effect on behavior.

We have already suggested that explicit pedagogy and imitation is, potentially at least, a more powerful means of transmitting information than observational learning or social facilitation. Pedagogy, however, requires that the tutor recognize that his own knowledge is different from his pupil's. Perhaps as a result, teaching in the strictest human sense is absent in most animals (Caro and Hauser, 1992).

Consider another example. Imagine a group of baboons in which individuals are extremely skilled at judging behavioral contingencies ("If I do X, Y is likely to happen"), but unable to identify the motives or knowledge of others or to recognize that these mental states can be different from their own. Imagine further that among these baboons, as among the baboons studied by Packer (1977), some pairs of males exchange reciprocal alliances: A and B regularly cooperate to chase other males from females, and each allows the other to mate with the female half of the time. In other pairs, however, support is not reciprocated: X helps Y gain access to a female but later, when X solicits Y, the latter refuses to help.

Under these conditions, males who are solicited as alliance partners will benefit if they can distinguish between an individual who seeks help and genuinely intends to reciprocate and one who seeks help but is unlikely to return the favor. This distinction, however, will not be an easy one to make, because selection will also favor males who solicit effectively, using the same patterns of behavior regardless of whether they intend to reciprocate or not. This is, in fact, what seems to occur in natural groups of baboons (e.g., Packer 1977; Noe 1986; Bercovitch 1988).

If male baboons are incapable of attributing beliefs and motives to others, then direct experience is the only means by which they will be able to acquire information about honest and dishonest solicitors. And in judging a solicitor's reliability, the more experiences they have had with him, the more accurate their assessment is likely to be; one interaction is unlikely to be sufficient. These baboons, in other words, will always be vulnerable to those who cheat on their first encounter or who continue to cheat at very low rates, and they will remain vulnerable until they have gained enough experience, through sometimes costly interactions, to make an accurate assessment of their partners' reliability.

Perhaps more important, because these baboons base their judgments of another individual's reliability on his actions rather than on his motives, one individual's knowledge about another will be specific to a particular type of interaction. If male X has learned that Y rarely reciprocates when forming alliances, he will know just that: Y rarely reciprocates when forming alliances. Such knowledge, however, will not necessarily make X skeptical about Y's reliability when their group is threatening another or when the males encounter a predator. Because their judgment of others rests on an assessment of behavioral contingencies, the knowledge they possess will be relatively context-specific. As a result, they will be vulnerable to individuals who cooperate in one situation only to cheat in another. Experiments with vervet monkeys suggest that, if one individual provides unreliable information about the presence of another group, others will cease responding to her calls as long as they concern other groups. Listeners will still pay attention, however, if the individual switches to an alarm call (Cheney and Seyfarth 1988). The vervets' failure to generalize their skepticism more broadly leaves them open to deception.

Now imagine that into this group of nonintentional baboons comes a mutant male capable of attributing states of mind to others and of recognizing that these states of mind may be different from his own. The first point to make about this individual is that he recognizes a distinction between an animal's behavior and the motives that underlie it. As a result, he recognizes that however much a solicitor seems likely to reciprocate, this may not actually be his intention. Such knowledge will not necessarily make the mutant male any less vulnerable to cheaters on his first interaction with them, but it is likely to make him more skeptical in subsequent interactions.

More important, because the new male is attentive to motives and not just behavioral contingencies, he is more likely to generalize his knowledge of different individuals to a variety of different contexts. When he encounters male Y cheating in the formation of alliances, for example, the mutant male's skepticism about this individual will extend not just to future alliances but also to Y's alarm calls, food calls, behavior during intergroup encounters, and so on. In short, the new male will have a competitive advantage over others because, by assessing his companions' motives, he can better predict their behavior.

We offer this hypothetical example not to argue for or against a theory of mind in any species, but instead to emphasize that the existence (or lack) of a theory of mind, long recognized as an important watershed in children's cognitive development, also has considerable evolutionary significance. Once an individual recognizes that his companions not only behave but also think, desire, and believe about behavior, he becomes a much better social strategist, and can use his knowledge much more skillfully to his own and his relatives' benefit.

SUMMARY

Research on the social strategies of group-living animals provides clues about the selective forces that have given rise to complex mental abilities. Group life favors individuals who can recognize others, remember past interactions, and modify their behavior appropriately. An individual gains a further selective advantage if he can classify relationships into types ("mother-offspring," "mates," "rivals"), organize them into a structure (in vervet monkeys, a ranked hierarchy of matrilines), deduce transitive relations (if A dominates B and B dominates C, then A dominates C), and attribute mental states to other individuals. Group life, in other words, favors individuals whose minds are predisposed to organize data on social interactions according to certain rules.

We argue that selection favoring a structured, representational view of social companions will be particularly strong when group size is large, when individuals interact with the members of other groups, and when alliances are common. Under these conditions, individuals must be able not only to predict their opponents' behavior but also to recognize and assess their opponents' social relationships. The ecological conditions that favor the evolution of alliances are not unique to nonhuman primates. Current data suggest, however, that nonhuman primates may be the only animals who assess the relative quality of potential alliance partners.

Other cognitive skills, like the attribution of mental states, are less widespread and, among primates, may be restricted to apes and humans. Nevertheless, there are clear adaptive advantages to having a "theory of mind." Individuals who are able to attribute mental states to others gain an advantage in deception, detecting cheaters, and transmitting information. By judging their companions' motives, animals that possess a theory of mind can better predict their behavior.

At present, evidence for a theory of mind in primates is inconclusive. Moreover, little is known about the mechanisms underlying social behavior in other large-brained mammals like dolphins and elephants. Documenting the existence (or lack) of a theory of mind in different species constitutes a central problem for ethologists.

REFERENCES

Allen, C. 1989. Philosophical issues in cognitive ethology. Ph.D. dissertation, University of California, Los Angeles.

Altmann, J. 1980. *Baboon mothers and infants.* Cambridge: Harvard University Press.

Andersson, M. 1980. Why are there so many threat displays? *Journal of Theoretical Biology* 86:773–81.

Astington, J. W., P. L. Harris, and D. R. Olson, eds. 1988. *Developing theories of mind.* Cambridge: Cambridge University Press.

Aureli, F., C. P. van Schaik, and J. A. R. A. M. van Hooff. 1989. Functional aspects of reconciliation among captive long-tailed macaques (*Macaca fascicularis*). *American Journal of Primatology* 19:39–51.

Bachmann, C., and H. Kummer. 1980. Male assessment of female choice in hamadryas baboons. *Behavioral Ecology and Sociobiology* 6:315–21.

Bercovitch, F. 1988. Coalitions, cooperation, and reproductive success among adult male baboons. *Animal Behaviour* 36:1198–1209.

Boesch, C. 1991. Teaching among wild chimpanzees. *Animal Behaviour* 41:530–32.

Boyd, R., and P. Richerson. 1985. *Culture and the evolutionary process.* Chicago: University of Chicago Press.

Byrne, R. W., and A. Whiten. 1988. Towards the next generation in data quality: A new survey of primate tactical deception. *Behavior and Brain Sciences* 11:267–73.

Byrne, R. W., and A. Whiten. 1990. Computation and mindreading in primate tactical deception. In A. Whiten, ed., *Natural theories of mind: The evolution, development, and simulation of everyday mindreading.* Oxford: Basil Blackwell.

Caro, T. M., and M. D. Hauser. 1992. Is there evidence of teaching in nonhuman animals? *Quarterly Review of Biology* 67:151–74.

Chapais, B., M. Girard, and G. Primi. 1991. Non-kin alliances and the stability of matrilineal dominance relationships in Japanese macaques. *Animal Behaviour* 41:481–91.

Charnov, E. L. 1976. Optimal foraging: The marginal value theorem. *Theoretical Population Biology* 9:129–36.

Cheney, D. L. 1983. Extra-familial alliances among vervet monkeys. In R. A. Hinde, ed., *Primate social relationships: An integrated approach,* 278–85. Oxford: Blackwell.

Cheney, D. L., and R. M. Seyfarth. 1980. Vocal recognition in free-ranging vervet monkeys. *Animal Behaviour* 28:362–67.

Cheney, D. L., and R. M. Seyfarth. 1982. Recognition of individuals within and between groups of free-ranging vervet monkeys. *American Zoologist* 22:519–29.

Cheney, D. L., and R. M. Seyfarth. 1983. Non-random dispersal in free-ranging vervet monkeys: Social and genetic consequences. *American Naturalist* 122:392–412.

Cheney, D. L., and R. M. Seyfarth. 1986. The recognition of social alliances among vervet monkeys. *Animal Behaviour* 34:1722–31.

Cheney, D. L., and R. M. Seyfarth. 1987. The influence of intergroup competition on the survival and reproduction of female vervet monkeys. *Behavioral Ecology and Sociobiology* 21:375–86.

Cheney, D. L., and R. M. Seyfarth. 1988. Assessment of meaning and the detection of unreliable signals by vervet monkeys. *Animal Behaviour* 36:477–86.

Cheney, D. L., and R. M. Seyfarth. 1989. Reconciliation and redirected aggression in vervet monkeys, *Cercopithecus aethiops. Behaviour* 110:258–75.

Cheney, D. L., and R. M. Seyfarth. 1990a. Attending to behaviour versus attending to knowledge: Examining monkeys' attribution of mental states. *Animal Behaviour* 40:742–53.

Cheney, D. L., and R. M. Seyfarth. 1990b. *How monkeys see the world: Inside the mind of another species*. Chicago: University of Chicago Press.

Cheney, D. L., R. M. Seyfarth, S. J. Andelman, and P. C. Lee. 1988. Reproductive success in vervet monkeys. In T. H. Clutton-Brock, ed., *Reproductive success*, 384–402. Chicago: University of Chicago Press.

Connor, R., R. A. Smolker, and A. F. Richards. 1992. Two levels of alliance formation among male bottlenose dolphins. *Proceedings of the National Academy of Sciences U.S.A.* 89:987–90.

Cowie, R. J. 1977. Optimal foraging in great tits, *Parus major*. *Nature* 268:137–39.

Dasser, V. 1985. Cognitive complexity in social relationships. In R. A. Hinde, A. Perret-Clermont, and J. Stevenson-Hinde, eds., *Social relationships and cognitive development*, 9–32. Oxford: Oxford University Press.

Dasser, V. 1988a. Mapping social concepts in monkeys. In R. W. Byrne and A. Whiten, eds., *Machiavellian intelligence: Social expertise and the evolution of intellect in monkeys, apes, and humans*, 85–93. Oxford: Oxford University Press.

Dasser, V. 1988b. A social concept in Java monkeys. *Animal Behaviour* 36:225–30.

Datta, S. B. 1983. Patterns of agonistic interference. In R. A. Hinde, ed., *Primate social relationships: An integrated approach*, Oxford: Blackwell Scientific Publications.

Davis, R. T., R. W. Leary, D. A. Stevens, and R. F. Thompson. 1967. Learning and perception of oddity problems by lemurs and seven species of monkey. *Primates* 8:311–22.

Dawkins, M. S., and T. Guilford. 1991. The corruption of honest signalling. *Animal Behaviour* 41:865–74.

Dennett, D. C. 1983. Intentional systems in cognitive ethology: The "Panglossian paradigm" defended. *Behavior and Brain Sciences* 6:343–55.

Dennett, D. C. 1987. *The intentional stance*. Cambridge, Mass.: MIT Press/Bradford Books.

Dickinson, A. 1980. *Contemporary animal learning theory*. Cambridge: Cambridge University Press.

Essock-Vitale, S., and R. M. Seyfarth. 1987. Intelligence and social cognition. In B. B. Smuts, D. L. Cheney, R. M. Seyfarth, R. W. Wrangham, and T. Struhsaker, eds., *Primate societies*, 452–61. Chicago: University of Chicago Press.

Galef, B. G. 1988. Imitation in animals: History, definition, and interpretation of data from the psychological laboratory. In T. R. Zentall and B. G. Galef, eds., *Social learning: Biological and psychological perspectives*, 3–28. Hillsdale, N.J.: Lawrence Erlbaum Associates.

Gallistel, C. R. 1989. *The organization of learning*. Cambridge, Mass.: Bradford Books/MIT Press.

Gallup, G. G. 1982. Self-awareness and the emergence of mind in primates. *American Journal of Primatology* 2:237–48.

Gillan, D. J., D. Premack, and G. Woodruff. 1981. Reasoning in the chimpanzee: I. Analogical reasoning. *Journal of Experimental Psychology: Animal Behavior Processes* 7:1–17.

Grafen, A. 1990. Do animals really recognize kin? *Animal Behaviour* 39:42–54.

Grafen, A. 1991. Modelling in behavioural ecology. In J. R. Krebs and N. B. Davies, eds., *Behavioural Ecology: An evolutionary approach*, 3d ed., 5–31. Cambridge, Mass.: Blackwell Scientific Publications.

Green, S., and P. Marler. 1979. The analysis of animal communication. In vol. 3, P. Marler and J. G. Vandenbergh, eds., *Handbook of behavioral neurobiology*. New York: Plenum Press.

Harcourt, A. H. 1989. Social influences on competitive ability: Alliances and their consequences. In V. Standen and R. Foley, eds., *Comparative socioecology*, 89–113. Oxford: Blackwell Scientific Publications.

Harcourt, A. H. 1992. Coalitions and alliances: Are primates more complex than nonprimates? In A. H. Harcourt and F. de Waal, eds., *Coalitions and alliances in humans and other animals*, 445–72. New York: Oxford University Press.

Harcourt, A. H., and F. B. M. de Waal, eds. 1992. *Coalitions and alliances in humans and other animals*. New York: Oxford University Press.

Harper, D. C. G. 1991. Communication. In J. R. Krebs and N. B. Davies, eds., *Behavioral ecology: An evolutionary approach*, 3d ed. Cambridge, Mass.: Blackwell Scientific Publications.

Hinde, R. A. 1976. Interactions, relationships, and social structure. *Man* 11:1–17.

Hinde, R. A. 1983. A conceptual framework. In R. A. Hinde, ed., *Primate social relationships: An integrated approach*, 1–7. Oxford: Blackwell Scientific Publications.

Humphrey, N. K. 1976. The social function of intellect. In P. P. G. Bateson and R. A. Hinde, eds., *Growing points in ethology*, 303–18. Cambridge: Cambridge University Press.

Judge, P. 1982. Redirection of aggression based on kinship in a captive group of pigtail macaques. *International Journal of Primatology* 3:301.

Krebs, J. R., and R. Dawkins. 1984. Animal signals: Mind reading and manipulation. In J. R. Krebs and N. B. Davies, eds., *Behavioural ecology: An evolutionary approach*, 380–402. Oxford: Blackwell Scientific Publications.

Krebs, J. R., and McCleery, R. H. 1984. Optimisation in behavioural ecology. In J. R. Krebs and N. B. Davies, eds., *Behavioural ecology: An evolutionary approach*, 2d ed., 91–121. Oxford: Blackwell Scientific Publications.

Kummer, H. 1968. *Social organization of hamadryas baboons*. Basel: S. Karger.

Kummer, H. 1982. Social knowledge in free-ranging primates. In D. R. Griffin, ed., *Animal mind-human mind*, 113–30. Berlin: Springer-Verlag.

Kummer, H., W. Goetz, and W. Angst. 1974. Triadic differentiation: An inhibitory process protecting pair bonds in baboons. *Behaviour* 49:62–87.

Mandler, G. 1967. Organization and memory. In K. W. Spence and J. T. Spence, eds., *The psychology of learning and motivation*, 303–47. New York: Academic Press.

Maynard Smith, J., and G. A. Parker. 1976. The logic of asymmetric contests. *Animal Behaviour* 24:159–75.

Moss, C. J. 1988. *Elephant memories*. Boston: Houghton Mifflin.

Nishida, T. 1987. Local traditions and cultural transmission. In B. B. Smuts, D. L. Cheney, R. M. Seyfarth, R. W. Wrangham, and T. T. Struhsaker, ed., *Primate societies*, 462–74. Chicago: University of Chicago Press.

Nishida, T., and M. Hiraiwa-Hasegawa. 1987. Chimpanzees and bonobos: Cooperative relationships among males. In B. B. Smuts, D. L. Cheney, R. M. Seyfarth, R. W. Wrangham, and T. T. Struhsaker, eds., *Primate societies*, 165–77. Chicago: University of Chicago Press.

Noe, R. 1986. Lasting alliances among adult male savannah baboons. In J. Else and P. C. Lee, eds., *Primate ontogeny, cognition, and social behavior*, 381–92. Cambridge: Cambridge University Press.

Packer, C. 1977. Reciprocal altruism in olive baboons. *Nature* 265:441–43.

Packer, C., and A. E. Pusey. 1982. Cooperation and competition within coalitions of male lions: Kin selection or game theory? *Nature* 296:740–42.

Povinelli, D. J., K. E. Nelson, and S. T. Boysen. 1990. Inferences about guessing and knowing by chimpanzees. *Journal of Comparative Psychology* 104:203–10.

Premack, D. 1976. *Intelligence in ape and man*. Hillsdale, N.J.: Lawrence Erlbaum Associates.

Premack, D. 1983. The codes of man and beast. *Behavior and Brain Sciences* 6:125–67.

Premack, D. 1986. *Gavagai*. Cambridge: MIT/Bradford Books.

Premack, D. 1988. "Does the chimpanzee have a theory of mind?" revisited. In R. W. Byrne and A. Whiten, eds. *Machiavellian intelligence: Social expertise and the evolution of intellect in monkeys, apes, and humans*, 160–79. Oxford: Oxford University Press.

Premack, D., and G. Woodruff. 1978. Does the chimpanzee have a theory of mind? *Behavioral and Brain Sciences* 1:515–26.

Ryle, G. 1949. *The concept of mind*. London: Hutchinson.

Seyfarth, R. M. 1978. Social relationships among adult male and female baboons. II. Behaviour throughout the female reproductive cycle. *Behaviour* 64:227–47.

Seyfarth, R. M., and D. L. Cheney. 1986. Vocal development in vervet monkeys. *Animal Behaviour* 34:1640–58.

Seyfarth, R. M., D. L. Cheney, and P. Marler. 1980. Vervet monkey alarm calls: Semantic communication in a free-ranging primate. *Animal Behaviour* 28:1070–94.

Sigg, H., A. Stolba, J. J. Abegglen, and V. Dasser. 1982. Life history of hamadryas baboons: Physical development, infant mortality, reproductive parameters, and family relationships. *Primates* 23:473–87.

Smuts, B. B. 1985. *Sex and friendship in baboons*. New York: Aldine.

Stammbach, E. 1987. Desert, forest, and montane baboons: Multilevel societies. In B. B. Smuts, D. L. Cheney, R. M. Seyfarth, R. W. Wrangham, and T. T. Struhsaker, eds., *Primate societies*, 112–20. Chicago: University of Chicago Press.

Strong, P. N., and M. Hedges. 1966. Comparative studies in simple oddity learning, 1: Cats, raccoons, monkeys, and chimpanzees. *Psychonomic Science* 5:13–14.

Strum, S. C. 1984. Why males use infants. In D. M. Taub, ed., *Primate paternalism*, 146–85. New York: Van Nostrand Reinhold.

Swartz, K. B., S. Chen, and H. S. Terrace. 1991. Serial learning by rhesus monkeys: I. Acquisition and retention of multiple four-item lists. *Journal of Experimental Psychology: Animal Behavior Processes* 17:396–410.

Terrace, H. S. 1987. Chunking by a pigeon in a serial learning task. *Nature* 325:149–51.

Thornhill, R. 1979. Adaptive female-mimicking behavior in a scorpion fly. *Science* 205:412–14.

de Waal, F. B. M. 1982. *Chimpanzee politics.* New York: Harper and Row.

de Waal, F. B. M., and L. M. Luttrell. 1988. Mechanisms of social reciprocity in three primate species: Symmetrical relationship characteristics or cognition? *Ethology and Sociobiology* 9:101–18.

de Waal, F. B. M., and R.-M. Ren. 1988. Comparison of the reconciliation behavior of stumptail and rhesus macaques. *Ethology* 78:129–42.

de Waal, F. B. M., and A. van Roosmalen. 1979. Reconciliation and consolation among chimpanzees. *Behavioral Ecology Sociobiology* 5:55–66.

de Waal, F. B. M., and D. Yoshihara. 1983. Reconciliation and redirected affection in rhesus monkeys. *Behaviour* 85:224–41.

Waldman, B., P. Frumhoff, and P. Sherman. 1987. Problems of kin recognition. *Trends in Ecology Evolution* 3:8–13.

Whiten, A., ed. 1990. *Natural theories of mind: The evolution, development, and stimulation of everyday mindreading.* Oxford: Blackwell Scientific Publications.

Whiten, A., and R. W. Byrne. 1988a. Tactical deception in primates. *Behavior and Brain Sciences* 11:233–73.

Whiten, A., and R. W. Byrne. 1988b. Taking (Machiavellian) intelligence apart: editorial. In R. W. Byrne and A. Whiten, eds., *Machiavellian intelligence: Social expertise and the evolution of intellect in monkeys, apes, and humans*, 50–67. Oxford: Oxford University Press.

Wilson, E. O. 1971. *The insect societies.* Cambridge: Harvard University Press.

Wimmer, H., and J. Perner. 1983. Beliefs about beliefs: Representation and constraining function of wrong beliefs in young children's understanding of deception. *Cognition* 13:103–28.

York, A. D., and T. E. Rowell. 1988. Reconciliation following aggression in patas monkeys, *Erythrocebus patas. Animal Behaviour* 36:502–09.

SEVENTEEN

Lanchester's Theory of Combat, Self-Organization, and the Evolution of Army Ants and Cellular Societies

NIGEL R. FRANKS AND LUCAS W. PARTRIDGE

LIVING IN GROUPS and societies opens up possibilities not available to the solitary individual. Two spectacular examples of biological societies are the army ants and cellular societies (by the latter we mean myxobacteria and cellular slime molds). Army ants attack prey that are either too numerous (the workers of social insect colonies) or too large (big arthropods) for single ants to kill; myxobacteria attack, digest, and consume their prey collectively. Such is the devastating efficiency of this collective activity that both kinds of society deplete local resources and must undergo periods of nomadism. Cellular slime molds, on the other hand, appear to feed separately but come together prior to dispersal.

One of the themes we discuss is the spectacular predatory power afforded by very large, literally overwhelming, numbers. A second, related, theme is how large numbers of rather simple units coordinate their collective activities through self-organization.

In this chapter we review Lanchester's theory of combat (Lanchester 1916), which explains the battle conditions under which large numbers of individuals in a fighting force are much more important than high individual fighting value. Next, we consider the constraints and behavioral mechanisms that have played a key role in the evolution of such societies, focusing on decentralized control and self-organization in particular. First, however, we outline the natural history of the best-known species of army ants and cellular societies.

THE ARMY ANT SYNDROME AS EXEMPLIFIED BY *Eciton burchelli*

In terms of their social integration and mutual interdependence, the members of army ant societies are an extreme in evolutionary ecology (Franks 1989a). Hundreds of thousands, or even millions, of army ant workers live together in extraordinarily cohesive societies (Schneirla 1971). Certain species form nests from their own living bodies. These bivouacs provide a protective envelope for the colony's developing brood. From the bivouac, the army ant colony stages gigantic group raids, forming exquisite recursive patterns, that may sweep over thousands of square meters in a single day and employ the efforts of hundreds of thousands of workers.

Thousands of these individuals may die every day during combat with their

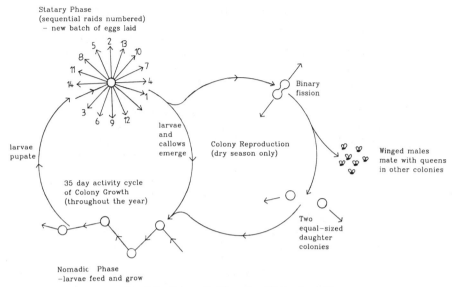

Fig. 17.1   Schematic life cycle of *Eciton burchelli.*

prey, which, in the case of *Eciton burchelli,* consist of other social insect colonies and large arthropods such as spiders, crickets, and scorpions. Relative to an individual army ant worker, these prey are collectively or individually large and dangerous. Only raiding by very large numbers of workers is likely to be effective (see below) and so the army ant colony needs to be huge.

The army ant raid is a series of dangerous battles. If only 1 or 2 percent of the 150,000 foragers are lost per day by an average *E. burchelli* colony, it will start to dwindle in size (Franks 1985). In many cases, individuals survive for only hours when separated from their nestmates. Members of certain castes literally become locked into a kamikaze role when they defend their colony. (*Eciton* majors can penetrate vertebrate flesh with their hooklike mandibles but are then unable to release themselves.)

The social insect prey of *E. burchelli* are slow to recover their numbers (Franks 1982). In order to find sufficient food, therefore, the army ant colony has to move from one feeding site to another. Colonies alternate rigidly between nomadic phases when they raid each day and much more quiescent statary phases in which the bivouac remains at one site and foraging occurs only intermittently. In the nomadic phase the army ant colony has its voracious larvae to feed, whereas in the statary phase all its brood are in the form of nonfeeding eggs and pupae. When colonies have grown sufficiently large, they split into two daughter colonies which are nevertheless considerably larger than most other colonies of social insects. The life cycle of *E. burchelli* is schematically illustrated in figure 17.1.

Although the exact number of times is controversial, the army ant syndrome appears to have evolved independently at least twice (Hölldobler and Wilson 1990; Bolton 1990). Remarkably, this extreme of social integration is repeated elsewhere at a completely different scale.

THE NATURAL HISTORY OF CELLULAR SOCIETIES

Here the term "cellular societies" refers to cells capable of existing as physically separate (but not necessarily independent) individuals that are also capable of aggregating to form collective entities. While such aggregations are often visible to the naked eye, the individual cells can be observed only through a microscope.

The potential for unicellular organisms to aggregate and form multicellular entities appears to have evolved on at least four separate occasions, namely, in the prokaryotic myxobacteria and in the following eukaryotes: the acrasid and dictyostelid amoebae and the ciliate protozoa (Bonner 1988). We shall restrict our attention to the myxobacteria (literally "slime bacteria") and the dictyostelid amoebae since these have been studied intensively.

Myxobacteria

Myxobacteria are strictly aerobic, Gram-negative, rod-shaped bacteria that are ubiquitous in soil, bark, decaying organic matter, dung, and fresh water from subarctic to tropical zones (Reichenbach 1986). In places they are extremely abundant: as many as 450,000 cells per gram may be found in certain soils (McCurdy 1969; and Singh 1947; cited in Shimkets 1990). The cells are usually embedded in a secreted matrix of macromolecules commonly referred to as slime (Shimkets 1990). The life cycle of a particularly well studied myxobacterium, *Myxococcus xanthus*, is illustrated in figure 17.2.

Throughout their entire life cycle these bacteria are undeniably social, since they move, feed, and lie dormant in organized multicellular masses (Dworkin and Kaiser 1985). Gliding through the soil in swarms of up to several million individuals, they prey upon and scavenge other bacteria as well as larger, eukaryotic organisms such as yeasts and nematodes (Zusman 1984).

Extracellular antibiotics, lytic enzymes, and digestive enzymes are secreted into the environment, and the liberated products of digestion are shared among the cells. While food is plentiful the myxobacteria grow and divide indefinitely by binary fission, but when food becomes scarce the bacteria switch from the feeding phase to the developmental phase. In *Myxococcus xanthus* tens of thousands of cells aggre-

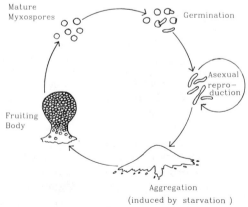

Fig. 17.2   Schematic life cycle of *Myxococcus xanthus*. (Adapted from Wireman and Dworkin 1975.)

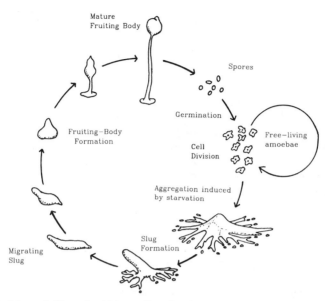

Fig. 17.3   Schematic life cycle of *Dictyostelium discoideum*. (Adapted from Alberts et al. 1989.)

gate to form discrete raised mounds of cells: the fruiting bodies. In other species, such as *Stigmatella aurantiaca*, the fruiting bodies are more elaborate and treelike. Many cells within these structures differentiate into dormant, spherical myxospores that are resistant to environmental extremes. Other cells form the stalk upon which the myxospores are borne in a mass. The spore-containing part of the fruiting body remains intact and helps protect the resting colony during its dispersal by water or passing animals. When favorable conditions return, the myxospores germinate simultaneously, and a new swarm of predatory rod-shaped myxobacteria emerge. (For a recent review of myxobacteria, see Shimkets 1990.)

Cellular Slime Molds

One of the most intensively studied eukaryotic cellular slime molds is *Dictyostelium discoideum*. Its life cycle is shown in figure 17.3.

Like the myxobacteria, cellular slime molds are ubiquitous in the soil (Webster 1980), and the individual amoebae (sometimes called "myxamoebae") also feed on bacteria and yeast cells. When food is plentiful, the amoebae reproduce by binary fission, but again, in the absence of food they stream together and form multicellular aggregates consisting of up to 100,000 or so cells (Bonner 1983). The aggregate gradually develops into a sluglike creature 1 to 2 millimeters long (Alberts et al. 1989). In contrast to the nonmotile fruiting body aggregations of the myxobacteria, these slugs are capable of migration. Eventually, the slug ceases to migrate and produces one or more fruiting bodies that bear spores on a stalk. Cells in the stalk gradually vacuolate and die, but the spore cells are dispersed by water, worms, or insects (Bonner 1983). The spores lie dormant until favorable conditions return, whereupon they germinate and release single amoebae. Unlike the myxobacteria, however, the spores do not necessarily remain together as an integral unit. The instantaneous formation of a swarm is therefore not guaranteed upon germination, in contrast to the myxobacteria.

OVERVIEW

Both army ants and cellular societies exhibit mass cooperative behavior in a nomadic predatory lifestyle. Here we briefly review the ecology and communication systems of these two types of societies, then focus on the fundamental similarities in the mechanisms underlying their behavioral repertoires. We suggest that such similarities (between army ants and myxobacteria) are likely to be the product of almost identical selection pressures acting on predatory power to produce an extreme form of convergent evolution.

The rationale of this comparative approach, which sweeps across phylogenetic boundaries to compare biological organization in societies with completely different progenitors, is to highlight the most fundamental processes in social selection and evolution. For example, as we detail below, the individual members of both army ant and cellular societies seem to have severely restricted cognitive fields that are dominated by inputs from other members of their society, particularly their immediate neighbors in space or time. In both cases the autonomy of the individual appears to have been lost during the evolution of the collective.

First, we consider a mathematical model that reveals the importance of concentrating large numbers in many combat situations.

THE ADVANTAGE OF LARGE NUMBERS: LANCHESTER'S THEORY OF COMBAT

A Deterministic Model

Lanchester was an engineer who devised a series of equations to describe the rates of attrition of two warring armies under different conditions. One of his original equations is as follows:

$$dm/dt = -\alpha n$$
$$dn/dt = -\beta m \tag{17.1}$$

Here, m and n are the numbers of combat units (e.g., troops, aircraft, or ships) engaged on each side, and α and β are constants representing the relative fighting values of the two types of unit. This equation is appropriate to a situation of concentrated aimed fire, in which individuals on the numerically weaker side may be the focus of concentrated attack by more than one individual on the opposing side. The disparity between the warring forces increases through time as the larger side brings progressively more and more of its firepower to bear on the dwindling opposition. Dividing the two equations to eliminate time and integrating them gives:

$$\beta(m_0^2 - m^2) = \alpha(n_0^2 - n^2) \tag{17.2}$$

where m_0 and n_0 are the number of units in the initial forces (Wallis 1968). Assuming equal combat effectiveness ($\alpha = \beta$), and rearranging for m when $n = 0$ (i.e., when m has won), we obtain:

$$m = \sqrt{(m_0^2 - n_0^2)} \qquad (17.3)$$

In other words, all else being equal, the average number of survivors in the winning army will be the square root of the difference in the squares of the initial numbers in the opposing forces.

A Numerical Example

Following Lanchester, we shall give a numerical demonstration of this theory using the battle data postulated by Lord Nelson in his memorandum of 9 October prior to the battle of Trafalgar in 1805 (Lanchester 1916). Nelson assumed that a British fleet of 40 would engage a larger combined French and Spanish fleet of 46 ships. Assuming equal firepower per ship ($\alpha = \beta$), equation 17.3 shows that the enemy would be expected to win with 23 surviving ships after dispatching all the British ones to a watery grave ($\sqrt{[46^2 - 40^2]} = 23$).

In his memo, Nelson spelled out the importance of concentrating his forces. He considered sending 8 ships to split the enemy fleet in half. His remaining 32 ships could then destroy the first half of the opposing fleet. Applying Lanchester's equations, Nelson's strategy would have left him 22 ships ($\sqrt{[32^2 - 23^2]} = 22$) (plus any of the original 8 that had survived the initial skirmish). He could then set about attacking the second half of the enemy fleet with the odds much more in his favor than one could imagine from an all-out battle at the outset of the conflict.

Equation 17.2 is known as Lanchester's square law. The power of numbers is so great that if you double the number of your troops, by making each half the size, with half the per capita firepower, you will suffer many fewer casualties (both in proportionate and absolute terms). Consider, if the French and Spanish had 92 half-size ships, each with half the virulence of each of the 40 English ships, then, in straight battle, on average, 73 of their half-size ships would survive (solving for m in equation 17.2 when $m_0 = 92$, $n_0 = 40$, $n = 0$, $\alpha = 1$, and $\beta = 0.5$). In terms of total tonnage this is a considerably better result than the 23 full-size survivors in the original script.

Is it then any surprise that army ant workers are not only extremely numerous but actually much smaller per capita than workers of their social insect prey, and that colonies reproduce by fission to maintain relatively large armies at all times? Small size is also a huge advantage in terms of modular growth. The smaller the individual, the faster it can be produced and the sooner it can actively contribute to the growth of other modules.

Two important points relating to risk and uncertainty can be made. First, Lanchester's original laws are deterministic and really give only the average numbers surviving at any one time. They give no idea of the variation in possible outcomes or of the probability of any one particular outcome. Thus if one side only slightly exceeds the other in numbers, then, although it would win in a deterministic system, it would not win every time in a stochastic situation. To increase the certainty of a given outcome, therefore, the disparity between the two sides should be increased. Second, if you face an enemy that is unpredictable in both size and strength, then a suitable strategy is to ensure that you have as much fighting strength as possible to increase the likelihood of winning. Both these points imply that, barring constraints, an army should be as large and strong as possible, especially since a more numerous army should suffer many fewer casualties.

Lanchester's square law implies that by increasing your numbers you will accrue disproportionately more benefit than by simply increasing the fighting value of your individual combat units (compare the square terms with the constants in equation 17.2 above).

If all battles are fought at the edge of the armies so that only a few troops are engaged at any one time, Lanchester's linear law is more appropriate (Wallis 1968):

$$\beta(m_0 - m) = \alpha(n_0 - n) \tag{17.4}$$

Franks and Partridge (1993) have shown that this linear law might explain the fighting strategy of slave-making ants that are less numerous but stronger per capita than the host workers they must fight. Such slave-makers use propaganda substances to ensure that they are not the subject of concentrated attack by their more numerous opponents.

This chapter is concerned with very large armies in which battles may begin at the edge of the swarm, but where the opponents are in most cases quickly engulfed and attacked from all sides. Thus even if a battle begins in the linear mode, it will most likely rapidly become an encounter of the square law variety. In other words, the importance of concentrating large numbers will still apply.

When it comes to applying Lanchester's theories to predator-prey systems in biology, it should be realized that the prey are by no means passive targets. They, too, have the potential to evolve corresponding countermeasures. For example, a defending prey social insect colony might also be selected to have larger numbers of smaller workers. Nevertheless, it should be remembered that many other selection pressures will act on worker size and number in such colonies, and, indeed, the army ants might be rare enemies (Dawkins 1982).

However, many of these selection pressures will also apply to the army ants: the production of an army with many small units is not without its costs and constraints. One such cost is increasing transport costs: energetic costs of transport per unit weight per unit distance are larger for smaller animals (Alexander 1982). To minimize such costs, one of the four worker castes in *Eciton burchelli*, the large submajor, is specialized for road-haulage and forms superefficient teams that lower the costs of transporting prey (Franks 1986). Just as we would predict, the two smaller and much more numerous worker castes in *E. burchelli* do the lion's share of the fighting with prey, whereas the much bigger and individually stronger submajors and majors have specialized roles (see Franks and Partridge 1993).

Finally, another cost that is sometimes associated with units of small size and low sophistication is a reduction in the cognitive field of an individual (Lythgoe 1979). How then does a colony operate in such a sophisticated collective manner? The answer may be self-organization, which will be considered below.

APPLYING LANCHESTER'S COMBAT THEORY TO ARMY ANTS AND CELLULAR SOCIETIES

Army Ants

The majority of the workers in an *E. burchelli* colony are individually much smaller than the workers of the defending prey colonies (Franks 1980). As we explain below, these army ants are virtually blind and may encounter a prey colony at any point along the front of a swarm raid, often without prior warning. Individual army ants can be rapidly killed by workers defending their colonies (N. R. Franks, pers. obs.), but the army ants generally break through the defenses of such colonies by sheer weight of numbers. Prey colonies might number from a few hundred to a few thousand, whereas the army ants have tens of thousands of workers per square meter at the swarm front. We can now explain the enormous numbers of army ants in the leading edge of the raiding system: Lanchester's theory of combat shows that having such overwhelming numbers should minimize casualties. These tens of thousands of workers are not redundant (even though the colony might only retrieve 30,000 prey items a day: Franks 1982) because the next battle with a dangerous prey colony may occur unpredictably both in space and time. Indeed, they may be waging battles simultaneously with more than one colony. The advantage of overwhelming numbers helps to explain why army ants reproduce by colony fission.

Cellular Societies

The traditional advantage ascribed to group living in cellular societies like myxobacteria is the benefit of cooperative feeding (e.g., Wireman and Dworkin 1975; White 1981; Burnham et al. 1981; Dworkin and Kaiser 1985; Reichenbach 1986; Shimkets 1990). Vast numbers of concentrated, closely interacting cells can secrete large amounts of digestive enzymes and antibiotics and absorb the liberated products of digestion with minimum loss to the environment in both cases. Experiments have shown that per capita growth rates, up to a point, are greater in collectives with higher densities of individuals (Burnham et al. 1981; Dworkin and Kaiser 1985).

Of course, population growth rates are affected by death rates as well as birth rates. By concentrating their numbers (referring again to Lanchester's theory of combat), death rates of cells due to prey resistance might be significantly reduced, and hence greater population growth rates assured. Theories for the evolution of cellular societies should not be neglectful of losses in combat situations. Similarly, it is important to reverse the popular but erroneous view of the army ants being the all-conquering denizens of the rainforest.

Although we can find little published evidence of the prey of cellular societies actively resisting and killing their predators, the prey's cell walls (and secreted defenses in many cases) provide some protection against attack. Such defensive cover is most efficiently and rapidly eroded under the pressure of vast numbers of individuals focusing their individual attacking abilities. The concept of cover was initially invoked (Lanchester 1916) possibly in reference to the old military adage of an attacking army requiring five times as many combatants as a besieged enemy under cover (since individual shots by the attackers are less likely to score direct hits).

Thus in army ants and cellular societies there is a clear advantage to overwhelming numbers acting in concert. In both cases the individual entities need not be acting totally unselfishly. Kin selection is the most likely explanation for the cooperation observed in both kinds of societies. The inclusive fitness of individuals may be enhanced by cooperating with their close relatives (Hamilton 1964; Grafen 1991). In both cases this cooperation has become so extreme that individual autonomy often appears to have been overridden by the needs of the society.

SELF-ORGANIZATION AND DECENTRALIZED CONTROL: CHEMICAL COMMUNICATION AND THE BLIND LEADING THE BLIND

We now examine the mechanisms that enable both army ants and cellular societies to concentrate their vast numbers to form a cohesive, nomadic group predator. One of the ways that army ants and cellular societies maintain their cohesion is through

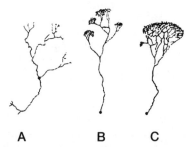

A   B   C

Fig. 17.4   The foraging patterns of (A) *Eciton hamatum*, (B) *E. rapax*, and (C) *E. burchelli*. The circles
represent the bivouac sites at the origin of the raids. (*A* and *C* are adapted from Rettenmeyer 1963;
B is adapted from Burton and Franks 1985.)

the use of trail substances, which may help to coordinate the mass attack of the
individuals on their prey. First, we will review recent work that shows how army
ant raids may be self-organizing, leading to tremendous collective efficiency based
literally on the blind leading the blind.

The power of self-organization in social organisms is exemplified in the swarm
raids of *E. burchelli* (fig. 17.4C). Such raids typically begin at dawn with thousands
of workers pouring out of the bivouac. At first, there are few signs of organization
or marked directionality in these mass movements. But then pattern begins to
emerge from the pandemonium and the raid develops a distinctive structure. Most
of the workers, and there may be up to 200,000 in a single raid, are concentrated in
a dense phalanx at the swarm front, which can be up to 20 meters across. Behind
the swarm front is a series of anastomosing columns forming loops that get progressively larger and less numerous away from the front. Perhaps 15 m or so behind
the swarm front the columns take the form of a single loop, which links with the
principal column of the raid. This is the main roadway for the ants and channels
dense high-speed traffic between the raid front and the bivouac. The principal column is the umbilical cord of the foraging system, and it grows longer and longer
as the raid front advances away from the bivouac. During this process, larger, well-
used loops in the fan of the raid outcompete smaller loops for traffic, and the smaller
ones decay away. Self-similarity therefore arises in the geometry of the branching
raid system, and a beautiful "fractal" pattern emerges.

In a single day such a raid can sweep out a rectangular area 10 to 15 meters
wide by 100 or even 200 meters long. During such a raid the army ants are able to
maintain much the same compass bearing (Franks and Fletcher 1983), even though
they are proceeding through the dim dark depths of the rainforest floor, where only
10% of the light reaching the forest canopy illuminates the scene.

Such army ant raids are one of the wonders of the insect world. Their beautiful,
highly dynamic structure is even more awe-inspiring when one realizes that the
workers who chart its path are virtually blind. Furthermore, they proceed unchecked
in spite of the complex and capricious environment through which the colony raids,
and in the absence of long-range communication or a specific caste of traffic con-
trollers.

Deneubourg et al. (1989) have shown how the blind can indeed lead the blind
in such a raid system, through the mutual guidance provided by chemical trails.
They produced a mathematical model that mimics the patterns of the army ant raid
based on the following assumptions. At first the army ants mill about at random.

But wherever they go they leave a sign of their passing in the form of a volatile chemical trail substance that is smeared on the ground. In the model, if an ant is presented with a choice to move ahead either left or right, it will choose either randomly if there is no trail present, or it will move in whichever direction bears the strongest scent. One feature of the model is that if an ant moves left because there is more scent to the left, it reinforces that trail with its own scent as it proceeds. This means that the next ant to reach that choice point is even more likely to go left, and it too adds its trail to the other ones. Strong positive feedback ensues: the blind leading the blind; or, as Dr. Samuel Johnson might have put it: I stink, you smell! (You follow my stink by using your powers to smell and reinforce my stink with your stink for others to smell.) In addition, the stronger the trail, the faster the ants run.

The ants form a strong principal trail simply because traffic begets traffic. They form a wide phalanx at the swarm front because here they venture into virgin terrain with no smell to guide them. Hence at the swarm front they mill about randomly and diffuse slowly outward like a two-dimensional cloud. The model structure formed by these simple rules has a convincing swarm front and principal trail, but lacks the fractal loops and branches found behind the front of a real raid (compare fig. 17.4C with fig. 17.5A). However, none of the model ants have yet had any instructions, or indeed reason, to find their way home to the bivouac.

Suppose the cue for ants to return home is the discovery and capture of food. An ant picking up a prey item at the leading edge of the swarm turns through 180° and blindly gropes its way home, choosing at each step to go where it perceives the strongest trail pheromones. On the way home the model ant also lays its own trail pheromone atop those of its nestmates, but now, due to its excited state, it lays much more per unit distance than on the way out. The model shows that it is these returning ants that initiate the system of branching and looped trails in the wake of the swarm raid. The trails of the returning ants are so strong that some of the outbound ants choose these trails and in turn reinforce them. In this way it is the interaction between outbound and inbound ants and their discovery of prey that generates the raid pattern (fig. 17.5B,C).

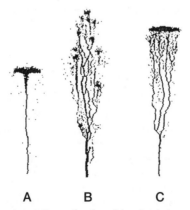

A B C

Fig. 17.5 Three distinct foraging patterns developed by simulations using the same model with three different food distributions: (A) No food. (B) Each point has a 1% probability of containing 400 food items (as might be the case for an army ant that is a specialist predator of other social insects, such as *Eciton rapax*). (C) Each point has a 50% probability of containing one food item (as might be the case for a generalist predator such as *E. burchelli*). (Adapted from Deneubourg et al. 1989.)

The key point is that this is a flexible, responsive, event-driven system. The location of prey feeds back on the geometry of the raid. The model predicts that if prey are scattered in small packets at random ahead of the raid, the system will have the geometry of an E. *burchelli* raid whose arthropod prey are indeed distributed in this way (N. R. Franks, pers. obs.) (fig. 17.5C). When the very same model armies are presented with prey distributed in scarce but large packets, such as the mostly or exclusively social insect prey of E. *rapax* and E. *hamatum*, respectively, they form the more dendritic raid system typical of these species (fig. 17.5B). The communication systems have not changed. The patterns change in response to the perceived distribution of the prey of the army ants.

Franks et al. (1991) tested both the assumptions and the predictions of the Deneubourg et al. (1989) model. They showed that the blind do indeed lead the blind with chemical trails laid by both outbound and inbound workers; and that running velocity is proportional to pheromone strength. They also showed that E. *burchelli* could be made to raid in the fashion of E. *rapax* by redistributing its prey in large discrete patches.

The important general point is that sophisticated patterns can flow from very limited local communication. It is entirely possible that similar simple communication and positive feedback could be involved in the formation of the army ants' bivouac and in its ability to thermoregulate; in the alternation of nomadic and statary phases; in the choice of a single new bivouac site at the front of a nomadic raid; and in the rapid switch to a full emigration at the end of a nomadic day's raiding.

Recently, self-organization has been recognized as being of central importance in the decision-making systems of an increasingly wide range of collective behavior patterns in social insects. Such patterns include rhythms and mutual exclusion (Franks and Bryant 1987; Franks et al. 1990; Hatcher et al. 1992); sorting algorithms (Franks and Sendova-Franks 1992); building behavior (Franks et al. 1992); foraging path choices (Stickland et al. 1992); and division of labor (Tofts and Franks 1992). Self-organization might also be the key to the extraordinary collective organization of the myxobacteria and cellular slime molds.

MECHANISMS OF SELF-ORGANIZATION IN THE MYXOBACTERIA

Swarming Behavior

When myxobacteria feed, thousands of cells move in a swarm (Shimkets and Kaiser 1982). Although the cells move independently of one another, the entire mass of cells migrates as a unit. Swarms move by establishing a loose reticulum of cells by means of "peninsulas" projecting from the swarm front and then looping back on themselves. The gaps are then filled in by individual cells or "rafts" of loosely associated cells (Kaiser and Crosby 1983). Swarms are highly dynamic, being able to divide or fuse with other swarms (White 1981).

Genetic studies have shown that the gliding of the cells is controlled by two main systems—one for "S" or Social motility, and the other for "A" or Adventurous motility (Dworkin and Kaiser 1985). The S system requires the cells to be close together (about one cell length apart) and is probably mediated by pili connecting the cells (White 1981; Kaiser and Crosby 1983; Dworkin and Kaiser 1985). The A system, meanwhile, controls the movement of isolated cells.

Single cells glide more slowly in virgin terrain and will retrace their paths back to the swarm when only about two cell lengths from it (Kaiser and Crosby 1983;

Stanier et al. 1986). This behavior pattern is strongly reminiscent of the so-called "rebound pattern" described by Schneirla (1940) for *Eciton* army ants. In this pattern, individual workers at the head of the swarm hesitantly venture a few centimeters ahead of the pheromone-saturated ground, laying their own pheromone trails, before quickly turning around and running back into the swarm. The swarm front thus advances as a result of the cumulative advance and recoil actions of many different workers.

Slime tracks are always left behind the gliding cells (Reichenbach 1986). The extracellular slime matrix reduces desiccation, protects against attack by several antimicrobial agents, provides a support medium for extracellular digestive enzymes, and is probably also involved in gliding motility (Gnosspelius 1978). From our studies of the patterns of organization in army ant swarm raids, we suspect that the extracellular matrix is analogous to the army ant pheromone trail in that it helps to maintain the cohesiveness of the colony. Furthermore, rather like the army ants that follow a pheromone trail on the way back to the bivouac, the bacterial cells may be able to use their slime trails to retrace their tracks toward their comrades during the rebound pattern or even during the process of aggregation mentioned below.

Cells at high densities tend to become less cohesive and show greater adventurous motility away from the colony (Shimkets 1990). Such a behavior pattern is strikingly similar to Schneirla's (1940) observation that individual army ant workers will move away from very forceful tactile stimuli (a surge of oncoming nestmates, for example). In both cases, such behavior patterns ease the local crowding pressure and facilitate expansion of the swarm into unexploited territory.

Colonial Spherule Formation

One of the most striking examples of collective organization in myxobacteria is the formation of colonial spherules in freshwater environments (Burnham et al. 1981; Burnham et al. 1984). Myxobacterial cells initially cluster around the filaments of their cyanobacterial prey (Daft and Stewart 1973) and eventually build up into spherical colonies consisting of thousands of bacteria, encapsulating the prey. The bacteria produce numerous fibrils that not only facilitate cell-to-cell adhesion but also provide a large surface area capable of ensnaring other filaments of prey in the aqueous medium. Burnham et al. (1981) hypothesize that the myxobacteria continually cover over the captured cyanobacteria, with the result that their prey become trapped in the core of the spherule. Within this core, extracellular digestive enzymes can be concentrated and the liberated products of digestion absorbed with minimum loss to and dilution by the surrounding aquatic environment. The possible importance of this concentrated fighting strength in relation to Lanchester battles and army ants was revealed earlier.

Rippling

Another remarkable behavior pattern shown by myxobacteria is "rippling"—rhythmic gliding movements that arise in dense fields of cells (Dworkin and Kaiser 1985). Ripples are most often observed spreading outward from localized areas in the swarm during aggregation into fruiting bodies (Shimkets and Kaiser 1982). Cells accumulate in the crests of ripples and are depleted in the troughs. Although it can be induced by the breakdown components of bacterial cell walls from lysed myxobacteria or their prey, the function, if any, of rippling remains obscure (Dworkin and Kaiser 1985; Zusman 1984). We have only mentioned it because it represents yet another striking example of a behavior pattern that has been coordinated across thousands of individual cells. Nevertheless, we speculate that rippling may be a means for facilitating efficient locomotion of the cells within the swarm (just as army ants facilitate rapid transport by building living bridges across difficult terrain). Alternatively, rippling may help circulate nutrients or signals among the genetically related cells.

Aggregation and Fruiting Body Formation

Three conditions are required for the switch from the feeding phase to the developmental phase: nutrient depletion, the presence of a solid surface for attachment, and a high cell density (Dworkin and Kaiser 1985). It appears that at least one diffusible molecule (possibly adenosine) is secreted and used by *Myxococcus* cells to monitor cell density and initiate aggregation if the density is sufficiently high (Dworkin and Kaiser 1985; Shimkets 1990).

Aggregation sites form in existing slime trails and sometimes begin as swirling populations of moving cells (White 1981). In contrast to the cellular slime molds, in which chemotaxis is known to play a major role in aggregation (see below), chemotaxis may not be the only mechanism involved in the aggregation of myxobacteria. For recent modeling work, see Stevens (1990, 1991).

O'Connor and Zusman (1989) describe the complex behavior of the cells in *Myxococcus xanthus* as they develop into a fruiting body composed of an unstalked mound of some 100,000 cells. At aggregation centers the swirling arrays of cells generate spirals, which become stacked into monolayers. These monolayers slowly align in register and form so-called "terraces." Mounds arise from the spirals of cells on the surface of such terraces. As the mounds grow, any distinction between the monolayers breaks down, and hollow tubular tunnels form within the mounds. Such tunnels may facilitate the passage of oxygen to the strictly aerobic cells deep within the mound, just as army ants form living ventilation or cooling ducts in their bivouacs (Franks 1989b).

Both aggregation centers and mounds can spontaneously disperse during development (White 1981). A mound may not become fixed in position until a substantial proportion of its cells have changed into immotile myxospores. This is analogous to the mechanism of bivouac formation in army ants, in which there may be dynamic competition for workers between the ant clusters at different potential sites for the bivouac. The final bivouac site is presumably not collectively "chosen" until a large proportion of the workers have interlocked their tarsi, stretched out, and become catatonically immobile.

As the mound forms, as many as 90% of the cells lyse in *Myxococcus xanthus*, while the remaining 10% turn into spores (Dworkin and Kaiser 1985). The extent of lysis appears to depend heavily on the culture method, and the role of lysis in development, if any, is highly controversial (e.g., contrast Wireman and Dworkin [1975] with Shimkets [1990]). In any case, the lysed cells release nutrients that can be taken up by any of the remaining cells developing into myxospores. A proportion of cells also die during development in species of myxobacteria such as *Stigmatella aurantiaca* that form fruiting bodies with stalks. Only the cells in the sporangia at the tip of the stalk develop into myxospores, while those making up the stalk eventually die (Shimkets 1990). Interestingly, no genetic exchange apparently occurs between the different cells that make up a fruiting body (Dworkin and Kaiser 1985; Shimkets 1990). The possibility that only cells of a very similar genotype will cooperate in the formation of fruiting bodies should be examined, for, as will be seen below, this appears to be the general rule for cellular slime molds (although genetic recombination prior to fruiting body formation would also facilitate genetic uniformity across cells).

During colony propagation in army ants, some workers will come to serve their parental queen in one of the daughter colonies, whereas their previous nestmates will end up with a sister queen in the other daughter colony. This will mean that certain workers raise nephews and nieces rather than brothers and sisters (Franks and Hölldobler 1987). Thus the fate of individuals during reproduction in both army ant and cellular societies can only be explained in terms of inclusive fitness (Hamilton 1964; Grafen 1991).

The intercellular interactions within a species of myxobacteria are so precise that the morphology of the fruiting bodies can be used as a consistent character in the taxonomic identification of a species (Dworkin and Kaiser 1985). Earlier in this chapter we described how the different raiding patterns of *Eciton burchelli*, *E. hamatum*, and *E. rapax* could be accounted for by means of the same mechanism simply responding to differences in the environment (as perceived through their different diets). At the simplest level, the differences between species in the morphology of their fruiting bodies could be interpreted as the products of species-specific differences in the fine-tuning of the same building and branching algorithm. For instance, there may be species-specific differences in the response thresholds, signal strengths, or levels of attraction (adhesion) between the cells. In a way, the fruiting body with its integral slime matrix can be viewed as a three-dimensional swarm raid, with slime trails instead of pheromone trails.

As was mentioned previously, the mature spore-containing fruiting body functions as a dormant collective, guaranteeing that a new feeding phase will be initiated by a population of cells able to feed cooperatively (Dworkin and Kaiser 1985). Furthermore, the developmental mechanism is flexible. If the cells are too starved to possess resources sufficient to form a fruiting body, then its formation can be bypassed altogether and sporulation initiated immediately (Shimkets 1990).

MECHANISMS OF SELF-ORGANIZATION IN CELLULAR SLIME MOLDS

We have already outlined the basic life cycle of *Dictyostelium discoideum*. Interestingly, it appears that when food is abundant, the individual amoebae feed in a more solitary fashion than the so-called "wolf packs" of swarming myxobacteria (Raper 1940; Dworkin and Kaiser 1985). The individual amoebae probably engulf their prey (Raper 1940) rather than secreting extracellular digestive enzymes, and are probably more efficient than myxobacteria at feeding alone. They certainly do not appear to attack prey larger than themselves, in contrast to the myxobacteria. Thus the primary selection pressure for social behavior in the case of cellular slime molds may have been enhanced powers of dispersal rather than predatory power.

In contrast to the myxobacteria, cellular slime molds use chemoattractants during aggregation. *D. discoideum*, for instance, uses cyclic adenosine monophosphate (cAMP). Different species appear to use different chemoattractants, thus enabling two species inhabiting the same habitat to sort themselves out independently of one another and form pure species aggregates (Bonner 1983).

Just as army ants exhibit positive feedback through the mutual reinforcement of their chemical pheromone trails as they run along them, so an individual myxamoeba responds to a cAMP gradient not only by moving up the gradient toward its source at the center of the nascent aggregate, but also by secreting its own pulse of cAMP. Thus the signal is effectively transmitted to cells at a greater distance from the center of the aggregate than would be expected from the original pulse of cAMP alone (for recent work, see Steinbock et al. 1991).

The first cells to secrete the cAMP signal become the potential centers of the future aggregates, and, in a manner analogous to the formation of aggregates in myxobacteria and to the competition between neighboring clusters of army ants during the process of bivouac formation, there is opportunity for competition between neighboring centers for amoebae (Alberts et al. 1989).

The amoebae also secrete a lectin (a carbohydrate-binding protein) called discoidin-1, which may help to guide the cells in streams toward aggregation centers. This mechanism is somewhat akin to the pathway guidance (Alberts et al. 1989) provided by the extracellular slime trails of myxobacteria and the pheromone trails of army ants.

Although there is no evidence for rippling or colonial spherule formation in cellular slime molds, they do develop into multicellular slugs that are capable of migration and differentiation into fruiting bodies composed of more than one cell type. The mechanisms of this complex pattern formation are still controversial and have yet to be fully elucidated (Williams et al. 1989). Nevertheless, they involve sorting and regulation of the proportions of the different cell types, as well as position-dependent induction or inhibition of differentiation (see the review by Schaap 1986). Also, there appears to have been little investigation into the mechanism by which the thousands of separate cells in the mass manage to coordinate their activities to produce a movement of the entire slug (but see Siegert and Weijer 1992).

Although the spores appear to be dispersed individually (Bonner 1983), thus causing a break in the multicellular nature of the life cycle that is not found in the myxobacteria, some cellular slime molds have been found to form "macrocysts" (Webster 1980). These arise from myxamoebae (not necessarily of the same genotype) that aggregate and appear to undergo a form of sexual reproduction and genetic recombination. When the macrocysts germinate, they release a swarm of up to 200 myxamoebae, which may be identical in genotype.

A maintenance of close proximity between cells of similar genotype throughout most of the life cycle (perhaps even during their somewhat "solitary wandering" through the soil [Bonner 1983]) would help explain the altruistic death of the stalk cells in the fruiting bodies. Through their closely related "sister" cells that form the spores, these stalk cells would have a genetic stake in the next generation. We have already seen that the same is true for the stalk cells fated to die in the fruiting bodies of myxobacteria such as *Stigmatella aurantiaca*. Unfortunately, few, if any, experiments have been performed to study the interactions between collectives or conspecific individuals of different genotypes in either the myxobacteria or the cellular slime molds (see also Wilson and Sober 1989).

SUMMARY AND CONCLUSION: A COMPARISON OF ARMY ANTS AND CELLULAR SOCIETIES

We have discussed the tremendous predatory power afforded by the extreme social cohesion in army ants and myxobacteria. Our application of Lanchester's theory of combat suggests the selection pressure for concentrated numbers, and shows that very large, poorly equipped armies might be more virulent than smaller, heavily armed forces in the situations covered here.

The example of cellular slime molds provides a good illustration of the perils of assuming that similar patterns have been generated by similar processes. Despite the striking similarity of many of the behavior patterns of myxobacteria and cellular slime molds to those of army ants, two different selection pressures are likely to be involved. Enhanced predatory power is a plausible selection pressure for the evolution of collective behavior in the army ants and myxobacteria, but it is much more likely that enhanced powers of dispersal have governed the evolution of collective behavior in the cellular slime molds.

On a more subtle level, the evolution of myxobacteria and army ants appears to have been driven by slightly different aspects of predatory power. In the case of army ants, group predation gives access to prey unavailable to an individual, thus creating a whole new niche for insect societies (similar to the niche occupied by mammalian anteaters). The colonial nature of myxobacteria, on the other hand, appears to allow them to consume the same prey as solitary bacteria, but with greater efficacy. In other words, they are merely exploiting a former niche more efficiently.

Large armies might, however, be difficult to coordinate with a potentially vulnerable centralized administration. Instead, army ants and cellular societies seem to use behavioral mechanisms based on decentralized self-organization.

In army ants the cohesiveness of the colony appears to be a result of both the chemotactile attraction of the workers to one another and the queen (and the repulsion of other colonies); and the attraction of the workers to the pheromone trails laid down by foregoing workers. The rebound pattern ensures that no worker ventures far from other workers or from the trails laid by its colony. It seems that all army ants have reduced eyesight compared with their ancestors (indeed, *Dorylus* army ants are completely blind). Thus the sensory range of army ants is largely restricted to their immediate environment. Information provided by good eyesight might be a distraction, whereas poor eyesight might facilitate the evolution of social cohesion.

In the case of cellular societies, individual cells can physically adhere to one another, and the extracellular slime matrix functions partly as a trail substance. Just as with the army ant rebound pattern, individual cells reverse direction if they venture too far from the colony (at least in the case of the myxobacteria).

In both types of societies individuals appear to be capable of sensing only their immediate neighborhood and of communicating with their closest neighbors. We hypothesize that such a constraint (when viewed from the point of view of more solitary, independent organisms) contributes to the cohesiveness of the society by forcing individuals to remain close together. Individuals are not distracted by information impinging from outside their immediate surroundings. Instead, they are maximally sensitive to the local conditions and in a position to respond rapidly to any changes in their immediate environment, which, simply because of their colonial nature, is dominated by other individuals like themselves.

ACKNOWLEDGMENTS

This chapter was greatly improved by the critical advice and encouragement of Mike Mogie, Alan Rayner, Tony Robinson, and Alan Wheals. We also wish to thank

Jean-Louis Deneubourg and Simon Goss for many very helpful discussions of the importance of self-organization in social behavior. N. R. F. acknowledges the support of a NATO Collaborative Grant (No. 880344) and a Venture Research Award from British Petroleum International. L. W. P. is grateful to the University of Bath for a Postgraduate Bursarship.

REFERENCES

Alberts, B., D. Bray, J. Lewis, M. Raff, K. Roberts, and J. D. Watson. 1989. *Molecular biology of the cell*. 2d ed. New York: Garland.

Alexander, R. McN. 1982. *Locomotion of animals: Tertiary level biology*. Glasgow: Blackie.

Bolton, B. 1990. Army ants reassessed: The phylogeny and classification of the doryline section (Hymenoptera, Formicidae). *Journal of Natural History* 24:1339–64.

Bonner, J. T. 1983. Chemical signals of social amoebae. *Scientific American* 248:106–12.

Bonner, J. T. 1988. *The evolution of complexity by means of natural selection*. Princeton: Princeton University Press.

Burnham, J. C., S. A. Collart, and M. J. Daft. 1984. Myxococcal predation of the cyanobacterium *Phormidium luridum* in aqueous environments. *Archives of Microbiology* 137:220–25.

Burnham, J. C., S. A. Collart, and B. W. Highison. 1981. Entrapment and lysis of the cyanobacterium *Phormidium luridum* by aqueous colonies of *Myxococcus xanthus* PCO2. *Archives of Microbiology* 129:285–94.

Burton, J. L., and N. R. Franks. 1985. The foraging ecology of the army ant *Eciton rapax*: An ergonomic enigma? *Ecological Entomology* 10:131–41.

Daft, M. J., and W. D. P. Stewart. 1973. Light and electron microscope observations on algal lysis by bacterium CP-1. *New Phytologist* 72:799–808.

Dawkins, R. 1982. *The extended phenotype*. Oxford: Oxford University Press.

Deneubourg, J. L., S. Goss, N. Franks, and J. M. Pasteels. 1989. The blind leading the blind: Modelling chemically mediated army ant raid patterns. *Journal of Insect Behavior* 2:719–25.

Dworkin, M., and D. Kaiser. 1985. Cell interactions in myxobacterial growth and development. *Science* 230:18–24.

Franks, N. R. 1980. The evolutionary ecology of the army ant *Eciton burchelli* on Barro Colorado Island, Panama. Ph.D. thesis, University of Leeds.

Franks, N. R. 1982. Ecology and population regulation in the army ant *Eciton burchelli*. In E. G. Leigh, A. S. Rand, and D. M. Windsor, eds., *The ecology of a tropical forest: Seasonal rhythms and long-term changes*, 389–95. Washington, D.C.: Smithsonian Institution Press.

Franks, N. R. 1985. Reproduction, foraging efficiency and worker polymorphism in army ants. In M. Lindauer and B. Hölldobler, eds., *Experimental behavioural ecology*, Karl von Frisch Memorial Symposium, Fortschritte der Zoologie, Bd. 31, 91–107. Stuttgart: G. Fischer Verlag.

Franks, N. R. 1986. Teams in social insects: Group retrieval of prey by army ants (*Eciton burchelli*, Hymenoptera: Formicidae). *Behavioral Ecology and Sociobiology* 18:425–29.

Franks, N. R. 1989a. Army ants: A collective intelligence. *American Scientist* 77:138–45.

Franks, N. R. 1989b. Thermoregulation in army ant bivouacs. *Physiological Entomology* 14: 397–404.

Franks, N. R., and S. Bryant. 1987. Rhythmical patterns of activity within the nests of ants. In J. Eder and A. Rembold, eds., *Chemistry and biology of social insects*, 122–23. Verlag J. Peperny.

Franks, N. R., and C. R. Fletcher. 1983. Spatial patterns in army ant foraging and migration. *Eciton burchelli* on Barro Colorado Island, Panama. *Behavioral Ecology and Sociobiology* 12: 261–70.

Franks, N. R., and B. Hölldobler. 1987. Sexual competition during colony reproduction in army ants. *Biological Journal of the Linnean Society* 30:229–43.

Franks, N. R., and L. W. Partridge. 1993. Lanchester battles and the evolution of combat in ants. *Animal Behaviour* 45:197–99.

Franks, N. R., and A. B. Sendova-Franks. 1992. Brood sorting in ants: Distributing the workload over the worksurface. *Behavioral Ecology and Sociobiology* 30:109–23.

Franks, N. R., S. Bryant, R. Griffiths, and L. Hemerik. 1990. Synchronization of the behaviour within nests of the ant *Leptothorax acervorum* (Fabricius): 1. Discovering the phenomenon and its relation to the level of starvation. *Bulletin of Mathematical Biology* 52:597–612.

Franks, N. R., N. Gomez, S. Goss, and J. L. Deneubourg. 1991. The blind leading the blind in army ant raid patterns: Testing a model of self-organization (Hymenoptera: Formicidae). *Journal of Insect Behavior* 4:583–607.

Franks, N. R., A. Wilby, B. W. Silverman, and C. M. N. Tofts. 1992. Self-organizing nest construction in ants: Sophisticated building by blind bulldozing. *Animal Behaviour* 44: 357–75.

Gnosspelius, G. 1978. Myxobacterial slime and proteolytic activity. *Archives of Microbiology* 16:51–59.

Grafen, A. 1991. Modelling in behavioural ecology. In J. R. Krebs and N. B. Davies, eds., *Behavioural ecology: An evolutionary approach*, 3d ed., 5–31. Oxford: Blackwell Scientific Publications.

Hamilton, W. D. 1964. The genetical evolution of social behaviour. *Journal of Theoretical Biology* 7:1–52.

Hatcher, M. J., C. M. N. Tofts, and N. R. Franks. 1992. Mutual exclusion as a mechanism for information exchange within ant nests. *Naturwissenschaften* 79:32–34.

Hölldobler, B., and E. O. Wilson. 1990. *The ants.* Cambridge, Mass.: Harvard University Press.

Kaiser, D., and C. Crosby. 1983. Cell movement and its coordination in swarms of *Myxococcus xanthus. Cell Motility* 3:227–45.

Lanchester, F. W. 1916. *Aircraft in warfare: The dawn of the fourth arm.* London: Constable.

Lythgoe, J. N. 1979. *The ecology of vision.* Oxford: Oxford University Press.

McCurdy, H. D. 1969. Studies on the taxonomy of the Myxobacterales. I. Record of Canadian isolates and survey methods. *Canadian Journal of Microbiology* 15:1453–61.

O'Connor, K. A., and D. R. Zusman. 1989. Patterns of cellular interactions during fruiting-body formation in *Myxococcus xanthus. Journal of Bacteriology* 171:6013–24.

Raper, K. B. 1940. Pseudoplasmodium formation and organization in *Dictyostelium discoideum. Journal of the Elisha Mitchell Scientific Society* 56:241–82.

Reichenbach, H. 1986. The myxobacteria: Common organisms with uncommon behaviour. *Microbiological Sciences* 3:268–74.

Rettenmeyer, C. W. 1963. Behavioral studies of army ants. *University of Kansas Science Bulletin* 44:281–465.

Schaap, P. 1986. Regulation of size and pattern in cellular slime molds. *Differentiation* 33:1–16.

Schneirla, T. C. 1940. Further studies of the army-ant behavior pattern. Mass organization in the swarm-raiders. *Journal of Comparative Psychology* 29:401–60.

Schneirla, T. C. 1971. *Army ants. A study in social organization.* H. R. Topoff, ed. San Francisco: W. H. Freeman.

Shimkets, L. J. 1990. Social and developmental biology of the myxobacteria. *Microbiological Reviews* 54:473–501.

Shimkets, L. J., and D. Kaiser. 1982. Induction of coordinated movement of *Myxococcus xanthus* cells. *Journal of Bacteriology* 152:451–61.

Siegert, F., and C. J. Weijer. 1992. Three-dimensional spiral waves in *Dictyostelium* slugs. *Proceedings of the National Academy of Sciences* 89:6433–37.

Singh, B. N. 1947. Myxobacteria in soils and composts: Their distribution, number, and lytic action on bacteria. *Journal of General Microbiology* 1:1–10.

Stanier, R. Y., J. L. Ingraham, M. L. Wheelis, and P. R. Painter. 1986. *General microbiology.* 5th ed. London: Macmillan.

Steinbock, O., H. Hashimoto, and S. C. Müller. 1991. Quantitative analysis of periodic chemotaxis in aggregation patterns of *Dictyostelium discoideum. Physica D* 49:233–39.

Stevens, A. 1990. Simulations of the gliding behaviour and aggregation of myxobacteria. Biological motion. In W. Alt and G. Hoffmann, eds., *Lecture notes in biomathematics* 89, 548–55. Berlin: Springer-Verlag.

Stevens, A. 1991. A model for gliding and aggregation of myxobacteria. In A. Holden, M. Markus, and H. G. Othmer, eds., *Nonlinear wave processes in excitable media*, 269–76. New York: Plenum Press.

Stickland, T. R., C. M. N. Tofts, and N. R. Franks. 1992. A path choice algorithm for ants. *Naturwissenschaften* 79:567–72.

Tofts, C. M. N., and N. R. Franks. 1992. Doing the right thing: Ants, honey bees and naked mole-rats. *Trends in Ecology and Evolution* 7:346–49.

Wallis, P. R. 1968. Recent developments in Lanchester theory. *Operational Research Quarterly* 19:191–95.

Webster, J. 1980. *Introduction to fungi.* 2d ed. Cambridge: Cambridge University Press.

White, D. 1981. Cell interactions and the control of development in myxobacteria populations. *International Review of Cytology* 72:203–27.

Williams, J. G., K. T. Duffy, D. P. Lane, S. J. McRobbie, A. J. Harwood, D. Traynor, R. R. Kay, and K. A. Jermyn. 1989. Origins of the prestalk-prespore pattern in *Dictyostelium* development. *Cell* 59:1157–63.

Wilson, D. S., and E. Sober. 1989. Reviving the superorganism. *Journal of Theoretical Biology* 136:337–56.

Wireman, J. W., and M. Dworkin. 1975. Morphogenesis and developmental interactions in myxobacteria. *Science* 189:516–23.

Zusman, D. R. 1984. Cell-cell interactions and development in *Myxococcus xanthus*. *Quarterly Review of Biology* 59:119–38.

EIGHTEEN

How Social Insect Colonies Respond to Variable Environments

Deborah M. Gordon

COLONY ORGANIZATION: TRACKING THE ENVIRONMENT AND RESPONDING TO CHANGE

ANY SOCIAL GROUP can be investigated at different levels of organization. In social insects, behavior at the group level stands out in stark relief. To explain the behavior of animals with which we readily identify, such as primates, it is easy to look to the motivation and decisions of individuals. But in social insects, while colonies behave in complex ways, the capacities of individuals are relatively limited. Bridging the gap between the simple behavior of individuals and the achievements of colonies is the central problem in understanding social insect organization.

We are steeped in the idea of hierarchical control, which we use to structure human organizations and computer algorithms, and see as characterizing natural processes such as genetic programs. Social insect colonies, however, are not hierarchically organized. No insect issues commands to another, or instructs it to do things in a certain way. Yet no individual is aware of what must be done to complete any colony task. Social insect colonies thus exemplify a type of organization that occurs in many biological systems, but which does not fit easily into our usual modes of explaining organizations. Another example is the brain: each neuron is incapable of more than a simple decision whether to fire or not, and there is no hierarchical control, yet brains manage to think and remember. How do such organizations work? If social insect colonies are not hierarchically organized, how *are* they organized?

Because social insects represent an unfamiliar type of system, it is important to be careful when framing questions about colony organization. There is a danger of falling back on familiar questions about better-known systems, and these can miss the point. For example, suppose a social insect colony were hierarchically organized, like an army, orchestra, or government. Then it would make sense to ask, Who is in charge? How are orders passed along the chain of command? How are the lower ranks persuaded to keep to their places? But in fact, such questions are not appropriate, except perhaps in the purely metaphorical sense adopted when we seek to explain why some individuals reproduce and others do not. From efforts to understand the day-to-day workings of a social insect colony, a completely different set of questions arises.

In considering how colonies are organized, the first step is to determine what the organization accomplishes. A colony responds to a changing environment by displaying a vast range of behavior, including foraging, nest construction, care of the brood, territorial interactions between colonies, swarming, and migration. When a new food source becomes available, or the need for food is great, more workers forage. When the nest is damaged, it is repaired or a new one is built. When interference by neighbors is intolerable, workers will fight; when more territory becomes available, a colony will expand its range. In some species, colonies that reach a certain size split to form a second, new nest.

Here I will consider two questions about a colony's response to its environment. The first is, how does a colony monitor its environment? The colony must keep track of events in order to respond to them. I will consider the organization of patrolling: how individuals that move around and communicate with one another create a network that continually relays information about the environment to the colony. The second question is, how does a colony respond appropriately to changing conditions? I will consider the allocation of workers to different colony tasks, and the dynamics that ensure that the numbers engaged in each task are appropriate for existing conditions.

Tracking a Changing Environment

One approach to explaining how colonies monitor a changing world is to ask how each individual perceives local events. How do individuals perceive that a food source is especially rich, that the larvae are especially hungry, that the nest needs repair, that the neighbors are encroaching, or that it is time to move to a new nest? These are fascinating questions, and we have begun to arrive at a few answers. Studies of communication show that social insects respond to a variety of chemical cues. For example, ants use chemicals from six or more glands, each of which secretes a different pheromone (Hölldobler and Wilson 1990). However, most pheromones are secreted in such tiny quantities that it is impossible for us to track them as they are emitted. We are thus reduced to indirect evidence in any attempt to link pheromones and behavior. We can examine how a colony reacts to a piece of paper that has been soaked in substance x, extracted from the glands of many workers, but it is much more difficult to establish when and where an individual would emit substance x, and how other individuals would react. Nevertheless, it is clear that chemical cues are very important in social insects' perception of one another and their environments.

Whatever the means of individual perception, colonies must find ways to track their environments effectively. They must monitor the nest and some area outside of it often enough, and thoroughly enough, not to miss any relevant events. For example, ants are famous for their tendency to appear at picnics. This raises the question of how they manage to find picnics so quickly and so often. In principle, for ants to find any picnic, anywhere, immediately, there would have to be an ant everywhere all the time. Even in a world packed with ants there are not enough for this, so colonies must employ some kind of patrolling system: that is, ants must partition the world so that some ant is at *some* possible picnic site *almost* all the time. How to search the world in a continuous and thorough way is the first important problem that a patrolling system must solve. There is, however, a second problem: maintaining contact among ants. We notice ants at picnics because they recruit other ants to help retrieve the food. In addition to alerting foragers to new food sources, patrolling ants communicate other kinds of events, such as disturbances and invasions by other species, to one another and to the rest of the colony. As patrollers search, they must come into contact with other ants often enough that communication can take place before the picnic is over.

Tracking the environment requires searching by individuals, but efficient tracking also requires interaction among individuals. Interaction determines how the insects partition space, and how they maintain contact with one another. How a colony monitors its environment is a crucial feature of colony organization.

The Allocation of Workers to Colony Tasks

Once a colony discovers what is going on around it, it must respond. A general feature of colony response to a changing environment is a change of worker allocation (Gordon 1989a). In any social insect colony, a variety of tasks are accomplished simultaneously: some workers forage, others remain in the nest and care for the brood, others clean and maintain the nest, and so on. In some species, individuals specialize in certain tasks. The best-known examples of specialization, perhaps, are from colonies of polymorphic species; that is, from colonies that contain adults in a variety of sizes. In such species, workers of a certain size tend to do certain tasks. It has been postulated that ants specialize in the task for which their size renders them most efficient (Oster and Wilson 1978), though this remains to be demonstrated. In any case, few ant genera contain polymorphic species (Oster and Wilson 1978), so few can have evolved any relation of worker task and worker size.

There are two ways that the numbers engaged in different tasks can change. First, individuals can switch from one task to another. Individuals change tasks from hour to hour in many ant species (Gordon 1989b). Bees, including bumblebees and honeybees, change tasks as they grow older, and ontogenetic changes of task can be altered or reversed to meet colony requirements (Michener 1974). In general, the task of a social insect worker is not fixed. Second, workers make transitions between active and inactive states. In any colony at any time, there will be some individuals apparently doing nothing. When conditions change, however, previously inactive individuals will begin to work. Active individuals will become inactive. Both these factors, task switching and transitions from inactivity to activity, cause changes in the numbers engaged in various colony tasks.

In general, we do not know how to explain the dynamics of worker allocation in terms of individual behavior, but in some cases, we can tell a coherent story. Recruitment to a newly discovered food source is perhaps the aspect of social insect behavior that has been most intensively studied. In some ant species, it happens more or less like this: A new food source appears. A patrolling ant worker finds it, and lays a chemical trail from the food source back to the nest. At the nest she encounters some inactive ants, who follow the trail back to the food source. Eventually a strong trail of foragers builds up, traveling to the food source and carrying food back to the nest. In honeybees, the returning forager performs a dance that informs other bees of the location of food, and active foragers then travel to the site to retrieve more.

Even in the case of recruitment, which is relatively well understood, many questions remain. Some workers must be waiting at the nest to be recruited. Will there be a sufficient number to retrieve all the food, and if so, how is the relationship between available foragers and food distribution maintained? How do the numbers waiting to be recruited depend on the colony's need for food?

We could hope for simple answers to questions about worker allocation if each individual did only one task and responded to simple cues about how intensively to pursue that task. However, we know that things are not so simple. Numbers engaged in one task depend on numbers engaged in another (Gordon 1987). In harvester ants, for example, the numbers foraging will vary, depending on numbers patrolling around the nest mound and also on numbers engaged in nest maintenance work. Even though individuals switch tasks, experiments with marked individuals show that task switching does not account for all the observed interactions among worker groups (Gordon 1989b). Instead, when some ants are recruited from inside the nest to do extra nest maintenance work, other ants, that day's foragers, remain inside the nest. The number of active foragers fluctuates, depending on the activities of distinct groups of workers.

Harvester ants provide evidence of further complexity of worker allocation in social insects. I performed a series of perturbation experiments, in which I changed the numbers of ants engaged in one task by creating a disturbance or condition to which just one worker group responded directly; this caused numbers of workers engaged in other tasks to change as well. For example, I increased numbers of active nest maintenance workers by putting out small piles of toothpicks near the nest entrance, which nest maintenance workers carried to the edge of the nest mound and then abandoned. I decreased numbers of foragers by placing small plastic barriers across the foraging trails.

In some experiments, I did more than one perturbation simultaneously. Combined perturbations had nonadditive effects (Gordon 1987). The response of large, old colonies was a homeostatic one: the more such a colony was disturbed, the more likely it was to behave like a normal, undisturbed one. When the behavior of undisturbed colonies was used as a baseline or equilibrium state, combined perturbations showed that the larger the disturbance, the more likely colonies were to return to this state. Faced with combined perturbations, colonies altered worker allocation less than would be expected if one were to sum the effects of each perturbation separately.

An example of the nonadditive, homeostatic response to combined perturbations is the following: I increased numbers engaged in nest maintenance, with the indirect effect that numbers foraging decreased. In a separate experiment, I decreased numbers foraging, with the indirect effect that numbers engaged in nest maintenance increased. Thus the expected effect of both perturbations, when combined, would be a larger decrease in foraging and increase in nest maintenance. However, expectations of additivity were not realized. Instead, numbers foraging decreased less than in either single perturbation experiment. This means that interactions between worker groups are modulated. The numbers that forage depend on the numbers engaged in nest maintenance work, and vice versa. But the relation is not linear. At some threshold, the "more nest maintenance, less foraging" rule breaks down. Beyond this threshold, the numbers foraging conform to another rule.

Models of worker allocation are needed that explain several empirical results. First, numbers engaged in different tasks change from one hour to the next. Second, distinct worker groups interact with one another: at any time, the numbers engaged in one task depend on numbers engaged in other tasks. Third, the effects of multiple interactions are homeostatic and nonlinear.

The organization of a colony determines how it monitors the environment and how it responds when conditions change. The rest of this chapter describes how one kind of cue—namely, rates of encounter among workers—may function in the organization of social insect colonies.

ARE RATES OF ENCOUNTER IMPORTANT IN COLONY ORGANIZATION?

Colony organization must be based on simple rules of individual behavior. For task allocation, such rules might be of two kinds: either individuals assess the amount of work to be done to complete a certain task, or they assess the numbers of workers currently engaged in the task. An individual decides whether to be active or inactive, and has a choice of tasks it might perform if it were active. Consider an inactive individual that begins to forage. Is the cue that causes this individual to forage actively based on the amount of food stored in the colony, or is it based on numbers of workers actively foraging? The results described above show that in a harvester ant colony, when numbers engaged in one task are changed, changes occur in numbers engaged in other tasks. For example, when numbers of active nest maintenance workers increase, numbers of active foragers decrease. Are the foragers responding to changes in numbers of active nest maintenance workers, or directly to the amount of nest maintenance work to be done?

Worker Number and Task Allocation in Harvester Ants

There are some indications that in harvester ants, numbers of workers, not the extent to which a task is accomplished, provide the cues that generate changes of worker allocation. First, it is easier to see how a worker could track numbers of workers than how it could appraise the status of various tasks. Workers engaged in different tasks tend to be spatially segregated (Gordon 1984). For example, harvester ant foragers leave the nest entrance and go directly to a foraging trail that can extend 20–30 meters. Nest maintenance workers often travel only a few centimeters from the nest entrance, where they put down debris collected inside the nest, and they rarely leave the nest mound. Debris near the nest entrance will be discovered by nest maintenance workers, but foragers will not encounter the debris unless it is located in the path from the entrance to the foraging trail. It seems unlikely that when more nest maintenance work is needed, foragers will apprehend the necessity directly.

While workers engaged in one task are unlikely to encounter cues that indicate the status of other tasks, they are very likely to encounter workers from other task groups. All exterior workers go in and out of the nest frequently. Colonies are active outside the nest for 6–7 hours each day (Gordon 1984). One trip by a forager may take half an hour (D. M. Gordon, unpubl.); nest maintenance workers, patrollers, and midden workers all go in and out of the nest even more often. Experiments with marked individuals showed that exterior workers rarely travel deep inside the nest (MacKay 1981). This means that workers of all task groups, as they move in and out, will mix in the upper chambers directly inside the nest entrance. Different task groups may be spatially segregated in work outside the nest, but encounters among them are frequent as workers go in and out of the nest.

A second line of evidence that numbers of workers are important in the dynamics of worker allocation in harvester ants comes from studies of colonies of different ages (Gordon 1987, 1991, 1992). Colony age is related to colony size: older colonies are larger (contain more workers) than younger ones. Colonies are founded by a single queen, and grow to a size of about 10,000 workers in 5 years, remaining at about this size until the queen dies at the age of 15 or 20. Workers, who live for up to a year, are all daughters of the single, founding queen.

Worker allocation operates differently in colonies of different sizes (Gordon

1987). Perturbation experiments were performed in young, 2-year-old colonies as well as in older ones (at least 5 years old). In older, larger colonies, the response to combined perturbations was more homeostatic than in younger, smaller ones. Older colonies compensate as the magnitude of disturbance increases, so that older colonies are less sensitive than younger ones to perturbations of foraging. In addition, the response of older colonies is more consistent, over repeated experiments, than that of younger ones.

We do not know how to account for increased stability of worker allocation in older, larger ant colonies. It cannot be due to the experience of older workers, since workers live at most a year and the colonies observed differed in age by at least 3 years. The most likely explanation is that individuals in old and young colonies operate according to the same dynamic rules, but that such rules specify qualitatively different outcomes when large and small numbers of workers are involved.

Colony age and size also influence interactions between neighboring colonies. Colonies compete for foraging area; when one colony is prevented from foraging, its conspecific neighbors will begin to use its foraging area within 10 days (Gordon 1992). In conflicts over foraging area, smaller 3- and 4-year-old colonies are more persistent and aggressive than older, larger ones at least 5 years old. This is apparent in two ways. First, younger colonies will enter into conflict with a neighbor over a small food source, like seed bait. It takes a more abundant food source to induce older, larger colonies to fight with their neighbors (Gordon 1991). Second, a temporary absence of one colony from a region of its foraging area permits its neighbors to encroach upon that region. When the colony resumes foraging there, however, the intruder will retreat. But older, larger colonies are more likely to retreat than younger, smaller ones (Gordon 1992). A smaller colony tends to persist in using a newly acquired foraging area even when the previous user attempts to regain it.

Younger colonies may be more persistent in conflicts with neighbors because they are growing quickly. A 3-year-old colony may contain 4,000 workers (Gordon 1992); since workers live only a year, those 4,000 workers will have to do the work necessary to produce the 6,000 workers it will contain the next year when the colony is 4 years old. A 5-year-old colony, containing 10,000 workers, will have to produce 10,000 workers over a year's time to maintain a stable size, but it has 10,000 workers to do the work. Most of a colony's food is devoted to feeding larvae. In a quickly growing colony, the ratio of larvae to foragers is high. Larval hunger may somehow stimulate foragers in small, quickly growing colonies to take greater risks and be more persistent in the use of contested foraging area. At the level of individual foragers, worker numbers—or more precisely, the ratio of larvae to workers—may influence decisions about whether to continue foraging in a particular direction, even if this means fighting with workers of a neighboring colony.

Colony size influences task allocation and foraging behavior in harvester ants. This suggests that in social insects generally, worker numbers may be an important component of the rules that determine individual decisions about which task to perform, and when.

Worker Number and Task Allocation in Other Social Insects

The organization of worker allocation is not yet understood in any social insect species. In a series of studies of the tropical social wasp *Polybia occidentalis*, Jeanne and colleagues have studied the allocation of workers to various tasks associated with nest building, and the interaction of nest repair and foraging (Jeanne 1987;

O'Donnell and Jeanne 1990). These studies suggest that individual wasps assess task status, rather than worker number (Jeanne 1986).

Recent models of worker allocation in honeybees and ants use cues based on worker number, and it is possible that large colonies may require different kinds of organization from those employed by smaller wasp colonies. In the study of task allocation, most attention has focused on the organization of work within one task group. By far the most frequently studied task is foraging. Deneubourg, Goss, and their colleagues have been concerned mostly with interactions among foraging ants (e.g., Deneubourg et al. 1986, 1987). The models assume that foragers imitate one another, for example, by following chemical cues. Once the process gets started, imitation can lead to complex, branching foraging trails without any individual having to assess the magnitude or distribution of food resources.

Seeley and colleagues have studied how honeybee foragers make transitions between the active and inactive states (Seeley 1989). When bees return to the hive with nectar, they pass it on to other bees that store it. The process of storing nectar takes time, so when more nectar is coming into the hive, returning foragers have to wait longer until they can unload it onto a nectar storer. Numbers of incoming foragers thus determine the length of delay at the hive, which in turn regulates the numbers of outgoing foragers. Seeley et al. (1991) model the allocation of foraging bees to two different nectar sources. Assuming that one forager can determine whether another is successful, and that unsuccessful foragers will follow successful ones, the rate of encounters between foragers from the two sources will determine the distribution.

The work described in the next section is concerned with the two different problems discussed so far, monitoring the environment and changes of worker allocation. The work centers on one fundamental question: How could relatively simple individual behavior generate the complex behavior we see in colonies? I will describe studies of the ways in which an ant's perception of the rate at which it encounters other ants could contribute to patrolling systems and to the dynamics of worker allocation.

THE FUNCTION OF ENCOUNTER RATE IN ANT COLONIES

There is much evidence to suggest that social insects respond to changes in the numbers of other workers present, both nestmates and non-nestmates (e.g., Nonacs 1990; Adams 1990). Worker allocation in harvester ants may be cued by changes of worker density. For example, if a forager can track changes in numbers of nest maintenance workers, it may act according to a rule such as, When density of nest maintenance workers is high, don't forage. How workers might assess density, that is, numbers of ants per unit area, is not known. No individual ant is capable of counting total numbers of workers. That is, an ant cannot decide, When there are more than 42 nest maintenance workers, don't forage.

Encounter rates could provide a simple cue to worker density. For a given path shape, the rate at which encounters occur among moving ants depends on worker number; this will be discussed in more detail below. As a simple generalization, the more ants in a given area, the more collisions will occur. A forager may be able to distinguish nest maintenance workers from other ants. Harvester ants engaged in different tasks respond differently to the same pheromone (Gordon 1983). This suggests that workers undergo physiological changes as they move from one task to another. It may be that such changes are perceived by other workers, and that somehow a nest maintenance worker has an odor different from that of a forager. The task itself may alter the odor of a worker. Suppose that a worker can perceive the interval between contacts with nest maintenance workers. When numbers of nest maintenance workers coming in and out of the nest increase, so will numbers of contacts with nest maintenance workers. A forager may have a threshold contact rate with nest maintenance workers, past which it becomes less likely to forage.

Ants Regulate Encounter Rate

One possible problem with the hypothesis that individual decisions about task allocation depend on contact rate is that contact rate may be too sensitive to changes of density. Suppose that contact rate were completely random, and that the dynamics of collisions between ants conformed to the principles of Brownian motion. Then the number of pairwise collisions that each of n ants experiences would increase as n^2 (Feynman et al. 1963). Contact rate would be extremely sensitive to worker density, increasing quadratically with an increase in numbers of ants per unit area. Small changes of density must be common as ants move around the nest and on and off foraging trails. If contact rate were to fluctuate wildly with small changes of density, this could diminish the usefulness of contact rate as a signal.

A recent theoretical study of the effects of group size on information transfer (S. W. Pacala, D. M. Gordon, and H. C. J. Godfray, unpubl.) suggests another reason why ants may regulate contact rate at high density. Suppose an ant's decision whether to perform a task depends in part on the rate at which it encounters other ants successfully performing that task. For example, an ant's decision whether to forage may depend in part on the rate at which it encounters foragers bring food back to the nest; empirical studies suggest this is true (Gordon 1991). As group size increases, so does encounter rate. The probability of encountering a successful ant thus increases. But success rate is limited by the amount of food available. In very large groups, or at very high densities, foraging could be elicited by a high encounter rate between inactive or unsuccessful foragers, and successful foragers. This would encourage foraging even when the amount of food in the environment does not warrant it. At high densities, suppression of encounters allows the group to track the environment more successfully.

In laboratory experiments with the ant *Lasius fuliginosus*, we tested whether contact rate is random (Gordon et al. 1993). We measured contact rate at a range of densities by varying both arena size and numbers of ants. The results show that in undisturbed conditions, contact rate is regulated, not random. Contact rate does not increase quadratically with density. Instead, it levels off as density increases. At very low densities, ants aggregate and maintain a relatively high contact rate. At very high densities, ants appear to avoid the other ants nearby, so that contact is suppressed. One ant can perceive another at a distance of about 1.2 centimeters, and thus decide whether to engage in contact. Encounter rates may be nonrandom in many social insect species. Why colonies adjust contact rates needs further empirical study.

Contact Rate as a Cue to Changes of Nestmate Density

While *L. fuliginosus* regulates contact rate in undisturbed conditions, the ants use contact rate as a cue to changes of nestmate density in disturbed conditions. When we added ants from another colony, contact rates increased significantly. The magnitude of the increase in contact rate depended on the proportion of non-nestmates present. By altering the numbers of host and added ants, we showed that proportions, not numbers, of alien ants determined the magnitude of the response. This rules out the possibility that the ants merely responded to the total quantity of foreign colony odor in the arena.

These results show that an ant tracks the interval between contacts with nestmates. When it meets non-nestmates too often, it increases the rate of contact with all other ants. By doing this, it can increase the rate at which it discovers the proportion of non-nestmates present. Such behavior would be useful for a foraging ant that begins to encounter non-nestmates, because a large proportion of non-nestmates would indicate that the forager had strayed into the territory of another colony. It would also be useful when one colony is invaded by another, as the proportion of non-nestmates would indicate the magnitude of the invasion. How contact rates are increased remains to be discovered. We tested whether increased speed might account for increased contact rates, but found that it did not. Recent theoretical work (Adler and Gordon 1992) shows how, if each ant adjusted the curviness, or turning angle, of its path, overall contact rate could increase.

How Do Encounter Rates Affect Task Allocation? A Neural Network Model

How might contact rate function in the allocation of workers to various colony tasks? One approach draws on the analogy of social insect colonies with brains, and uses neural network models to describe worker allocation (Gordon et al. 1992). We simulated a neural net (Hopfield 1982) in which each unit, or ant, could be in one of eight possible categories. Each ant could be either active or inactive, and could belong to one of four possible task groups: midden workers, patrollers, nest maintenance workers, or foragers. The state of the kth ant is defined by $(\mathbf{a}_k, \mathbf{b}_k, \mathbf{c}_k)$, where **a, b,** and **c** can each take on the values of 1 or -1. Each of the three vectors (**a, b,** or **c**) corresponds to one decision: whether to be active or not (**a**), whether to belong to a patroller/forager subclass (**b**), or whether to belong to a nest maintenance/midden worker subclass (**c**). An ant changes task, and changes from the active or inactive category, by changing the relevant vector from 1 to -1, or vice versa.

Consider the transition from the active to the inactive state. How the jth ant will behave is described by a threshold function:

$$h_j = \Delta \alpha_{jk} \mathbf{a}_k - \theta_j > 0$$

Whether the jth ant is active or not depends on the sum, over all ants j not equal to k, of its interactions with all other ants. What is summed is the product of two elements: (1) a constant, taken from the matrix α, which specifies the interaction weights between each pair of categories (e.g., active-inactive), and (2) the contribution, 1 or -1, specified by the vector $(\mathbf{a}_k, \mathbf{b}_k, \mathbf{c}_k)$ that defines the state (task group, active/inactive) of the kth ant. For example, when $\mathbf{a}_k = 1$, the kth ant is active. Suppose the jth ant is also active. The matrix α_{jk} sets a value or weight on the interaction between two ants both in the active category. The interaction between the jth and kth ant will be the product of this interaction weight, and 1, the value of $\mathbf{a}_k$. All such interactions are summed. If the sum of the interactions is greater than a threshold, θ, the ant becomes or remains active. If not, the ant becomes or remains inactive.

Each decision about activity and about task thus depends on the sum of pairwise encounters with other ants. To model the perturbation experiments described above, we made two assumptions. First, ants interact with active and inactive ants in their own task category. Second, active ants also interact with active ants in all other task categories. It should be noted that this model differs from conventional Hopfield-like nets in that when a unit or ant changes category, by changing task or changing between activity and inactivity, the ant alters its interaction matrix.

We used the network to simulate interactions among 200 ants, 25 in each category. The interaction matrices were defined so that the system had a global attractor, tending to return to the initial state of equal numbers in each of the eight categories (active/inactive, four tasks). To represent perturbations, simulations began with a change in numbers active in one task. The results reflected empirical observations. An initial increase or decrease in one category propagated to other tasks, until eventually the system returned to the stable state.

The neural network model shows that the observed interactions among task groups could result from accumulated pairwise encounters between individuals. Each ant must be capable of perceiving the task group and activity status of the other ants it encounters. Combining this information according to some simple rule, such as the sum of interactions, the model colony allocates workers in a way that reflects some of the behavior of real colonies: events changing numbers engaged in one task will change the numbers engaged in other tasks.

Encounter Rate and Spatial Patterns of Movement

Could real ants behave like a neural network? In the model, worker allocation depends on numbers of encounters between ants in different task groups. When used to describe brains, neural networks usually assume fixed spatial relationships among the units, or neurons (Rumelhart and McClelland 1986). Unlike neurons, however, social insects move around. The pattern of encounters will depend on the spatial patterns of movement.

Ant species differ in characteristic movement patterns (Harkness and Maroudas 1985; Leonard and Herbers 1986). Some ants, such as the formicines in the *Formica rufa* group, tend to walk in relatively straight lines. Other ants follow more convoluted paths; for example, the ant *Conomyrma insana* probably gets its name from the fact that its typical path resembles that of a coiled spring. The shapes of the paths of moving individuals determine not only how often they meet, but how exhaustively, and how often, they cover the ground. Encounter rates may be important in the ways a colony monitors its environment, as well as in the organization of worker behavior. I have begun to investigate how path shape and encounter rates interact in the behavior of patrollers.

I studied path shapes in patrollers of the fire ant, *Solenopsis invicta* (Gordon 1988). This ant is a very effective invader and competitor on the ecological scale, which suggests that patrollers may explore very effectively on a smaller scale, such as that of a colony's territory. I traced the paths of workers filmed as they entered a region newly available to the colony. The path shapes of these patrolling workers could be classified into four categories, each of which was performed by workers of a particular size. The four paths tended to be used in the same temporal sequence in many replicated trials. The first type of worker to enter the new region moved rapidly and engaged in frequent antennal contacts with other workers. Ants of the second type came into the new region and then remained motionless for long intervals, acting perhaps as sentries. Ants of the third type traversed slow, convoluted paths, and tended to contact stationary ants of the second type. Hours later, when the new region had perhaps become incorporated into the ordinary foraging range of the colony, ants of a fourth type walked quickly and directly through the region without engaging in contact. Further work is needed to determine the role of path shape and antennal contact in the exploration tactics of fire ants. The existence of specialized individuals that explore a new region in characteristic spatial and temporal patterns is strongly suggestive. At the level of the group—that is, the group of patrollers in a colony—fire ants seem to employ a network of path shapes that may be especially effective in determining how the colony will exploit a newly available region.

In a recent theoretical study, we considered the relation between spatial patterns of movement and the efficiency of a network of patrolling ants (Adler and Gordon 1992). Patrolling ants face two problems: they need to search the world effectively, and they need to maintain contact with one another so that they can communicate if they find something. The contact may be direct, or there may be a lag. For example, if two ants are to communicate through antennal contact, they must be close enough together at the same time for their antennae to meet. If they communicate with a volatile pheromone, they must be within range of each other within the time it takes for the pheromone to disperse.

These two problems, searching and maintaining contact, impose conflicting constraints on the spatial behavior of patrollers. How individual foragers should move in order to search effectively is an important and growing field of research (Pyke 1978; Zimmerman 1979; Dusenberry 1989). Such work shows that an effective search method may be a convoluted path, close to a random walk, in which an individual may circle over the same ground for a long time. But a group of individuals on random walks will not encounter each other often, because each will tend to get stuck in one place. To maintain contact with one another, individuals should walk in paths approaching straight lines. The problem we set for networks of patrolling ants is thus different from that of individual foragers. An individual forager's task involves discrete episodes of finding food and eating it, or bringing it back to a central nest. But patrollers need to monitor a region continually while staying in contact with one another.

Our model described the behavior of a network of n patrolling ants, using both simulation and analytical approximation. In a given network, all ants used the same shape of path, analogous to the species-specific movement patterns of ants mentioned above. The straightness, or lack of straightness, of the ants' paths was described in terms of the variation in direction between successive path steps. Higher variation in step direction produced a more convoluted path. We tested how path shape and number of ants affect the efficiency of the network. Events were scattered randomly throughout the region occupied by the ants, and the duration of the events varied. Ants were considered to disseminate information when they came into contact; that is, within a specified small distance of each other. Efficiency was measured in two ways: what proportion of events were discovered, and the rate at which ants became informed about an event discovered by one of them. Efficiency at covering space depended on the shape of the ants' paths. Efficiency at disseminating information depended on the rate and distribution of encounters.

The results show that the effects of number of ants and of path shape interact with each other. As numbers of ants increase, ants cover more ground and contact one another more often. Larger networks are more efficient, both at event discovery and at disseminating information, than smaller networks. Straighter paths also enhance the rate at which randomly occurring, scattered events are discovered. However, network size has a stronger effect than path shape, leading to an interaction of the two variables. When networks are large enough, path shape becomes less important. Large colonies can thus patrol efficiently while patrollers transfer information to one another, even when individual patrollers follow apparently inefficient paths.

Ecological differences among species may be reflected in differences of path shape and encounter rate. The model predicts that patrollers will adopt straighter paths when patroller numbers are limited, because path shape contributes more to network efficiency when numbers of ants in the network are small. Colony size is a species-specific characteristic, and it is interesting that some of the most ecologically active species have large, easily mobilized colonies.

Species-Specific Encounter Rates

To investigate species differences in contact rates, we examined three, ecologically very different, species in laboratory arenas (Gordon et al. 1993). The first, *Myrmica rubra*, has small colonies, and its distribution tends to be limited by the presence of other species (Elmes and Wardlaw 1982). The second, the fire ant *Solenopsis invicta* Buren, is an extremely effective competitor that invades disturbed habitats. The third, *Lasius fuliginosus*, has large, long-lived colonies that collect honeydew from aphids, and in many ways the species appears to play the ecological role of the red wood ant group that is abundant in the conifer forests of northern Europe (Dobrzanska 1966). To examine species differences in the use of contact rate as a signal, we compared the extent to which each species changes contact rate in response to food. We further considered whether contact is spatially or temporally patterned in the three species.

The three species differed significantly in the extent to which contact rate responds to changes in the environment. For example, in the fire ant *S. invicta*, contact rate is more differentiated than in *M. rubra*. Contact rate in *S. invicta* varied significantly with time of day, and depended strongly on location in the arena and on the presence of food (as in Gordon 1988). In *M. rubra*, rates of contact were uniformly low. Contact rates were much higher in *L. fuliginosus* than in the other two species, and did not seem to depend on location or the presence of food. In *L. fuliginosus*, as explained above, contact rate appears to be regulated.

These results suggest that ecological differences among species reflect characteristic features of colony organization: how fully the colony monitors its environment, and the magnitude and intensity of its response to changed conditions. Many empirical questions remain. Do patrolling colonies use encounters to disseminate information? Does the efficiency of a patrolling network reflect the colony's ability to respond to changes in its environment? Under what conditions will a colony regulate contact rates, and when will such rates be allowed to vary?

CONCLUSIONS

In many ways, encounter rates may provide the cue that links changes of colony behavior and changes of colony environment. Rates of encounter among workers of different task groups may underlie the organization of worker allocation. Movement patterns of patrollers and the rate at which they transfer news of a discovery may determine how quickly and thoroughly a colony can find out about and respond to its environment.

Further work is needed to investigate the evolution of the behavior considered here. Natural selection requires variation, and variation among social insect colonies is not well understood. Further studies are needed on the population biology of social insects. The reproductive success of a colony will depend on what it accomplishes. For example, if fitness depends on the amount of food a colony can obtain, what will matter is not merely how many foragers the colony has, but how many forage in particular conditions, and how effectively they retrieve food.

The evolution of colony organization will depend on variation among colonies in their ability to utilize resources and maintain a stable internal environment. To study how natural selection is acting on colony organization, we will need to examine variation among colonies in the dynamics of worker allocation; that is, variation among colonies in the rules that determine how many workers do each task, and when. Colonies may also vary in how well they find resources and how appropriately they react when the environment is disturbed. How quickly a colony must find events and respond to them depends on the temporal grain of its environment and on the rapidity of adjustment its ecology demands. The evolution of colony organization entails the tuning of this link. To investigate how colony behavior evolves, we will need to relate variation in colony behavior to variation in reproductive success.

The evolution of colony organization in social insects suggests analogies with that of brains or multicellular organisms. Rather than small groups of individual units, each of which is elaborate enough to make complex decisions about its own welfare, social insect colonies contain large numbers of relatively dispensable individuals whose survival and reproduction depends on a complex organization. In this sense, social insects are like cells, or neurons, and this is the aspect of social insect biology that led Wheeler (1911) to refer to colonies as "superorganisms." Understanding such organizations requires us to ask clear questions about how they operate at the colony level.

ACKNOWLEDGMENTS

I thank J. R. Gregg, C. Boggs, and K. Human for comments on the manuscript.

REFERENCES

Adams, E. S. 1990. Boundary disputes in the territorial ant *Azteca trigona*: Effects of asymmetries in colony size. *Animal Behaviour* 39:321–28.

Adler, F. R., and D. M. Gordon. 1992. Information collection and spread by networks of patrolling ants. *American Naturalist* 140:373–400.

Deneubourg, J. L., S. Aron, S. Goss, J. M. Pasteels, and G. Duerinck. 1986. Random behavior, amplification processes and number of participants: How they contribute to the foraging properties of ants. *Physica* 22D:176–86.

Deneubourg, J. L., S. Goss, J. M. Pasteels, D. Fresneau, and J.-P. Lachaud. 1987. Self-organization mechanisms in ant societies (II): Learning in foraging and division of labor. In J. M. Pasteels and J. L. Deneubourg, eds., *From individual to collective behaviour in social insects*, 177–96. Basel: Birkhauser.

Dobrzanska, J. 1966. The control of the territory by *Lasius fuliginosus* Latr. *Acta Biologica Experientia* (Warsaw) 26:193–213.

Dusenberry, D. B. 1989. Ranging strategies. *Journal of Theoretical Biology* 136:309–16.

Elmes, G. W., and J. C. Wardlaw. 1982. A population study of the ants *Myrmica sabuleti* and *Myrmica scabrinoidis*, living at two sites in the south of England. (I.) A comparison of colony populations. *Journal of Animal Ecology* 51:651–64.

Feynman, R. P., R. B. Leighton, and M. Sands. 1963. *The Feynman lectures on physics.* Reading, Mass.: Addison-Wesley.

Gordon, D. M. 1983. The dependence on social context of necrophoric response to oleic acid in the harvester ant, *Pogonomyrmex badius. Journal of Chemical Ecology* 9:105–11.

Gordon, D. M. 1984. Species-specific activity patterns in harvester ants. *Insectes Sociaux* 31: 74–86.

Gordon, D. M. 1987. Group-level dynamics in harvester ants: Young colonies and the role of patrolling. *Animal Behaviour* 35:833–43.

Gordon, D. M. 1988. Group-level exploration tactics in fire ants. *Behaviour* 104:162–75.

Gordon, D. M. 1989a. Caste and change in social insects. In P. Harvey and L. Partridge, eds., *Oxford surveys in evolutionary biology,* vol. 6, 55–72. Oxford: Oxford University Press.

Gordon, D. M. 1989b. Dynamics of task switching in harvester ants. *Animal Behaviour* 38: 184–204.

Gordon, D. M. 1991. Behavioral flexibility and the foraging ecology of seed-eating ants. *American Natualist* 138:379–411.

Gordon, D. M. 1992. How colony growth affects forager intrusion in neighboring harvester ant colonies. *Behavioral Ecology and Sociobiology* 31:417–27.

Gordon, D. M., B. Goodwin, and L. E. H. Trainor. 1992. A parallel distributed model of the dynamics of ant colony behaviour. *Journal of Theoretical Biology* 156:293–307.

Gordon, D. M., R. Paul, and K. Thorpe. 1993. What is the function of encounter patterns in ant colonies? *Animal Behaviour* 45:1083–1100.

Harkness, R. D., and N. G. Maroudas. 1985. Central place foraging by an ant (*Cataglyphis bicolor* Fab.): A model of searching. *Animal Behaviour* 33:916–28.

Hölldobler, B., and E. O. Wilson. 1990. *The ants.* Cambridge, Mass.: Harvard University Press.

Hopfield, J. K. 1982. Neural networks and physical systems with emergent collective computational abilities. *Proceedings of the National Academy of Sciences U.S.A.* 79:2554–58.

Jeanne, R. L. 1986. The evolution of the organization of work in social insects. *Monitore Zoologico Italiano* 20:119–33.

Jeanne, R. L. 1987. Do water foragers pace nest construction activity in *Polybia occidentalis?* In J. M. Pasteels and J. L. Deneubourg, eds., *Behavior in social insects: Experientia Supplementum,* vol. 54. Basel: Birkhauser Verlag.

Leonard, J. G., and J. M. Herbers. 1986. Foraging tempo in two woodland ant species. *Animal Behaviour* 34:1172–81.

MacKay, W. P. 1981. A comparison of the nest phenologies of three species of *Pogonomyrmex* harvester ants. *Psyche* 88:25–75.

Michener, C. D. 1974. *The social behavior of the bees.* Cambridge, Mass.: Belknap Press of Harvard University Press.

Nonacs, P. 1990. Death in the distance: Mortality risks as information for foraging ants. *Behaviour* 112:23–34.

O'Donnell, S., and R. L. Jeanne. 1990. Forager specialization and the control of nest repair in *Polybia occidentalis* Olivier (Hymenoptera: Vespidae). *Behavioral Ecology and Sociobiology* 27:359–64.

Oster, G., and E. O. Wilson. 1978. *Caste and ecology in the social insects.* Princeton: Princeton University Press.

Pyke, G. H. 1978. Optimal foraging: Movement patterns of bumblebees between inflorescences. *Theoretical Population Biology* 13:72–98.

Rumelhart, D. E., and J. L. McClelland. 1986. *Parallel distributed processing.* Cambridge, Mass.: MIT Press.

Seeley, T. D. 1989. Social foraging in honey bees: How nectar foragers assess their colony's nutritional status. *Behavioral Ecology and Sociobiology* 24:181–99.

Seeley, T. D., S. Camazine, and J. Sneyd. 1991. Collective decision-making in honey bees: How colonies choose among nectar sources. *Behavioral Ecology and Sociobiology* 28:277–90.

Wheeler, W. M. 1911. The ant-colony as an organism. *Journal of Morphology* 22:307–25.

Zimmerman, M. 1979. Optimal foraging: A case for random movement. *Oecologia* 43:261–67.

NINETEEN

Chaos and Behavior: The Perspective of Nonlinear Dynamics

BLAINE J. COLE

IN THIS CHAPTER I will explore the relation between nonlinear dynamics and animal behavior. This chapter is not just about chaos but rather the full spectrum of nonlinear dynamics that may occur in animal behavior. I will touch on phase-locking, bifurcations, and other subjects that are closely related to chaotic dynamics. While chaos is an appealing topic, it is only one end of a spectrum of possible dynamical behavior. The importance of chaos itself to animal behavior is, at this time, problematic; if chaos occurs in behavior, it may or may not be enormously important. The importance of nonlinear dynamics, more broadly cast, is not really in doubt. My goal here is to explore the dynamics of behavior, including the production of chaos. I will use the pattern of activity in ant colonies as a case study, and I will suggest some future directions for research.

Chaotic dynamics appear in a number of theoretical contexts relevant to biology, as well as in several empirical situations. The implication of chaotic dynamics in the behavior of animals may have profound consequences for our ability to model the mechanisms of behavior, for our ability to predict the occurrence of behavior, and for our ability to understand the evolution and ecological consequences of behavior. I do not pretend to give a thorough treatment of nonlinear dynamics here. Excellent treatments of the subject, with special references to biological problems, are found in Schaffer and Kot (1985a), Glass and Mackey (1988), Godfray and Blythe (1990), and Denton et al. (1990). An excellent, more technical, yet readable, reference is Grassberger et al. (1991).

One of the most recognizable characteristics of behavior is its variability. This variability both generates diversity in behavioral phenomena and limits our ability to make generalizations. To describe the behavioral phenotype of an organism, we often use methods that increase the repeatability of a measurement, such as using long-term averages, binning the data, or pooling data from a number of individuals. But behavior also explicitly exists in time, and by averaging or pooling data and neglecting the temporal component of behavior we may lose an important dimension of the phenomena. We can use some of the tools of nonlinear dynamics to examine behavior from a new perspective and gain insights into its operation and evolution.

Chaotic dynamics are produced by deterministic, nonlinear systems that are characterized by "sensitive dependence on initial conditions." If two replicates of an experiment (either a biological experiment or a numerical experiment on the computer) are performed, there will be differences in the initial conditions that are less than the measurement error of the investigator. In a chaotic process the outcomes of the two replicates will diverge from one another exponentially with time. The lack of repeatability in the long-term properties of processes that are initially indistinguishable means that this deterministic process appears indeterminate. Technically this condition requires that there be at least one positive Lyapunov exponent in the system of equations that underlie the process (about which I will have some more to say below). Although the trajectories or replicates initially diverge from one another, they are also bounded and thus do not diverge forever. The deterministic process is confined to an attractor. If a process has an equilibrium point (a point attractor), then it will return to the equilibrium if it is perturbed. When a process is chaotic, the attractor behaves in the same manner; if the trajectory is perturbed from the attractor, it will return to it. In many cases, chaotic processes have an attractor that has a noninteger or fractal dimension and is often called a strange attractor.

Animal behavior has characteristics that invite us to imagine that chaos is one plausible outcome. Behavior consists of an extremely complex collection of nonlinear processes. Part of the nonlinearity of behavior results from the essential nonlinearity of payoffs or utility for an organism making economic decisions. The payoff to a forager searching for food for one time unit is unlikely to be half of the payoff of foraging for two time units. Even if the payoffs are linear, the utility of twice as much food is unlikely to be twice as great. An additional form of nonlinearity is that caused by the nonlinearity of responses to stimuli or to interactions between individuals. For example, twice as much aggression between two individuals is unlikely to have exactly twice the effect of any single amount of aggression. The effect on the endocrine system of a subordinate in a dominance interaction may be much greater than twofold if the dominance encounters are twice as frequent. The second feature of behavior that leads us to expect chaos is its complexity. The internal state of an organism has a large number of interacting components, including neural, nutritional, hormonal, and sensory components. Behavior results from the interactions of these internal components as well as interactions with other organisms. These interactions among the component subsystems of behavior are themselves often nonlinear.

A complex system with a large number of interacting, nonlinear components is prone to complex nonlinear dynamics, including chaos. This line of reasoning has only the status of a plausibility argument for behavioral chaos, but I find it to be a compelling reason to look further. Certainly there is no theoretical reason to suppose that chaos could not occur in behavior. One could easily construct theoretical models of behavior in which the outcome of an animal's actions would be chaotic. For example, an organism forages when it reaches a certain threshold level of hunger and becomes sated with some time delay after it has eaten. By adjusting the rate of success in foraging, the rate at which the gut empties, and the time delay between foraging success and satiation, one could generate periodic or chaotic foraging bouts. The critical point is that there is nothing about behavior, as a complex nonlinear system, that makes it inherently more or less likely than any other system to exhibit chaos. I will return later to the sorts of behavior in which chaos might plausibly occur.

The existence of behavioral chaos is therefore largely an empirical issue. In order to discover chaos in behavior, we must examine behavioral data for evidence of chaos. Analyzing behavior for evidence of chaos requires numbers of data. With the rapid development of image analysis systems, the complexity of behavior that can be automatically monitored will increase dramatically. The techniques for measuring the characteristics of chaos (or as Schaffer calls them, the "fingerprints of chaos") have been developed largely in the last few years. The development of new techniques and the refinement of older techniques is a subject of intense effort and will doubtless lead rapidly to more powerful techniques. The detection of evidence for chaos has largely settled on techniques for determination of Lyapunov exponents, the measurement of the dimension of the attractor, and forecasting methods, including the construction and interpretation of return maps and nonlinear forecasting. The best evidence for the existence of behavioral chaos would be an experimental manipulation of characteristics of the behavioral system to manipulate the onset of chaos, but this awaits the development of suitable experimental systems.

CHAOS IN BEHAVIORAL DATA

In this section I will describe the evidence for chaos in the activity of ants as an example of the application of nonlinear dynamics to the study of behavior. I have studied the activity of *Leptothorax allardycei*, a small ant that inhabits hollow grass stems in the Florida Keys (Cole 1991a, 1991b, 1991c). The activity patterns of colonies and of single worker ants have been studied in laboratory observation nests. Activity has been quantified using an automatic digitizing camera that produces a simplified picture of a colony, and software that compares successive images and counts the number of pixels that have changed between images.

The activity patterns of colonies of *L. allardycei* are apparently periodic (Cole 1991c). In figure 19.1A and B I show two colony data records. The activity varies rhythmically with a period of about 25 minutes. The activity of single individual worker ants is not periodic, however. In figure 19.1C and D I show the activity record for two isolated worker ants. Although the worker ants become active spontaneously and stop activity spontaneously, they do not exhibit the pattern of activity characteristic of the colony. Physical contact among the workers (Cole 1991b, 1991c) is responsible for coupling their activity rates to one another. Workers also vary tremendously in their level of spontaneous activity (Cole 1992). These sorts of colony dynamics are known from other species (*L. acervorum:* Franks et al. 1990; *L. muscorum, Zacryptocerus varians, Tapinoma littorale, Pseudomyrmex elongatus, P. cubaensis:* B. J. Cole, pers. obs.), but they are not universal among ants (e.g., *Tapinoma sessile* does not show periodic colony activity).

Below I will describe methods for detecting evidence of chaos and the state of current evidence for the existence of chaos in ant behavior.

Lyapunov Exponents

The best evidence for chaos, at least one positive Lyapunov exponent, is problematic. The Lyapunov exponent is an average measure over the attractor of how rapidly nearby trajectories diverge or converge. Two nearby trajectories on the attractor will diverge from one another, while they will converge toward the attractor if displaced from it. The methods that are currently widely used to measure Lyapunov exponents (e.g., the Fixed Evolution Time algorithm of Wolf et al. [1985]) are very sensitive to minor changes in the parameters (see Schaffer et al. 1988). It is often difficult to verify that the Lyapunov exponent is positive unless there are large numbers of data (typically 10^3–10^4 data points) that have less noise than biological systems characteristically have. Recently developed methods for nonlinear noise reduction (Kostelich and Yorke 1990) combined with nonlinear mapping to determine the Lyapunov exponents (Bryant et al. 1990) may prove more useful for biological data.

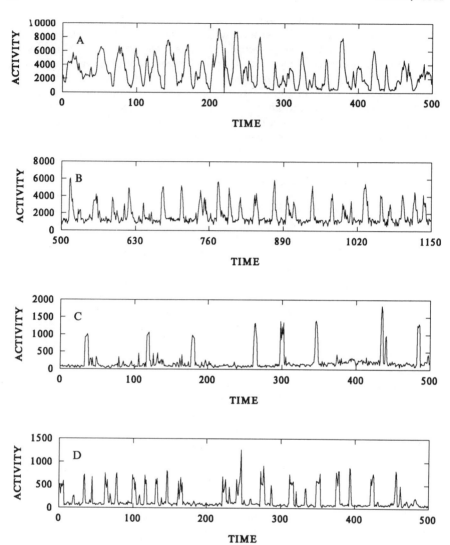

Fig. 19.1   Activity records for two colonies of *Leptothorax allardycei* (*A, B*) and for two isolated single workers (*C, D*). The time scale is in half-minute intervals; the ordinate is the number of pixels that change between successive images. The colony activity records are strongly periodic (Cole 1991a, 1981b), whereas the single ant activity records show no clear-cut periodicity.

The Dimension of the Attractor

Chaotic systems are deterministic ones that are constrained to an attractor. Random fluctuations, by contrast, are not constrained to a surface of low dimension, but are space-filling. Finally, periodic or quasiperiodic trajectories take place on surfaces of integer dimension. If there is a single period, the trajectory is a stable limit cycle that occurs on a one-dimensional periodic orbit. If there are two periods that form a rational fraction, then the trajectory occurs on the surface of a torus (a two-dimensional surface) and repeats after an integer number of cycles. If the two periods do not form a rational fraction, the entire surface of the torus will be filled. The trajectory is, however, still confined to the two-dimensional toroidal surface, and is not chaotic but quasiperiodic. We can reconstruct the attractor of movement activity and measure its dimension in order to make inferences about the underlying dynamics of the system.

Dimension measurements are frequently made using the method of Grassberger and Procaccia (e.g., Olsen and Schaffer 1990; Schaffer et al. 1988). This method is in widespread use and gives reliable results for certain types of data; however, it can give misleading results for other types of data (see, e.g., Albano et al. 1988; Möller et al. 1989; Sugihara and May 1990). Grassberger-Procaccia can give misleading results when the data are noisy. When the signal-to-noise ratio (SNR) of the data is less than about 20 dB (measured as $10\log_{10}$ [var(signal) ÷ var(noise)]), then Grassberger-Procaccia will not give an accurate result (Hediger et al. 1990) because there is not a well-defined scaling region in the calculations of the correlation integral. There is instrumental noise in the measurement of ant activity. For the activity records of intact colonies, I have estimated the SNR to be about 35 dB, or well within the range of what Grassberger-Procaccia can tolerate. However, the SNR of individual ants is in the neighborhood of 10 dB. This level of noise makes it inappropriate to use Grassberger-Procaccia.

To measure the dimension of the attractor of individual ant activity records, I have used the Local Intrinsic Dimensions (LID) method of Hediger et al. (1990). Their procedure, as does Grassberger-Procaccia, makes use of Takens's (1981) embedding theorem. If a complex process is governed by a collection of differential equations, then we could determine the behavior of the system by studying these governing equations. In practice we may have only the ability to measure a single output variable. Takens's theorem says that we can study the properties of a reconstructed attractor. For any data point at time T, the lagged data points at $T + t, T + 2t, \ldots,$ $T + (r - 1)t$ form the coordinates of the point in an r-dimensional space. Figure 19.2 shows the reconstructed attractor for a single ant activity record embedded in three dimensions. Takens shows that the dynamical properties of this reconstructed attractor are the same as the properties of the attractor derived from the unknown, but underlying, set of governing equations. Takens's result applies exactly to an infinte number of noise-free data, and an empirical question of great interest is how many noisy data are required to provide good estimates. The embedding dimension must be greater than $2m + 1$, where m is the dimension of the attractor, to ensure that the properties of the attractor are preserved (Takens 1981; Broomhead and King 1986).

Using the LID method, one estimates the dimension as the number of orthogonal directions along which the data are arrayed in the local neighborhood of a randomly selected point. The mean of a number of such points around the attractor is the LID. If the embedding dimension is r, then we take some number, n, closest neighbors of a randomly chosen point (n is typically three times the embedding dimension: Albano et al. 1988; Hediger et al. 1990). The matrix of the coordinates of the n points in r dimensions forms the data matrix, **M**. The eigenvalues of the standardized variance-covariance matrix ($\mathbf{M^{-1}M}$), are related to the fraction of variation that occurs along principal axes; essentially we do a principal components analysis of the data in the coordinates of points in the local neighborhood of a randomly chosen point. The contribution of Hediger et al. (1990) was to use a result from signal processing that makes use of an information theoretic criterion, the Minimum Description Length (MDL: Wax and Kailath 1985), to estimate, with a maximum

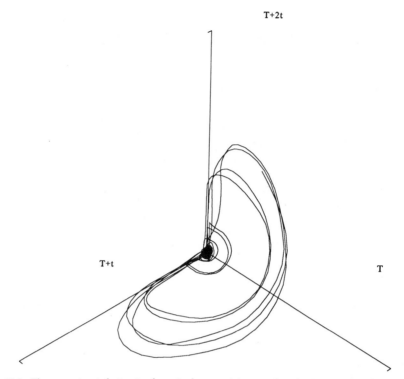

Fig. 19.2   The reconstructed attractor for a single ant activity record, embedded in three dimensions.

likelihood method, which of the axes are associated with the signal and which are associated with noise (Cole 1991a). In other words, the variation along the first several axes is associated with the dispersion of data around the attractor; the remaining axes are associated with random variation around the attractor. The important advantage of the LID technique is that it can be used even when the SNR in the system is as high as 5 dB; that is, when the noise is as much as 30 times as great as can be tolerated by Grassberger-Procaccia. The method will overestimate the dimension of the attractor when the SNR exceeds about 15 dB, since the method concludes that more of the axes are "real."

Since the SNR of the activity records of ant colonies is high enough to be analyzed by Grassberger-Procaccia, and is too high to be analyzed by LID, I have calculated the correlation dimension using the implementation of Schaffer et al. (1988) and the multidimensional tree searching algorithm of Bingham and Kot (1989). Since it is difficult to obtain any sort of estimate of the error of any particular Grassberger-Procaccia or LID estimate, I have used the replicate estimates of the dimension of a number of activity records to estimate the distribution of the dimension. The dimension of colony activity records (mean = 3.09 ± 0.24) is not significantly different from a value of 3 dimensions ($t = 0.24$, df $= 9$, $P > .4$). Because the dimension of the attractor is an integer, the activity of colonies is consistent with nearly periodic motion about a toroidal attractor.

Does a measurement of 3.09 for the dimension of the attractor of a colony of ants guarantee that the activity of colonies is periodic? Certainly there is a strong periodic component to the activity, but we already knew that from the Fourier analysis. It is not possible to exclude the possibility that colony activity is chaotic. A cautionary point should be raised here. The dimension of the attractor of one of the best-studied chaotic systems, the Lorenz system, is 2.06. Given a limited number of noisy data, it would be nearly impossible to measure the dimension with sufficient accuracy to falsify the appropriate null hypothesis that the attractor has an integer dimension. However, during long records of colony activity, the period of colony activity often changes rather abruptly, a phenomenon that is associated with chaos. Certainly the weight of evidence favors colony periodicity, but the empirical question of chaos in colony activity will not be resolved by measuring dimensions.

Obtaining a reliable estimate of dimension using the LID method requires a high density of points around the attractor. If one uses an embedding dimension of 10, it is necessary that the 30 nearest points to a chosen point be in the local neighborhood of that point of interest. If the data density is low around the attractor, the 30 nearest points will be some distance away, and the cloud of points will not be locally linear. This will result in an overestimate of the dimension of the attractor. The data for single ant activity records involve larger numbers of data points (about 7,200 points per record). I embed each single ant activity record in ten dimensions. The lag interval, t_l, that I use is 33. Within broad limits, the choice of the particular time lag does not seem to affect the results (e.g., Cole 1991a).

The mean dimension of the attractor of single ant activity records is 2.43 ($\pm$ 0.30 = 2 SE). This value is significantly different from the integer values of either 2 ($t = 2.90$, $P < .01$, one-tailed) or 3 ($t = 3.84$, $P < .005$, one-tailed) as well as from that of colony records ($t = 3.49$, df = 18, $P < .01$). The fractal dimension of the attractor provides some evidence that the activity of single ants is chaotic.

Forecasting

Prediction of the long-term course of a chaotic process is not possible because nearby trajectories diverge exponentially with time (there is a positive Lyapunov exponent). However, it may be possible to make short-term predictions about the dynamics of a chaotic system by constructing a first-return map of successive excursions around the attractor. The nonlinear forecasting techniques for producing predictions from chaotic dynamics will not be discussed here (see Farmer and Sidorowich 1988; Sugihara and May 1990; Sugihara et al. 1990). The return map of a chaotic process will typically show a nonrandom pattern. Observation of a nonrandom pattern does not guarantee the presence of chaos, but in combination with other evidence, such as the noninteger dimension of the attractor, is strongly indicative. Poincaré sections are produced from the reconstructed attractor by cutting through the attractor (when the attractor is embedded in three dimensions, it is sliced with a plane, e.g., fig. 19.3). The plot of successive values of the trajectory against one another as they pass through this plane is the first-return map. In figure 19.4 I show two first-return (Poincaré) maps of two activity records of intact colonies. The return maps show no indication of pattern in successive peaks; this is to be expected with noisy periodicity. However, the first-return maps of single ant activity records (fig. 19.5) seem to show clearly nonrandom patterns.

Because of our ability to perceive pattern in random data, the return maps must be tested to determine whether the apparent nonrandomness is significant. Quantifying the extent of the nonrandomness in a return map is not straightforward. There are at least two reasons for this difficulty. The first is that the return map has an unspecified form. Therefore, the null hypothesis against which the data are to

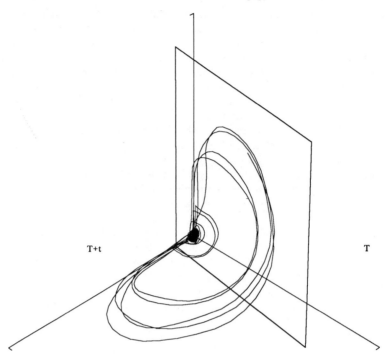

Fig. 19.3 A Poincaré section produced from the attractor in figure 19.2.

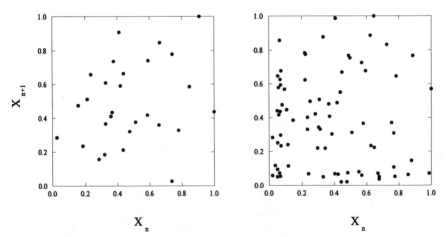

Fig. 19.4 First-return maps for two colony activity records. There is little evidence of pattern in the level of activity in successive excursions around the attractor.

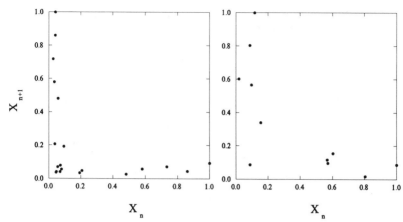

Fig. 19.5   First-return maps for two single ant activity records.

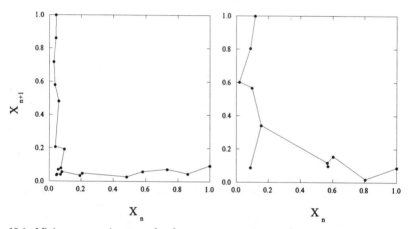

Fig. 19.6   Minimum spanning trees for the return maps given in figure 19.5. The length of the minimum spanning trees is significantly less than that of those from randomized return maps.

be tested is not known in advance. Second, the expected form of the return map will likely not have an integer dimension. This makes standard curve fitting meaningless. I have used several methods to test for patterns in the return maps. The methods are all based on randomizations and themselves only begin to delimit the possibilities. The first, following a suggestion by W. Schaffer, is to use as a test statistic the sum of the product of the x–y coordinates of the points of the return map. This statistic is tested against a randomized distribution. The null distribution is formed by randomizing the x-coordinates (equivalent to randomizing the successive passes through the Poincaré section), creating the randomized return maps, and calculating the statistic. This procedure is sensitive to either an approximately hyperbolic return map (as in fig. 19.5) or to a return map in which large excursions around the attractor frequently succeed one another.

A second method is to use the length of the minimum spanning tree as the test statistic. The minimum spanning tree (fig. 19.6) connects the points such that the sum of the distances is as small as possible. The order of the points is randomized and the length of the minimum spanning tree calculated. The distribution of the lengths of the minimum spanning trees of randomized points is used to estimate the significance of the observed statistic. In patterned data the lengths of the minimum spanning trees may be less than in randomized data. Other tests are outlined in Cole (1991a).

The results of several of these tests indicate that the return maps for single ant activity records are nonrandom. It is possible that a unimodal map may even provide a good initial approximation of the data in figure 19.5. Given that the estimated dimension of single ant activity is about 2.4, we might expect that the first-return maps would be nearly one-dimensional.

DISCUSSION

The Origins of Behavioral Chaos

The Production of Chaos from Internal Networks. If the physiological and neurobiological mechanisms underlying the production of behavior are chaotic, then behavior may be chaotic as well. Chaos has, for example, been suggested as the normal waiting state of brain activity. Babloyantz and Destexhe (1986), Babloyantz (1986), Albano et al. (1986), and Skarda and Freeman (1987) have suggested that the electroencephalogram (EEG) pattern of normal brain function is chaotic. The complexity of the EEG, as measured by its dimension, increases with certain cognitive tasks and decreases under anaesthesia, for example (e.g., Mayer-Kress and Layne 1987; Nan and Jinghua 1988; King 1991).

Models of neural networks (e.g., Clark 1991; Freeman 1987; Lewis and Glass 1991; Kepler et al. 1990; Kürten and Clark 1986; Sompolinsky et al. 1988) often have chaotic dynamics as one of the outcomes. The chaos of a neural network model is deduced by examining the temporal fluctuations of state in the output of the interacting neural components. If the chaotic neural output has output to a motoneuron, then chaotic movement patterns are a natural result (Mpitsos et al. 1988).

Physiological networks of other types can also generate chaotic dynamics. The best-known example is the control of white cell production, examined by Mackey and Glass (1977). At the intersection of neural networks and the sensory system of an organism, Siegel (1990) discusses the relation of visual processing to nonlinear dynamics. One could imagine delayed, nonlinear responses to sensory, nutritive, or reproductive input, which could induce complex feeding, movement, or social behavior.

There are two characteristics of networks of internal control that must be taken into account in discussing the probability of complicated nonlinear results. The first is the number of interacting control variables: as their number increases, the likelihood of chaotic outcomes increases. An important determinant of the dynamical outcome of network control is the type of interaction between the process being controlled and the control variables. Glass and Malta (1990) considered the special case of a network of negative feedback loops operating on a single response variable. Each of the negative feedback loops was randomly constructed to have a unique and characteristic time scale for the lag of the response, the rate of the response, and the strength (or gain) of the response. In a large number of simulations, such randomly constructed negative feedback networks typically exhibited periodic output, often with long periods. Chaos was an infrequent result; Glass and Malta estimated that chaos would result in approximately 2% of the cases. Although their results apply specifically only to a single nonlinear function and over a specific range of parameter values, nevertheless, they do suggest that chaos is unlikely simply to happen when there is pure negative feedback.

The situation is different when there is a network of mixed feedback. When there are regions of the controlled variable over which the control variables exert positive feedback and other regions over which they exhibit negative feedback, the outcome is far more likely to be chaotic (Oliveira and Malta 1987; Glass et al. 1988; Sompolinsky et al. 1988). The implication for animal behavior seems clear. Behavioral functions often have mixed feedback. The interface of physiology and behavior provides many examples of behavior having a stimulatory physiological effect over one range of values and an inhibitory effect over another range. The occurrence of positive feedback in biological systems has been reviewed by DeAngelis et al. (1986).

This is the *deus ex machina* solution to the production of behavioral chaos: If the underlying proximate mechanism of behavior is chaotic, then behavior will be chaotic as well. This may be the major method by which chaos is produced in behavior, and the major contribution of nonlinear dynamics to animal behavior may be toward the understanding of how complex network models generate chaotic output.

Periodically Driven Behavior. Systems that are periodically driven can couple to the forcing period in simple or very complex phase-locked patterns. They can also be driven by periodic stimulation into chaotic regimes. Periodic forcing of epidemiological models has been discussed in relation to the apparent chaos of measles epidemics by Schaffer and others (Schaffer 1985; Schaffer and Kot 1985b; Olsen and Schaffer 1990). Here periodic changes in the infectiveness of a disease result in chaotic epidemics. The most likely candidates for a periodic driver in a behavioral system are the circadian and circannual rhythms. I will give two examples of the behavioral effects of periodic forcing in such rhythms. Periodic forcing of integrate-and-fire models or nonlinear oscillators and the dynamical behavior that can result is discussed, in a physiological context, by Glass and Mackey (1988). The description given earlier of an animal that begins to forage when it reaches a critical level of hunger is an integrate-and-fire model.

Let us explore this possibility further. Suppose we have the situation summarized in figure 19.7. An animal eats when it reaches a certain threshold level of hunger. However, this threshold itself changes periodically. We let the threshold: $T(t) = K + \sin t$; in other words, it varies sinusoidally in the simplest manner possible. For a mammal, the threshold may be low at night and much higher during the day; for birds, the situation is typically reversed. The simplest model that we can construct is one in which there is a linear increase in hunger and the animal is completely satiated as soon as it begins foraging. Here we let foraging occur when the hunger level exceeds the threshold.

In this case the "hunger trace" follows the sawtooth curves in figure 19.7 for various rates of becoming hungry. Over a broad range of rates of becoming hungry, the organism phase-locks into a pattern of daily foraging. As the rate increases beyond a critical value, there is a dramatic change in the dynamics of foraging. The critical point here is that there is a smooth increase in the rate at which an animal becomes hungry as a result of the relationship between the size of an organism and various parameters that are related to foraging, such as the metabolic rate and the time required for gut clearance (for reviews see Peters 1983, and especially Calder 1984). The rate of becoming hungry is a function of body size, which requires that

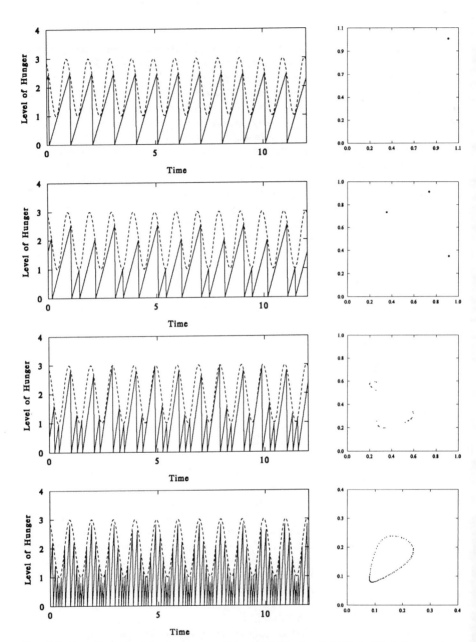

Fig. 19.7 Periodic forcing of an integrate-and-fire model. The threshold (dotted line) changes sinusoidally, K = 2.0. In the left half of the figure I show the "hunger trace" (solid line), in which the slope indicates the rate at which the organism becomes hungry. From top to bottom the rates are 0.04, 0.045, 0.08, and 0.2. When the hunger level exceeds the sinusoidally changing threshold, the organism eats. In the right-hand portion of the figure I show the relation of successive intervals between foraging bouts. Note that there is a sudden change in the pattern of foraging near a rate of 0.045.

organisms that decrease in size below a critical point exercise one of two behavioral and physiological options: either become torpid during one fraction of the day (the typical strategy used by small birds) or begin foraging at other times of the day (the strategy employed by small mammals). In figure 19.7 I also plot the interval between successive foraging bouts as the mammal decreases in size. Note that there is an abrupt, qualitative change in the dynamics of foraging with a small, quantitative change in the rate at which an animal becomes hungry. This sort of periodically forced integrate-and-fire model generates $M:N$ phase-locking for all integers that are relatively prime, as well as chaotic dynamics (Glass and Bélair 1986).

Periodic forcing also plays a role in determining the timing of reproduction. The timing of reproduction obviously influences a wealth of important behavior associated with mating systems and parental care. Here we suppose that there is seasonal forcing of the reproductive cycling of organisms. Organisms with a gesta-tion-lactation-parental care period of longer than one year may have difficulty in phase-locking to seasonal cycles. Periodic forcing of a process that is itself periodic may lead to complicated dynamics, including chaos. Surprisingly, there may even be some evidence for complicated dynamics in the seasonal timing of reproduction. Kiltie (1988) has shown that there is a relation between the gestation period of mammals and the degree to which they have sharply defined seasonal reproduction patterns. Furthermore, there is a difference between tropical mammals (in which there is a low amplitude of seasonal forcing, and in which there is less phase-locking of the breeding season to the seasonal cycle) and extratropical mammals (in which there is a higher amplitude of seasonal forcing and in which there is greater phase-locking to the seasonal cycle). It is a reasonable hypothesis that the timing of repro-duction could be illuminated by considerations of nonlinear dynamics. A clear first step would be to produce a return map by plotting successive intervals between reproductive episodes (or perhaps more appropriately, the successive seasonal phases at which reproduction occurs for a number of individuals).

Social Networks. Interactions in social groups may alter the dynamics of the behav-ior of the individuals within the group or the collective behavior of the group. If individuals within a group are capable of influencing the timing of subsequent be-havior of other individuals, then the range of possible outcomes is as broad as that produced by internal networks. One potential outcome of these collective interac-tions is chaos, but it is not the only outcome, or even the most interesting one. In addition to temporal patterns that result from interactions in social networks, the production of spatial patterns in the expression of a particular behavior is a possi-bility.

A rich source of information on social network interactions comes from socially displaying organisms. Sismondo (1990) studied the interactions between singing katydids. He showed that katydids respond to one another's stridulatory calls, form-ing synchronously calling pairs, and also demonstrated a variety of other sugges-tive nonlinear phenomena, including various modes of phase-locking and apparent period-doubling bifurcations. This study does not claim to demonstrate that the stridulation of katydids is chaotic, but it takes the first steps in an experimental demonstration of the transition between nonchaotic and chaotic behavior. Other organisms that have synchronized calls or displays, such as frogs (Schwartz 1991; Brush and Narins 1989) and fireflies (Buck 1988; Carlson and Copeland 1985), clearly interact with one another in complex ways and may exhibit synchrony in display (i.e., phase-locking), or period doubling of intercall frequency in response to artificial or natural stimuli.

Interactions in the social insects will be another fruitful source of material for nonlinear dynamics in behavior. The theoretical structures that are most appropriate for describing the dynamics of social insects are likely to be mobile cellular automata (MCA) models. In cellular automata models (e.g., Ermentrout and Edelstein-Keshet 1993; Nowak and May 1992, 1993) the state of the subunits that constitute the system is determined at a set of lattice points by internal dynamics and interactions between the subunits. Under some circumstances it would be appropriate to consider a densely packed array of individuals; in other cases, only a fraction of available sites need to be occupied. The subunits are connected in some fashion. The two extremes are globally connected (all subunits are connected to all other subunits) or locally connected (only the nearest neighbors influence one another's dynamics). When there are nonlinear interactions between the subunits, the dynamics of the cellular automata can be of any of the kinds that we have encountered before in network models (see, e.g., Clark 1991; Matthews et al. 1991). Furthermore, there is the strong possibility that spatial patterns will emerge from the interactions among the automata (e.g., Satoh 1990, or in the case of nonlinear behavioral interactions, Nowak and May 1992, 1993).

In MCA models, the cells have internal dynamics and interact locally, but are capable of movement between lattice points as well. In a model colony of ants, for example, each worker occupies a single lattice point, with activity being a function of both the internal dynamics of the ants and the interactions among them. Solé et al. (1992) and Miramontes et al. (1993) have modeled the production of cycles of activity in ant colonies using a cellular automata model in which the internal dynamics of the worker ants are either random or produced by a chaotic process. They show that the emergence of periodic colony activity depends on the characteristics of the interaction matrix, J. The interaction matrix determines the outcome of interactions between ants, which are either active or inactive. In particular, one necessary condition for oscillation is that interactions between two active ants cause changes in the subsequent activity patterns of both ants.

The crucial point is that the characteristics of the interaction matrix are themselves subject to natural selection. While periodicity is one of the possible outcomes of the model, it is not guaranteed by the structure of the model. Selection, operating at the colony level, can produce different dynamical outcomes based on alterations of the terms of the interaction matrix. We can gather the same sort of data with *Drosophila* as with ants. We expect the collective dynamics of an aggregate of solitary individuals such as *Drosophila* to be different from the dynamics of ants because the terms of the interaction matrix have had a fundamentally different evolutionary history. In figure 19.8 I show an activity record of a single *Drosophila* and of a group of twenty *Drosophila*. Rather than the predictable modulation of activity in the ant colonies, the activity record of the group of *Drosophila* is, as far as I can determine, random.

Solé et al. (1992) show that periodicity emerges with the addition of more subunits to their cellular automata model. Does the emergence of a periodic component to activity indicate that the activity is truly periodic? After all, a chaotic process is often nearly periodic and will frequently have significant periodic components. Furthermore, in cellular automata models, there are often complicated transitions between periodicity and chaos, frequently involving locally chaotic phenomena (Li et al. 1990). Miramontes et al. (1993) show that chaos can be one of the outcomes

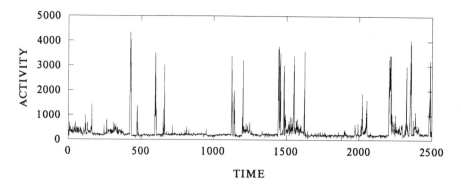

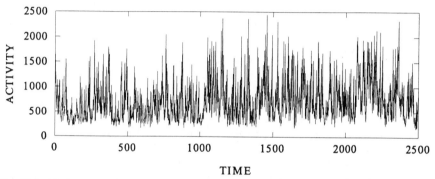

Fig. 19.8 (A) An activity record for a single *Drosophila melanogaster*. (B) The activity record of an aggregate of 20 *D. melanogaster*. The pattern of activity of this aggregate is apparently random. The flies used here are a curly-plum stock in a Hochi-R background and were obtained from James Jacobson. The time scale is in 3-second increments.

in an MCA model. The frequency and likelihood of chaos in this situation poses a theoretical question, which, like the corresponding empirical one, is unresolved.

The Consequences of Behavioral Chaos

Regardless of the origins of chaotic behavior, observations of behavioral chaos and the perspective of nonlinear dynamics can have a number of consequences for how we study animal behavior and for the theoretical description of behavior. Behavioral chaos can in turn have ecological effects and, particularly, consequences for the evolution of behavior.

As behavioral biologists we have all had the experience of attempting to replicate an experiment under conditions that are controlled for the time of day, physiological state of an animal, social situation, experience, and so forth, only to find that the response of the animals is completely different. We chalk the variation up to the essential randomness of behavior. I will take the view in this section that it is possible that some of the variation that we perceive as fundamental to behavior may be due to the sensitive dependence on initial conditions that is characteristic of chaos, rather than simply to the chance vagaries of behavior. The existence of chaos in animal behavior has a number of consequences for the study of behavior, the interpretation of behavior, and the evolution of behavior.

The perspective of nonlinear dynamics in animal behavior changes how we approach the description of behavior, changes our ability to predict behavior, and even changes what we mean by behavior. Chaotic systems are inherently simpler to describe and study theoretically than systems that are dominated by a stochastic component. Incorporating stochastic variation into a theoretical description of behavior requires the addition of terms that are difficult to manipulate and the abandonment of attempts to account for this variation. Systems with large numbers of nonlinear interactions are precisely those that would most benefit from some simplification. If we could discover chaos in complex animal behavior, it would simplify the task of describing its dynamics. We could reduce the large number of complex variables and interactions to a more tractable number that are sufficient to describe the dynamics. This reduction would result in a simplification of the task of building theoretical models to describe behavior. Rather than a model that contains tens of variables, an adequate model might need to employ only as many variables as there are dimensions in the attractor.

Using the perspective of nonlinear dynamics changes what we mean by behavior. Behavior is a process, but it can also be viewed as an object: the attractor. The activity records in figure 19.1 are not any better or more complete a description of the movement activity of an ant than are the attractors in figure 19.9. We can study these geometric objects in the same way that we study any sort of object, by examining its morphology. A behavioral process need not be chaotic in order for us to exploit this technique. As long as the data exist in time, they can be treated in this way.

Because chaos is deterministic, prediction is possible. It may be true that long-term predictions are not possible; however, short-term predictions can be made. The return maps of figures 19.5 and 19.6 are short-term predictions of successive heights of the activity peaks as they pass through the Poincaré section. If we hold the view that the variation we observe in behavior is due simply to chance variation, then such predictions are not possible. To the extent that chaotic behavior is predictable, at least over the short term, then we are able to predict something that we could not before.

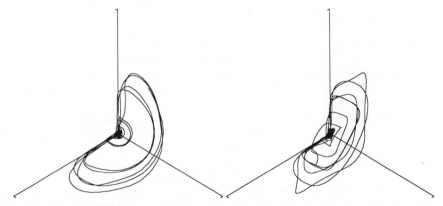

Fig. 19.9   The reconstructed attractors of two single ant activity records. Apart from minor transients, the shapes of the two attractors are virtually the same.

The effects of behavioral chaos can resonate in ecological interactions with other organisms. In discussing the chaotic orbit of Pluto, Sussman and Wisdom (1988) made the following remarkable statement: "Pluto's irregular motion will chaotically pump the motion of the other members of the solar system and the chaotic behavior of Pluto would imply chaotic behavior of the rest of the solar system." Since all planets are gravitationally interacting with one another, chaotic behavior in one part of the solar system will induce chaos everywhere. We could make a similar argument for ecological interactions. If the behavior of an organism is chaotic, and this organism interacts with other organisms, then the effect of chaos could be pervasive. In practice, what we need to know is the extent to which the possible effects of behavioral chaos on an animal's prey, its predators, or other members of its social unit can persist.

How long can a chaotic signature be detected in a noisy world? This seems a fertile field for theoretical exploration. We know that the answer, in principle, will be a function of the sorts of interactions that take place between animals. We have already seen that one kind of interaction leads to apparent periodicity (in *Leptothorax* ants) and another leads to apparent randomness (in *Drosophila*). Changes in the number of individuals that interact with one another (the connectedness) and the type of interactions between individuals will govern the outcome in any network of individuals. If we overlay random variation on a set of coupled interactions that produces chaos, how many data do we need in order to distinguish chaos from randomness?

Admitting chaos into animal behavior will have a profound effect on the way that behavioral evolution is studied. The essential reason is that the study of evolution is intimately bound up with the study of variation. If we perceive some fraction of behavioral variation as being illusory, as having little to do with the informative behavioral phenotype, then we will have a different perspective on the pervasiveness and power of natural selection. We may see enormous variation in a behavior both among individuals and in a single individual from one observation time to the next. If we attempt to modify this variation or to probe the genetic basis of the variation experimentally by a selection experiment, we may conclude that there is no detectable genetic basis to the variation. Our conclusion would be that there is no heritable variation and that natural selection is not possible; the behavior has low repeatability. If we conclude that much of the apparent variation is simply due to iterations of the same chaotic process, then our attempts to modify behavior may not be acting on a piece of behavior that in any meaningful sense exists. The level of variation indicated by the difference between the attractors of figure 19.9 is not large; apart from an initial transient, they appear nearly identical. If we identify the attractor and repeat our selection experiment on the shape of the attractor, we may have an enormous effect on the behavior and reach the conclusion that it is highly heritable and has tremendous sensitivity to natural selection. If we perceive the relevant phenotype as being the geometry of the attractor itself, then our perception of behavioral variation will be fundamentally altered.

One of the more intriguing possible consequences of behavioral chaos is that it may influence the evolution of social behavior. It is possible that a social unit (such as an ant colony) constructed of subunits (ant workers) that are individually chaotic can respond more flexibly than a social unit constructed of components that have different dynamics. This idea is outlined in general terms by Ott et al. (1990). A chaotic process has embedded within it an infinte number of unstable periodic orbits. If a chaotic process is repeatedly perturbed, it can have a periodic output. In fact, the range of possible periodic outputs that can be achieved with small perturbations is far larger in a perturbed chaotic system than in a system constructed of periodic components. Indeed, parallel processing with chaotic components has recently been shown to be highly efficient (Inoue and Nagayoshi 1991).

From the perspective of the colony of ants, it may be advantageous to have behavior that can be predicted over a long time scale, but which is adaptable to changing demands. Periodic colony activity is predictable over fairly long time periods, although the period of colony activity sometimes changes for little apparent reason. If a colony is constructed of periodic or randomly behaving components, the range of possible social phenotypes may be smaller than if the colony is composed of chaotic components.

CONCLUSION

Bringing nonlinear dynamics into the study of animal behavior may have revolutionary consequences for our view of animal behavior. I have tried to make several points in this chapter. The first is that chaos is a likely outcome in behavior due to its complex and nonlinear nature. Chaos and other forms of complicated nonlinear dynamics may arise because (1) complicated internal networks are responsible for the production of behavior; (2) because periodic forcing of behavior is likely to occur; or (3) because social networks may yield complicated collective behavioral dynamics. The second point is that because chaos is deterministic, prediction is possible. The extent to which we are able to make predictions that were previously impossible, even in principle, is a measure of progress. The final point is that the bounded variability of chaos will have a major effect on how we perceive behavioral evolution. The study of variation in behavior, which is fundamental to the study of behavioral evolution, could become associated with the study of variation in the form of a behavioral attractor rather than variation in an excursion around it.

ACKNOWLEDGMENTS

I would like to thank Bill Schaffer for introducing me to the study of nonlinear dynamics and Diane Wiernasz for reading manuscripts and discussing ideas, and for critical commentary.

REFERENCES

Albano, A. M. et al. 1986. Lasers and brains: Complex systems with low-dimensional attractors. In G. Mayer-Kress, ed., *Dimensions and entropies in chaotic systems*, 231–40. Berlin: Springer-Verlag.

Albano, A. M., J. Muench, C. Schwarz, A. Mees, and P. Rapp. 1988. Singular-value decomposition and the Grassberger-Procaccia algorithm. *Physical Review* A 38:3017–26.

Babloyantz, A. 1986. Evidence of chaotic dynamics of brain activity during the sleep cycle. In G. Mayer-Kress, ed., *Dimensions and entropies in chaotic systems*, 114–22. Berlin: Springer-Verlag.

Babloyantz, A., and A. Destexhe. 1986. Low-dimensional chaos in an instance of epilepsy. *Proceedings of the National Academy of Sciences U.S.A.* 83:3513–17.

Bingham, S., and M. Kot. 1989. Multidimensional trees, range scaling and a correlation dimension algorithm of reduced complexity. *Physics Letters* A 140:327–30.

Broomhead, D., and G. King. 1986. Extracting qualitative dynamics from experimental data. *Physica* 20D:217–36.

Brush, J., and P. Narins. 1989. Chorus dynamics of a neotropical amphibian assemblage: Comparison of computer simulation and natural behaviour. *Animal Behaviour* 37:33–44.

Bryant, P., R. Brown, and H. Abarbanel. 1990. Lyapunov exponents from observed time series. *Physical Review Letters* 65:1523–26.

Buck, J. B. 1988. Synchronous, rhythmic flashing of fireflies. II. *Quarterly Review of Biology* 63:265–89.

Calder, W. A. III. 1984. *Size, function, and life history.* Cambridge, Mass.: Harvard University Press.

Carlson, A. D., and J. Copeland. 1985. Flash communication in fireflies. *Quarterly Review of Biology* 60:415–56.

Chay, T., and J. Rinzel. Bursting, beating and chaos in an excitable membrane model. *Biophysical Journal* 45:357–66.

Clark, J. W. 1991. Neural network modelling. *Physics in Medicine and Biology* 36:1259–1317.

Cole, B. J. 1991a. Is animal behaviour chaotic? Evidence from the activity of ants. *Proceedings of the Royal Society of London* B. 244:253–59.

Cole, B. J. 1991b. Short-term activity cycles in ants: A phase response curve and phase-resetting for worker activity. *Journal of Insect Behavior* 4:129–37.

Cole, B. J. 1991c. Short-term activity cycles in ants: Generation of periodicity by worker interaction. *American Naturalist* 137:244–59.

Cole, B. J. 1992. Short-term activity cycles in ants: Age-related changes in tempo and colony synchrony. *Behavioral Ecology and Sociobiology* 31:181–87.

DeAngelis, D. L., W. M. Post, and C. C. Travis. 1986. *Positive feedback in natural systems.* New York: Springer-Verlag.

Denton, T., G. Diamond, R. Helfant, S. Khan, and H. Karagueuzian. 1990. Fascinating rhythm: A primer on chaos theory and its application to cardiology. *American Heart Journal* 120: 1419–40.

Ermentrout, G., and L. Edelstein-Keshet. 1993. Cellular automata approaches to biological modeling. *Journal of Theoretical Biology* 160:97–133.

Farmer, P., and J. Sidorowich. 1988. Predicting chaotic dynamics. In J. Kelso, A. Mandell, and M. Schlesinger, eds., *Dynamic patterns in complex systems*, 248–64. Singapore: World Scientific Press.

Franks, N., S. Bryant, R. Griffiths, and L. Hemerik. 1990. Synchronization of the behaviour within nests of the ant *Leptothorax acervorum* (Fabricius)—I. Discovering the phenomenon and its relation to the level of starvation. *Bulletin of Mathematical Biology* 52:597–612.

Freeman, W. J. 1987. Simulation of chaotic EEG patterns with a dynamic model of the olfactory system. *Biological Cybernetics* 56:139–50.

Glass, L., and J. Bélair. 1986. Continuation of Arnold Tongues in mathematical models of periodically forced biological oscillators. In H. G. Othmer, ed., *Nonlinear oscillations in biology and chemistry*, 232–43. New York: Springer-Verlag.

Glass, L., and M. Mackey. 1988. *From clocks to chaos: The rhythms of life.* Princeton: Princeton University Press.

Glass, L., and C. P. Malta. 1990. Chaos in multi-looped negative feedback systems. *Journal of Theoretical Biology* 145:217–23.

Glass, L., A. Beuter, and D. Larocque. 1988. Time delays, oscillations and chaos in physiological control systems. *Mathematical Biosciences* 90:111–25.

Godfray, H. C. J., and S. P. Blythe. 1990. Complex dynamics in multispecies communities. *Philosophical Transactions of the Royal Society of London* B 330:221–33.

Grassberger, P., T. Schreiber, and C. Schaffrath. 1991. Nonlinear time sequence analysis. *International Journal of Bifurcation and Chaos.* 1:521–47.

Hediger, T., A. Passamante, and M. Farrell. 1990. Characterizing attractors using local intrinsic dimensions calculated by singular-value decomposition and information-theoretic criteria. *Physical Review* A 41:5325–32.

Inoue, M., and A. Nagayoshi. 1991. A chaos neuro-computer. *Physics Letters* A 158:373–76.

Kepler, T., S. Datt, R. Meyer, and L. Abbott. 1990. Chaos in a neural network circuit. *Physica D* 46:449–57.

Kiltie, R. 1988. Gestation as a constraint on the evolution of seasonal breeding in mammals. In Mark Boyce, ed., *Evolution of life histories of mammals*, 257–90. New Haven: Yale University Press.

King, C. C. 1991. Fractal and chaotic dynamics in nervous systems. *Progress in Neurobiology* 36:279–308.

Kostelich, E. J., and J. A. Yorke. 1990. Noise reduction: Finding the simplest dynamical system consistent with the data. *Physica D* 41:183–96.

Kürten, K., and J. Clark. 1986. Chaos in neural systems. *Physics Letters* 114A:413–18.

Lewis, J., and L. Glass. 1991. Steady states, limit cycles and chaos in models of complex biological networks. *International Journal of Bifurcation and Chaos* 1:477–83.

Li, Wentian, N. H. Packard, and C. G. Langton. 1990. Transition phenomena in cellular automata rule space. *Physica D* 45:77–94.

Mackey, M., and L. Glass. 1977. Oscillation and chaos in physiological control systems. *Science* 197:287–89.

Matthews, P. C., R. E. Mirollo, and S. H. Strogatz. 1991. Dynamics of a large system of coupled nonlinear oscillators. *Physica D* 52:293–331.

Mayer-Kress, G., and S. P. Layne. 1987. Dimensionality of the human electroencephalogram. In J. Koslow, S. A. Mandell, and M. Schlesinger, eds., *Perspectives in biological dynamics and theoretical medicine*, 62–87. Annals of the New York Academy of Sciences, no. 504.

Miramontes, O., R. V. Solé, and B. C. Goodwin. 1993. Collective behaviour on random-activated mobile cellular automata. *Physica D* 63:145–60.

Möller, M., W. Lange, F. Mitschke, N. Abraham, and U. Hübner. 1989. Errors from digitizing and noise in estimating attractor dimensions. *Physics Letters* A 138:176–82.

Mpitsos, G. J., H. C. Creech, C. S. Cohan, and M. Mendelson. 1988. Variability and chaos: Neurointegrative principles in self-organization of motor patterns. In J. Koslow, A. Mandell, and M. Schlesinger, eds., *Dynamic patterns in complex systems*, 162–90. Hong Kong: World Scientific.

Nan, X., and X. Jinghua. 1988. The fractal dimension of EEG as a physical measure of conscious human brain activities. *Bulletin of Mathematical Biology* 50:559–65.

Nowak, M. A., and R. M. May. 1992. Evolutionary games and spatial chaos. *Nature* 359:826–29.

Nowak, M. A., and R. M. May. 1993. The spatial dilemmas of evolution. *International Journal of Bifurcation and Chaos* 3:35–78.

Oliveira, C. R. de, and C. P. Malta. 1987. Bifurcations in a class of time-delay equations. *Physical Review* A 36:3397–4001.

Olsen, L., and W. Schaffer. 1990. Chaos versus noisy periodicity: Alternative hypotheses for childhood epidemics. *Science* 249:499–504.

Ott, E., C. Grebogi, and J. A. Yorke. 1990. Controlling chaos. *Physical Review, Letters* 64:1196–99.

Peters, R. H. 1983. *Ecological implications of body size*. New York: Cambridge University Press.

Satoh, K. 1990. Single and multiarmed spiral patterns in a cellular automation model for an ecosystem. *Journal of the Physical Society of Japan* 59:4204–7.

Schaffer, W. M. 1985. Can nonlinear dynamics elucidate mechanisms in ecology and epidemiology? *IMA Journal of Mathematics Applied in Medicine and Biology* 2:221–52.

Schaffer, W., and M. Kot. 1985a. Do strange attractors govern ecological systems? *BioScience* 35:342–50.

Schaffer, W., and M. Kot. 1985b. Nearly one-dimensional dynamics in an epidemic. *Journal of Theoretical Biology* 112:403–27.

Schaffer, W., G. Truty, and S. Fulmer. 1988. *Dynamical software*. Dynamical Systems, Inc., Tucson, Ariz.

Schwartz, J. J. 1991. Why stop calling? A study of unison bout singing in a Neotropical treefrog. *Animal Behaviour* 42:565–77.

Siegel, R. M. 1990. Non-linear dynamical system theory and primary visual cortical processing. *Physica D* 42:385–95.

Sismondo, E. 1990. Synchronous, alternating, and phase-locked stridulation by a tropical katydid. *Science* 249:55–58.

Skarda, C., and W. J. Freeman. 1987. How brains make chaos in order to make sense of the world. *Behavior and Brain Science* 10:161–95.

Solé, R. V., O. Miramontes, B. C. Goodwin. 1992. Oscillations and chaos in ant societies. *Journal of Theoretical Biology* 159:269–80.

Sompolinsky, H., A. Crisanti, and H. J. Sommers. 1988. Chaos in random neural networks. *Physical Review Letters* 61:259–62.

Sugihara, G., and R. M. May. 1990. Nonlinear forecasting as a way of distinguishing chaos from measurement error in time series. *Nature* 344:734–41.

Sugihara, G., B. Grenfell, and R. M. May. 1990. Distinguishing error from chaos in ecological time-series. *Philosophical Transactions of the Royal Society of London* B. 330:235–51.

Sussman, G. J., and J. Wisdom. 1988. Numerical evidence that the motion of Pluto is chaotic. *Science* 241:433–37.

Takens, F. 1981. Detecting strange attractors in turbulence. In D. Rand and L.-S. Young, eds., *Dynamical Systems of Turbulence*, 366–81. Lecture Notes in Mathematics, no. 898. Berlin: Springer.

Wax, M., and T. Kailath. 1985. Detection of signals by information theoretic criteria. *IEEE Transactions in Acoustics, Speech and Signal Processing*. 33:387–92.

Wolf, A., J. Swift, H. Swinney, and J. Vastano. 1985. Determining Lyapunov exponents from a time series. *Physica* 6D:285–317.

CONTRIBUTORS

Arthur P. Arnold
Department of Physiological Science
University of California, Los Angeles

Stevan J. Arnold
Department of Ecology and Evolution
University of Chicago

Dorothy L. Cheney
Departments of Psychology and
    Biology
University of Pennsylvania

Blaine J. Cole
Program in Evolutionary Biology and
    Ecology
Department of Biology
University of Houston

Fred C. Dyer
Department of Zoology
Michigan State University

Nigel R. Franks
School of Biology and Biochemistry
University of Bath

Todd M. Freeberg
Department of Biology
Indiana University

Deborah M. Gordon
Department of Biological Sciences
Stanford University

Alastair J. Inman
Department of Biology
Wesleyan College

Alan C. Kamil
School of Biological Sciences, Depart-
    ment of Psychology, and Nebraska
    Behavioral Biology Group
University of Nebraska
Department of Psychology
University of Nebraska

Ellen D. Ketterson
Department of Biology
Indiana University

Andrew P. King
Department of Psychology
Indiana University

John R. Krebs
Department of Zoology
University of Oxford

Val Nolan Jr.
Department of Biology
Indiana University

Daniel R. Papaj
Department of Ecology and Evolution-
    ary Biology and Center for Insect
    Science
University of Arizona

Lucas W. Partridge
The Macaulay Land Use Research
    Institute
Craigiebuckler, Aberdeen

Leslie A. Real
Department of Biology
Indiana University

Michael J. Ryan
Department of Zoology
University of Texas

Robert M. Seyfarth
Departments of Psychology and
    Biology
University of Pennsylvania

Michael C. Singer
Department of Zoology
University of Texas

Earl E. Werner
Department of Biology
University of Michigan

Meredith J. West
Department of Biology and Psychology
Indiana University

R. Haven Wiley
Department of Biology and Curriculum
   in Ecology
University of North Carolina

Marlene Zuk
Department of Biology
University of California, Riverside

NAME INDEX

Abarca, N., 105, 111, 112
Abrahams, M. V., 60
Abrams, P. A., 298–301, 305, 306, 314
Abramson, C. I., 23, 32
Adams, E. S., 415
Ader, R., 354, 365
Adkins, E. K., 224
Adkins-Regan, E., 223, 224, 226, 231, 233
Adler, F. R., 417, 419
Adler, N. T., 328
Akutagawa, E., 221, 224, 229
Albano, A. M., 427, 432
Alberch, P., 267
Alberts, B., 393, 404
Alcock, J., 134, 146, 365
Alexander, J., 355, 359
Alexander, R. M., 396
Alkon, D. L., 2
Allais, M., 117–18
Allee, W. C., 359
Allendorf, F. W., 363
Altmann, J., 375
Alvarez-Buylla, A., 229–30
Amsel, A., 67, 69–70
Andersson, M., 179, 181, 190, 191, 192, 205, 211, 334, 380
Andrews, E. A., 18
Angerilli, N. P. D., 149
Anholt, B., 300, 301
Arak, A., 202–3, 210
Archawaranon, M., 329
Arnold, A., 4, 219–33, 329
Arnold, S. J., 6, 258–74, 292
Arrow, K., 114
Ascenzi, M., 226, 233
Astington, J. W., 381
Atchley, W. R., 260, 263, 267, 268
Atkinson, R. C., 140
Aureli, F., 375
Austin, J. L., 161

Babloyantz, A., 432
Bachmann, C., 27, 374
Baerends, G. P., 87, 88, 89
Bailey, W. J., 166
Baker, A. G., 69
Balda, R. P., 33, 36, 38–39, 101
Baldwin, J. M., 133, 140
Ball, G. F., 328, 329, 334
Balthazart, J., 224, 328, 329, 334

Barlow, G. W., 190
Baron, M. A., 18
Barto, A. G., 125
Basolo, A. L., 208–9, 210–11
Bateman, A., 5
Bateson, P., 202, 316
Battalio, R. C., 118
Baum, W. M., 121
Beach, F. A., 16
Becker, P. H., 158
Bélair, 435
Beletsky, L. D., 328, 329, 336
Bellwood, J. J., 180
Bender, E. A., 298, 314, 315
Bennett, J., 162
Bennet-Clark, H. C., 205
Bercovitch, F., 379, 384
Bernays, E. A., 279
Bernoulli, D., 103
Bernstein, C., 61–62, 63
Berven, K. A., 316
Bessemer, D. W., 25
Beutler, B., 360
Bickerton, D., 127
Biederman, I., 78, 79, 91
Bingham, S., 428
Bitterman, M. E., 16–17, 20, 23, 26, 32, 111–12, 148
Blyth, B., 223
Blythe, S. P., 423
Boag, D. A., 329
Boake, C. R. B., 258
Boesch, C., 383
Bogdany, F. J., 86
Böhner, J., 227
Boice, R., 15, 33
Boinski, S., 332
Bollens, S. M., 312
Bolles, R. C., 15, 18, 23, 29, 31–32, 34, 110
Bolton, B., 391
Bond, A. B., 26, 36
Bonner, J. T., 267, 393, 403, 404
Bottjer, S. W., 228, 229, 230
Bottoni, L., 329
Boughton, D., 286
Bowers, M. D., 291
Boyd, R., 140, 146, 149, 383
Boyse, E. A., 361, 362
Bradbury, J. W., 190, 205, 332
Brandes, C., 137, 150